S. MIHOJEVICH
D1432727

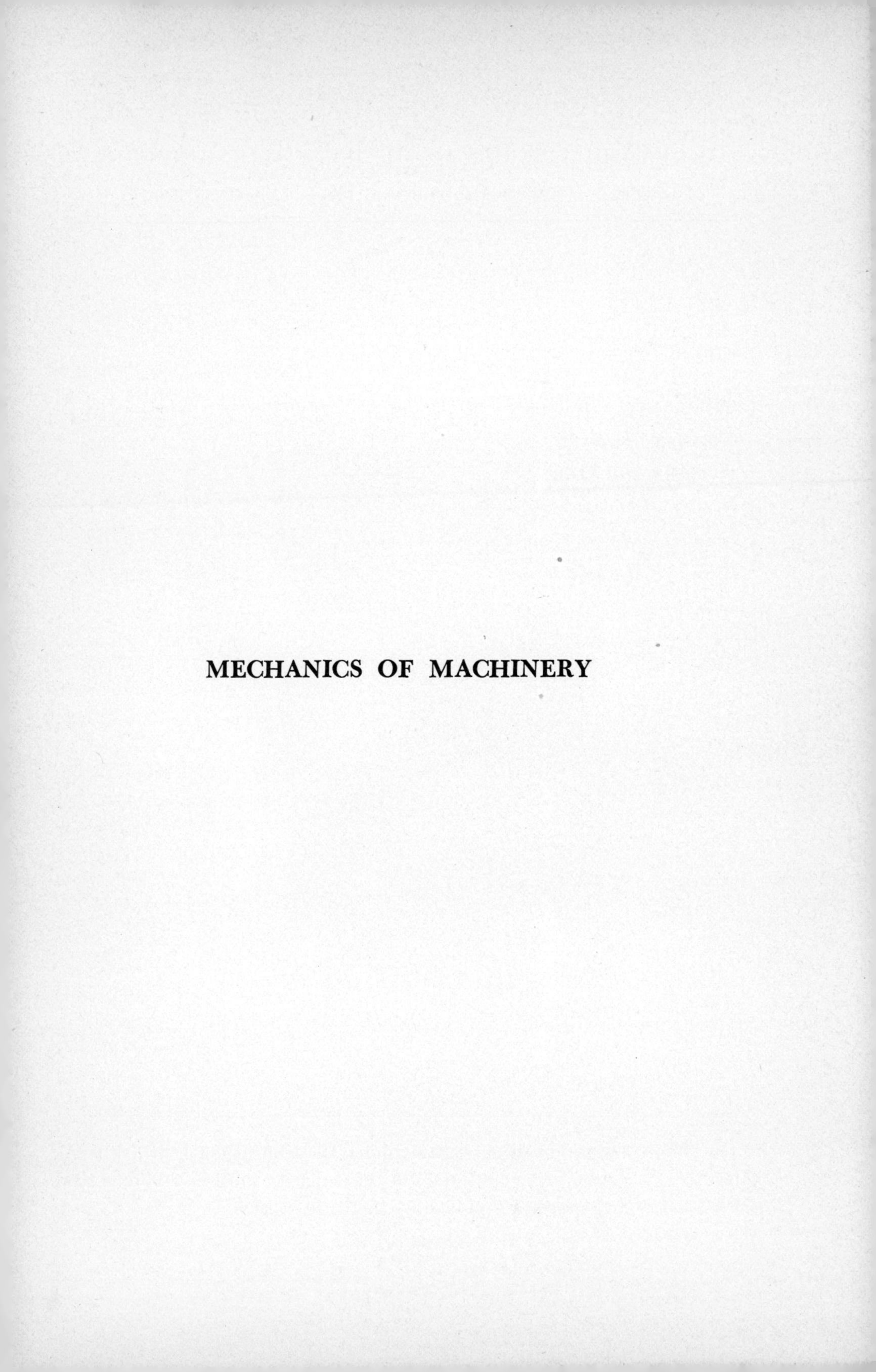

MECHANICS OF MACHINERY

McGRAW-HILL SERIES IN MECHANICAL ENGINEERING

Robert N. Drake and Stephen J. Kline, *Consulting Editors*

BEGGS · *Mechanism*

CAMBEL AND JENNINGS · *Gas Dynamics*

HAM, CRANE, AND ROGERS · *Mechanics of Machinery*

HARTMAN · *Dynamics of Machinery*

JACOBSON AND AYRE · *Engineering Vibrations*

PHELAN · *Fundamentals of Mechanical Design*

SABERSKY · *Engineering Thermodynamics*

SHIGLEY · *Machine Design*

STOECKER · *Refrigeration and Air Conditioning*

The Series was established in 1954 under the Consulting Editorship of Richard G. Folsom, who continued in this capacity until he assumed the Presidency of Rensselaer Polytechnic Institute in 1958.

MECHANICS OF MACHINERY

C. W. HAM, M.E.

Professor Emeritus of Mechanical Engineering, University of Illinois
Member of American Society of Mechanical Engineers

E. J. CRANE, M.E.

Western Electric Company
Member of American Society of Mechanical Engineers

W. L. ROGERS

Associate Professor of Mechanical Engineering, Northwestern University
Member of American Society of Mechanical Engineers

FOURTH EDITION

McGRAW-HILL BOOK COMPANY, INC.

New York Toronto London

1958

MECHANICS OF MACHINERY

Copyright © 1958 by the McGraw-Hill Book Company, Inc.

Copyright, 1927, 1938, 1948, by the McGraw-Hill Book Company, Inc. Printed in the United States of America. All rights reserved. This book, or parts thereof, may not be reproduced in any form without permission of the publishers.

Library of Congress Catalog Card Number 57-6395

THE MAPLE PRESS COMPANY, YORK, PA.

II

PREFACE

The first edition of this book, published in 1927, was written by the two senior authors, C. W. Ham and E. J. Crane, while they were associated in teaching the subject at the University of Illinois. This edition was revised by these authors and published as the second edition in 1938, followed by the third edition in 1948. W. L. Rogers of Northwestern University has joined in the authorship of the fourth edition, and the responsibility for the present revision has devolved upon him, since Professor Ham has retired from active teaching, and Mr. Crane is in active engineering practice.

In the fourth edition of this book, the basic fundamental approach and the general sequence and arrangement which were features of the previous editions have been retained. Mechanism is treated in the early part of the book under Part I, followed by kinematics and dynamics of machinery under Part II. This arrangement offers inherent flexibility; for example, it allows the instructor to assign work in mechanism or in motion analysis (kinematics) independently or concurrently.

In this new edition a brief discussion of velocity and acceleration has been included in Chap. 1, with the intent that it will be helpful in simple problems concerning linkages. However, the full treatment of velocity and acceleration has been reserved for Part II, as in previous editions. The general intent during the revision has been to shorten or modify sections of the text where it was felt that some improvement in approach or content might be achieved; to amplify and expand certain parts in order to include appropriate new topics or recent developments; to supply new examples and applications of current interest; and to adopt and conform to the symbology and terminology of the American Standards Association where applicable. Photographs have been added at selected locations with the view of stimulating interest in the text material and giving the student a clearer concept of the devices and their functioning than could otherwise be obtained. In the problems set there have been replacements and additions.

The scope and objectives of the book remain essentially unchanged. The text is not offered in any sense as a complete treatise. As before, the material has been limited to that necessary to bring out fundamental principles and methods.

In carrying on the work of this revision, the new coauthor had available, in addition to his own teaching experience, the comments, suggestions, and helpful criticism of many past users of the book. Many of these suggestions have been taken into account in the revision, where logical and feasible and in keeping with the inherent limitations of space and scope. The original authors critically reviewed the entire manuscript, and their suggestions and contributions were incorporated in bringing the manuscript to its final form.

The authors wish to express their appreciation to the manufacturers who generously supplied photographs and explanations. They also acknowledge that they have freely consulted other books on the subject, publications of manufacturers, engineering periodicals, etc., and have attempted to give specific acknowledgment where suitable and where material has been consciously used.

The book presupposes a background of basic analytical mechanics, although Part I may be given without this requirement. The authors have attempted to keep in mind the varying requirements of instructors who use the book, and the material in Part II is so arranged that, if time is limited to a brief course, the more complex examples and methods in the latter part of each chapter may be omitted without interfering with the sequence of the work. On the other hand, the material may be adapted to more advanced work by emphasis on the more complex problems or applications to problems of the type outlined in Chap. 13.

The problems for student solution, which are placed at the end of the text, have been selected with the view of illustrating, as far as possible, practical applications of the fundamental theory covered in classroom discussion. Answers to many of the problems are given at the end of the problem section. Most of the problems are of a practical engineering nature in that the mechanisms and data have been selected from manufacturers' blueprints and catalogues, engineering periodicals, and similar sources. This set of typical problems will be of particular value to the instructor using the book for the first time and will serve as a guide in developing additional problems.

The book is particularly suitable for use in a 5-hour semester course of three discussion periods and two 3-hour problem-solving periods per week for a term of 16 weeks. However, Part I may be given as a 2- or 3-hour course in mechanism, or kinematics (by incorporating the chapters on velocities and accelerations from Part II, if desired), followed by Part II as a 3-hour course in dynamics of machinery. The book also is easily adapted to courses given on the quarter system, with terms of 11 or 12 weeks.

In this era of high speeds in industrial machinery, the subjects of kinematics and dynamics of machinery are of increasing importance to engineers. The intention of the authors is to include in one textbook just the amount of material necessary to give the student an understanding of the fundamentals of these subjects in the amount of time available in the usual mechanical engineering curriculum, and to serve as a sound basic background for solving many practical problems or for pursuing more advanced or specialized studies.

C. W. Ham
E. J. Crane
W. L. Rogers

CONTENTS

PART I

MECHANISM

CHAPTER 1

INTRODUCTORY CONSIDERATIONS

Engineering is based on the sciences of *mathematics*, *physics*, and *chemistry*. These three sciences are more or less interdependent, this fact being particularly true of mathematics and physics. *Mechanics* is one of the several branches of physics. It deals with the relations that exist between masses, motions, and forces and is usually considered under three main divisions, namely, *kinematics*, *statics*, and *kinetics*. The first division deals with motion alone, the second with forces considered apart from motion, and the third with forces in conjunction with motion. Since both statics and kinetics deal with the action of forces on bodies, that part of mechanics embraced in these two subdivisions of the subject is sometimes called *dynamics*.

1-1. Scope of the Subject. *Mechanics of machinery* deals with the study of masses, motions, and forces in machines. The subject is restricted so as not to include the study of elastic forces and deformations in machine parts, these lying in the province of *mechanics of materials;* neither does the subject include the determination of the sizes and shapes of machine parts necessary for carrying the loads and transmitting the forces imposed upon them, this lying in the province of *machine design*.

The subject of mechanics of machinery as presented in this book will be considered under two main divisions, as follows: Part I, Mechanism; and Part II, Kinematics and Dynamics of Machinery.

1-2. Definitions. A *machine* is a combination of rigid or resistant bodies, so formed and connected that they move upon each other with definite relative motions and transmit force from the source of power to the resistance to be overcome. Within itself the machine thus has the two functions of transmitting definite relative motion and of transmitting force. When attention is concentrated upon the first of these—strength and rigidity needed for transmitting force being assumed—the term *mechanism* is applied to the combination of geometrical bodies constituting the machine or part of the machine. A *mechanism* may therefore be defined as a combination of rigid or resistant bodies, so formed and connected that they move upon each other with definite relative motions. The term *mechanism*, used in a broad sense as in the title of Part I of this book, is concerned primarily with a study of the functions, geometrical properties, and relative motions in various mechanisms in common use. The study of motions in a broader sense—to include velocities and accelerations—is reserved for Part II of the text. Although the general scheme is *mechanism* first, as defined above, and then *kinematics and dynamics of machinery*, the treatment of the separate divisions is not exclusive, elementary principles of mechanics being introduced in Part I as the occasion demands.

Kinematics of machinery treats of the motions of machine parts without considering the manner in which the influencing factors (force and mass) affect the motion. It deals with the fundamental concepts of space and time, and the quantities velocity and acceleration derived therefrom. *Dynamics of machinery*, embracing the two subdivisions *statics of machinery* and *kinetics of machinery*, treats of machine parts that are acted upon by both balanced and unbalanced forces, taking into account the masses and accelerations of the parts as well as the external forces.

1-3. Machine Motion. The motion of a machine member is usually such that all its points move in parallel planes. This type of motion is called plane motion.

A *mechanism* in which all points move in parallel planes is called a *plane mechanism.* The true motions of all points of this type of mechanism can be projected onto a single plane, allowing a motion analysis to be made in one plane. Plane motion sometimes can advantageously be considered to be a combination of two simpler types of motion, namely, rotation and translation. *Translation* of a machine member occurs when any line drawn on the member remains always parallel to itself during the motion. *Rotation* is present when such a line does *not* remain parallel to itself, or, in other words, the line changes its direction during the motion. The motion of the driving member of machines is usually either rotation about some fixed axis or translation, and this is true also for the motion of the final driven member which overcomes the useful resistance.

Plane mechanisms are a special case of *space mechanisms*, in which points do not necessarily move in parallel planes. For instance, two examples of *space motion* (as distinguished from plane motion) which are of importance are helical motion and spherical motion. A point on a screw thread which is advancing through a threaded hole is an example of helical motion. It is of interest to note that helical motion is a combination of rotation about an axis and a translation along that axis. When a body has spherical motion, all points in the body move in the surfaces of spheres having a fixed point for a common center, as in the case of the flyball governor, the universal joint, ball-and-socket joint, etc.

A body that has no material connection with other bodies, so that the path of its motion varies with the external forces acting upon it, is said to have *free motion.* The planets, for example, have free motion. A body that has a material connection with another body, so that the path of its motion relative to that body must be of a definite character, regardless of the external forces that may act upon it, is said to have *constrained motion.* The crankpin of an engine, for example, has constrained motion. All machine members are dependent upon constraint for their proper action.

1-4. Cycle, Period, and Phase of Motion. When, after starting from some simultaneous set of relative positions, the parts of a machine have passed through all the possible positions they could assume and have returned to their original relative positions, they are said to have completed a *cycle* of motion. The time required for a cycle is called a *period.* The simultaneous relative positions occupied by the parts of a machine at any instant during the cycle constitute a *phase.*

In the ordinary steam engine, in many diesel engines, and in some small-sized gasoline engines, the energy cycle and also the motion cycle correspond to one revolution of the crankshaft (two-stroke cycle). In the ordinary gasoline engine, however (and in many diesels also), each energy cycle requires two revolutions of the crankshaft or four strokes of the piston: the intake, compression, expansion, and exhaust strokes. Hence, for each energy cycle of the engine, the main parts, namely, crank, connecting rod, and piston, complete two motion cycles. The camshaft and valves, however, complete only one motion cycle during this period; so that, considering the engine as a whole, the moving parts return to their initial relative positions only after two revolutions of the crankshaft or four strokes of the piston. For this reason engines of this type are referred to as four-stroke-cycle engines, or more simply as four-cycle engines.

1-5. Continuous, Intermittent, and Reciprocating Motion. The motion of a machine member is *continuous* if during each successive cycle it neither stops nor reverses; *intermittent* if during each cycle it stops for a finite interval of time; and *reciprocating* if during each cycle its motion is reversed. For example, the crankshaft of the gasoline engine has continuous motion, the valve tappet rods an intermittent reciprocating motion, and the piston a purely reciprocating motion.

The rotation of a member may be continuous, intermittent, or reciprocating, the last being more commonly designated as oscillating. The translation of a body, which may be either rectilinear or curvilinear, also may be continuous, intermittent, or reciprocating, with the exception that, obviously, rectilinear translation cannot be continuous. The motion of the side rods of a locomotive moving on a straight track is a good example of curvilinear translation. When the term translation is used without qualification it is understood to mean rectilinear translation.

1-6. Pairing Elements. Pairs. The resistant parts of a machine must be so placed and so connected to adjacent members that the relative motion between any two is of a definite character. The geometrical forms placed upon two bodies so that they may be connected in this manner are called *pairing elements.* Two elements thus joined constitute what is called a *pair*, and the process of connecting them is called *pairing.*

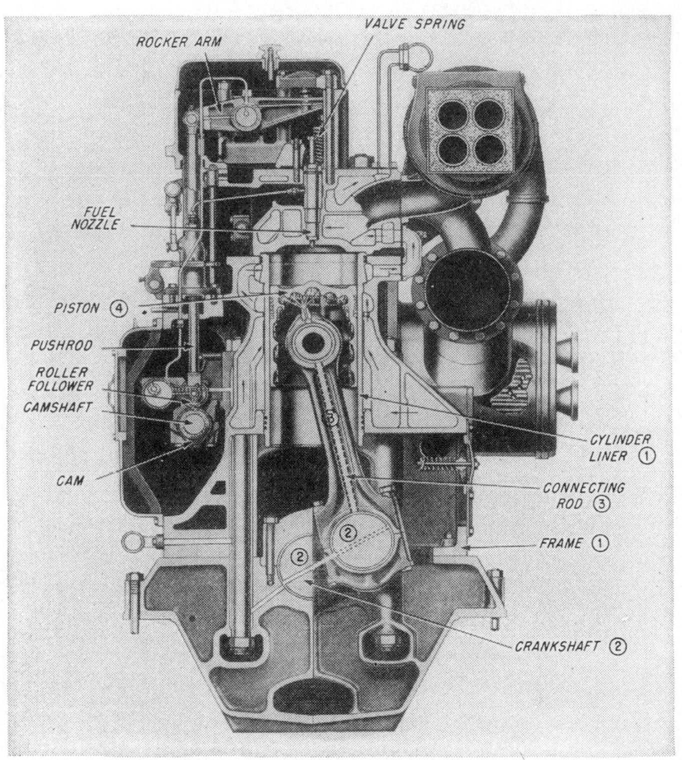

FIG. 1-1. Cross section through power cylinder of a diesel engine. (*Courtesy of Nordberg Manufacturing Co.*)

Pairs which permit surface contact are called *lower* pairs; those which permit only point or line contact are called *higher* pairs. The most common examples of lower pairing are the turning pair and the sliding rectilinear pair.

The diesel engine illustrated in Fig. 1-1 contains examples of the various types of pairs. The connection between the connecting rod (3) and the crankshaft (2) is a *lower* pair of the *turning* variety, with sliding occurring at the contact surfaces of the journal and bearing. A *sliding rectilinear* pair is exemplified by the piston and cylinder liner. A higher pair is formed by the cam-and-roller-follower combination, with contact along a line only.

Other examples of higher pairs are ball and roller bearings and the contact surfaces of a pair of mating gear teeth. These examples are so familiar as to require no explanatory illustrations. Usually, a lower pair completely constrains the motion of the pair of links connected, whereas in the case of a higher pair the constraint is often incomplete.

It frequently occurs that a single lower pair may have several contact surfaces. An example of a turning pair with multiple contact is the engine crankshaft and its bearings. An example of a sliding pair with multiple contact is the piston link of a steam engine, consisting of piston, piston rod, and crosshead, with sliding contact between piston and cylinder, piston rod and stuffing box, and crosshead and guides. Thus a single sliding rectilinear pair may have any number of contact surfaces, provided that these surfaces are parallel, and a single turning pair may have any number of journals and bearings, provided that they all have the same axis.

1-7. Links. A *link* may be defined as a rigid body having two or more pairing elements by means of which it may be connected to other bodies for the purpose of transmitting force or motion. A link is sometimes defined in a broader sense to include bands and fluid connectors. In ordinary practice, however, the name is applied to a rigid connector, and it is in this sense that the term will be used in this work.

In every machine there is one link that either occupies a fixed position relative to the earth or carries the machine as a whole along with it during motion. This link is essentially the frame of the machine, and is called the *fixed* link. As an example showing the use of the term *link*, let the engine shown in Fig. 1-1 be considered. Taking into account only the main members of the machine, it is seen that the number of links is four. Link 1 is the fixed link and includes the frame and all other stationary parts such as the cylinder, crankshaft bearings, and camshaft bearings. Link 2 includes the crankshaft, flywheel, etc., all of which have a motion of rotation about the axis of the main bearing. Link 3 is the connecting rod and, being an intermediate or "floating" link, has general plane motion. Link 4, the piston, has a motion of reciprocating rectilinear translation. It is evident that a link includes all the auxiliary parts that are rigidly fastened to the main part of the link and move with it as a unit.

1-8. Kinematic Chains. When a number of links are connected by means of pairs, the resulting system is called a *kinematic chain*, or simply a chain. A chain may be *locked*, *constrained*, or *unconstrained*. If a chain is locked, no relative motion is possible between the links, and any motion that takes place is that of the chain as a whole. A locked chain is shown at *a* in Fig. 1-2. If a chain is constrained, definite relative motion between the links is possible; i.e., if one of the links is held fixed and another put in motion, all points of the remaining links will move in certain definite paths and will always move in these same paths no matter how many times the motion is repeated. A constrained chain is shown at *b* in Fig. 1-2. If a chain is unconstrained, the relative motions of the links are not definite; i.e., if one link is held fixed and the motion of another link is repeated, the points of the remaining links will not, in general, follow the same paths. An unconstrained chain is shown at *c* in Fig. 1-2.

The locked chain has its application in structures. The constrained chain is the one with which kinematics is chiefly concerned and is the basis of all machines. The unconstrained chain is of little use practically.

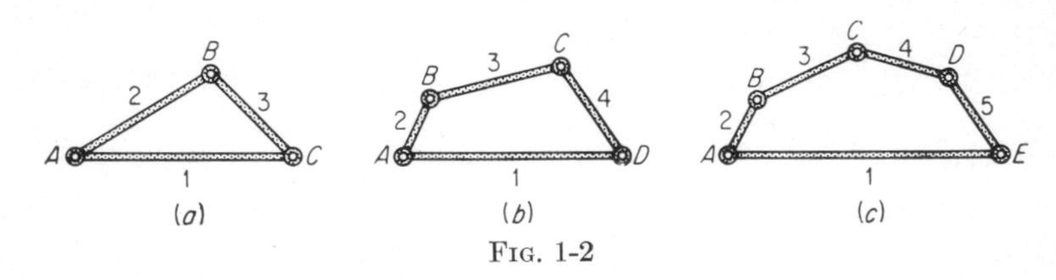

Fig. 1-2

It should not be inferred that only four-link chains are constrained. For example, the shaper mechanism of Fig. 1-4 contains *six* links or parts. A sliding-follower-and-cam mechanism can have as few as *three* parts.

1-9. Mechanism. Machine. Structure. If one of the links of a constrained kinematic chain is made a fixed link, the result is a *mechanism.* If a different link of the same mechanism is made the fixed link, the result is a different mechanism. Evidently there are as many different mechanisms for the same chain as there are links in the chain. Some confusion may arise as to the distinction between the term *mechanism* and the term *machine*, both of which are fundamentally a constrained chain with one link fixed. As stated in Art. 1-2, the primary function of a mechanism is to transmit or modify motion, whereas that of a machine is to modify energy and do work. If, therefore, the chain is considered solely from the viewpoint of motion modified or transmitted, it should be referred to as a mechanism. If, on the other hand, the chain is considered as an agent for modifying energy or doing useful work, it should be thought of and referred to as a machine. There is a large class of instruments or apparatus, such as clockwork, adding machines, typewriters, engineers' instruments, etc., that may seem to be on the border line between mechanism and machine, according to the above definitions. But they are more properly classified as mechanisms, since the force is no greater than necessary to produce relative movement. In other words, change of position of members is the useful effect, and there is no performance of external work. When a kinematic chain is analyzed as a mechanism, no special consideration need be given to the forms or proportions of the links except in so far as concerns the location of the pairing elements. As a machine, however, the requirements of strength, stiffness, clearances, etc., make it imperative to consider the links in all their details.

The leading distinction between a *machine* and a *structure* is that the former serves to modify and transmit energy, or force and motion, while the latter modifies and transmits force only.

1-10. Machine Representation. Skeleton Outlines. For the purpose of kinematic analysis, a mechanism may be represented in an abbreviated or skeleton form called the *skeleton outline* of the mechanism. The skeleton outline should give all the geometrical information necessary for determining the relative motions of the links. In Fig. 1-3 the skeleton outline has been drawn for the engine shown in Fig. 1-1. It will be observed that in this skeleton outline everything necessary for determining the relative motions of the main links has been shown, namely, the length O_2A of the crank; the length AB of the connecting rod; O_2, the location of the axis of the main bearing; and the path O_2B of point B, which represents the wrist-pin axis.

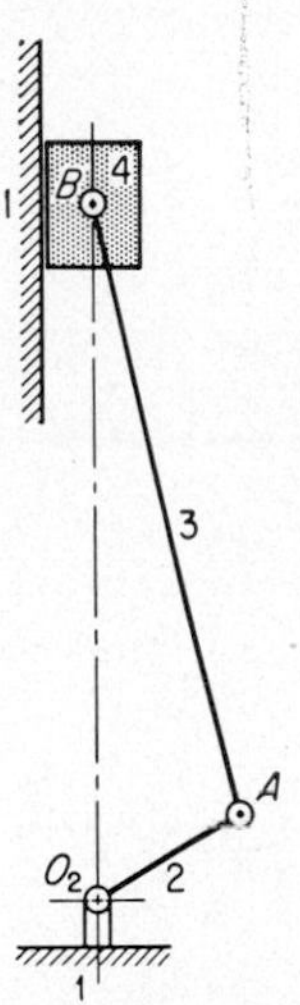

FIG. 1-3. Skeleton outline of the main parts of the diesel engine of Fig. 1-1.

When drawing the skeleton outline, a link that has only two pairing elements may be represented by a straight line joining the axes of the two elements, as shown by link 2 in Fig. 1-3, for example. The turning pairs themselves may be represented by small circles, often with dots at their centers, as shown at O_2, A, and B. The fixed link need not be shown except in the vicinity of its pairing elements, where crosshatching may be used, as shown in Fig. 1-3, as a conventional way of indicating that it is the fixed link. A sliding link may be represented in the conventional manner shown as 4 in the figure, i.e., by a rectangle pivoted on the axis of the pairing element of the link with which it has a turning connection. A simple method of representing a turning connection with the fixed link is that shown at O_2 in the figure. That shown at O_2 or O_4 in Fig. 1-4*b* or 1-15*a* may be used, if preferred. A link with three or more pairing elements is represented by the geometric figure formed by connecting the axes of the pairing elements, as shown by link 3 in Fig. 1-15, for example.

An example of the skeleton outline of a more complex mechanism is illustrated by the shaper mechanism in Fig. 1-4. In this mechanism, gear 2 is driven by a pinion on the shaft of the drive pulley (not shown) and carries a pin A on which is pivoted the block 3. Through the sliding action of the block in the slot in the vibrator (link 4) and the rotation of gear 2, link 4 is caused to oscillate. Since the ram, link 6, is so constrained

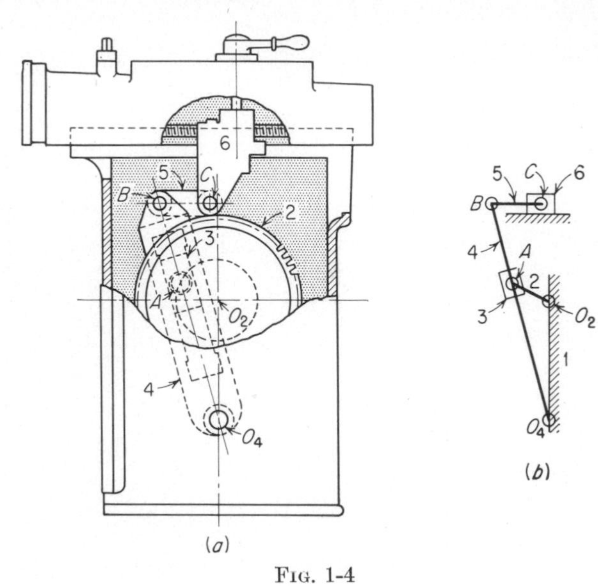

FIG. 1-4

as to have only reciprocating motion in a horizontal line, whereas the end of link 4 moves in the arc of a circle as it oscillates, it becomes necessary to have link 5 interposed between links 4 and 6. It is seen that link 2, the main driving link, is in effect a crank, and is so represented in the skeleton outline shown at b. It is also evident that link 6 is completely represented in skeleton outline by a rectangle pivoted on the end of link 5 and guided in a horizontal direction by link 1. The simplest representation of the action between links 3 and 4 is a rectangle sliding on a line as shown.

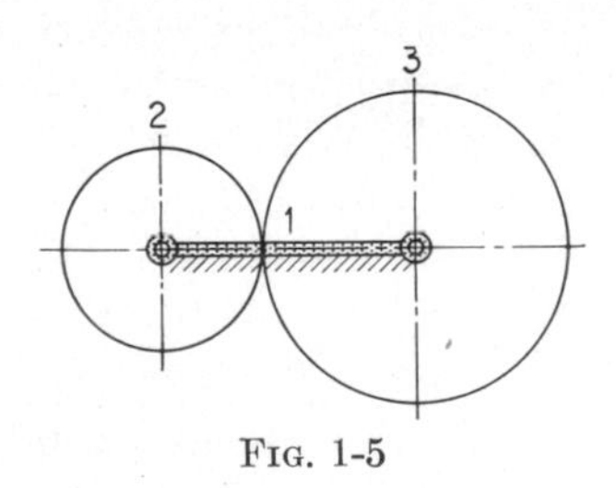

FIG. 1-5

For links containing higher pairs, it is usually impossible to make much reduction from the actual form of the link when drawing the skeleton outline; this is particularly true for certain types of cams. Two spur gears in mesh may, nevertheless, be represented by their pitch circles as shown in Fig. 1-5. Sometimes the existence of higher pairing in a mechanism may be disregarded, as for example, in a mechanism in which a ball bearing is used in place of the ordinary pin-and-eye joint.

The ball bearing is distinctly a case of higher pairing in so far as concerns its elements, but it acts as a whole exactly like an ordinary turning joint.

1-11. Expansion of Pairs. The size of the pairing elements has no effect on the relative motion of the links connected by the pair. Sometimes by simply expanding pairing elements the appearance of a machine may be changed beyond recognition without altering in the slightest degree the character of the motion. Thus it is often the case that mechanisms differing widely in appearance have the same or similar skeleton outlines, as will be made clear by consideration of the mechanisms shown in Figs. 1-6 to 1-11.

In each of these figures the skeleton outline has been superimposed on the drawing of the actual mechanism at the place where the expansion has been made. An inspection of the several skeleton outlines will make it evident that they are identical, thus showing that the mechanisms are all the same from a kinematic point of view. The differences in appearance are chiefly due to modifications in the detail (but not in the character) of the pairs.

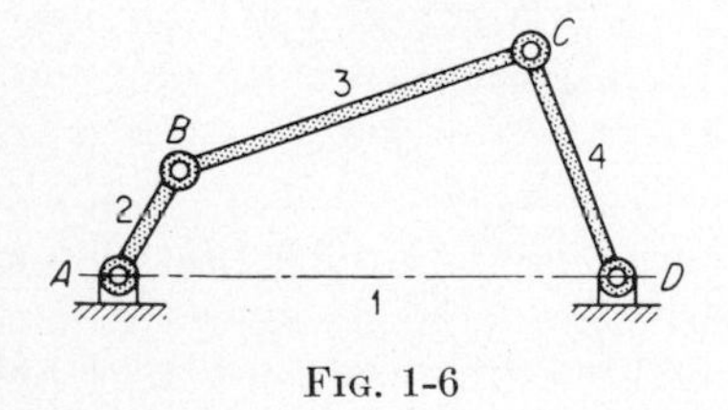

FIG. 1-6

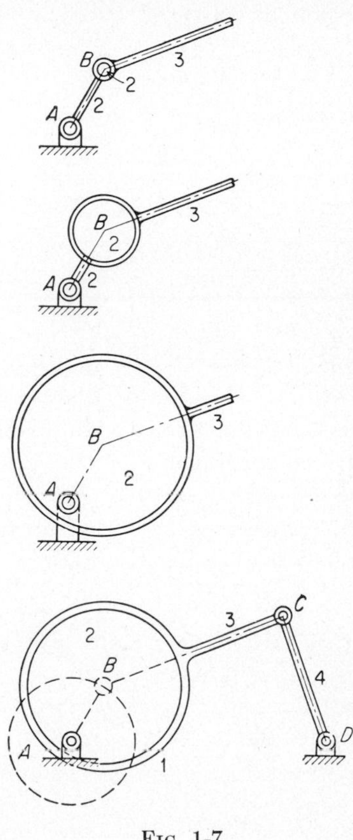

FIG. 1-7

With the fundamental mechanism in Fig. 1-6 as a starting point, link AB was replaced in Fig. 1-7 by an eccentric disk by the simple process of enlarging or expanding the pairing elements at B to include pair A. In Fig. 1-6 the pairing elements were represented by a small circle with the center at B. In Fig. 1-7 they are represented by the enlarged circle (the circumference of the disk) with the position of center B unchanged. The relative positions or motions of the two links affected, 2 and 3, have, therefore, suffered no change whatever. It is evident from a comparison of Figs. 1-6 and 1-7 that an eccentric is equivalent to a crank whose length is equal to the eccentricity.

In Fig. 1-8 the pair D has been enlarged to include C. Evidently the motion of DC will be equally constrained if the sides of the circular disk are cut away, leaving the zone indicated by the solid lines. A further modification is shown in Fig. 1-9, where the link CD is given the form of an annular ring instead of a disk. The next step in simplification is to cut away all the ring except a small block containing the pair C.

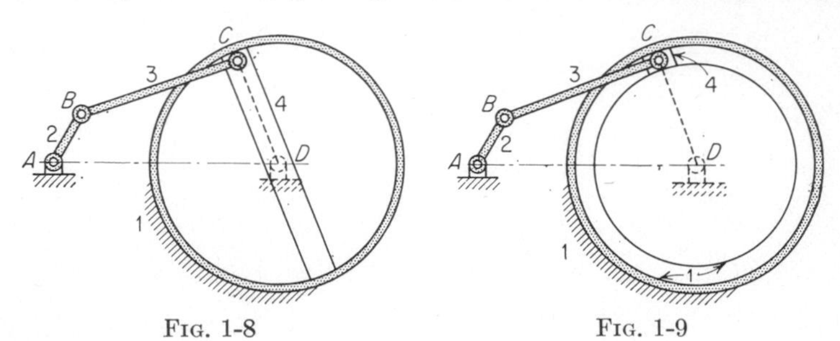

FIG. 1-8 FIG. 1-9

Finally, since by the rotation of crank AB the pair C is forced to reciprocate in a limited circular arc, the slot in which the block moves may as well be shortened to the length actually required as shown in Fig. 1-10. By comparing Fig. 1-10 with Fig. 1-6, it is seen that the sliding connection C in Fig. 1-10 is equivalent to a turning pair whose center of curvature is the center of curvature D of the slot. It is also evident that the block containing the point C is equivalent to a bar of length CD equal

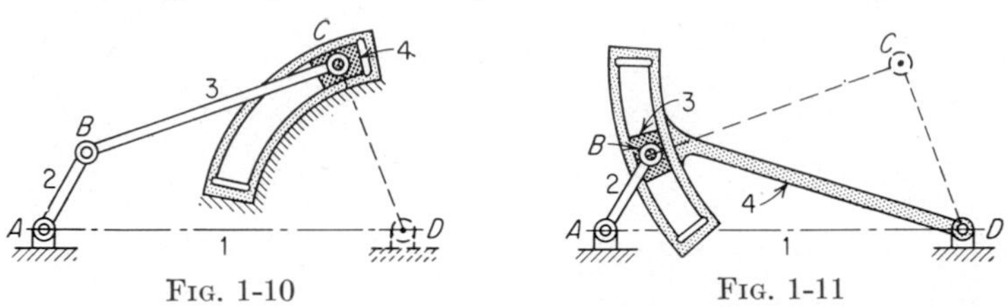

FIG. 1-10 FIG. 1-11

to the radius of curvature of the slot and pivoted at the center of curvature D. By increasing the radius of curvature of the slot, the equivalent of very long links may be obtained without increasing the dimensions of the machine. Thus the ordinary sliding rectilinear pair, consisting of a block moving in a straight slot, is equivalent to a turning pair whose axis is at an infinite distance.

By expanding pair C in a series of modifications similar to those described above, the mechanism shown in Fig. 1-11 has been derived from Fig. 1-6, the relative motions of the links remaining unchanged. The solution is left to the reader.

1-12. Inversions. *Inversion* is a term used in kinematics for a reversal or interchange of form or function, as applied to kinematic chains and mechanisms.

It was stated in Art. 1-9 that from a given kinematic chain as many mechanisms may be derived as the chain has links, by fixing each link in turn. These mechanisms are called *inversions* of the original mechanism, and the exchange of one fixed link for another is known as *inversion of the mechanism.*

If, for example, link 2 of the four-link mechanism shown at *a* in Fig. 1-12 is fixed and link 1 made free to move, the resulting mechanism, shown at *b*, is said to be an inversion of the original mechanism. It is

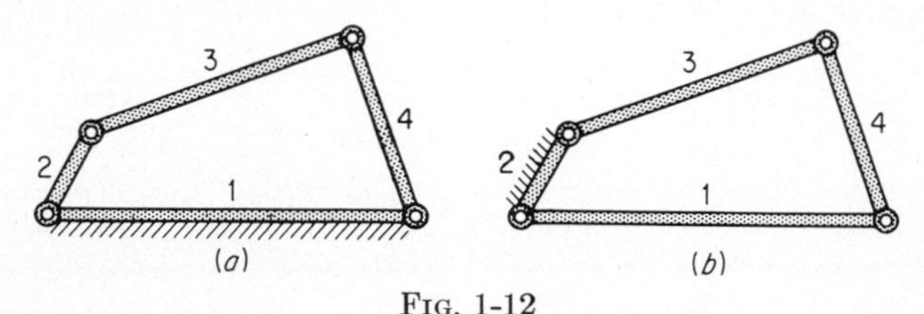

FIG. 1-12

important to keep in mind that the inversion of a mechanism does not change the motions of its links relative to each other but does change their absolute motions. As another example of the inversion of a mechanism, consider the gear train at *a* in Fig. 1-13. The epicyclic gear train at *b* is an inversion of the original train at *a*. In making an inversion of a mechanism it may often be necessary to make an inversion of a pair or change proportions of links to suit requirements.

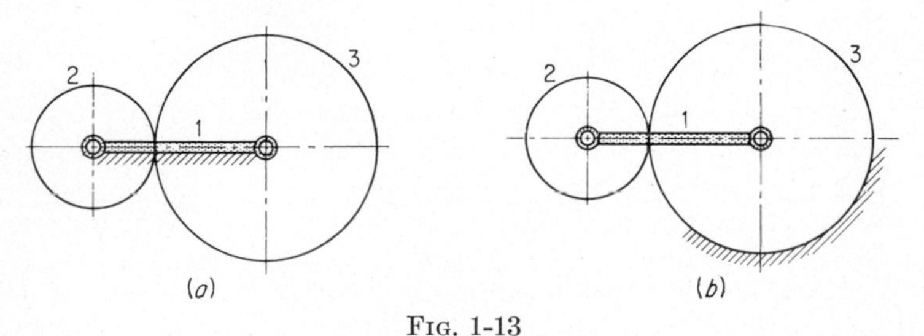

FIG. 1-13

Inversion of function may be noted in the case of such machine tools as the shaper and planer, lathe and boring mill, etc. Whereas in the shaper the work is held stationary while the cutting tool moves, in the planer the work moves and the cutting tool is stationary. Similarly, in drilling and boring on a lathe the tool is stationary, while in a drilling machine the work is stationary. Functional inversion occurs in a machine, also, when reversal of motion causes the original driving member to become the driven member.

1-13. Point Paths—Displacement Diagrams. In the kinematic study of mechanisms, it is frequently desirable to plot the paths of the motion of certain points and also to construct a diagram of the displacement of some point corresponding to the motion or displacement of some other point. In most cases the points considered are those that represent the axes of the moving pairs. In constructing a displacement diagram, the usual scheme is to plot displacements of some point on the last driven link against displacements of some point on the first driving link, employing any system of coordinates that may be found convenient. Usually rectangular coordinates are the most convenient to use. Taking first a simple illustration, in the engine mechanism (Fig. 1-14*a*), it is known that one end A of the connecting rod describes a circle and that the other

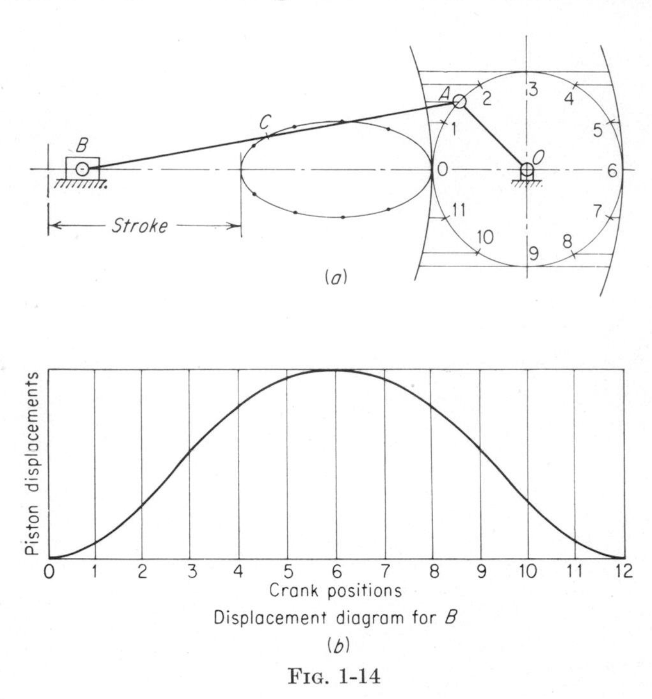

FIG. 1-14

end B describes a straight line. By assuming, say, 12 equally spaced positions of the crankpin A in the complete cycle of its motion, a corresponding number of positions of any other point such as C on the rod may be plotted, giving the elliptical curve shown in the figure. In a similar manner a diagram of the displacements of the piston B corresponding to the displacements of crankpin A may be plotted. In Fig. 1-14*b* displacements of the crankpin A, or angular displacements of the crank, have been plotted as abscissas, and piston displacements from top dead center position as ordinates. It is important to note that the length of the base line, representing abscissas, has been taken as an arbitrary length and is not necessarily equal to the circumference of the crank circle. Since the information sought is piston displacements correspond-

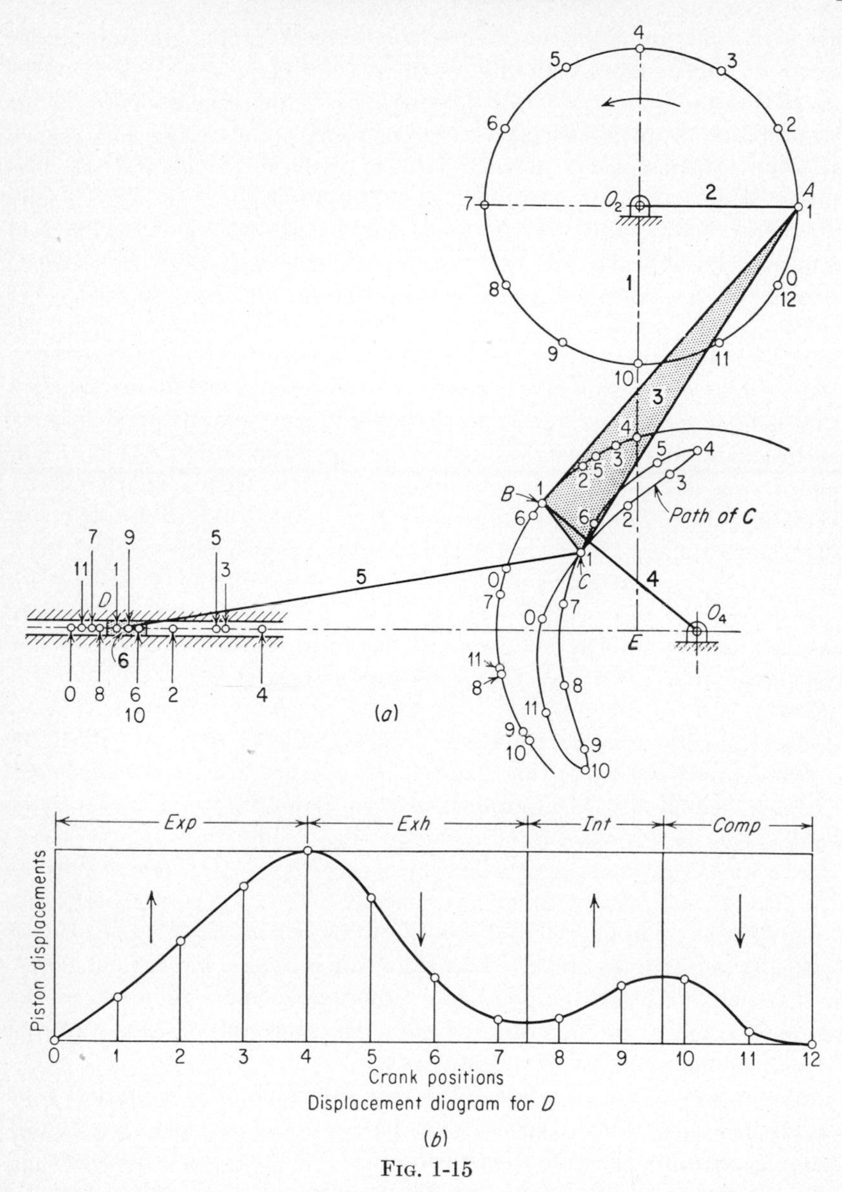

FIG. 1-15

ing to equally spaced crank positions, it is evident that in such a diagram any convenient length may be chosen for the base line. The use that is made of this particular diagram will become evident after studying Chap. 2.

As a further example of problems in plotted movements, let it be required to plot the paths of the motion of points B and C and the displacement diagram of piston D for the complete cycle of the Atkinson gas-engine mechanism, shown in skeleton outline in Fig. 1-15*a*. This type of engine is now obsolete because of its mechanical complexity, although the characteristics of a short compression stroke and a long expansion stroke are desirable even today from the standpoint of high thermal efficiency. For example, a combination of diesel engine and gas turbine which is driven from the engine exhaust gas is a means of obtaining the same general characteristics, but with different machinery.

The Atkinson mechanism is ingenious, and therefore of kinematic interest. The engine was of the four-stroke-cycle variety and was designed so that the four strokes of the cycle occurred in one revolution of the crank O_2A, with longer strokes for expansion and exhaust than for intake and compression. Just how well this end was attained may be seen by examining the piston-displacement diagram at Fig. 1-15*b*. This diagram was obtained by determining the piston displacements corresponding to the 12 crank positions indicated and plotting these displacements as ordinates on the arbitrarily chosen length of base line 0-12. The plotting of point paths of B and C is an incidental part of the work of obtaining the piston displacements but useful for crankcase design. This mechanism affords an excellent example of the value of such a diagram as that shown at Fig. 1-15*b*, as an aid in studying machine motions.

1-14. Velocity and Acceleration. These aspects of motion will be introduced only briefly in this chapter, as they are discussed extensively in Chaps. 8 and 9. The purpose here is primarily to review certain concepts for readers who probably already have some knowledge of them. *Velocity* (V) of a point (also called linear velocity) is the rate of change of position of the point with respect to time. Velocity has the attributes of magnitude and direction, and so it is a *vector* quantity. The *magnitude* of velocity is called the *speed.* If an equation is available which expresses the distance s moved by a point as a function of time, the speed at any instant is equal to the derivative ds/dt. Examples of velocity units are feet per second or inches per minute.

Angular velocity (ω) of a line is the rate of change of the angular position of the line with respect to time. Mathematically, $\omega = d\theta/dt$. Examples of units are radians per second or revolutions per minute.

Acceleration (A) of a point (also called linear acceleration) is the rate of change of velocity of the point with respect to time. It is a vector quantity. There are two types of acceleration. The time rate of change of the *magnitude* of velocity is called the *tangential* acceleration A^t, and the time rate of change of the *direction* of velocity is called the *normal* acceleration A^n. Examples of acceleration units are feet per second squared (ft/sec^2) or inches per minute squared.

Angular acceleration (α) of a line is the rate of change of angular velocity of the line with respect to time. Mathematically, $\alpha = d\omega/dt$. Typical units are radians per second squared (radians/sec^2).

A simple example will be reviewed which is probably familiar to most readers. Figure 1-16 shows a rigid body rotating about a fixed axis. At a of the figure, assume point P on the body to move a distance s (along the arc) to position P' while the body turns through the angle θ. From the definition of a radian, the angle θ in *radians* is equal to s/r (dimensionless). Therefore,

$$s_P = r\theta \tag{1-1}$$

in which θ must be in *radians* and s and r must have the same length units.

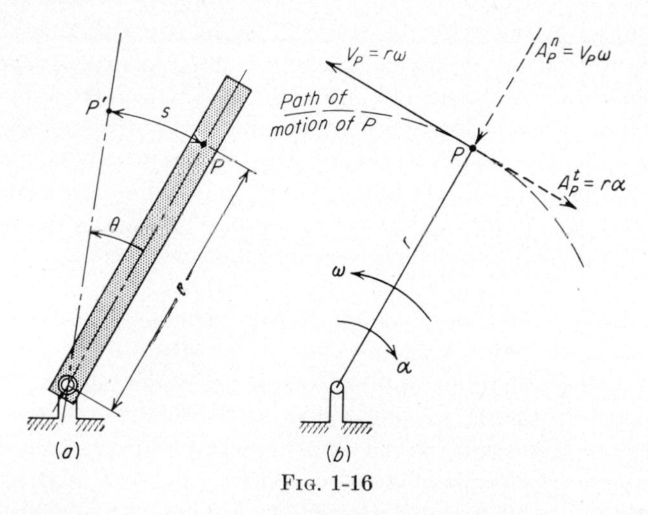

FIG. 1-16

The magnitude of the velocity (the speed) of P at any instant is

$$V_P = \frac{ds}{dt} = r\frac{d\theta}{dt} = r\omega \tag{1-2}$$

Here, ω must have units of *radians* per unit time.

The rate of change of *magnitude* of V_P is the *tangential* acceleration A_P^t.

$$A_P^t = \frac{d(r\omega)}{dt} = r\frac{d\omega}{dt} = r\alpha \tag{1-3}$$

The term α must have units of *radians* per unit time squared. The rate of change of *direction* of V_P is the *normal* acceleration A_P^n.

$$A_P^n = V_P\omega = \frac{V_P^2}{r} = r\omega^2 \tag{1-4}$$

The proof of Eq. (1-4) is given in Chap. 9 and will not be duplicated here. Most readers of this book probably will recognize Eq. (1-4) from previous studies.

Figure 1-16*b* shows the proper corresponding directions of the various velocities and accelerations which have been reviewed.

1-15. General Remarks. All mechanisms derived from one kinematic chain have been shown to be inversions of that chain. These inversions frequently differ very widely in general appearance and in purpose.

The *motions* in a machine depend entirely on the relative positions of the pairs and not at all upon the size and shape of the links. The *appearance* of a machine depends largely on the size and shape of its component links. Links are of diverse shapes; a crank and eccentric, for example, which are kinematically identical but quite different in appearance, are shown in Figs. 1-6 and 1-7.

In the study of mechanics of machinery it is necessary to eliminate from consideration problems of the design and manufacture of machine parts and confine attention largely to the skeleton outlines of the moving parts, since the motions of the parts are of chief concern. Once the kinematic relations among the links of a kinematic chain are known, those for all mechanisms obtained by inversions or expansions are also known. This fact is of great importance, for *no matter which link is fixed, the relative angular motions of the constituent links of a kinematic chain are always the same.* The study of kinematics of machinery is, therefore, greatly simplified and needless repetition avoided if the fundamental kinematic chains are studied in place of the many mechanisms which are kinematically identical.

Furthermore, the study of the subject may be approached from two different points of view, namely, the analysis of forces and motions in existing machines, and the creation of machines to produce desired motions and forces. The usual procedure is to study and analyze existing machines, and that method will be followed in this book.

CHAPTER 2

LINKAGES AND FLEXIBLE CONNECTORS

In Art. 1-7 a link was defined as a rigid body having two or more pairing elements by which it could be connected with other bodies for the purpose of transmitting force or motion. The term *linkage* refers to a mechanism made up of such links as cranks, levers, and rods, with turning and rectilinear-sliding pairs, thus excluding from the classification cams, gears, and flexible connectors such as belts, ropes, and chains.

The topics of cams and gears are taken up in later chapters. The first part of the present chapter is concerned with various representative *linkages*, and the latter part with *flexible connectors* involving belts, ropes, and chains.

2-1. Functions of a Link Mechanism. The function of a link mechanism is to produce rotating, oscillating, or reciprocating motion from the rotation of a crank, or the reverse. Stated more specifically, linkwork may be used to convert:

1. Continuous rotation into continuous rotation, with a constant or variable angular velocity ratio.
2. Continuous rotation into oscillation or reciprocation (or the reverse), with a constant or variable velocity ratio.
3. Oscillation into oscillation, or reciprocation into reciprocation, with a constant or variable velocity ratio.

2-2. The Four-link Mechanism. One of the simplest examples of constrained linkwork is the *four-link* mechanism. A great variety of useful mechanisms may be formed from the ordinary four-link mechanism by means of slight modifications such as changing the character of the pairs, proportions of the links, etc. Furthermore, it may be observed that many complex link mechanisms may be made up of a combination of two or more such mechanisms. A typical case is shown in Fig. 2-6. A great majority of four-link mechanisms fall into the two following classes: (1) the four-bar mechanism, and (2) the slider-crank mechanism.

2-3. The Four-bar Mechanism. This linkage, in its simplest form, has four bar-shaped links and four turning pairs, as shown in Figs. 2-1 and 2-2, which represent *open* and *crossed* linkages respectively. The fixed link may actually be bar-shaped, but more frequently it represents the frame of a machine and in that case is usually a massive casting of irregular shape.

In order to facilitate discussion, one of the rotating members of the four-bar mechanism will be called the crank, or driver, and the other will be called the rocker or follower. The floating link that connects the crank and the rocker will be called the connecting rod, and the fixed link will be called the frame.

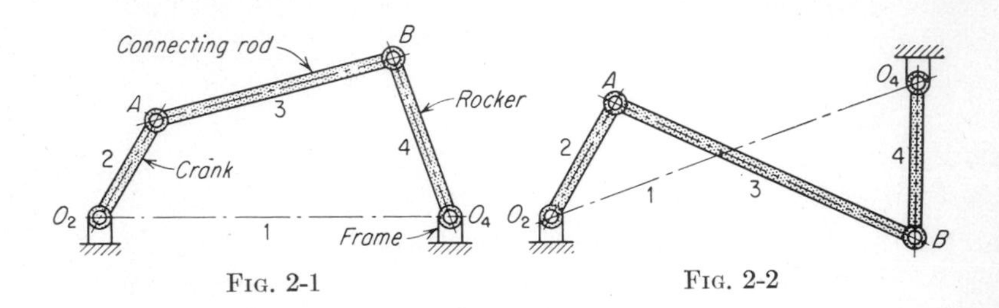

FIG. 2-1 FIG. 2-2

In some of the applications of the four-bar mechanism, the crank has a motion of oscillation, whereas in other applications it must be capable of making a complete rotation. The same statement may be applied to the follower. In cases of less frequent occurrence, both the driver and follower must make complete rotation. In order to meet the requirements of certain types of motion, it is necessary to hold the relative lengths of the links within certain well-defined limits. In proportioning the links, consideration must also be given to the avoidance or the neutralization of dead points in the motion of the driven link.

2-4. Dead Points. Referring to Fig. 2-3, let 2 be the driving link and 4 the driven link. The lengths are such that neither of the links

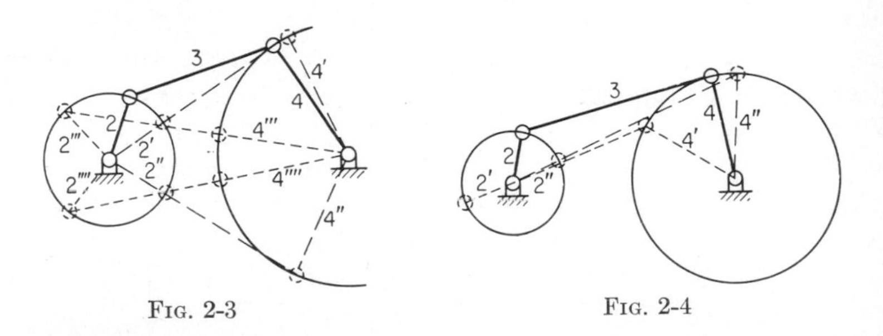

FIG. 2-3 FIG. 2-4

may make a complete rotation. Link 2 may oscillate between the extreme positions $2'''$ and $2''''$, and link 4 oscillates between the extreme positions $4'$ and $4''$. The position $4'''$ of the follower is called a dead-point position, for it is obvious that any driving force transmitted from the driver by the connecting rod 3 will be radial and hence can have no effect on the turning of the link 4. In practical applications of such a linkage, the range of action required is usually much less than the maximum. This is a desirable condition, since the action is not smooth when the follower is near a dead point.

In Fig. 2-4 is shown a four-link mechanism in which the driver may make a complete rotation, and the follower 4 oscillates from 4′ to 4″. It can be seen that if link 4 should become the driver there would be two dead points in the cycle. When link 4 is the driver and complete rotation of the crank 2 is required, it becomes necessary to provide external means, as for example, a flywheel to carry the crank 2 past the dead points.

2-5. Four-bar Mechanisms in Combination. With suitable proportioning and proper choice of relative positions of the links the four-bar mechanism may be made to produce a great variety of relative motions of driver and follower. The arrangement shown in Fig. 2-5 is just one example of the many possibilities. If the driver (link 2) is turned clockwise with uniform motion through the angle θ_2, the follower (link 4) will turn in the same sense through the much smaller angle θ_4 and with decreasing speed, which will become zero when link 2 reaches the position 2′. Reversing the motion of link 2 will cause link 4 to return toward its initial position, its motion being slow at first and then gradually increasing. This type of motion has been used in the Corliss valve gear on steam engines to secure quick opening and closing of the valves. In this particular application neither of the links 2 or 4 makes a complete rotation but they oscillate through comparatively small angles, generally not exceeding these shown as θ_2 and θ_4 in Fig. 2-5.

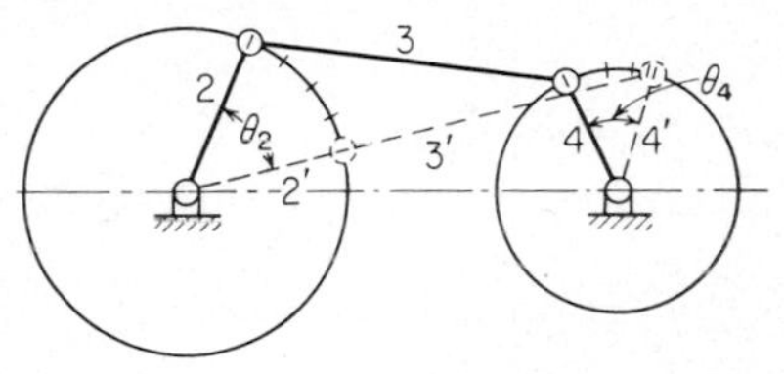

FIG. 2-5

When links 2 and 4 and 1 and 3 (Fig. 2-5) are equal, forming a parallelogram, the driver and follower have simultaneous dead points, and if the driver makes a complete rotation special means must be resorted to for complete constraint of the follower.

The application of these proportions and many others to a variety of mechanisms will be discussed in the following articles of this chapter.

The Corliss valve gear of Fig. 2-6 is one example of many that might have been chosen to illustrate a device incorporating combinations of four-bar linkages. The wrist plate 6 is given an oscillatory motion by means of the complete rotation of driving link 2, which, in the actual mechanism, is in the form of an eccentric. The linkage 1, 6, 13, 14, actuating one of the exhaust valves, will give to the link 14 a very slow motion when M is near M_1, i.e., when the valve is closed, but between M and M_2, when the valve is opening or closing, the motion is much faster. The same statement applies in the case of the admission valves, as shown by the linkage 1, 6, 7, 8.

It is interesting to note that the main part of the valve gear represented in Fig. 2-6 is made up wholly of four-bar mechanisms, six in number, and also that it furnishes examples of four distinct types of these mechanisms. Unit 1, 2, 3, 4 has in eccentric 2 a short crank arm (in skeleton outline) that makes a complete rotation and through rod 3 gives to the rocker arm 4 a symmetrical oscillation about O_4C. In unit 1, 4, 5, 6 this oscillation is transmitted to wrist plate 6, which, so far as its function

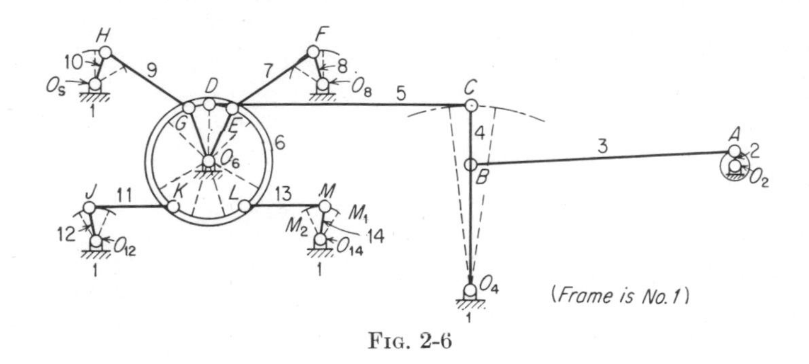

FIG. 2-6

in this section of the mechanism is concerned, is sufficiently represented by radius O_6D. Unit 1, 6, 7, 8 is an example of an *open* four-bar linkage, and unit 1, 6, 13, 14 is an example of a *crossed* four-bar linkage. No attempt is made here to show the remainder of the valve mechanism, namely, the releasing gear whereby arms 7 and 9 alternately take hold and let go of their valves.

2-6. The Drag-link Mechanism. A common application of the four-bar linkage, in which there is a continuous rotation of both the driver and follower, is found in the *drag-link* mechanism shown in Fig. 2-7*a* where it takes the form of a quick-return mechanism for a slotting machine. In Fig. 2-7*b*, only the skeleton outline of the main rotating links is shown. This mechanism has for its object the transformation of a constant angular velocity of the crank 2 into a varying angular velocity of the follower 4, the latter making one revolution in the same time as the crank but at a varying angular velocity during the revolution. For the successful operation of the four-bar linkage in the drag-link mechanism, the requirements are that both the crank and the follower must be able to make a complete rotation and that no dead points shall occur. These requirements are met when the following relations,

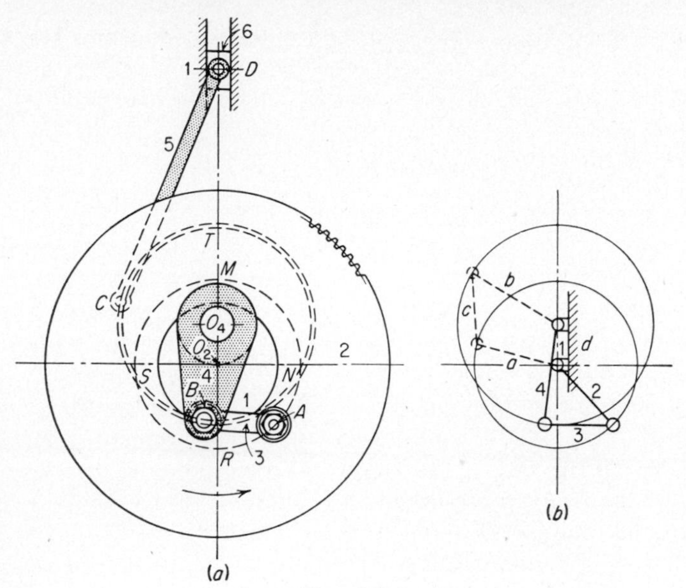
6
1
D
5
T
M
C
O4
O2
2
S
4
N
B
1
A
R
3
(a)
b
c
d
1
a
4
2
3
(b)

Fig. 2-7

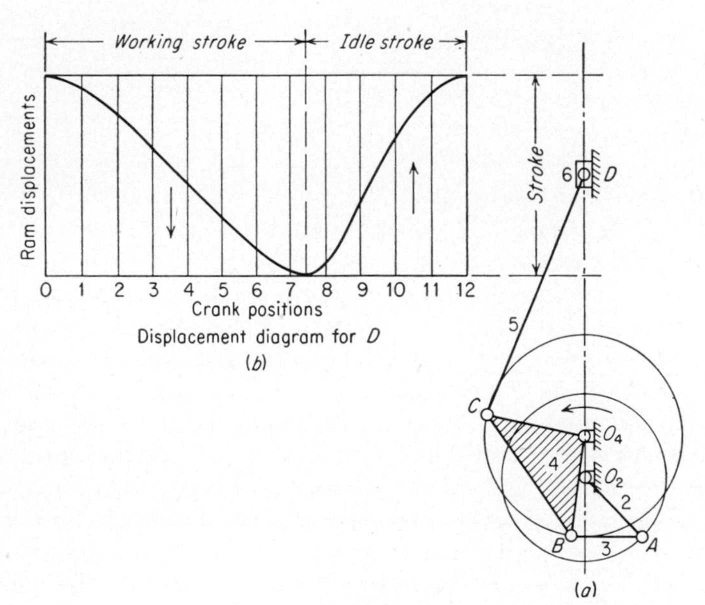
Working stroke
Idle stroke
Ram displacements
0 1 2 3 4 5 6 7 8 9 10 11 12
Crank positions
Displacement diagram for D
(b)
Stroke
6
D
5
C
O4
4
O2
2
B
3
A
(a)

Fig. 2-8

referring to the skeleton outline of the mechanism shown in Fig. 2-7*b*, are true:

$$c > d + b - a$$

and $$c < a + b - d$$

Hence $$d + b - a < a + b - d$$

Therefore $$d < a$$

The action of the drag-link mechanism can best be understood by plotting a displacement diagram, as shown in Fig. 2-8. The complete skeleton outline at *a* shows the proportions of the various links of this mechanism as used in a slotting machine of the type represented in Fig. 2-7*a*. Link 2 is a driving crank, which rotates uniformly, 3 is the drag link, and 4 is a follower link that drives the tool head 6 by means of connecting rod 5. At *b*, uniform angular displacements of the crank (link 2) are represented as abscissas, and corresponding displacements of the slide (link 6) are represented as ordinates. Since angular displacements are directly proportional to time, Fig. 2-8*b* shows that the return stroke of the slide (link 6) is made in about one-half the time required for the working stroke.

2-7. The Slider-crank Mechanism. This four-link mechanism, which has a well-known application in gasoline, diesel, and steam engines, is simply a special case of the crank-and-rocker mechanism. In order to make this clear, let it first be noted that if the rocker arm 4 is of considerable length, as shown in Fig. 2-9, it may be desirable to replace it by a block sliding in a curved slot or guide as shown. If the length of the rocker arm is infinite, the guide and block are no longer curved, but are straight, as shown in Fig. 2-10, and the linkage takes the form of the ordinary slider-crank mechanism.

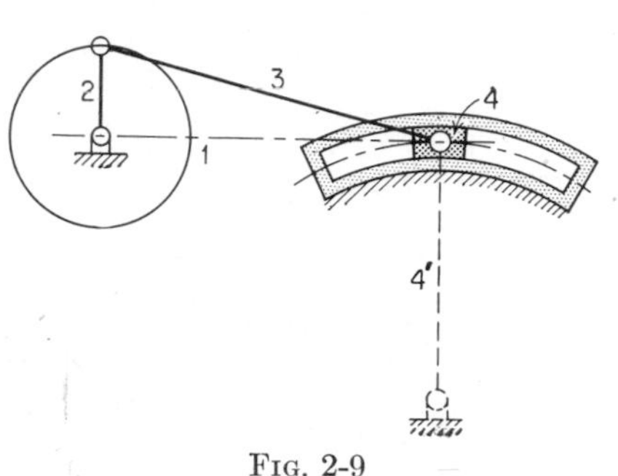

Fig. 2-9

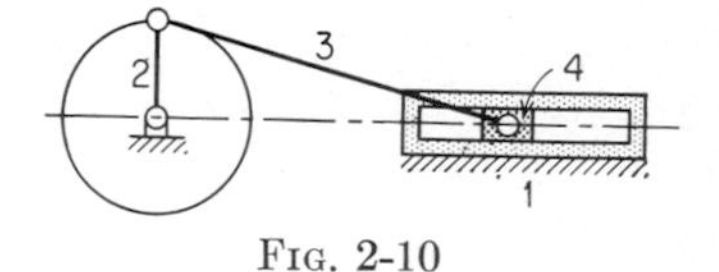

Fig. 2-10

2-8. Variations of the Slider-crank Mechanism. There are numerous variations of the slider-crank linkage, only a few of which will be discussed here. In Fig. 2-11 the connecting rod has been replaced by a sliding block by the process of expanding pair *C* to include pair *B*. The skeleton outline of the mechanism (in dotted lines) shows, however, that although the forms of the links have been greatly modified, the kinematic essentials of the mechanism have not been changed. The relative motions of the links will, therefore, be unchanged. When the mechanism shown in Fig. 2-11 is only slightly modified, it has an application in punching and shearing machines. In Fig. 2-12 part of the head of such a machine is shown. Uniform rotation is imparted to the eccentric 2 by the main shaft, which turns about the axis O. The eccentric transmits its motion through the connecting rod 3, called the pendulum in

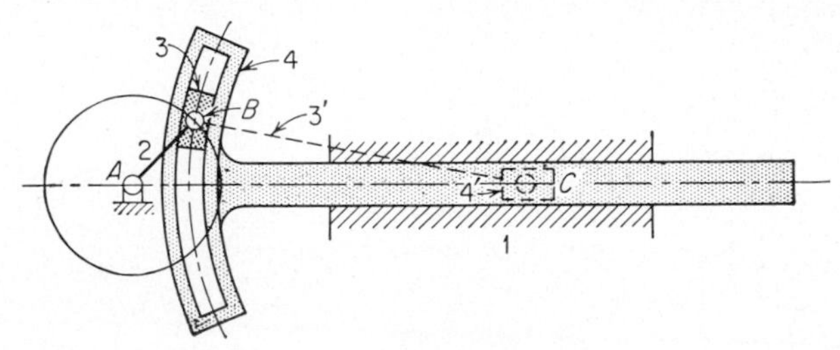

FIG. 2-11

this type of mechanism, to the ram 4, giving to the latter a motion of vertical reciprocation, the magnitude of which is equal to twice the length of the crank 2. The punch is fastened to the ram as shown at P and is forced down through the work W by the motion of the ram. The skeleton outline of the mechanism is shown at b, bringing out very clearly the fact that kinematically the linkage is nothing more or less than the ordinary slider-crank mechanism with the pair C expanded to include both B and A.

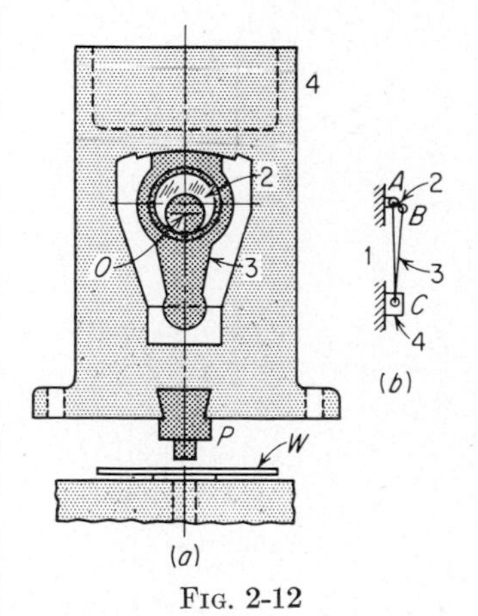

FIG. 2-12

2-9. Connecting Rod of Infinite Length. When the slot in the mechanism shown in Fig. 2-11 is made straight, as shown in Fig. 2-13, the linkage is equivalent to a slider-crank mechanism having a connecting rod of infinite length. This mechanism is known as the *Scotch yoke*. Uniform rotation of the crank 2 gives simple harmonic motion of the slotted crosshead 4. This can be shown as follows: Referring to Fig. 2-13, let the displacement of the crosshead be measured by the displacement of the center line of the slot. When the crank angle θ is zero, the center line of the slot passes through the point B_0 which lies at a distance R from the center O_2, R being the length AO_2 of the crank. As θ increases, the center line of the slot will move to the left. Then, letting the distance of the center line of the slot to the left of B_0 be represented by x, the displacement of B is

$$x = R - R \cos \theta = R(1 - \cos \theta) \tag{2-1}$$

The velocity of B is

$$V_B = \frac{dx}{dt} = R\omega \sin \theta \tag{2-2}$$

where $\omega = d\theta/dt$.

The acceleration of B is (for constant ω)

$$A_B = \frac{d^2x}{dt^2} = R\omega^2 \cos \theta \tag{2-3}$$

A point is said to have simple harmonic motion if it moves in a straight line with an acceleration that is proportional to its distance from an origin and is directed toward the origin. The point B therefore has simple harmonic motion, since its acceleration is proportional to its distance $R \cos \theta$ from O_2 and is directed toward O_2. A useful conception of simple harmonic motion is the following:

If a point moves along the circumference of a circle with uniform magnitude of velocity, the projection of the point on a diameter of the circle moves with simple harmonic motion.

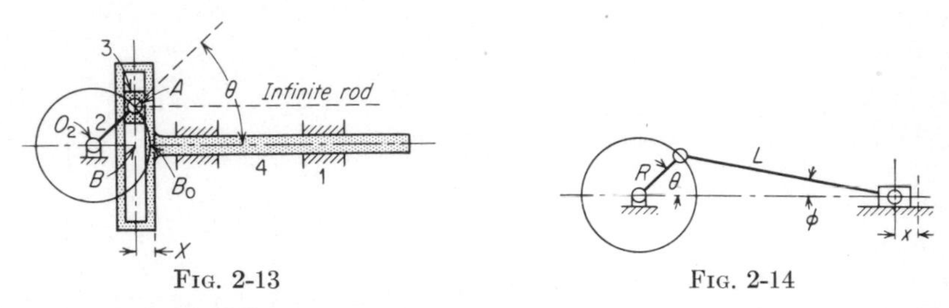

FIG. 2-13　　　　FIG. 2-14

2-10. Connecting Rod of Finite Length. In the case of the ordinary slider-crank mechanism, where the connecting rod must be of finite length, the motion of the crosshead is obviously not a simple harmonic motion. The deviation from simple harmonic motion, owing to what is known as the *angularity* of the connecting rod, may be indicated by comparing displacements. Referring to Fig. 2-14, let R be the length of the crank, L the length of the connecting rod, and x the displacement of the crosshead to the left of its extreme right-hand position. Then

$$\begin{aligned} x &= R + L - R \cos \theta - L \cos \phi \\ &= R(1 - \cos \theta) + L(1 - \cos \phi) \\ &= R(1 - \cos \theta) + L(1 - \sqrt{1 - \sin^2 \phi}) \\ &= R(1 - \cos \theta) + L\left[1 - \sqrt{1 - \left(\frac{R}{L}\right)^2 \sin^2 \theta}\right] \end{aligned} \tag{2-4}$$

This is the exact formula. A simplified formula that is sufficiently accurate for the usual values of R/L (1/3.5 or less) can be obtained by expanding the quantity under the radical in accordance with the binomial theorem and discarding the terms beyond the square in the series thus obtained. Then

$$x = R(1 - \cos\theta) + \frac{R^2}{2L}\sin^2\theta \qquad \text{approx} \tag{2-5}$$

Comparison of this expression with that of Eq. (2-1) shows that the deviation of motion of the slider-crank mechanism from simple harmonic motion is indicated by the second term of the expression.

In Fig. 2-15*b* displacements of the crosshead (slider) have been plotted against crank angles, both for the Scotch yoke (infinite rod) and for a

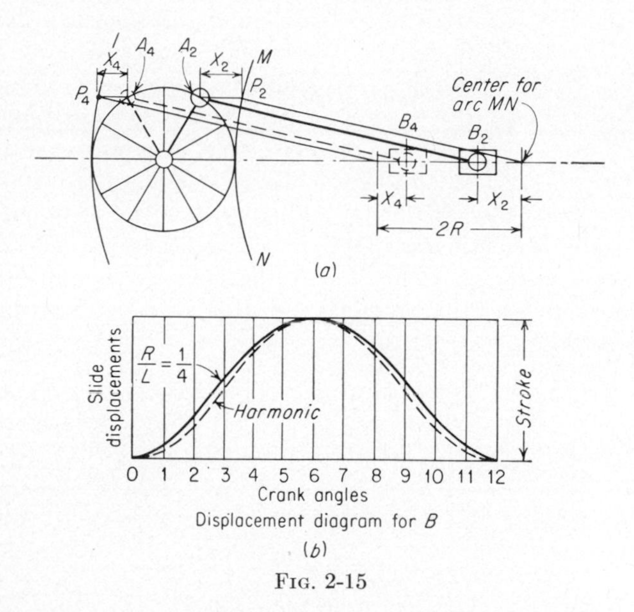

FIG. 2-15

value of $R/L = \frac{1}{4}$ for the slider-crank mechanism. Attention is called to the fact that, for the same length of crank, as the length of the connecting rod becomes smaller the deviation of the crosshead from simple harmonic motion becomes greater.

A convenient graphical method for obtaining the displacements of the slider corresponding to given crank positions of the ordinary slider-crank mechanism is shown in Fig. 2-15*a*. For any crank position, such as A_2, the displacement X_2 of the slide B from the right end of the stroke is equal to the horizontal intercept A_2P_2 between the crank circle and the arc MN, whose radius is the length of the connecting rod. In like manner the displacement X_4 of the slide from the left end of the stroke is A_4P_4.

2-11. The Scotch Yoke. Figure 2-16 shows an application of the Scotch yoke as it has occasionally been used on small engines and steam pumps. P indicates the pump cylinder and S the steam cylinder. The flywheel is carried on the crankshaft Q. The practical value of this modification of the slider-crank linkage is to provide a more compact mechanism than could be obtained with a connecting rod of finite length. Arranged as in Fig. 2-16, it gives a very compact means of introducing flywheel control into a steam pump or compressor, but it is an awkward arrangement, and the rubbing surfaces between block and slide are difficult to keep in proper lubrication.

Proportioned as in Fig. 2-17 with a large block surrounding an eccentric of small throw, the mechanism thus formed is of common occurrence in punching and shearing machines, its function being the same as that of the similar arrangement shown in Fig. 2-12. It should be understood, however, that it is only the basic form shown in Fig. 2-16 that may be designated as the Scotch yoke.

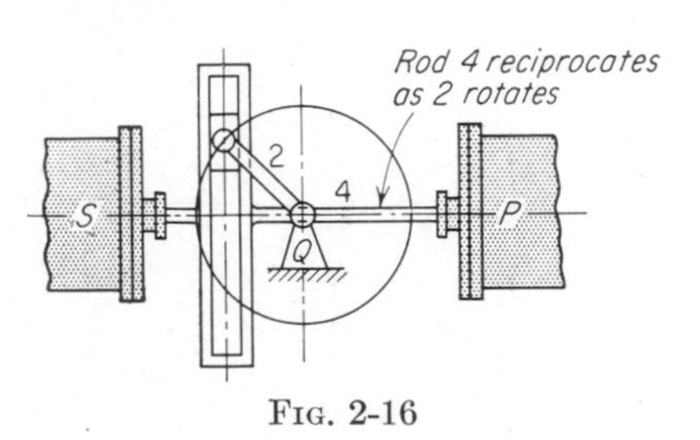

FIG. 2-16

2-12. Toggle Mechanisms. When the slider-crank mechanism approaches dead center, as indicated by the positions of links 2, 3, and 4 in Fig. 2-18, there is a rapid rise in the ratio of the resistance Q to vertical driving force P. Let Q be the resistance at C, P the vertical force

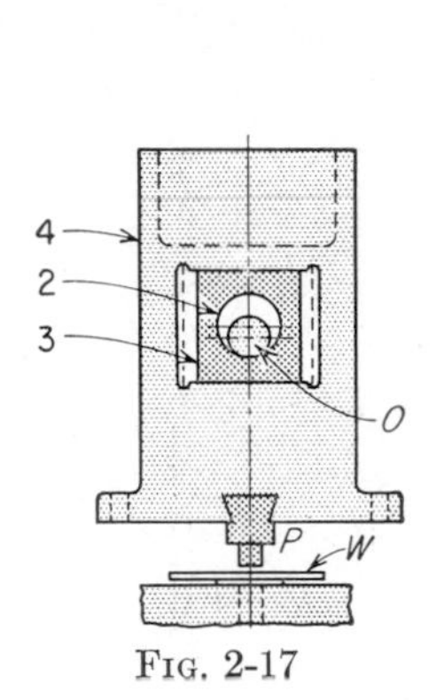

FIG. 2-17

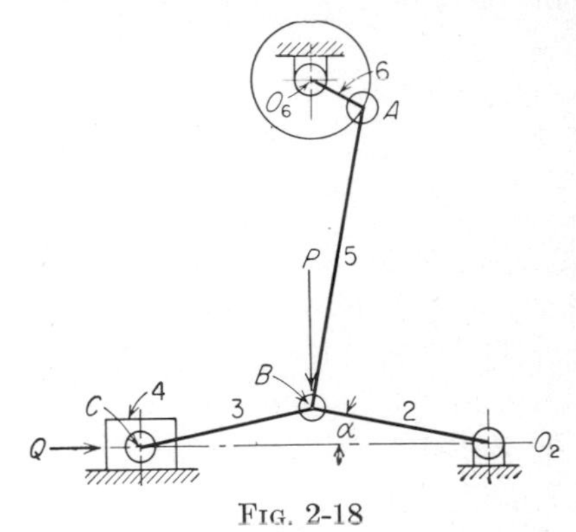

FIG. 2-18

applied at B, and α the angle that each link makes with the line of motion of C. Then, if 2 and 3 are equal in length,

$$\frac{P}{Q} = 2 \tan \alpha$$

This means of overcoming a very great resistance with a small force by straightening out a flat angle between two links is called a toggle effect. The toggle mechanism is applied in a variety of forms in stone crushers, presses, pneumatic riveters, pliers, friction clutches, etc. Figure 2-19 is a diagram showing an example of a toggle mechanism as applied in one type of stone crusher. Figure 2-20 represents a pair of toggle pliers which exert a powerful gripping action.

2-13. The Isosceles Linkage. Another modification of the slider-crank mechanism that is of interest is that obtained by making the crank

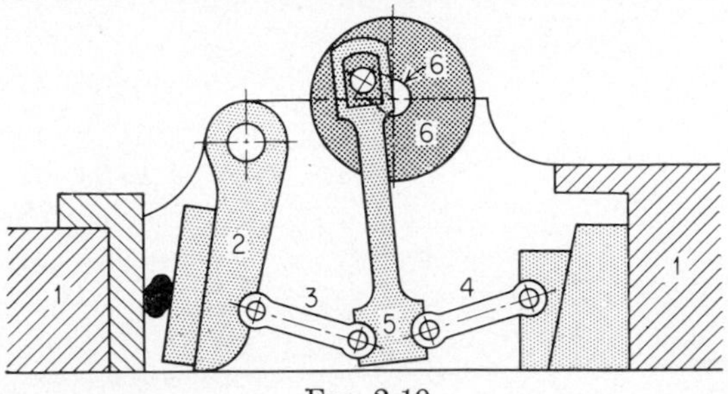

FIG. 2-19

and connecting rod of equal length, as in Fig. 2-21. In this arrangement, these two members always form an isosceles triangle ABC with the line of travel DE of the sliding member. For this reason the arrangement is sometimes called the *isosceles linkage.* It is evident that, for a complete rotation of crank 2, the length of stroke of the sliding member C is $DE = 4AB$. With the arrangement ABC, the slide C cannot have positive motion at the mid-point A of its stroke. In practical applications of this linkage, positive action is secured by the simple expedient of extending link CB to F where another sliding member is attached to it and guided in exactly the same manner as at C. With this addition, the motions of the several links become independent of AB and hence the link AB may be dispensed with as far as constraint is concerned.

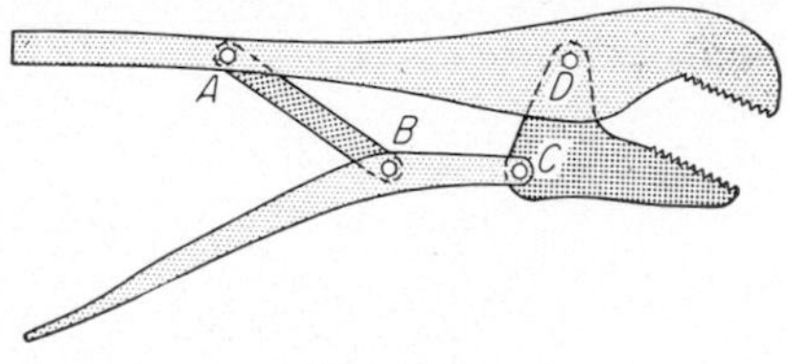

FIG. 2-20

The isosceles linkage is the basis for several useful devices, among these being the *ellipsograph,* an instrument for drawing ellipses. The principle of the ellipsograph is illustrated in Fig. 2-22. If a bar AC has the two points A and C guided in perpendicular straight lines, any other point on the bar such as D will trace an ellipse.

Let $$DC = b \quad \text{and} \quad DA = a$$
then $$DE = y = b \sin \theta$$
and $$DF = x = a \cos \theta$$
whence $$\frac{x^2}{a^2} + \frac{y^2}{b^2} = \cos^2 \theta + \sin^2 \theta = 1 \tag{2-6}$$

the equation of an ellipse. For the particular point B midway between A and C, the ellipse becomes a circle. This can be easily seen by referring back to Fig. 2-21.

The isosceles linkage is also the basis for straight-line motions as applied in certain types of indicator mechanisms.

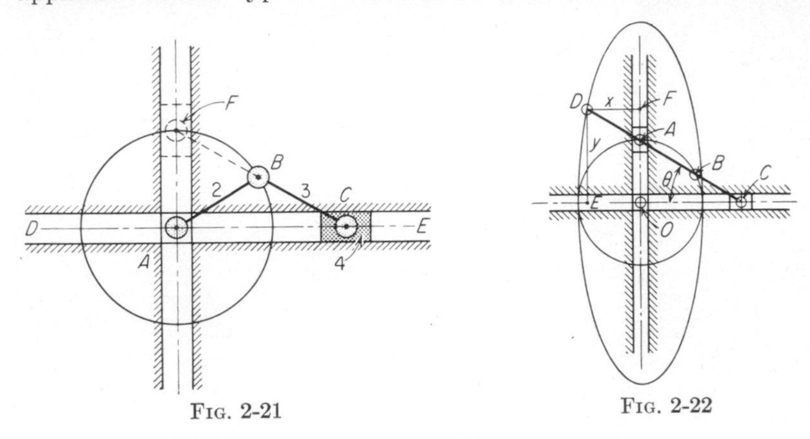

Fig. 2-21

Fig. 2-22

2-14. Inversions of the Slider-crank Mechanism. The meaning of the term *inversion* as applied to *mechanisms* was explained in Chap. 1. The inversions of the slider-crank linkage give rise to several well-known mechanisms, probably the most important of which is the engine mechanism.

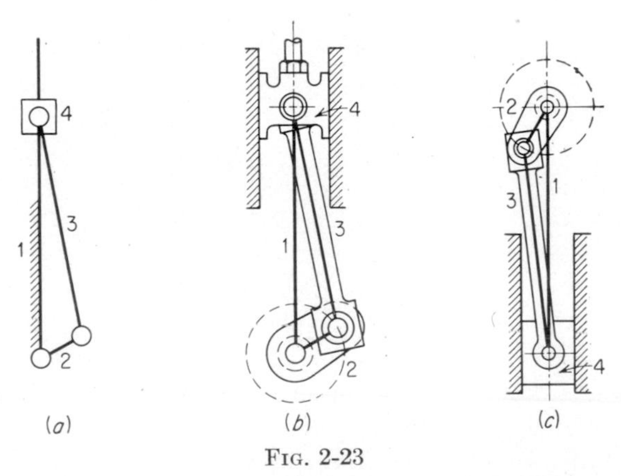

Fig. 2-23

Figures 2-23, 2-24, and 2-25 have been selected and arranged with the purpose of showing not only common examples of inversions of the slider-crank linkage, but also typical examples of the kinematic identity of machines that, at first sight, appear to be very different.

In each figure the skeleton outline is shown at a, and b and c represent two kinematically identical mechanisms that have been derived from it. The links in the actual mechanisms are numbered similarly to those in the skeleton outlines, so that the figures are almost self-explanatory.

1. *The First Inversion.* In the first inversion (Fig. 2-23) with link 1 fixed, *b* represents the ordinary engine mechanism in which the piston 4 is the driver, and *c* represents a pump mechanism where the crank 2 is the driver. It should be noted that in machines of this type a flywheel is needed if the crank is to rotate at substantially constant angular velocity.

2. *The Second Inversion.* The second inversion (Fig. 2-24) results from fixing connecting rod 3 of the original mechanism. At *b* is shown a form of oscillating engine in which sliding block 4 has been elaborated to

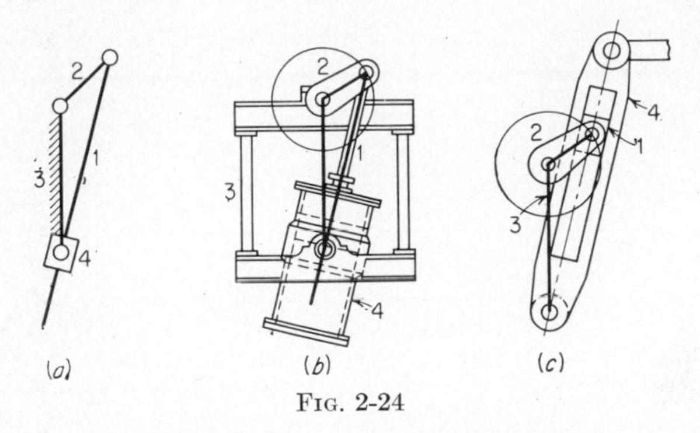

FIG. 2-24

form a cylinder on trunnions and the original frame 1 changed into a piston and rod. This type of engine has had limited application as a hoisting engine. Its chief advantage is cheapness of construction, since it permits of a very simple scheme for supplying steam to the cylinder. This is also a usual form of toy steam engine. A more useful inversion is the crank-shaper mechanism shown at *c*. Suitable extension of the linkage is required beyond link 4 for connection to the ram of the shaper. The object of this mechanism as applied to the shaper is to secure a slow cutting stroke and a quick return stroke of the tool.

3. *The Third Inversion.* In the third inversion (Fig. 2-25), crank 2 of the original mechanism becomes the fixed link, or frame. The Whitworth quick-return mechanism shown at *b* is derived from this inversion and is used in shapers and other machine tools. Gear 3, rotating on axis O_3 and carrying pin A, on which slide 4 is pivoted, corresponds with the connecting rod 3 of the original mechanism and acts as the driver. Link 1 rotating on axis O_1 is driven with a variable angular velocity and carries crankpin B, which can be adjusted to various radial distances from O_1, to change length of stroke of slide to which it is connected by an additional link. The rotary type of engine shown at *c*, at one time used in air

service, is another interesting example of this inversion in which the crank is the fixed member. The cylinders are mounted upon a rotating frame that turns on axis O_1, and the connecting rods are pivoted on fixed crankpin A.

4. *The Fourth Inversion.* The fourth inversion, obtained by fixing the sliding block (link 4) of the original mechanism, does not give a mechanism of sufficient practical value to warrant discussion.

2-15. Quick-return Mechanisms. It was shown above that two very important quick-return mechanisms were derived from inversions of the slider-crank mechanism, namely, the ordinary crank-shaper mechanism (Fig. 2-24*c*) and the Whitworth quick-return motion (Fig. 2-25*b*), which is used in shapers, slotting machines, etc.

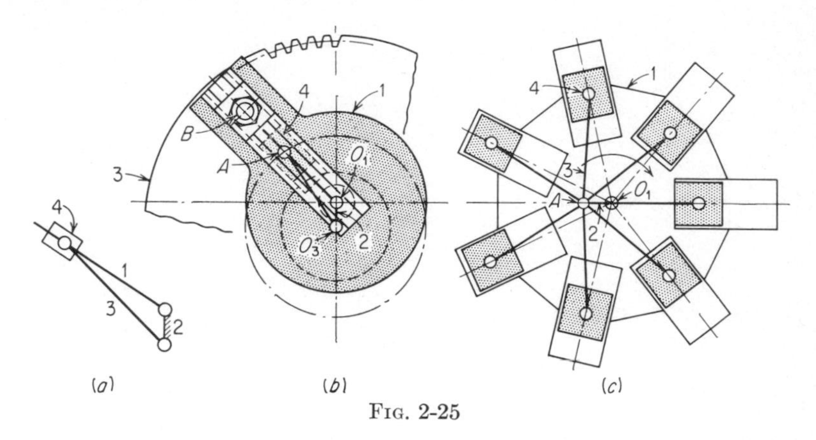

FIG. 2-25

A quick-return mechanism may be defined as one in which a slow working stroke of the tool and a fast idle or return stroke is produced from uniform rotation of a crank. Just how well this end is attained in case of the crank shaper may be determined by plotting the displacement diagram shown in Fig. 2-26. At a is shown the skeleton outline of the mechanism of Fig. 2-24*c*, with suitable extension of the linkage beyond link 4 for connection to the ram of the shaper, which is represented by link 6. In the displacement diagram at b, angular displacements of crank 2 have been plotted as abscissas, and corresponding displacements of ram 6 as ordinates.

In the study of quick-return mechanisms, considerable use is made of what is called the time ratio. This is simply the ratio of the time required for the cutting stroke to the time required for the return stroke, both on the assumption that the driving link has a motion of uniform rotation. Referring to Fig. 2-26, it is evident that, in the case of the crank-shaper mechanism, the ram is at the end of its stroke when the crank 2 is perpendicular to the rocker arm 4. The cutting stroke takes place while the crank is turning through the angle α, and the return stroke while the crank is returning through the angle β. Since the crank rotates uniformly, it is evident that the time required for the cutting and return strokes of the tool must be proportional to those angles; that is, the time ratio is equal to α/β.

Another type of quick-return mechanism, in which there are no sliding pairs between the first driving link and the last driven link, is the drag-link mechanism shown in Figs. 2-7 and 2-8.

Elliptical gears were formerly employed to some extent in shaper mechanisms to secure the quick-return motion. Although abandoned for this application, they are occasionally used in other mechanisms where a quick-return effect is desired.

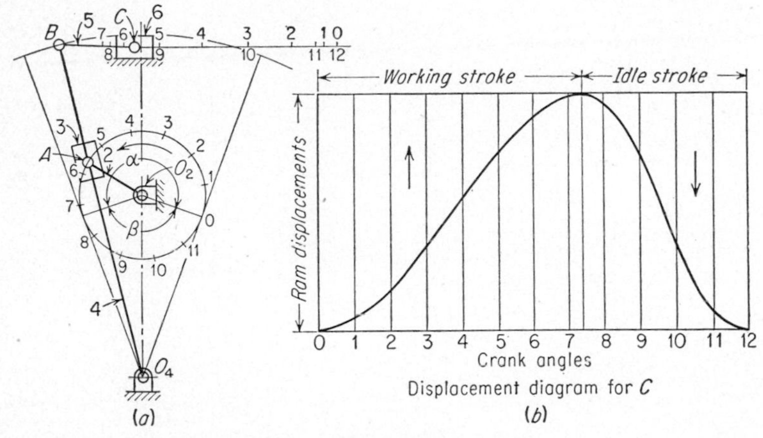

FIG. 2-26

2-16. The Oldham Coupling. A striking example of inversion is found in the Oldham coupling illustrated in Fig. 2-27*a*. This is a well-known form of flexible coupling used to transmit uniform angular velocity between parallel shafts whose axes may not coincide. As shown in the skeleton outline in Fig. 2-27*b* it is seen to be an inversion of the mechanism of Fig. 2-21 (in its practical form) with link 3 instead of 1 as the fixed member. With link 1 fixed (Fig. 2-27*b*) the locus O of the center of the cross is a circle of diameter AB. Let O and O' be two positions of the center of the cross. Then angles OAO' and OBO', as may be seen from the geometry of the figure, are equal. Hence, if link 2 turns through an angle OBO', 3 will turn through an equal angle OAO'. If, therefore, 2 and 3 are mounted on shafts as shown in the actual form of the coupling in Fig. 2-27*a* rotary motion may be transmitted from one to the other with an angular velocity ratio equal to unity. Evidently the center O of the cross will travel around a circle whose diameter AB is the distance between the axes of the shafts. Disk 4, instead of being shown with two slots at right angles as in the diagram (Fig. 2-27*b*), is in its actual form sometimes made with a slot on one side and with a guide strip on the other

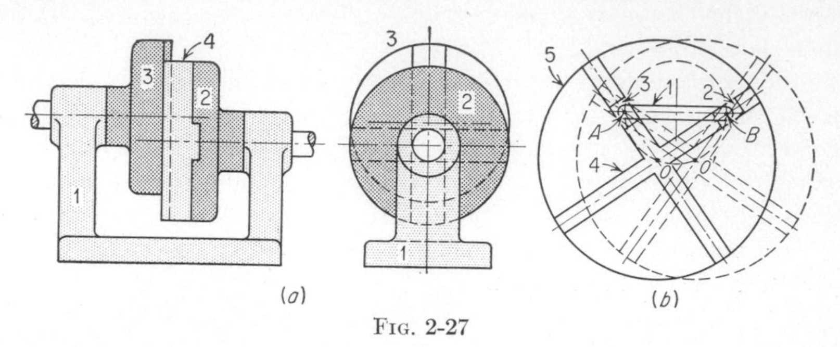

Fig. 2-27

side at right angles to the slot, as shown in Fig. 2-27*a*. Member 4 also can be simply a square block fitting in grooves cut in members 2 and 3, as indicated in Fig. 2-28.

2-17. Reducing Motions. In Fig. 2-29 the three mechanisms that may be used for the same purpose show interesting kinematic variation in the slider-crank mechanism. The crank 2 is arranged here for oscillating motion only, and the purpose is to give a cord *C*, which actuates the drum of an engine or compressor indicator, a miniature copy of the motion of the crosshead 4.

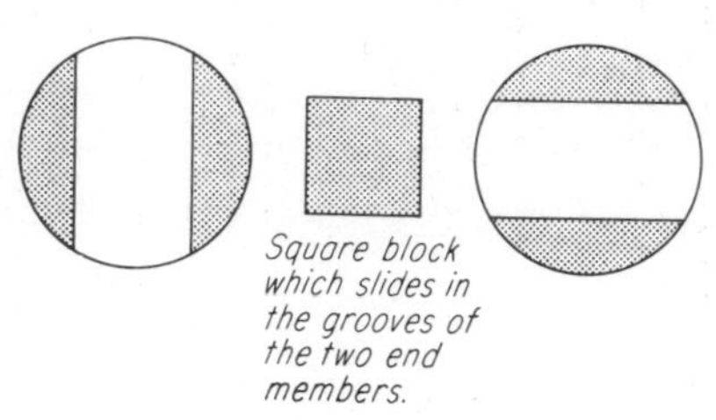

Fig. 2-28

2-18. Chamber Trains. The term "chamber train" covers mechanisms of the type shown in Fig. 2-30, used as pumps for comparatively low pressures and as blowers delivering against moderate pressures.

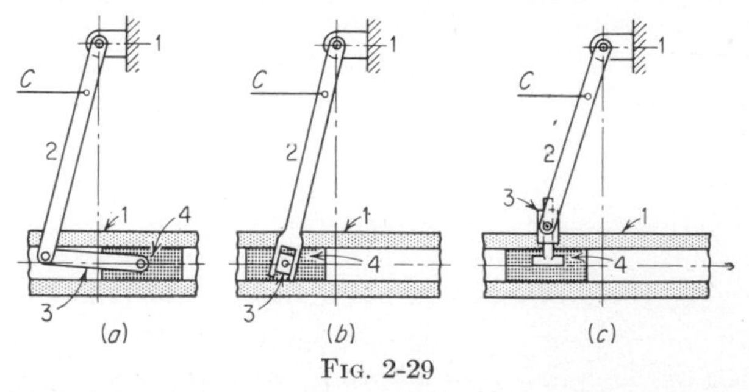

Fig. 2-29

Although chamber trains are made in a great variety of constructions, they really fall into two general divisions: (1) those which have a single rotor or impeller placed eccentrically within a cylindrical casing as in

Fig. 2-30 (it will be found that this type is usually built around some form of the slider-crank mechanism) and (2) those which have two rotors, each concentric with its part of the casing (Fig. 2-32), the two working together like a pair of gears, though actuated by a positive drive from outside the casing.

In Fig. 2-30 the fact that the mechanism is essentially of the slider-crank form is clearly brought out by the skeleton form outlined on the figure. The crank and connecting rod are embodied in the inner eccentric 2 and the outer ring 3, respectively, and the slide 4 in the reciprocating partition. It is interesting to observe that eccentric 2 may be considered as formed by expansion of the pairing elements at A (Art.

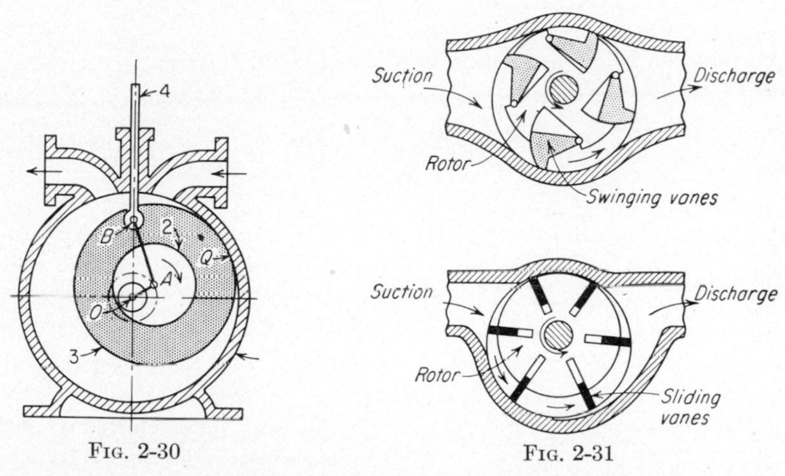

FIG. 2-30 FIG. 2-31

1-11), thus replacing crank OA; and that likewise ring 3 may be considered as formed by a further expansion of the outer, or hollow, element at A, thus replacing the connecting rod AB. The variable displacement is contained between contact point Q of rotor with casing 1 and the sliding partition 4. In Fig. 2-31 are shown two types of pumps which operate similarly but have moving elements on the rotor which are held against the inside of the circular hollow casing by centrifugal force as the rotor revolves.

The arrangements in Fig. 2-32 are regularly used in large positive blowers. Correct relative motion between the two rotors in each case is maintained by the pair of equal gears (indicated by the dotted circles) outside the casing. Since these chamber-wheel mechanisms are toothed wheels, discussion of their kinematic properties should be taken up under the subject of toothed gears, rather than under the subject of linkwork.

2-19. Straight-line Mechanisms. A straight-line mechanism is a linkage designed to make a point on one of the links move exactly or approximately in a straight line. In order to avoid the friction and also the structural difficulties arising from the use of straight guides, links with turning pairs are almost invariably employed in a straight-line mechanism. A large number of linkages have been devised for the purpose, only a few of which will be considered here.

1. *Approximate Straight-line Mechanisms.* The Watt straight-line mechanism, one of the earliest and best known of the approximate

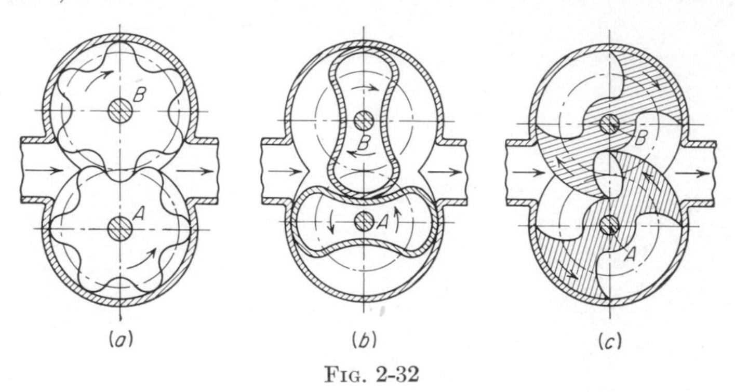

Fig. 2-32

straight-line mechanisms, is shown in Fig. 2-33 in its simplest form. Links 2 and 4 are pivoted at O_2 and O_4, and their free ends are connected by link 3. When in mid-position, links 2 and 4 are parallel and connecting link 3 is shown perpendicular to both. It is not absolutely necessary, however, for link 3 to be perpendicular to links 2 and 4. P is a point on link 3 that traces an approximate straight line, within certain limits of the motion. The complete path of P is the *figure* 8 shape of curve shown. The path of P is a function not only of arms 2 and 4, but also the location of P on the intermediate link 3. It can be seen that the principle of action is based on the fact that, as the tracer bar is deflected by the arm at one end, it receives a compensating deflection at the other end. If 2 and 4 are of the same length, P is located at the mid-point of link 3. In general, the segments AP and BP are inversely as the lengths of the adjacent arms.

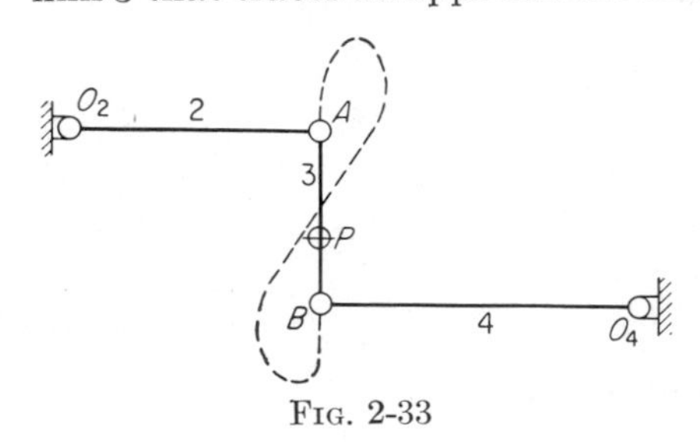

Fig. 2-33

2. *Exact Straight-line Mechanisms.* The best-known of the exact straight-line mechanisms is the Peaucellier mechanism, shown in Fig. 2-34. Two equal links 2 and 3 have a fixed center O. The links 4, 5, 6, and 7 are equal, and link 8 with a fixed center at Q equals the distance OQ. P is constrained to move in an *exact* straight line mn, within the working limits of the construction. This statement can be proved by showing that for any phase of the mechanism PP' is perpendicular to OP' at P',

which is a fixed distance from O. Considering the phase in solid outline.

$$(OA)^2 - (OD)^2 = (AP)^2 - (PD)^2$$

or

$$\begin{aligned}(OA)^2 - (AP)^2 &= (OD)^2 - (PD)^2\\ &= (OD + PD)(OD - PD)\\ &= (OP)(OC)\end{aligned}$$

Therefore, the product $(OP)(OC)$ is a constant. Now consider the linkage in its mid-position, where C falls at C', and P falls at P'. Then $(OC')(OP') = (OC)(OP)$, or

$$\frac{OC'}{OC} = \frac{OP}{OP'}$$

and $OQ = QC = QC'$, so that O, C, and C' lie on a semicircle, which makes the angle OCC' a right angle.

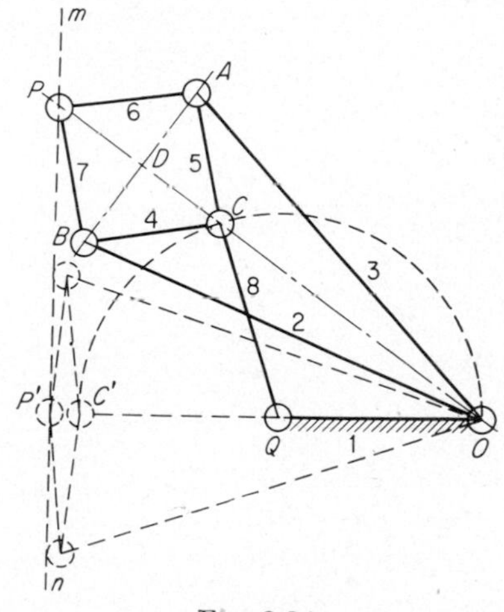

FIG. 2-34

In the triangles COC' and POP', the angle POP' is common, and since the sides are proportional, the triangles are similar.

Thus, since the angle OCC' is a right angle, the angle $PP'O$ is also a right angle.

In the same way it can be shown for any other phase of the linkage that the line PP' is perpendicular to OP' at P', and therefore P moves in a straight line.

If the length of CQ is made adjustable so that it is not necessarily equal to OQ, then point P traces a *curve.* Such a device forms the basis of operation of a compass for the drawing of very gradual curves such as for railroads.

2-20. Copying Devices. The Pantograph. This class of mechanism is composed mostly of variations of the pantograph. The pantograph is a four-bar linkage so arranged as to form a parallelogram in which, if one point is made to travel in any path, some other point will be constrained to describe a similar path, enlarged or reduced.

Figure 2-35 shows one such arrangement in which the links form a parallelogram with one corner fixed at O. A point p on ab that lies on the line connecting P and O will move in a path similar to that traced by P. Assume that P moves to P' along the path indicated by the dotted line; then p moves to p' along a similar path. From the figure it is evident that

$$\frac{OP}{Op} = \frac{cP}{cb}$$

also

$$\frac{OP'}{Op'} = \frac{c'P'}{c'b'}$$

But $cP = c'P'$ and $cb = c'b'$

Therefore

$$\frac{OP}{Op} = \frac{OP'}{Op'}$$

Hence the ratio of the distances of p and P from O is a constant. Since OcP and pbP are similar triangles in all positions, point p always lies on a line drawn from O to P. Hence the angular motions of points p and P about O are the same. Any path of P is determined by its angular motion about O and its radius vector to O as a pole. Since the angular motions of P and of p about O are equal for any motion of either of these points, and since the radius vector of p bears a constant ratio to that of P, the path of p is similar to that of P.

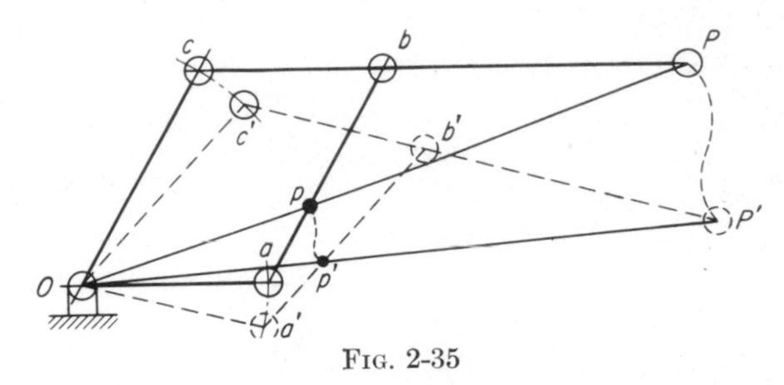

Fig. 2-35

The pantograph is the standard instrument for copying and reducing or enlarging drawings, maps, engravings, etc. It is also frequently used to increase or reduce motion in some definite proportion, as in the indicator rig on an engine where the motion of the crosshead is reduced proportionally to the desired length of the indicator diagram. The practical objection to its use as an indicator rig is the large number of joints and consequent likelihood of lost motion from wear. Another application is in cutting operations, where a cutting tool can be guided by a tracing point which traces an enlarged pattern of the shape to be cut.

2-21. Applications of Straight-line Mechanisms. In the early days of the steam engine, the straight-line and parallel mechanisms possibly were of greater importance than now. These mechanisms represented the method whereby the extremity of the piston rods of the beam engine of that time (shown in Fig. 2-36) were constrained to move in a straight-line path. The modern device, that of crosshead and guides, was not feasible because of structural and machining difficulties. The straight-line link motion was, therefore, an ingenious solution providing suitable constraint of the piston rod of a beam engine. Watt combined the pantograph with his straight-line mechanism so that the steam piston rod and pump rod were guided in straight lines by one combination of links as shown diagrammatically in Fig. 2-36. The links CD, DE, and EF form the Watt straight-line mechanism, and DE is divided at G so that $DG/GE = EF/DC$. CD is one arm of the oscillating beam of the engine, and it is extended to the point J. On this link CDJ, the pantograph $CDJHE$ is constructed, the lengths of DJ and EH being such that the point H traces a path similar to that of G. From previous discussions of the pantograph, it will be recognized that the condition for the motion is that C, G, and H lie on a straight line. The main piston rod is then connected at H and the pump rod at G. Although straight-line mechanisms are no longer required for the original purpose for which they were designed, they still have an important use in connection with multiplying gear for recording instruments, of which the engine indicator rig is typical.

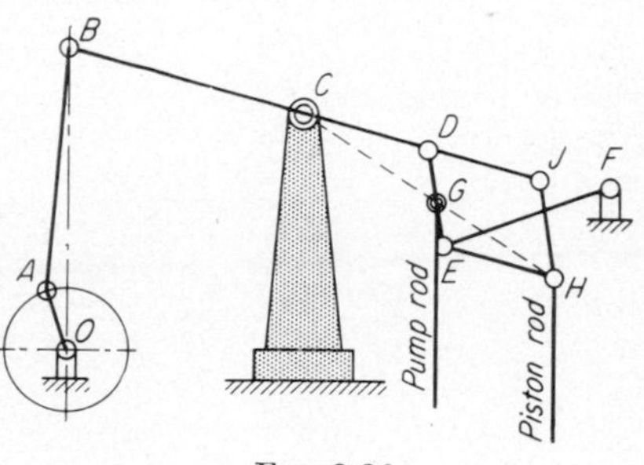

FIG. 2-36

2-22. Indicator Mechanisms. The function of the indicator is to give a magnified copy of the movement of the indicator piston, as variable pressure in the cylinder (against resistance of a spring) causes that piston to move up and down. To the kinematic requirements of straight-line path and constant-travel ratio for the pencil is added the necessity for minimum mass, in order that the instrument may be quick acting without serious disturbance by inertia.

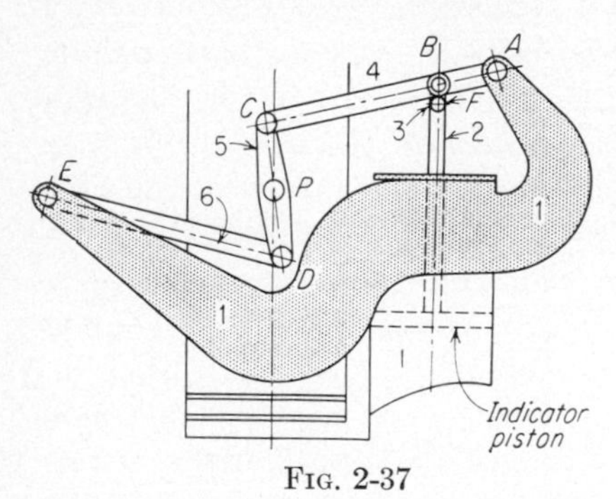

FIG. 2-37

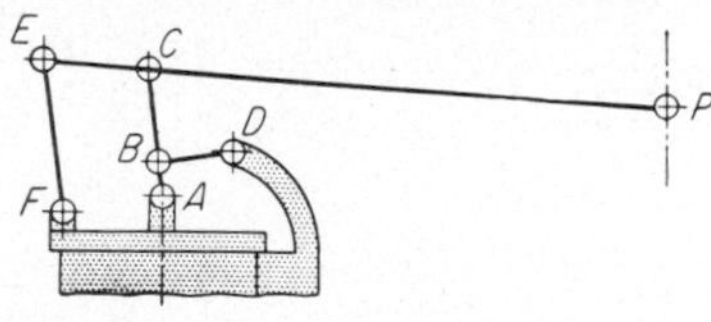

FIG. 2-38

The Richards indicator (Fig. 2-37), one of the earliest of the modern types, used the Watt linkage $ACDE$ to guide the pencil point at P in a straight line.

The Crosby indicator, a more modern type, is shown in Fig. 2-38.

This mechanism is a modification of the pantograph. The guiding is all concentrated on member CA and dependent upon radius rod BD.

2-23. Parallel Linkages. These are commonly applications of the four-bar linkage in which the links form a parallelogram. The *parallel rule* (Fig. 2-39), a piece of drafting equipment, furnishes an example of its simplest application. The *universal drafting machine* (Fig. 2-40), a modern substitute for the T square and triangle, is an interesting example of its present-day use. This instrument consists primarily of a combination of parallelograms $ABCD$ and $EFGH$ in which the coupling ring $CDEF$ is an important feature.

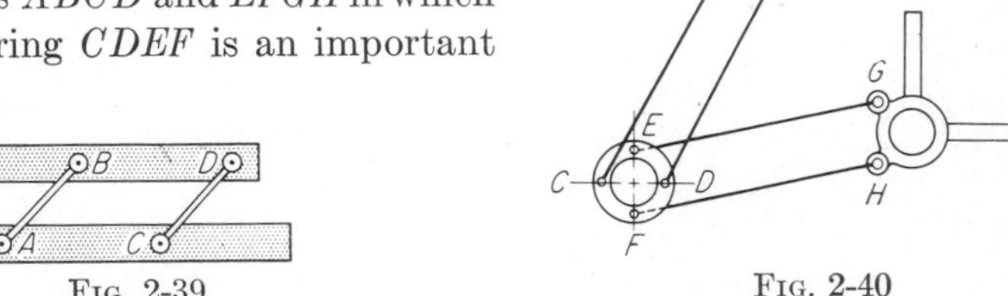

Fig. 2-39

Fig. 2-40

2-24. The Universal Joint. The universal joint (Fig. 2-41a), also known in the older literature as *Hooke's coupling*, is a spherical linkage in common use for connecting two shafts that intersect. Regardless of how constructed or proportioned for practical use, it has essentially the form shown in Fig. 2-41b, consisting of two semicircular forks 2 and 4, pin-jointed to a right-angle cross 3. If only one-half of each fork is considered, as eb of 2 and fd of 4, and these are assumed to be connected by a spherical link db equal to the fixed distance between the two adjacent points of the cross 3, a four-bar conic linkage is produced in which the axes of all the turning pairs intersect in O. With this arrangement, the fork could be omitted and there would be, as the kinematic equivalent of the original mechanism, the arrangement shown in Fig. 2-41c.

The driver 2 and the follower 4 make complete revolutions in the same time, but the velocity ratio is not constant throughout the revolution. The following analysis will show how complete information as to the relative motions of driver and follower may be obtained for any phase of the motion.

1. *Analysis of a Universal Joint.* If a plane of projection is taken perpendicular to the axis of 2, the path of a and b will be a circle $AKBL$ (Fig. 2-41d). If the angle between the shafts is β, the path of c and d will be a circle that is projected as the ellipse $ACBD$, in which

$$OC = OD = OK \cos \beta = OA \cos \beta$$

If one of the arms of the driver is at A, an arm of the follower will be at C; and if the driver arm moves through the angle θ to P, the follower arm will move to Q; OQ will be perpendicular to OP; hence $COQ = \theta$.

But COQ is the *projection* of the real angle described by the follower. Qn is the real component of the motion of the follower in a direction parallel to AB, and line AB is the intersection of the planes of the driver's and the follower's paths. The true angle ϕ described by the follower, while the driver describes the angle θ, can be found by revolving OQ

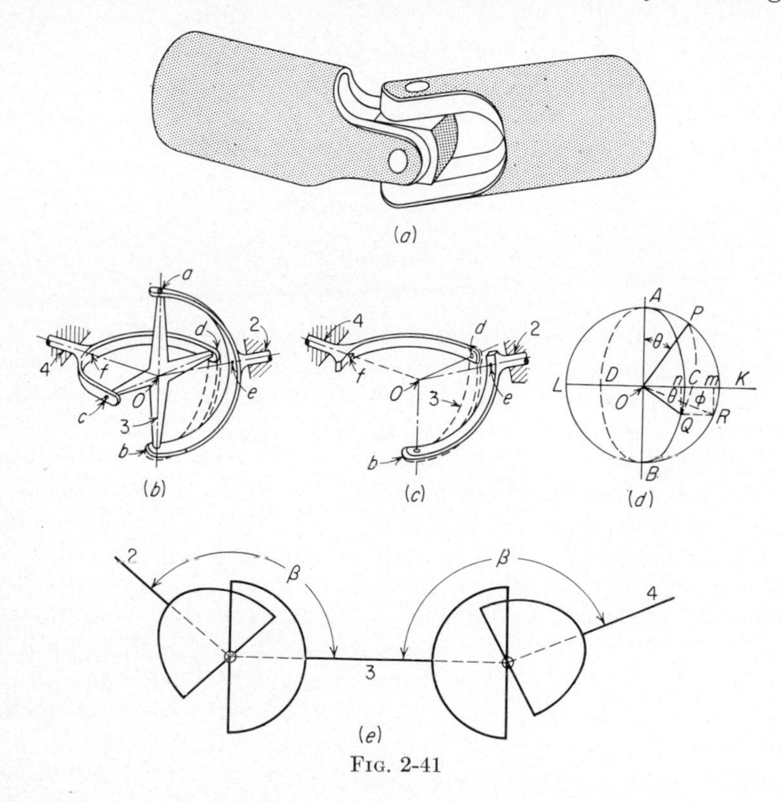

FIG. 2-41

about AB as an axis into the plane of the circle $AKBL$. Then OR = the true length of OQ, and $ROK = \phi$ = the true angle that is projected as $COQ = \theta$.

Now
$$\tan \phi = \frac{Rm}{Om}$$

and
$$\tan \theta = \frac{Qn}{On}$$

But
$$Qn = Rm$$

Hence
$$\frac{\tan \theta}{\tan \phi} = \frac{Om}{On} = \frac{OK}{OC} = \frac{1}{\cos \beta}$$

Therefore
$$\tan \phi = \cos \beta \tan \theta \qquad (2\text{-}7)$$

The ratio of the angular motion of the follower to that of the driver is found as follows, by differentiating Eq. (2-7), remembering that β is constant

$$\frac{\omega_4}{\omega_2} = \frac{d\phi}{d\theta} = \frac{\cos\beta \sec^2\theta}{\sec^2\phi} = \frac{\cos\beta \sec^2\theta}{1 + \tan^2\phi} \tag{2-8}$$

Eliminating ϕ by means of Eq. (2-7)

$$\frac{\omega_4}{\omega_2} = \frac{\cos\beta \sec^2\theta}{1 + \cos^2\beta \tan^2\theta} = \frac{\cos\beta/\cos^2\theta}{(\cos^2\theta + \sin^2\theta \cos^2\beta)/\cos^2\theta}$$

$$= \frac{\cos\beta}{\cos^2\theta + \sin^2\theta(1 - \sin^2\beta)} = \frac{\cos\beta}{1 - \sin^2\theta \sin^2\beta} \tag{2-9}$$

By a similar process θ could be eliminated, giving

$$\frac{\omega_4}{\omega_2} = \frac{1 - \cos^2\phi \sin^2\beta}{\cos\beta} \tag{2-10}$$

It is seen from Eq. (2-9) that ω_4/ω_2 is a minimum where $\sin\theta = 0$, or when $\theta = 0, \pi$, etc., which corresponds to a value of $\phi = 0, \pi$, etc. The same thing is seen from Eq. (2-10), which gives a minimum value of ω_4/ω_2 when $\cos\phi = 1$, or $\phi = 0, \pi$, etc. Also ω_4/ω_2 is a maximum when $\sin\theta = 1$, or $\cos\phi = 0$, corresponding to $\theta = 90$ deg, etc.; $\phi = 90$ deg, etc. To summarize the above, when the driver has a uniform angular velocity, the *ratio* of angular velocities varies between extremes of $\cos\beta$ and $1/\cos\beta$, where β is the angle between the connected shafts. For an angle β of 15 deg, the velocity ratio lies between 0.966 and 1.037; for $\beta = 30$ deg, the ratio varies between 0.866 and 1.155. These variations in velocity give rise to inertia forces, torques, noise, and vibration which would not be present if the velocity ratio were constant. The follower has a minimum angular motion when the driving arm is at A or B (Fig. 2-41*d*) and the follower arm is at C or D. The follower has a maximum angular motion when the driving arm is at K or L and the follower arm is at A or B.

2. *The Effect of a Double Universal Joint.* By using a double joint (Fig. 2-41*e*), the variation of angular motion between driver and follower can be entirely avoided. This compensating arrangement is to place an intermediate shaft 3 between the two main shafts 2 and 4, making the same angle β with each. The two forks of this intermediate shaft must lie in the same plane. If the first shaft rotates uniformly, the angular motion of the intermediate shaft will vary according to the law deduced above. This variation is exactly the same as if the last shaft rotated uniformly, driving the intermediate shaft. Since, therefore, the variable motion transmitted to the intermediate shaft by the uniform rotation of the first shaft is exactly compensated for by the motion transmitted from the intermediate to the last shaft, the uniform motion of either of

these shafts will impart, through the intermediate shaft, uniform motion to the other. Universal joints, particularly in pairs, are used in many machines. One common application is in the drive shaft which connects the engine of an automobile to the rear axle.

3. *Constant-velocity Universal Joints.* There are available several types of universal joints which transmit constant velocity (or a velocity ratio of unity) by means of a single joint. One example (the Rzeppa joint) is shown in Fig. 2-42. The mechanical construction is such that the steel balls which transmit the motion from one shaft to the other always automatically lie in a plane which bisects the angle between the shafts. Thus velocity is transmitted through any one ball at the same radius from both shaft center lines, resulting in identical angular velocities for the two shafts.

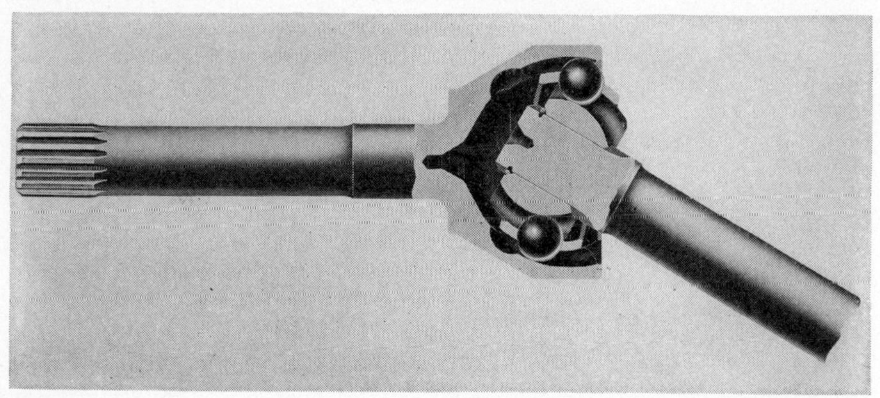

FIG. 2-42. (*Courtesy of the Gear Grinding Machine Co.*)

Applications of constant-velocity joints include the front-wheel automotive drive and drives on road-building machinery, tanks, street-railway cars, and mining equipment. Shaft angles of the order of 35 deg can be handled successfully.

2-25. Automobile Steering Gear. Figure 2-43 shows an automobile during a turn to the left. All four wheel centers should have motion perpendicular to radii from one center such as c if sideways skidding and sliding of wheels, with resulting tire wear, are to be avoided. This requires that the front wheels turn in a manner which places the intersection of their extended axes on the line which is the extended axis of the rear wheels, as illustrated. The restraint given by the four-bar linkage shown fulfills the requirements *approximately.* During a turn to the left, the left member 2 turns through a greater angle than the right member 4. Thus the left wheel, the axle of which turns with member 2, is rotated through a greater angle than the right wheel (which turns with 4). This is necessary if the extended axes are to intersect as required. The four-bar linkage lengths must be proportioned to achieve the closest approximation to the desired motion within the limits of the assigned turning angles.

Modifications of the simple steering linkage of Fig. 2-43 have been adopted in some variety, but the one example serves to indicate one way in which the basic requirements have been fulfilled with reasonable accuracy.

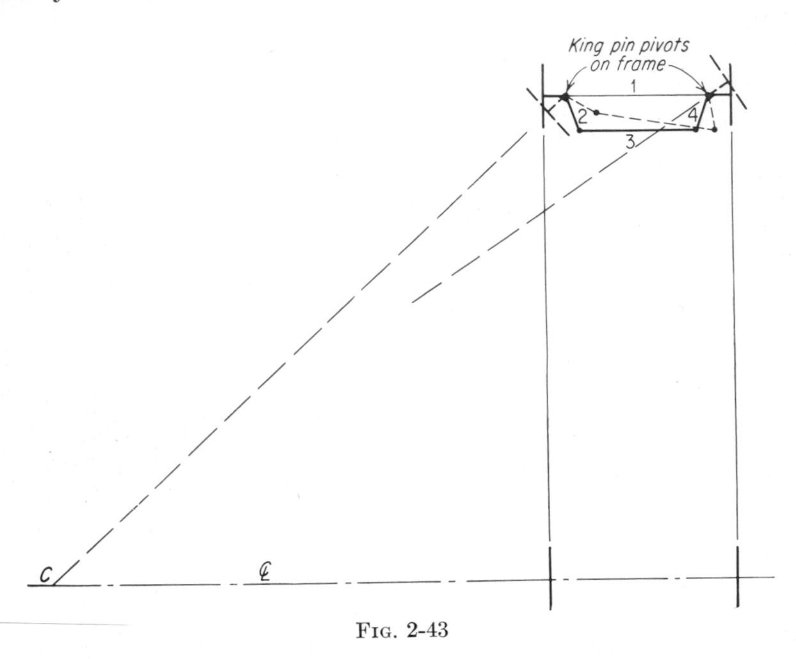

Fig. 2-43

2-26. Valve Gears. The linkages so far considered have, in general, been made up of not more than four to six links. Linkwork of a much greater degree of complexity is found in great variety in many mechanisms, classical examples being those linkages which have been used to actuate the valves of steam, gasoline, and locomotive engines. Linkages of this class are called *valve gears*. Although in some cases, particularly in gasoline engines, the actuating motion for the valve is derived from or mostly applied through cams, a great many valve gears are link motions that appropriately come under the heading of this chapter and merit extended discussion. It is not practicable, however, to consider more than a few examples here.

Kinematic study of valve-gear linkages usually involves problems relating to the determination of point paths, displacement of valve corresponding to displacement of crank or piston, and relative motions of various links.

2-27. Steam-engine Valve Gears. The simplest form of valve gear, used on the ordinary steam engine, adds to the main slider-crank linkage a secondary linkage of the same kind. A skeleton outline of the arrangement is shown in Fig. 2-44. Since the crankshaft, crank, and eccentric form one rigid link, they are represented by the compound link OAC, OA representing the crank and OC the eccentric. Whereas in the steam engine the length of the main connecting rod 3 is usually from five to seven times that of the crank arm OA, the length of eccentric rod 5 varies from twenty to forty times the throw OC of the eccentric. For this reason, in analysis of the motions, it is usual to neglect the effect on the valve movement of the finite length of rod, except in case of close analysis.

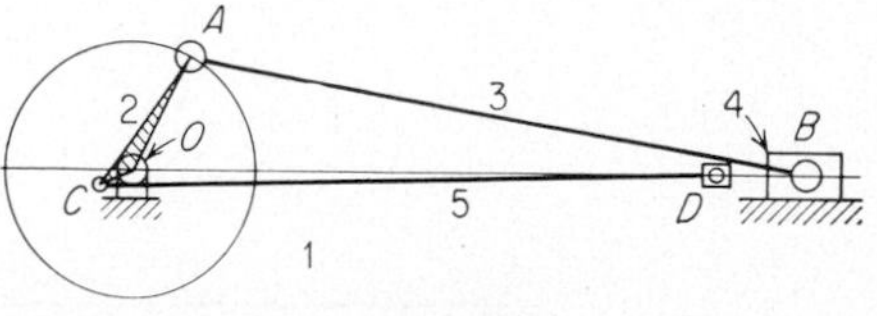

Fig. 2-44

The simple form of the regular Corliss type of valve gear used on steam engines (Fig. 2-6) has been discussed briefly in Art. 2-5. Because of the time element in the dashpot releasing gear, the full Corliss type of valve gear represented in Fig. 2-6 is suitable only for engines running at comparatively low speeds. For high-speed steam engines there have been developed modifications of the Corliss gear in which the dashpot control has been eliminated. A valve gear of this type, known as the Corliss nonreleasing gear, is outlined in Fig. 2-45. Through the linkwork shown, the eccentrics at A actuate the steam-valve mechanisms at B and C and the exhaust-valve mechanisms at D and E. The steam valves rotate through the same angles as links a and b, to which they are rigidly connected. In like manner the exhaust valves rotate through the same angles as their driving links c and d.

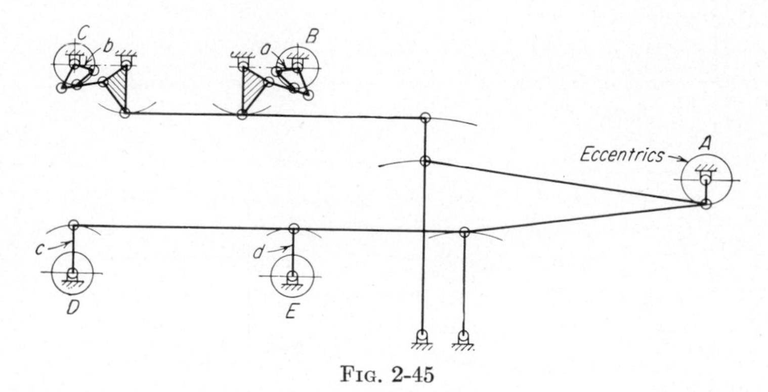

Fig. 2-45

2-28. Locomotive Valve Gears. Although the steam locomotive is being rapidly replaced by the diesel locomotive, two of the best-known steam-locomotive valve gears will be described here because of their kinematic and historic interest. The two are the *Stephenson valve gear*, which for a long time held predominant place in locomotive design, and the *Walschaert valve gear*, the predominant type in later steam locomotives.

The Stephenson gear is shown in Fig. 2-46. Located between the wheels, this gear was successful up to the time when later developments

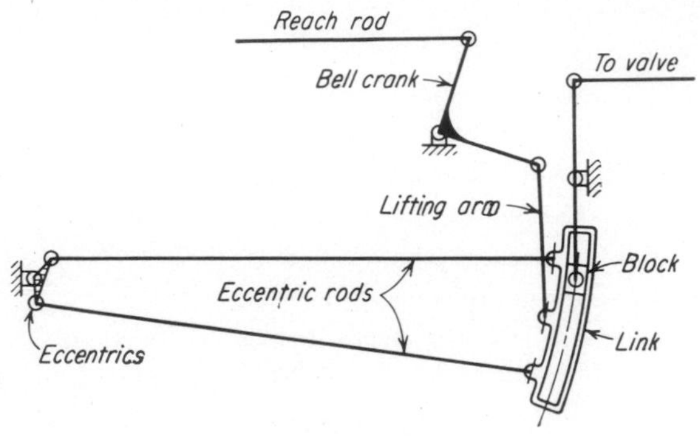

FIG. 2-46

in locomotive practice began to crowd uncomfortably the space between the wheels. When this condition became serious, the Stephenson gear was abandoned in favor of the Walschaert type (Fig. 2-47), which, in various modifications, has been adopted on practically all the more modern steam locomotives.

The principal advantage of the Walschaert gear lies in the accessibility of its parts, which are placed entirely outside the driving wheels. This facilitates inspection, oiling, and cleaning.

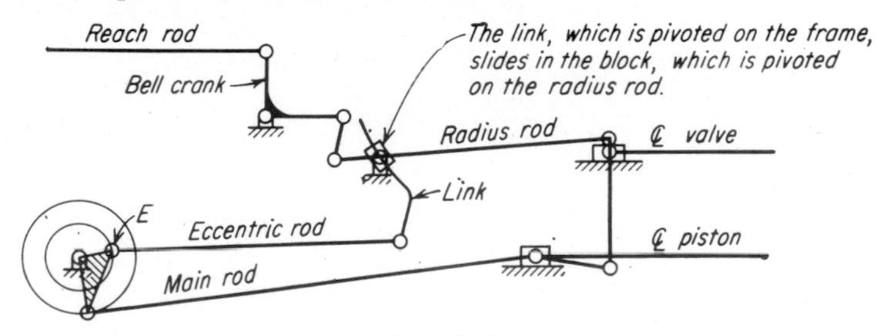

FIG. 2-47

In heavy locomotives employing the Stephenson valve gear, the eccentrics must be of large diameter to secure the required throw. This increases the velocity of the rubbing surfaces and the tendency to heat. Also, the inertia effects of the heavy slotted links are objectionable.

The various parts of the Walschaert gear are pin-connected and are easily lubricated; hence troubles due to overheating are reduced to the minimum. The Walschaert gear also transmits the moving force of the valve in practically a straight line; consequently there are less springing and yielding of the parts than in the Stephenson gear. By removal of the valve gear from between the wheels, a better opportunity is afforded to introduce frame bracing, thus reducing the possibility of frame breakage. The Walschaert type of valve gear also possesses advantages as regards steam distribution and operation.

The general adoption of the Walschaert valve gear has been accompanied by a number of variations which, however, do not eliminate its essential elements. The Baker, the Young, and the Southern valve gears, other well-known modern locomotive valve gears, are simply variations of the Walschaert gear.

2-29. Governors. A governor may be defined as a device for regulating automatically the output of a machine. The governor of a steam engine, steam turbine, gas or diesel engine, gas turbine, or other prime mover is a linkage that automatically regulates the supply of steam, gas, or other fluid so as to keep the driving force exerted by the working fluid constantly adjusted to the resistance to be overcome.

The output of a prime mover is continually varying. If, for example, a steam turbine is driving a generator used for lighting purposes, the number of lights in use varies from time to time. The same condition exists in the case of an engine that furnishes power for a machine shop, since the power demands from the various machine tools in the shop are continually changing.

The governor partakes of the motion of the machine to which it is attached so that a change in the speed of the machine due to a change in the load causes a corresponding change in the moving parts of the governor, which, in turn, cause, by means of a suitable mechanism, a change in the pressure of the fluid or a change in the quantity of the fluid delivered to the machine.

The regulating action of a governor should not be confused with that of the flywheel of an engine. The flywheel that acts as a reservoir of energy is useful in regulating speed throughout the very short period of a revolution or engine cycle, whereas the function of the governor is to regulate speed over greater periods, maintaining a balance between the energy supplied to the engine and the resistance to be overcome.

2-30. Types of Governors. Classified according to disposition of the revolving masses and the method of attaching them to the prime mover, governors are of two general types: flyball governors and shaft governors.

Flyball governors are of two kinds:

1. Gravity loaded (Fig. 2-48), in which the centrifugal force due to the revolving masses is largely balanced by gravity.

2. Spring loaded (Fig. 2-49), in which the centrifugal force is largely balanced by springs.

Shaft governors are of two kinds:

1. Centrifugal governors (Fig. 2-50), in which centrifugal force plays the major part in the regulating action.

2. Inertia governors (Fig. 2-51), in which the inertia effect predominates.

2-31. The Gravity-loaded Flyball Governor. This type of governor is older than the shaft governor. It is the original type used by Watt on his engines and has been much used since on low-speed steam engines. Modifications of the earlier designs have been adapted to gas-engine and steam-turbine practice.

The older forms of steam-engine flyball governors are gravity loaded and run at moderate speeds. A typical gravity-loaded flyball governor of the conic-pendulum form used on a Corliss engine is illustrated in Fig. 2-48.

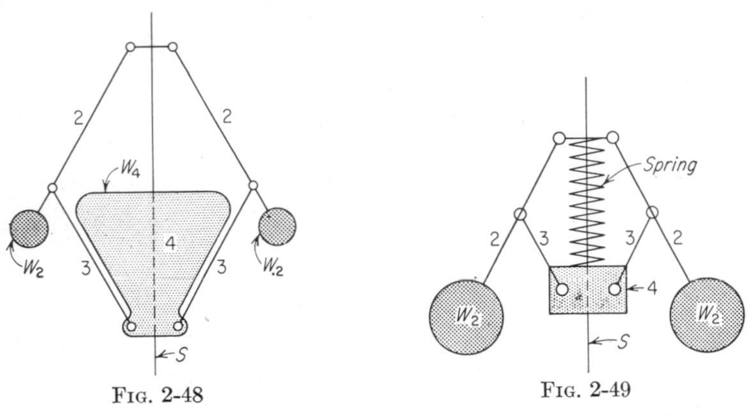

Fig. 2-48 Fig. 2-49

In its simplest form it has two or more weights W_2 carried by arms 2 which in turn are pivoted on a rotating spindle S. The spindle is driven at a speed proportional to the speed of the engine. As the weights revolve with the spindle, they tend to move outward under the action of centrifugal force. If this force is great enough, the weights will move outward, and by means of links 3 will move the weighted slide 4 along the spindle S. Through a suitable linkage this slide actuates the valve mechanism, and thus the supply of working fluid is controlled by the position of 4, decreasing as 4 rises.

2-32. The Spring-loaded Flyball Governor. More recent designs of the flyball governor such as are used for steam turbines, diesel engines, etc., are spring loaded, as illustrated in Fig. 2-49. A spring-controlled governor may be considered a special case of the gravity-loaded flyball governor, in which a spring is substituted for the dead weight for the purpose of rendering the governor more effective. In large turbines or engines the flyball governor may act indirectly. That is, it may take only the light service of moving a small pilot valve of an auxiliary device that operates the main valve gear or valve.

2-33. The Shaft Governor. This governor is a later development than the flyball type, having come into use in this country mainly as an adjunct to the high-speed steam engine, which requires a governor of greater effort and power than is usually possible with the flyball type. Builders of high-speed steam engines have governed them almost entirely by varying the point of cutoff of the steam. This has been accomplished by changing the angle of advance and also the throw of the eccentric by means of a governor that causes the center of the eccentric to vary in position relative to the crank, according to the load, resulting in a change in the events of the stroke. Since the governor usually acts directly upon the eccentric, it is necessary to have a powerful governor in order properly to overcome the resistance of the valve. Again, since the governor acts directly on the eccentric, it is convenient to have the governor on the crankshaft and hence running at the engine speed. The term *shaft governor* is therefore appropriate for this type of governor.

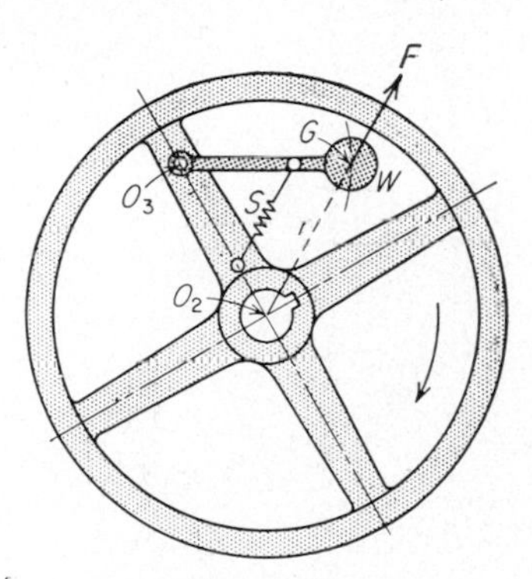

FIG. 2-50

2-34. Principle of the Centrifugal Shaft Governor. The elements of a governor of this type are shown in Fig. 2-50. The governor parts are usually mounted inside the flywheel and rotate with it on the engine power shaft O_2. The weight W is pivoted at O_3 to one of the arms of the flywheel so that it can assume various positions with respect to the flywheel. As the flywheel rotates, the weight W tends to move outward under the action of centrifugal force. This motion is opposed by the tension of spring S. If the speed increases beyond a certain point, weight W will assume a new position, and by means of suitable linkage (not shown) the eccentric will be shifted relative to the shaft O_2, thus regulating the supply of working fluid.

In addition to the centrifugal force, there is another force tending to move weight W relative to the flywheel. Let is be assumed that the flywheel is rotating about shaft center O_2 with angular velocity ω. If the load is decreased, the engine tends to speed up and the flywheel receives an angular acceleration α. Owing to its inertia the governor weight W resists this acceleration and thus tends to rotate about the center O_3 in a sense opposite that of ω. This action is known as the inertia effect. In this case, however, the inertia effect is relatively small and the centrifugal effect very powerful.

When a shaft governor is so proportioned that the centrifugal action predominates as in Fig. 2-50, the governor is known as a *centrifugal shaft governor*. It is not practicable to construct a shaft governor that operates by inertia alone, but when the inertia effect is relied upon to furnish the greater part of the regulating force, the governor is called an *inertia governor*.

2-35. Principle of the Inertia Shaft Governor. One of the simplest forms of inertia governor is outlined in Fig. 2-51. It consists of a single piece of cast iron in the general form of a bar pivoted to the flywheel at P, with the center of mass very close to the shaft center O. The eccentric that operates the valve is attached to this bar, or weight, by means of a suitable linkage. Neglecting the small centrifugal force that will be set up, it is evident that, as long as the flywheel rotates with a uniform angular velocity, there will be no relative motion between the weight and the flywheel. Now, if the flywheel is given an angular acceleration, the weight, because of its inertia, will resist this acceleration. As a result, there will be a relative motion between the flywheel and weight that will cause the eccentric to shift its position relative to the shaft. Although the centrifugal force has a certain desired effect in regulation, owing to the fact that the mass center is placed a short distance from the shaft center O, it is evident that in the action of this governor inertia force plays decidedly the major part.

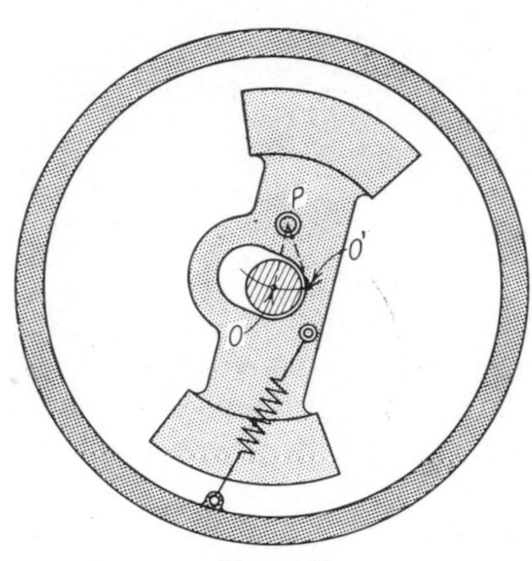

FIG. 2-51

2-36. Action of the Two Types of Shaft Governor Compared. It has been shown that shaft governors operate as a result of combined centrifugal force and inertia force and are classified according to the force that predominates. It is important to have a clear understanding of the principle of action of each type.

In the centrifugal type the governing force depends upon the *amount* of change in the angular velocity of the flywheel. Hence there must be a definite amount of change in the speed before the change in the centrifugal force is great enough to move the valve gear against the frictional forces.

In the inertia governor, on the other hand, the governing force depends on the *rate* of change of angular velocity, which is usually greatest just at the beginning of the change of speed. Hence it begins to act just as soon as the flywheel begins to change its speed. It follows, therefore, that the inertia governor is more sensitive than the centrifugal governor.

FLEXIBLE CONNECTORS

The remainder of this chapter is concerned with examples of flexible connectors which are used to transmit motion (and force) in machines. The main categories to be considered are belts, ropes, and chains.

2-37. Belts, Ropes, and Chains. When the distance between shafts becomes so great as to render the use of gear trains impracticable, rotary motion is transmitted by means of belts, ropes, or chains. In these cases the belt, rope, or chain is passed in a continuous band over pulleys, sheaves, or sprockets, and the motion is transmitted as though the wheels rolled directly upon each other. Owing to conditions that will be explained later in this chapter, the first two of the above methods of driving cannot be used where exact velocity ratio is required. These methods, nevertheless, have a wide range of employment in the transmission of motion and power, because of the numerous applications where exact velocity ratio is not necessary.

2-38. Flat Belts. When a belt is in the form of a thin flat band designed to run on a cylindrical surface it is known as a *flat belt.* The flat-belt drive has been one of the most widely used methods of transmitting power. Although very largely superseded by more modern methods, it still has applications. The use of belts is subject to certain limitations. For example, there is a limit to the distance over which power can be transmitted with economy by this method. No specific figures can be given for this limit, as each installation is a special case and must be treated as such. The distance between centers for a great number of flat-belt drives, however, varies from 15 to 30 ft. Another limitation upon the use of belts results from the fact that the driving force transmitted depends upon the friction between the belt and the faces of the pulleys it connects. There is always the possibility of some slipping between the belt and the faces of the pulleys, and hence the character of the motion transmitted is not positive. Where positive action is required, gears or chains must be used.

2-39. Velocity Ratio of Belt Drives. When the rotation of the driven pulley of a belt drive must have the same sense as that of the driving pulley, the scheme shown in Fig. 2-52 is employed. When, on the other hand, the sense of rotation of the driven pulley must be opposite that of the driving pulley, the belt is arranged in the manner shown in Fig. 2-53. These two arrangements are called *open* and *crossed* belt drives, respectively.

The velocity ratio of a belt drive is defined as the ratio of the number of turns of the driving pulley to the number of turns of the driven pulley in a unit of time. This may be determined with sufficient accuracy in the following manner: Let it be assumed that the center of the belt section has a constant linear velocity, whether it is between the pulleys or passing around one of them. Let this velocity in feet per second be V; the thickness of the belt t; the diameters of the driving and driven pulleys D_1 and D_2, respectively; and the rpm of the driving and driven pulleys n_1 and n_2, respectively. If there is no slip, the linear velocity of the belt, while passing around one of the pulleys, will be equal to the angular velocity of that pulley multiplied by the sum of its radius and half the thickness of the belt. Hence,

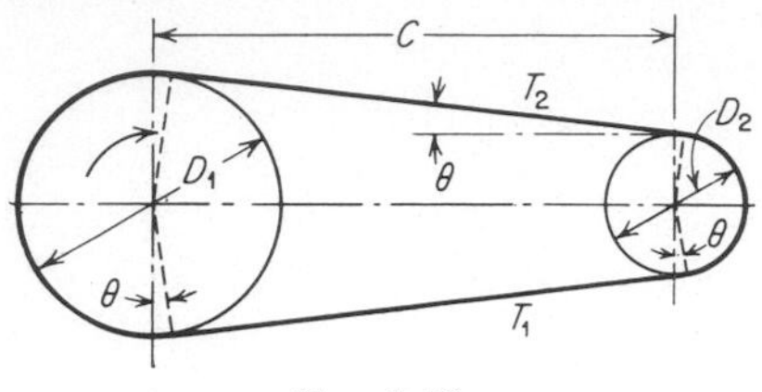

FIG. 2-52

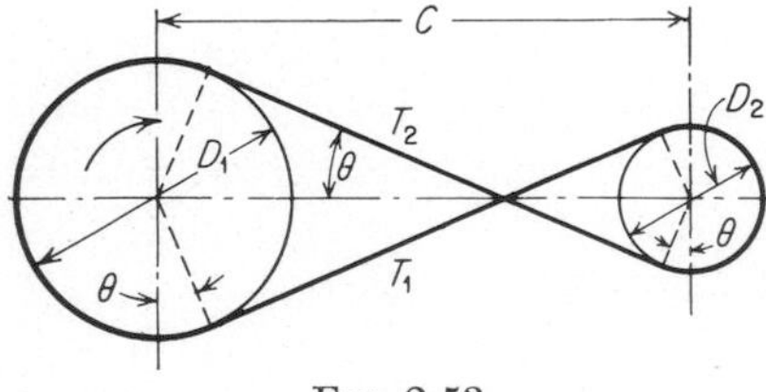

FIG. 2-53

$$V = \frac{2\pi n_1}{60}\left(\frac{D_1}{2} + \frac{t}{2}\right) = \frac{2\pi n_2}{60}\left(\frac{D_2}{2} + \frac{t}{2}\right)$$

$$n_1(D_1 + t) = n_2(D_2 + t)$$

$$\frac{n_1}{n_2} = \frac{D_2 + t}{D_1 + t} \qquad (2\text{-}11)$$

Since the thickness of the belt is usually small compared with the radii of the pulleys, the approximate formula

$$\frac{n_1}{n_2} = \frac{D_2}{D_1} \qquad (2\text{-}12)$$

is sufficiently accurate, except for rare cases where a relatively thick belt runs on a comparatively small pulley.

2-40. Belt Tensions. Referring to Fig. 2-53, it is evident that, if the belt is to stay on the pulleys, both the part marked T_1, called the tight side, and the part marked T_2, called the slack or loose side, must exert a pull on the pulleys; that is, the slack side is only relatively slack. If the tensions in pounds are represented by T_1 and T_2 as shown, the effective pull or the force transmitted from one pulley to the other is

$$T = T_1 - T_2$$

If D_1 is the diameter of the driving pulley in feet, and n_1 is the number of revolutions it makes in a minute, the horsepower transmitted is

$$\text{hp} = \frac{\pi D_1 n_1}{33{,}000} (T_1 - T_2) \tag{2-13}$$

Methods of determining the effective pull, or as it is often called, the net tension, will be found in works on machine design. The thickness of the belt is disregarded in the formula for the horsepower given above, the error resulting therefrom being negligible in view of the incompleteness of available information regarding the net tension.

Frequently an idler pulley is used to increase the arc of contact, especially on the smaller pulley, and also to produce more tension in the belt. When a swinging idler is used, as shown in Fig. 2-54, the belt tension can be regulated and may be released altogether, if desired.

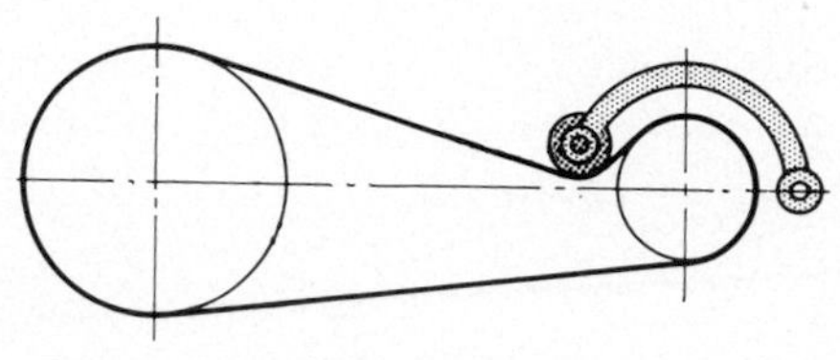

FIG. 2-54

2-41. Length of Belt. The determination of the correct length of belt for a given installation is a somewhat complicated problem, even when the effect of sag is not taken into account. Formulas for this length will be deduced, first for the crossed and then for the open type of belt drive.

1. *Crossed Belt.* Referring to Fig. 2-53, let D_1, D_2, and C be known. The total length of the belt is made up of the part around the smaller pulley, the part around the larger pulley, and the part between the pulleys.

Part around the smaller pulley $= (\pi + 2\theta)D_2/2$.
Part around the larger pulley $= (\pi + 2\theta)D_1/2$.
Part between the pulleys $= 2C \cos \theta$.

The total length is then

$$L = \left(\frac{\pi}{2} + \theta\right)(D_1 + D_2) + 2C \cos \theta \tag{2-14}$$

$$\sin \theta = \frac{D_1 + D_2}{2C} \qquad \text{and} \qquad \cos \theta = \sqrt{1 - \left(\frac{D_1 + D_2}{2C}\right)^2}$$

$$L = \frac{\pi}{2}(D_1 + D_2) + (D_1 + D_2) \arcsin \frac{D_1 + D_2}{2C} + 2C\sqrt{1 - \left(\frac{D_1 + D_2}{2C}\right)^2} \tag{2-15}$$

Equation (2-15) may be simplified by making use of the series obtained by expanding the arcsine and the quantity under the radical as follows:

Let
$$\frac{D_1 + D_2}{2C} = x$$

Then,
$$\arcsin x = x + \frac{x^3}{6} + \frac{3x^5}{40} + \cdots$$

and
$$\sqrt{1 - x^2} = 1 - \frac{x^2}{2} - \frac{x^4}{8} - \frac{x^6}{16} - \cdots$$

Substituting the above quantities in Eq. (2-15), an expression is obtained for the length of the belt that involves only the diameters and center distance of the pulleys, thus

$$L = 2C + \frac{\pi}{2}(D_1 + D_2) + \frac{(D_1 + D_2)^2}{4C} + \frac{(D_1 + D_2)^4}{192C^3} + \frac{(D_1 + D_2)^6}{2{,}560C^5} + \cdots \qquad (2\text{-}16)$$

The number of terms to be taken in calculating the length of a crossed belt by means of Eq. (2-16) is dependent upon the degree of accuracy desired. Not more than three terms are necessary, all powers above the square being negligible. The final expression for the length of a crossed belt then becomes

$$L = 2C + \frac{\pi}{2}(D_1 + D_2) + \frac{(D_1 + D_2)^2}{4C} + \cdots \qquad (2\text{-}17)$$

For rough calculations, as in making a cost estimate, the first two terms will give sufficiently accurate results. The values of the quantities D_1 and D_2 in the above equation are usually taken as the diameters of the pulleys under consideration. To be strictly accurate the term D_1 (or D_2) should be taken as the diameter of the pulley plus twice the distance from the face of the pulley to the neutral plane of the belt. The latter distance is approximately equal to one-half the thickness of the belt, thus making the value of D_1 equal to the diameter of the pulley plus the thickness of the belt. Such accuracy is seldom required.

2. *Open Belt.* The case for open belts is similar. Referring to Fig. 2-52 it may be seen that the component parts of the total length are as follows:

Part around the smaller pulley $= (\pi - 2\theta)D_2/2$.
Part around the larger pulley $= (\pi + 2\theta)D_1/2$.
Part between the pulleys $= 2C \cos \theta$.

By a process of reasoning similar to that employed for crossed belts, it may be shown that the length of an open belt is

$$L = 2C + \frac{\pi}{2}(D_1 + D_2) + \frac{(D_1 - D_2)^2}{4C} + \cdots \tag{2-18}$$

From a comparison of Eqs. (2-17) and (2-18), it is evident that for relatively large center distances the same belt may be used for either crossed or open drives.

2-42. Cone Pulleys. It is sometimes desirable so to construct a belt drive that its velocity ratio may be changed. This may be accomplished by means of a single belt and a pair of stepped cone pulleys as shown in Fig. 2-55. This change may be accomplished in a similar manner by the use of a pair of *speed cones* which are merely cones tapered down smoothly instead of by steps. It is customary to construct cone-pulley drives in such a manner that the speed of the driven cone for a constant speed of the driving cone increases in geometric ratio for each step as the belt is shifted from one side of the pair to the other. Referring to the figure, let ω_1 be the constant angular velocity of the driving cone, and let ω_2 be the lowest angular velocity of the driven cone. Then, if the speeds are to increase in geometric ratio, the next pair of pulleys on the cones should give the driven cone an angular velocity of $k\omega_2$; the succeeding pair should give it a velocity of $k^2\omega_2$; the next $k^3\omega_2$; etc. The corresponding velocity ratios are ω_1/ω_2, $\omega_1/k\omega_2$, $\omega_1/k^2\omega_2$, etc. After the magnitudes of the velocity ratios for the various steps of a cone-pulley drive have been determined, the diameters of the pulleys for one step may be chosen at will, subject, of course, to the requirements of space and satisfactory operation. The choice of the diameters of the other pairs of pulleys is complicated by the fact that the same belt is used for all the pairs. The problem will first be taken up for the crossed belts and then for the open belts.

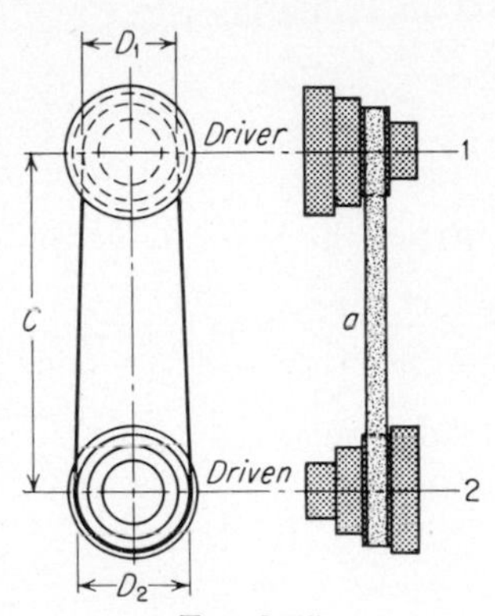

Fig. 2-55

1. *Cone Pulleys with Crossed Belt.* Referring to Fig. 2-55, let n_1/n_{2a} be the velocity ratio for the position of the belt shown. The length of the belt for the diameters D_{1a} and D_{2a} is the same as for any other pair of diameters D_{1x} and D_{2x} to which the belt may be shifted. It follows from Eq. (2-16) that

$$D_{1a} + D_{2a} = D_{1x} + D_{2x} \tag{2-19}$$

For the velocity ratio n_1/n_{2x}, the diameters D_{1x} and D_{2x} must also satisfy the relation

$$\frac{D_{2x}}{D_{1x}} = \frac{n_1}{n_{2x}} \tag{2-20}$$

These two relations may be solved simultaneously for D_{1x} and D_{2x}, and the diameters of any pair of pulleys of a crossed-belt cone-pulley drive thus determined.

2. *Cone Pulleys with Open Belt.* The case for open belts is as follows. Referring again to Fig. 2-55, the length of belt required for pulleys a is given by Eq. (2-18) as

$$L = 2C + \frac{\pi}{2}(D_{1a} + D_{2a}) + \frac{(D_{1a} - D_{2a})^2}{4C} + \cdots \tag{2-21}$$

This value may be calculated and substituted in the equation

$$L = 2C + \frac{\pi}{2}(D_{1x} + D_{2x}) + \frac{(D_{1x} - D_{2x})^2}{4C} + \cdots \tag{2-22}$$

From the equation expressing the velocity ratio,

$$D_{2x} = \frac{n_1}{n_{2x}} D_{1x} \tag{2-23}$$

Substituting,

$$L = 2C + \frac{\pi}{2}\left(1 + \frac{n_1}{n_{2x}}\right) D_{1x} + \frac{[1 - (n_1/n_{2x})]^2}{4C} D_{1x}^2 + \cdots \tag{2-24}$$

This equation may easily be solved for D_{1x}, and from the preceding equation D_{2x} may then be found.

Convenient graphical methods of obtaining the results expressed by the above analytical expressions may be found in several well-known treatises on mechanism.

2-43. Belt Connections between Nonparallel Shafts. Belts may be used for drives between shafts that are not parallel, provided the pulleys are so located as to conform to a fundamental principle that governs the operation of all belt drives, namely: *The center line of that part of the belt approaching a pulley must lie in the central plane of that pulley.* The angle at which the belt leaves the pulley is immaterial. This will be made clear by reference to Fig. 2-56. In this case the shafts 1 and 2 are turning in the directions indicated by the arrows. The center line of the belt leaves the driving pulley A at the point m. The driven pulley B is in such a position on shaft 2 that a plane through the middle of its face and perpendicular to its axis contains the center line mn. The belt will, therefore, run properly on to pulley B. It is evident that, if the direction of motion were reversed, the belt would immediately leave the pulleys. Changes in direction of rotation thus necessitate corresponding changes in relative positions of the pulleys. In Fig. 2-56 the axes of the pulleys are at 90 deg with each other. The belt would run equally well if the pulleys were turned at any angle about xx as an axis. It can be seen that the extreme practical limits of motion about this axis would in one direction result in an open-belt drive between parallel shafts, and in the other direction a crossed-belt drive between parallel shafts. Belts arranged as in Fig. 2-56 are called *quarter-turn* belts.

If the drive between two nonparallel shafts is to be such that the pulleys may run in either direction and still deliver the belt properly, in accordance with the fundamental principle already explained, it is necessary to make use of intermediate pulleys to guide the belt into the proper plane. Such pulleys are called *guide* pulleys. It is impossible to arrange a reversible belt drive between nonparallel shafts using only two pulleys. An example of a reversible belt drive is shown in Fig. 2-57.

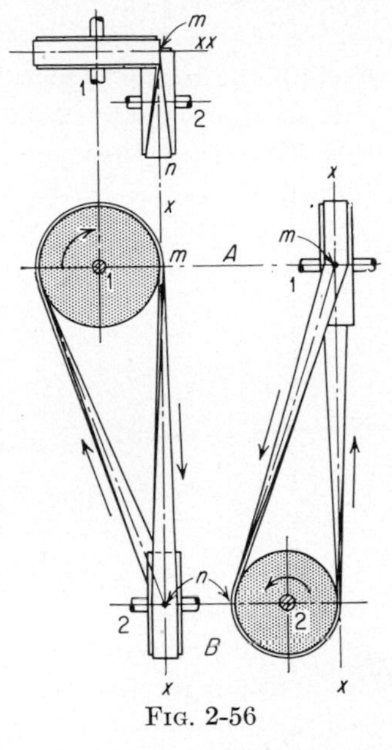

FIG. 2-56

2-44. Crowning of Pulleys. Pulleys are crowned to prevent the belt from running off their sides. By crowning is meant making the pulley

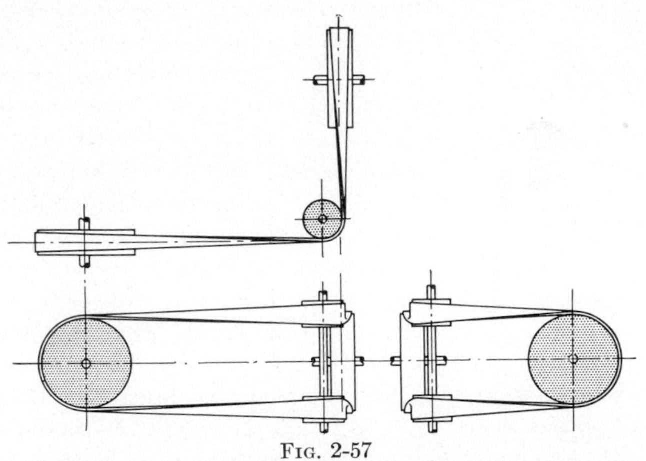
FIG. 2-57

slightly conical or spherical, as shown in Fig. 2-58a and b. Just how the crowned pulley tends to center the belt may be seen from the following considerations: Referring to Fig. 2-58c, if a belt is led on to a double conical pulley as shown, it will be immediately subjected to two influences, the first of which is to lie flat, thus causing the belt to be bent laterally as shown. The second is to resist this lateral bending and, as a result, that part of the belt that is approaching the pulley will be thrown over to the right. Once in contact there is not much tendency for the belt to slip sideways, and, since the first point of contact tends to be thrown toward the center of the pulley, the belt as a whole will run centrally.

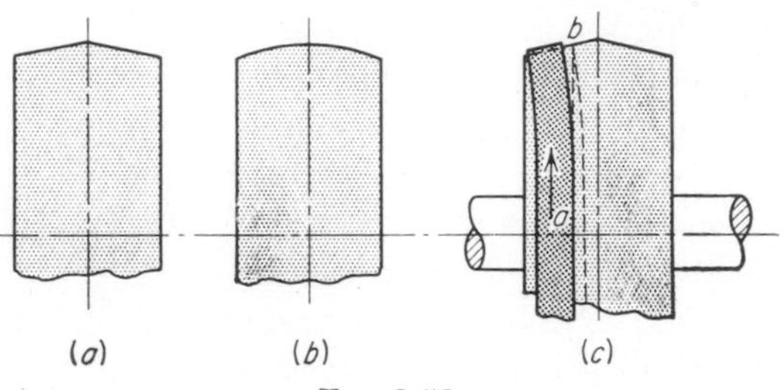

FIG. 2-58

2-45. V Belts. When a belt is trapezoidal in section, designed to run in a V-shaped groove, it is known as a *V belt.* The usual form of V belt and pulley is shown in section at a in Fig. 2-59, where D indicates the nominal pitch diameter of the pulley. This form of belt has long been in use for special purposes, the ordinary automobile fan belt being perhaps the most familiar form of this type of drive. The modern V belts, made of fabric and vulcanized rubber with a cotton-cord tension element, run on sheaves having a varying number of grooves to suit the horsepower ratings. They are in certain cases used with a grooved driving pulley and a flat driven pulley. Also, these belts may be adapted to quarter-turn drives and, with modifications in design, as shown at b in the figure, to reversed bending about sheaves. Where the distance between centers of shafts is small the V belt is rapidly replacing the older flat belt drive, since it transmits a larger amount of power from a pulley of a given width of face. Also, being almost positive and slipless in action, the V-belt drive is displacing chain and gear drives for many short-center-distance machine drives.

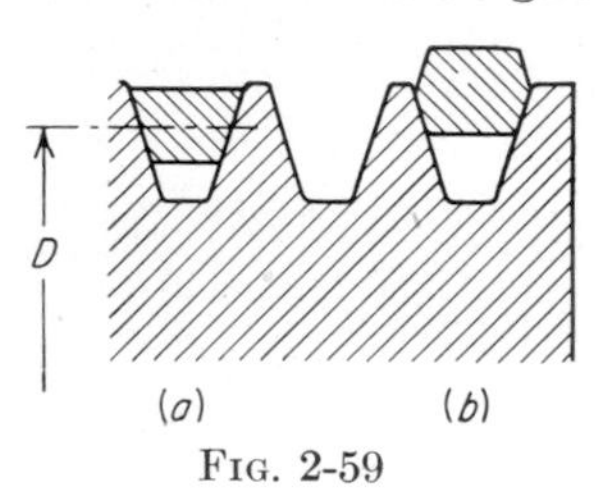

FIG. 2-59

2-46. V-belt Drives. The discussion in the preceding articles relating to belts applies specifically to flat-belt drives. It should be noted, however, that the part pertaining to velocity ratio and length of belt may be applied to V-belt drives by using the proper diameters.

2-47. Ropes and Cables. Ropes used in engineering operations usually are made of a fibrous material such as manila, hemp, and cotton, or of strands of iron or steel, in which case they may be called wire ropes or cables. As to the kind of service, rope and cable may be classified as follows: (1) for hoisting and transporting of loads; (2) for transmitting motion (and force) as a tension element in a machine or mechanism; and (3) for the transmission of power.

Fibrous rope is used chiefly for hoisting tackle, whereas present-day application of wire rope is chiefly to hoisting, haulage, and transporting service. The development of electrical transmission has gradually crowded out the rope for power transmission.

Examples of hoisting and transporting applications of rope and cable include elevators and mine hoists; cranes, draglines, and power shovels; and package and message conveyors in buildings.

Rope and cable are much in evidence in machines and devices in which motion can be transmitted by a tension element. Control cables on aircraft, dial cords on radio receivers, and emergency-brake cables on automobiles are familiar examples. The flexibility of rope or cable is a definite asset in obtaining desired motions with relative ease in applications such as these.

2-48. Chains. Chain drives are used where positive action is required and where it is desired to obtain a more compact arrangement than would be possible by the use of a flat belt. In general, the power is transmitted over a shorter distance in a chain drive than is the case for a flat-belt drive. The wheels over which chains run are called sprockets and have their surfaces shaped to conform to the type of chain used. Chains and sprockets are made in a large variety of forms, according to the service to be performed.

The various types of chains found in engineering practice may be classified as follows:

- Hoisting chains
  - Coil
  - Stud-link
- Conveying and elevating chains
  - Detachable, or hook joint
  - Closed joint
- Power-transmission chains
  - Block
  - Roller
  - Inverted-tooth

2-49. Hoisting Chains. The coil chain (Fig. 2-60*a*) is used on hoists, cranes, dredges, etc., and the stud-link chain (Fig. 2-60*b*) is used mainly in marine work in connection with anchors and moorings. The chief advantage of the stud-link chain is that it will not kink or entangle so readily as a coil chain.

2-50. Conveying and Elevating Chains. Conveying chains are used for conveying and elevating all kinds of materials under a wide range of conditions. The detachable or hook-joint chain (Fig. 2-61) is used extensively for conveying and elevating but is not well adapted for conveying gritty bulk material. It is also used to some extent for power transmission where the speeds are low. The closed-joint chain, one type of which is shown in Fig. 2-62, is well adapted to the elevating and conveying of gritty bulk material, as well as the transmission of power at moderate speeds.

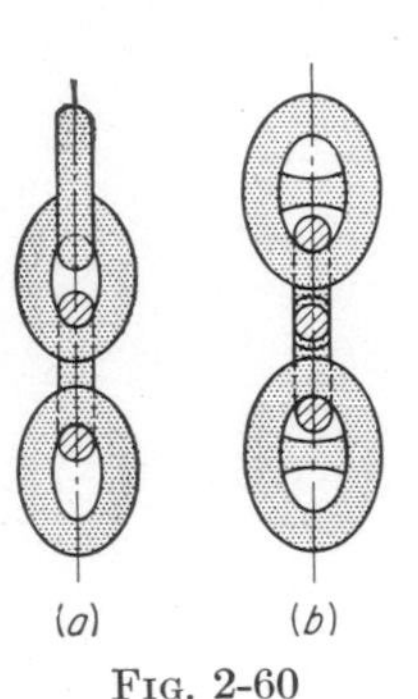

Fig. 2-60

2-51. Power-transmission Chains. The types of chains discussed in the preceding paragraphs are not well adapted to any service requiring speeds above 600 fpm and consequently are not suitable for the transmission of power where the speed exceeds this limit. For this class of service the special forms of power-transmission chains, all of which are machined fairly accurately, have been devised.

Figure 2-63 illustrates a block chain. As the name indicates, the block chain consists of solid steel blocks, shaped like the letter B or the figure 8, to which the side links are fastened by hardened steel rivets. Chains of the block type are less expensive to make than roller or inverted-tooth chains, and have proved satisfactory for light power transmission where the speeds do not exceed 800 to 900 fpm.

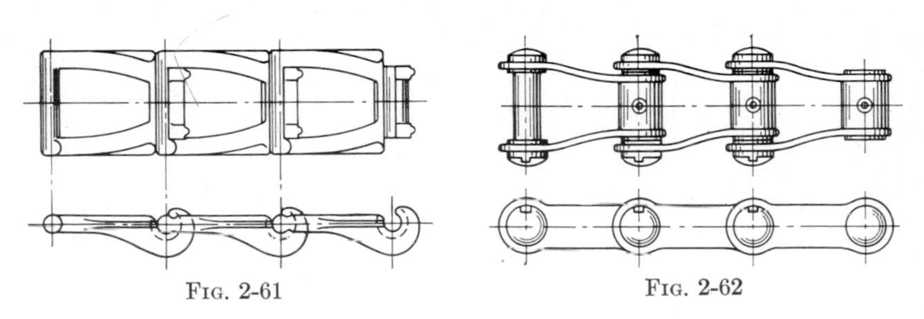

Fig. 2-61 Fig. 2-62

A typical roller chain is shown in Fig. 2-64. This type of chain is used to some extent in motor-vehicle service, especially on trucks, as well as for general power transmission. A chain speed of 1,500 fpm with proper lubrication is about the limit for general use.

The most widely used type of chain for transmitting power at relatively high linear speeds is the inverted-tooth chain. There are now in use several well-known makes of inverted-tooth chain that run satisfactorily at high speeds. One example (Fig. 2-65) will serve to illustrate this type of chain, since the several designs have, in general, the same form of link, differing only in the form of joint used. If properly designed and constructed, an installation of an inverted-tooth chain will be just

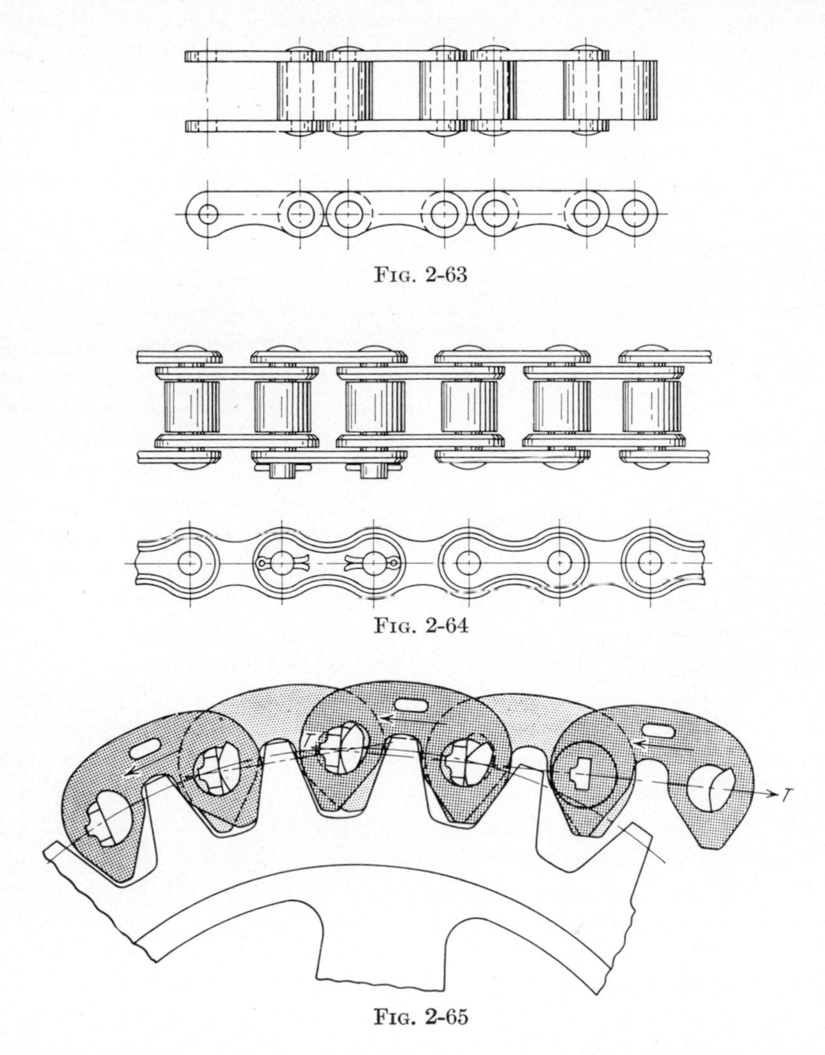

Fig. 2-63

Fig. 2-64

Fig. 2-65

as efficient, it is claimed, as a gear drive for the same conditions of operation. Where a positive drive is essential, and where the shafts are too far apart for gearing, inverted-tooth chains are used extensively. Inverted-tooth chains are well adapted for transmitting power economically at speeds of 1,200 to 1,500 fpm, and the smaller sizes of chain may be run as high as 2,000 fpm. The higher limit of speed, however, is attained at a cost of reduced life of the chain. This type of chain is also known as the *silent chain*.

2-52. Variable-speed Transmissions. It has been observed (Art. 2-42) that, with the flat belt, change of speed is accomplished by moving the belt along the axes of the cone pulleys or speed cones to give a different pair of working diameters on the driving and driven pulleys. These may be regarded as the elementary mechanisms through which variable-speed transmission is accomplished with flexible connectors.

One of the best-known devices for accomplishing variable-speed transmission through the use of a belt is the *Reeves variable-speed transmission*. It consists essentially of a pair of pulleys connected by a V-shaped belt in the manner indicated in Fig. 2-66. Each of the pulleys consists of a pair of beveled disks (cones) keyed to the shaft so that disks and shaft must turn together but the disks are free to be moved along the shaft. By means of suitable adjusting devices the two conical disks on one shaft are caused to approach each other at the same time that the two disks on the other shaft separate. Bringing the two disks nearer together causes the V-shaped belt to be in contact with them farther out from the axis, and separating them causes the belt to be in contact nearer the axis. Thus, the effective radius of one pulley is increased at the same time that the effective radius of the other pulley is decreased. In this manner, the ratio of driving diameter to driven diameter is readily and quickly changed, thus securing any desired speed, between the maximum and minimum, without the necessity of stopping the machine. The device is in the form of a compact unit that can readily be interposed between the source of power and the driven machine.

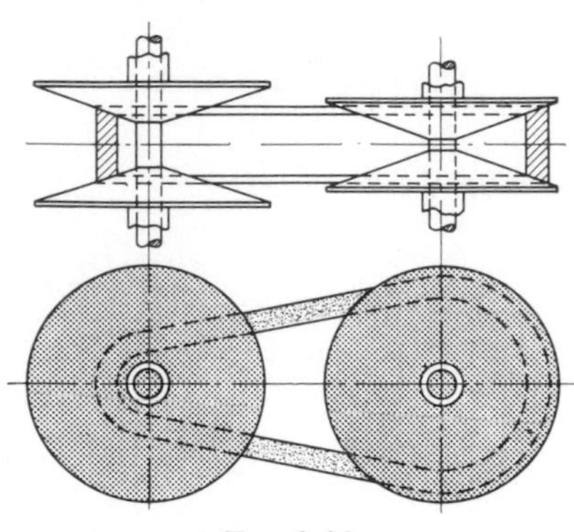

Fig. 2-66

Variable-speed transmission in the case of the chain drive may be secured by a device similar to that described above for the V-shaped belt. One such device known as the *PIV* (*positive, infinitely variable*) *gear* consists essentially of a V-shaped link chain with teeth on the sides that engage with radial teeth in a pair of axially adjustable conical disks on the driving and driven shafts. Each pair of disks thus becomes an adjustable-diameter wheel in the transmission of power, so that an infinite number of speeds between the minimum and maximum are available.

2-53. Hoisting Tackle. Under the title of *hoisting tackle* may be grouped all those combinations of ropes and chains with pulleys and pulley blocks, the purpose of which is to overcome a considerable resistance that acts through a relatively small distance by means of a relatively small force that acts through a considerable distance.

2-54. The Block and Tackle. Probably the simplest example of hoisting tackle is the ordinary block and tackle illustrated in Fig. 2-67. For purposes of analysis it is more convenient to use the diagrammatic representation of the block and tackle, as shown in Fig. 2-68. For the case shown, a force P acting through a distance p will, if friction is disregarded, lift the load Q, which is equal to $4P$, through the distance $\frac{1}{4}p$. The *mechanical advantage* of a block and tackle, as in the case of any other machine, is the ratio of the resistance overcome to the force that would have to be applied if friction were disregarded. There are a number of methods for determining the mechanical advantage of a block and tackle, but probably the most convenient is simply to count the ropes supporting the load. This number is equal to the mechanical advantage.

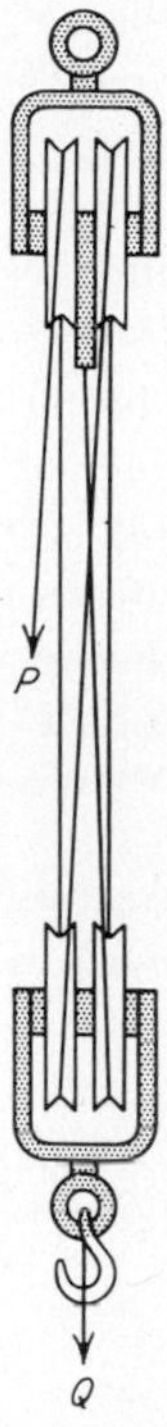

2-55. The Differential Chain Block. Although the hoisting mechanism shown in Fig. 2-69 differs somewhat in principle from the ordinary block and tackle, it is used for the same purpose, namely, the raising of a heavy load by means of a small applied force. The upper block contains two sheaves of slightly different diameters that are fastened together so as to move as a unit. The lower block has but one sheave and is attached to the upper block by means of the endless chain, as shown. The force is applied at P, and the load Q is fastened to the hook on the lower block. The relation between P and Q may be determined

P

Q

Fig. 2-67

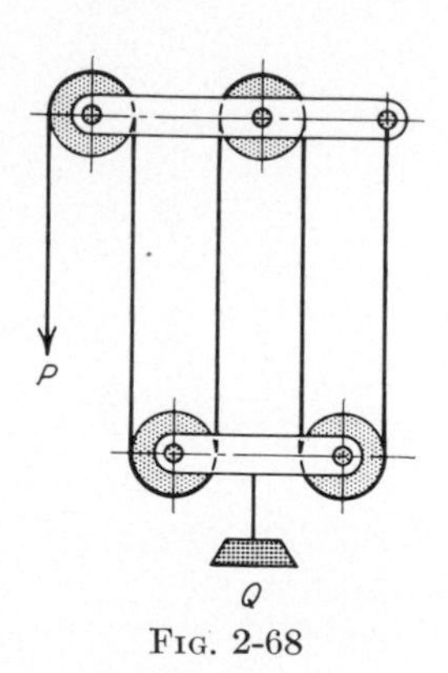

Fig. 2-68

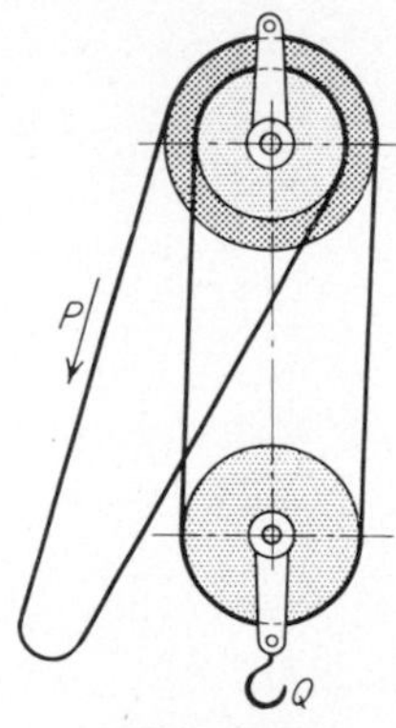

Fig. 2-69

as follows: Let the diameter of the larger sheave in the upper block be D_a, and let the diameter of the smaller sheave be D_b. If these sheaves are rotated in a counterclockwise sense, the chain that is run off the smaller sheave will increase the total length of chain between the upper and lower blocks, and that which is run on to the larger sheave will decrease this length. The net change of the total length of chain between the blocks will be the difference of the lengths run on and off the upper sheaves. This change of length is equally divided between the two parts of the chain that connect the upper and lower blocks, and hence the change of center distance of the two blocks is equal to one-half the difference of the lengths of chain run on and off the sheaves in the upper block. Let the sheaves in the upper block be turned through one revolution under the action of the force P. The load Q will be raised through a distance

$$h = \frac{\pi}{2}(D_a - D_b) \tag{2-25}$$

Assuming no friction, the amount of work delivered to the hoist by P must equal the amount done in raising Q. Therefore,

$$P\pi D_a = Qh = Q\frac{\pi}{2}(D_a - D_b)$$

$$P = \frac{Q}{2}\frac{D_a - D_b}{D_a} \tag{2-26}$$

CHAPTER 3

CAMS

The transformation of one of the simple motions, such as rotation, into any other motion is often conveniently accomplished by means of a cam mechanism, consisting usually of two moving elements, the cam and the follower, mounted in a fixed frame. Of course, a system of pin-connected linkages also acts to transform motion from one kind to another. However, cam devices are extremely versatile, and almost any arbitrarily specified motion can be obtained. In some instances, it is certainly true that they offer the simplest and most compact transformation system possible.

A *cam* may be defined as a machine element having a curved outline or a curved groove, which, by its oscillating or rotating motion, gives a predetermined specified motion to another element called the *follower*. The cam has a very important function in the operation of many classes of machines, especially those of the automatic type, such as printing presses, shoe machinery, textile machinery, gear-cutting machines, and screw machines. In any class of machinery in which automatic control and accurate timing are paramount, the cam is an indispensable part of the mechanism. The possible applications of cams are unlimited, and their shapes occur in great variety. Some of the most common forms will be considered in this chapter.

3-1. Cam Motions. Two simple examples of a *disk* cam and follower are shown in Figs. 3-1 and 3-2, one with a roller follower and one with a flat-faced follower. If the cam is given a motion of rotation, it will first lift the follower by direct contact to its extreme position, will next permit its return to its initial position, and will then let it remain at rest in its initial position during a given interval of time, after which the cycle of motion will be repeated. In cams of this type the return motion of the follower must be brought about by some force external to the cam, by means of a spring, for example. The roller, by means of which contact is made between cam and follower (Fig. 3-1), serves to reduce friction to a minimum, but it is sometimes desirable or necessary to use a flat-faced follower, such as represented in Fig. 3-2. That part of the follower face which comes in contact with the cam is usually provided with a hardened surface to prevent excessive wear.

3-2. Fundamental Procedure in a Cam Layout. In the layout of a cam, the initial position, length of stroke, direction, and character of the motion of the follower are, as a rule, completely known. The angular motion of the cam and the location of the axis of the cam relative to the follower are usually known also. The problem, therefore, is to determine the shape of the cam profile that will impart the required motion to the follower.

The fundamental principles to be observed in the layout of a cam will be brought out by means of the examples given in Figs. 3-1 and 3-2. In Fig. 3-1 is shown a disk cam A whose clockwise rotation is to drive the roller follower B radially outward along the line OC. The cam is assumed to rotate clockwise at a constant rate.

The outline of the cam is to be such that for each 30 deg of rotation of the cam for the first 120 deg the axis of the roller follower is to move progressively from its initial position at 0 to 1, 2, 3, and 4 and then return progressively from its extreme position at 4 to 3, 2, 1, and 0 during each 30-deg rotation of the cam for the second 120 deg, remaining at rest in its initial position during the remaining 120 deg of the cycle.

The first step in the solution is to draw the four radial lines dividing the first 120-deg rotation of the cam into the required 30-deg intervals. The next step is to locate the centers 1′, 2′, 3′, and 4′ of the roller follower along these radial lines at distances from the center O of the cam successively equal to the distances $O1$, $O2$, $O3$, and $O4$ measured on the initial radial line OC, and draw in the outline of the roller follower in these positions. The final step in the solution is to determine the required outline of this part of the cam by drawing a smooth curve tangent to the initial position and the four constructed positions of the roller follower.

In determining the constructed positions 1′, 2′, 3′, and 4′ of the roller follower, as described above, it is convenient to apply the principle of *inversion:* that is, to assume that the cam is the fixed link and that the frame that guides the follower and carries the bearing supporting the camshaft is the moving link. Obviously, fixing the cam and rotating the frame in a sense opposite that assumed for the cam will give the same relative motion between the follower and frame as would be given by the actual setup, i.e., rotation of the cam with the frame fixed. In the present instance, e.g., when the cam has rotated clockwise through an angle α, which in this case is equal to 60 deg, the axis of the roller follower is to have moved from 0 to 2; or, if the cam is fixed and the roller and frame are rotated counterclockwise through the angle α to the dotted position shown in Fig. 3-1, the roller will move a distance equal to that of 0 to 2 relative to the frame. Hence point 2′ is where an arc of radius $O2$ intersects the radial line $O2'$. Obviously, if a circle of radius equal to that of the roller follower is drawn from point 2′ and a part of the cam outline drawn tangent to this curve and the initial position O of the roller, and the cam is then rotated clockwise through 60 deg, the center of the roller follower will be positively driven from O to 2. Having determined the remaining points 1′, 3′, and 4′ in a similar manner, the outline of the working surface of the cam for the first 120-deg rotation was determined as shown in the figure. Since the method of determining the working surface of the cam for the second 120 deg of its motion is exactly the same as for the first 120 deg, the outline has been drawn in without showing the construction. Since the follower is to remain stationary while the cam rotates through the last 120 deg, this part of the cam outline will be a circular arc drawn from O tangent to the initial position of the roller follower.

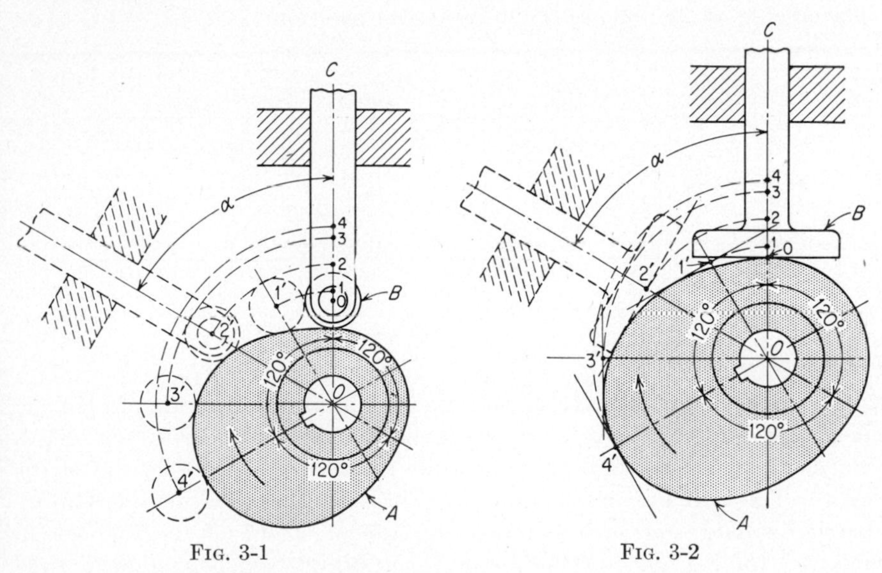

Fig. 3-1

Fig. 3-2

In order to avoid confusion in a cam layout, the constructed positions of the follower, such as 0, 1′, 2′, 3′, and 4′ in Fig. 3-1, should always be numbered to correspond to the numbered positions 0, 1, 2, 3, and 4, which define the motion of the follower along its path.

Figure 3-2 illustrates the layout when the roller follower of Fig. 3-1 has been replaced by a flat-faced follower. The specifications are otherwise exactly the same as for the layout of Fig. 3-1 and the solution is made in the same manner. The working surface of the cam is a smooth curve tangent to the several constructed positions of the face of the follower as in the previous case.

It should be noted that in this case the face of the follower makes an angle of 90 deg with the initial radial line OC, and, therefore, in its constructed positions the face should make the same angle with the corresponding radial lines, as illustrated by the dotted position of the follower at position 2′.

3-3. Displacement Diagrams. In the layout of a cam mechanism it is usually desirable to make use of a *displacement diagram.* Such a diagram is a specification of the motion desired for the follower as a function of the motion of the driver (cam). For example, in Fig. 3-3, the length of the diagram represents one revolution of the cam, and the height of the diagram the total displacement of the follower from its lowest position. If the cam rotates with constant angular speed, which is usually the case, the time of a revolution is divided into any convenient number of equal parts, called time periods, which appear as equal spaces on the base line. The ordinates corresponding to the several time periods represent displacements of the follower from its initial position.

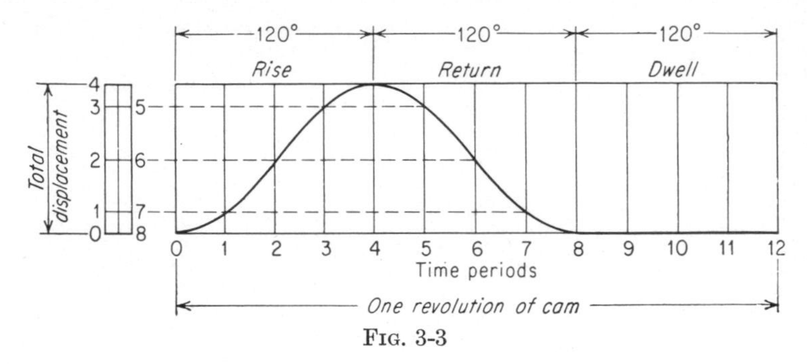

Fig. 3-3

Figure 3-3 represents the displacement diagram for the cam mechanism in Fig. 3-1 (or Fig. 3-2) as it would be plotted on an arbitrary length of base line divided into 12 equal time intervals for one rotation of the cam. The ordinates in the diagram represent the correspondingly numbered displacements shown along the line of stroke of the follower in Fig. 3-1 (or Fig. 3-2). The law of variation of displacement with time during the rise and return of the follower is clearly pictured by the curves in the diagram.

In Figs. 3-1 and 3-2 the construction for the return motion of the follower during the second 120 deg of motion of the cam was omitted to avoid confusion in the drawings. But since the return motion is exactly the same as the rise, except in reverse order, the return motion is represented in the displacement diagram (Fig. 3-3) by the reversed curve shown from positions 4 to 8. The remaining 120-deg period of the cam motion, where the follower is at rest in its initial position, is represented in the diagram by positions 8 to 12. Hence, positions 0 to 4 in the displacement diagram represent the rise of the follower, 4 to 8 the return, and 8 to 12 a rest, or dwell, in the initial position until the beginning of the next cycle.

In addition to providing a very convenient means for studying the character of the motion of the follower, a displacement diagram may also be of value in showing the proper sequence of events in a mechanism. In general the diagram may be drawn on any suitable length of base line, since this dimension is arbitrary.

Cam outlines should be designed so as to avoid shock at the beginning and end of the follower stroke. In order to secure smooth action no abrupt changes should appear anywhere in the displacement diagram. One method sometimes used in the design of a cam is to lay out a displacement diagram such as shown in Fig. 3-3, drawing in with the compass or irregular curve an outline that will represent a smooth motion for the follower. The intersection of this outline with the ordinates of the diagram gives a scale for the motion of the follower similar to that shown at the left of the diagram in Fig. 3-3. In general, however, the travel of the follower conforms to some definite law of motion, such as is discussed in the next article, in which case the curve showing the variation of the motion between its extreme positions is a specific calculated curve.

It is not always *necessary* that a displacement diagram such as that of Fig. 3-3 be drawn for every problem, of course. Some of the examples given later in this chapter illustrate that sometimes a quicker solution is obtained by constructing the follower positions directly on a drawing of the mechanism (see Figs. 3-13 and 3-15), particularly when a relatively simple law of motion is specified. The desirability or necessity of making a displacement diagram in any given case is thus governed by the requirements of the problem.

In cam devices in which all points on the follower do not have the same motion, it is necessary to choose a particular *reference point* on the follower. The motion of the follower is specified, then, by specifying the motion of the reference point. Pivoted followers are representative examples and will be discussed later in the chapter.

3-4. Base Curves. The curves in the displacement diagram that represent the motion of the follower are called *base curves.* The motion of the follower may be made to conform to a great variety of base curves, but only three simple forms that are commonly used will be considered in detail here; namely, the modified straight-line base curve, the harmonic base curve, and the parabolic base curve. An analysis of these types will serve to indicate the methods of studying the characteristics of any type of curve which might be chosen.

For purposes of comparison the base curves shown at *a* in Figs. 3-4 to 3-7 have been drawn on base lines of equal length, and for equal total displacements. Each of these base curves, therefore, may be considered as applying to a cam which imparts to the follower the same total displacements during the same angular motion of the cam. The difference lies in the law of motion of the follower. The slope of the displacement diagram at any point is the magnitude of velocity ds/dt of the follower at that point. Similarly, the slope of the velocity diagram gives the rate of change of the magnitude of velocity, which is the tangential acceleration along the direction of motion. The velocity and acceleration diagrams shown directly beneath the base curves, at b and c, respectively, in Figs. 3-4 to 3-7, give a comparison of the characteristics of the follower motion for the base curves shown at a in these figures.

1. *Straight-line Base Curve.* If the motion of the follower were represented by the straight line, Fig. 3-4a, it would have equal displacements in equal units of time, i.e., uniform velocity from the beginning to the end of its stroke, as shown at b. The acceleration except at the ends of the stroke would be zero, as shown at c. The diagrams indicate very

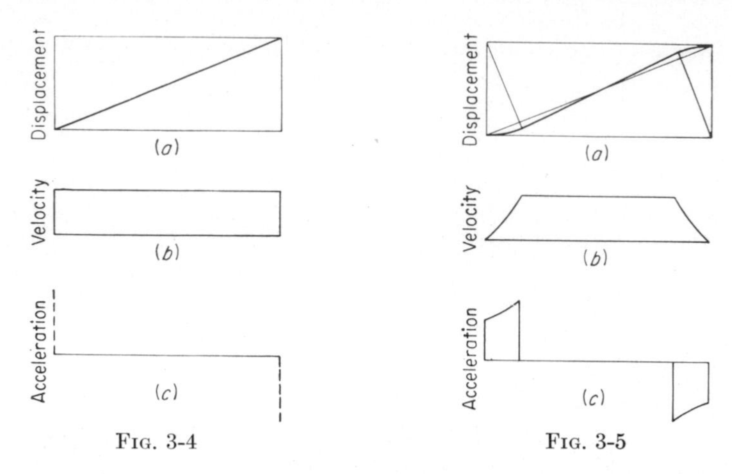

FIG. 3-4 FIG. 3-5

clearly the abrupt changes of velocity, with the consequent large forces, at the beginning and end of the stroke. This condition is undesirable in any case and especially so when the cam rotates at high velocity. The straight-line base curve is, therefore, only of theoretical interest.

2. *Modified Straight-line Base Curve.* To avoid the very undesirable condition described in a above, the law of motion of the follower should be such as to give gradually increasing velocity at the beginning of the stroke and then gradually decreasing velocity at the end of the stroke. This can be accomplished in the case of the straight-line base curve of Fig. 3-4a by easing off the corners of the diagram as shown in Fig. 3-5a, giving what is known as the modified straight-line base curve. The lengths of the periods for acceleration and retardation at the ends of the stroke, and for constant velocity between, depend entirely on the "easing-off" radius. This radius may be taken as any convenient length, within reasonable limits. The shorter the radius the nearer the approach to the

undesirable conditions of the straight-line base curve. The longer the radius the more gradual the action at the ends of the stroke, but the more abrupt at the middle of the stroke. A radius equal to the follower displacement is often used in practice. The diagrams of Fig. 3-5 are based on this radius. Although the velocity and acceleration diagrams at b and c in this figure show a decided improvement in the character of the follower motion as compared with the similar diagrams at b and c in Fig. 3-4, the motion is not so smooth as that of Figs. 3-6 and 3-7, which will be discussed next.

3. *Harmonic Base Curve.* A cam with the base curve at a in Fig. 3-6 will impart simple harmonic motion to the follower. This curve is sometimes called the *crank curve*, owing to the fact that when a crankpin moves with constant speed in a circular path the motion of the projection

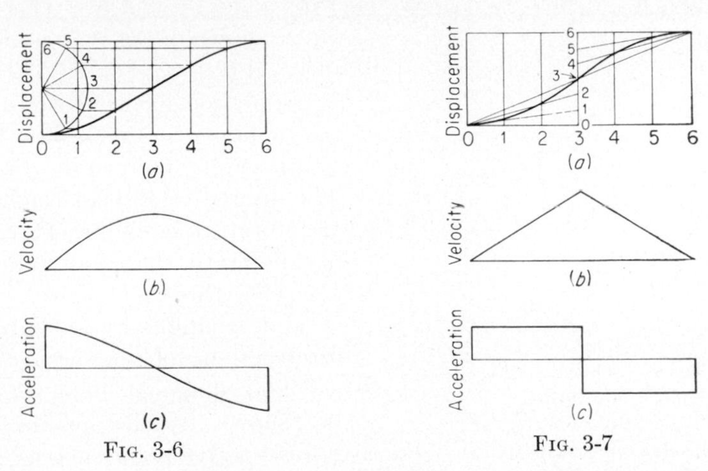

FIG. 3-6

FIG. 3-7

of its center on the diameter of the circle is, by definition, a simple harmonic motion. The velocity diagram at b indicates smooth action. The acceleration, as shown at c, is a maximum at the initial position, zero at the mid-position, and a negative maximum at the final position. The method of constructing the harmonic base curve follows from the definition, and is shown in the figure. Mathematically, the displacement diagram indicating simple harmonic motion is a sine or cosine type of curve.

4. *Parabolic Base Curve.* The parabolic base curve is shown in Fig. 3-7a. As indicated at b, the velocity increases at a uniform rate during the first half of the motion and decreases at a uniform rate during the second half of the motion. The acceleration is constant and positive throughout the first half of the motion, as shown at c, and is constant and negative throughout the second half. Because of the constant acceleration this base curve is sometimes called the *gravity curve.* This

type of curve gives the follower the smallest value of maximum acceleration along the path of motion for a given total displacement in a given time. In high-speed machinery this is particularly important because of the forces which are required to produce the accelerations.

Mathematically, for the condition that tangential acceleration A^t is constant, dV/dt = constant = A^t.

$$\text{Velocity } V = \int A^t\,dt = A^t t + c_1 = \frac{ds}{dt}$$

$$\text{Displacement } s = \int V\,dt = \tfrac{1}{2}A^t t^2 + c_1 t + c_2$$

In these equations c_1 and c_2 are constants of integration. If $s = 0$ and $V = 0$ when $t = 0$, c_1 and c_2 are each 0, and $s = \frac{1}{2}A^t t^2$. This is the curve shown in Fig. 3-7*a* in the time interval 0 to 3.

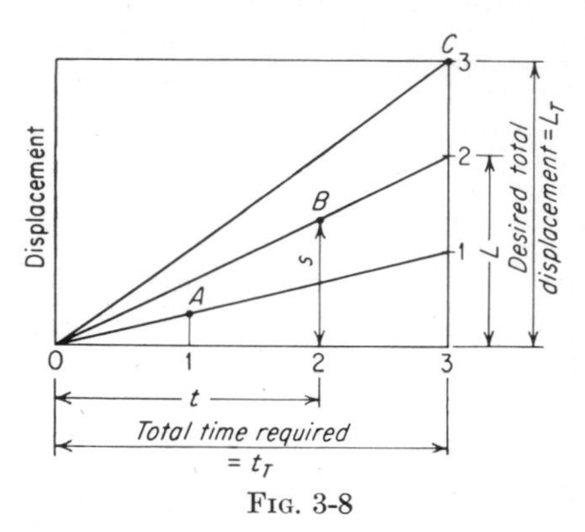

Fig. 3-8

A method of constructing the parabolic curve is shown in the figure, and enlarged in Fig. 3-8. The desired total displacement and the total time required are each divided into the same number of intervals (three in Fig. 3-8). Points *A* and *B* (Fig. 3-8) are located at intersections of the proper vertical and diagonal lines, and will be shown to be points on a parabola drawn from 0 to *C*. For a representative point such as *B*, and from the geometry of the figure, $s/t = L/t_T$. But because there are equal intervals along L for equal intervals of time, L is proportional to t, or $L = Kt$. Therefore, $s/t = Kt/t_T$ and $s = (K/t_T)t^2$, or $s = K't^2$, which is the equation of a parabola with origin at 0. Therefore, point *B*, having lift s, must lie on this parabola. That the parabola passes through *C* can be seen by noting that K can be evaluated as L_T/t_T (from $L = Kt$), and that, at *C*, $t = t_T$. The equation for s then shows $s = L_T$, which indicates that *C* is on the same parabola as *B*.

In some applications, the rate of change of acceleration (often called "jerk" or "snap") is of importance, because it is a measure of the rate of change of force producing acceleration. To minimize the effects of jerk, there should be no abrupt changes in the acceleration diagram. Often in design, the logical starting place is the acceleration diagram, with the velocity and displacement diagrams obtained by successive integrations, which can be done either analytically or graphically with reasonable accuracy. Space limitations do not allow extended discussion here of further details. The motions already discussed serve as examples.

3-5. Disk Cam with Radial Roller Follower. 1. *Layout of the Cam.* A disk cam rotating clockwise at constant angular velocity is to impart harmonic motion to a roller moving in a radial path as follows:

Up ¾ in. while the cam turns through 120 deg.
Rest (or dwell) while the cam turns through 60 deg.
Down ¾ in. while the cam turns through 90 deg.
Rest (or dwell) while the cam turns through 90 deg.

The motion of the follower is fully represented by the displacement diagram shown in Fig. 3-9*a*, and its lowest position with respect to the

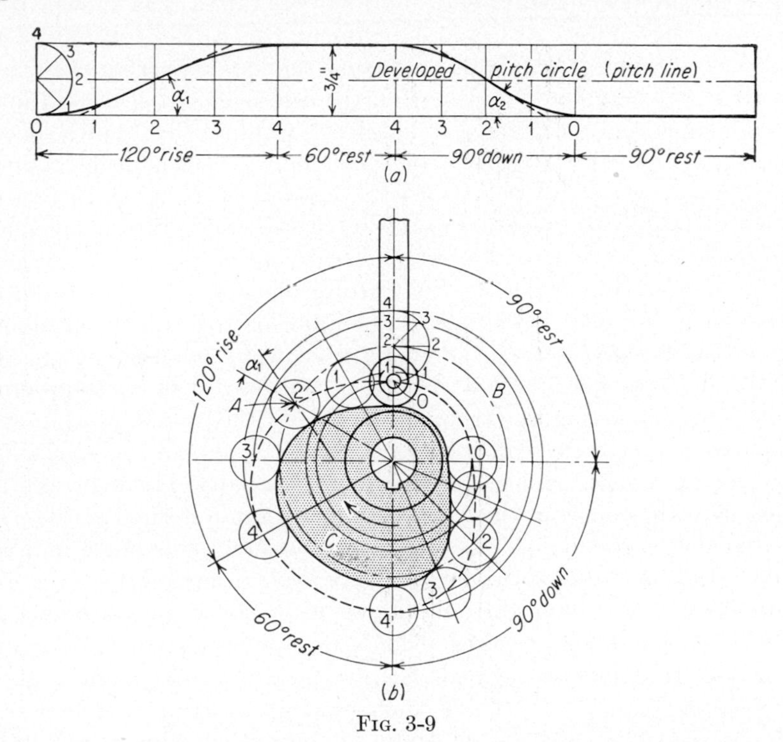

Fig. 3-9

axis of the camshaft is shown in Fig. 3-9*b*. The first step in the solution is to lay out the scale of motion of the follower *along the path of motion of the axis of the roller* between its initial and final positions, 0 to 4, respectively. This may be accomplished either by transferring points from the displacement diagram or by laying out a semicircle of ¾ in. diameter, dividing it into a number of equal parts (four were assumed in this case), and projecting to the diameter to obtain the division 0 to 4 shown along the line of the follower stroke.

Since the follower is to complete its cycle of motion during one revolution of the cam, the next step is to lay out around the axis of the cam the four angular divisions, 120, 60, 90, and 90 deg, required for the motion of the follower. In the solution of this problem, the same expedient is employed as in Art. 3-2, the cam being assumed as fixed and the frame and follower rotated in a sense the reverse of that specified for the cam. Since the cam motion is clockwise, the angles referred to above will therefore be laid out in a counterclockwise sense from the path of motion of the follower, as shown in the figure. Now, since the displacement scale has been constructed on the basis of four equal time intervals for both the 120-deg up and the 90-deg down motion, these angular spaces are divided into four equal divisions and numbered 0 to 4 as shown. The radial lines through 1, 2, 3, and 4 represent the *paths of motion* of the center of the roller. Arcs with radii equal to distances from the axis of the cam to the numbered divisions on the displacement scale will intersect correspondingly numbered radial lines at points that represent constructed positions of the axis of the roller follower. The roller follower is then drawn at each of these positions, and the working surface of the cam is obtained by drawing a smooth curve which is the internal envelope of all the roller positions. The cam must touch (be tangent to) all positions of the roller follower if it is to force the follower to the desired constructed positions. For greater accuracy of the cam profile, more positions of the follower can be constructed than are used in Fig. 3-9. The working surfaces of the cam for the periods of rest of the follower are obtained by connecting the working surfaces for the up and down motion of the follower with circular arcs drawn from the axis of the camshaft.

2. *Pressure Angle.* For a roller follower, the angle between the line of motion of the center of the roller and the common normal at the point of contact of cam and roller is the *pressure angle.* This angle indicates the deviation of the direction of the force of cam against follower from the direction of motion of the follower center. The larger this angle, the larger is the side thrust tending to bind the follower stem in its guides. In Fig. 3-9*b* the pressure angle at A on the upward stroke is the angle α_1.

In general, the length of the displacement diagram is immaterial. However, for the special case shown in Fig. 3-9 in which a roller follower moves in a radial path, if the length of the displacement diagram is made equal to the circumference of the circle ABC, the diagram will represent a true development of the pitch surface of the cam. This circle, with radius equal to the distance from the axis of the cam to the mid-point of the travel of the follower, is called the *pitch circle.* In Fig. 3-9*a* the length of the displacement diagram was made equal to the circumference of the pitch circle. The developed pitch circle in the displacement diagram is called the *pitch line.*

A feature of such a displacement diagram is that the tangent to the base curve at any point makes an angle with the horizontal base line which is the pressure angle at that point. Thus the pressure angles α_1 and α_2 are seen to be the maximum pressure angles during the rise and down motions, respectively. It is interesting to note that the velocity of the follower is proportional to the tangent of the pressure angle.

The maximum limit for the pressure angle on the working stroke is governed by the character of the design.

3. *Size of Cam.* For a given stroke, or travel, of the follower, the maximum size of the cam is governed only by the space available. The minimum size may be determined by construction details such as size

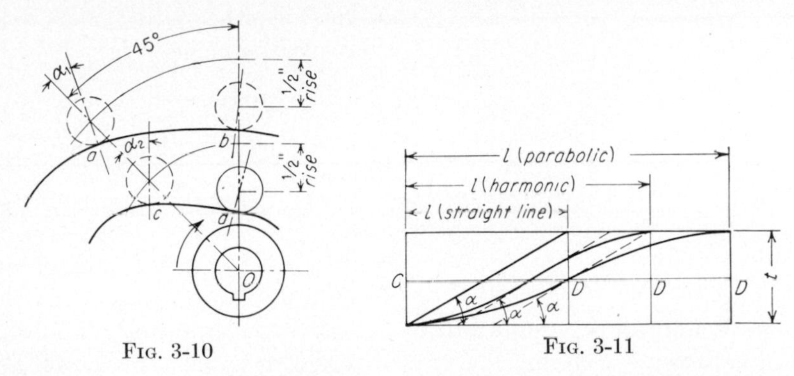

Fig. 3-10

Fig. 3-11

of camshaft, or by the consideration of pressure angle. The size chosen should be large enough to ensure an easy motion of the follower. The larger the cam the easier will be the motion of the follower for any given travel in any given angular motion of the cam. This is true because the pressure angle is reduced as the cam is made larger, as illustrated in Fig. 3-10.

Although the parabolic base curve combines two of the most important theoretical considerations, namely, smoothest motion (constant acceleration and therefore constant force producing the acceleration) and least power for operation, the harmonic curve permits the use of a smaller cam for a given pressure angle than does the parabola, provided the follower is a roller moving in a radial direction. In other words, for the same size of cam, the pressure angle is smaller for the harmonic curve, and therefore the action of the harmonic curve is easier in this respect than the action of the parabolic curve.

This fact is evident from Figs. 3-11 and 3-12. The length of the displacement diagram has been chosen to the scale already discussed, in which the full length is equal to the circumference of the pitch circle of the cam. For the fraction of the full diagram shown, the length CD (or l) in Fig. 3-11 is equal to the fraction cd of the circumference shown in Fig. 3-12. The angles α shown in Fig. 3-11 thus are true pressure angles. For the same maximum pressure angle α and the same rise t, the length l of the base curve will bear a definite relation to the rise t.

It will now be shown that the radius r of the pitch circle cde of Fig. 3-12 is less for the harmonic base curve than for the parabolic base curve, when the maximum pressure angles α are the same. Let it be assumed that the rise t of the follower is to take place while the cam turns through an angle β radians. It should be recalled that radians are dimensionless, the number of radians being the arc length divided by the radius length. If cd and r are measured in inches,

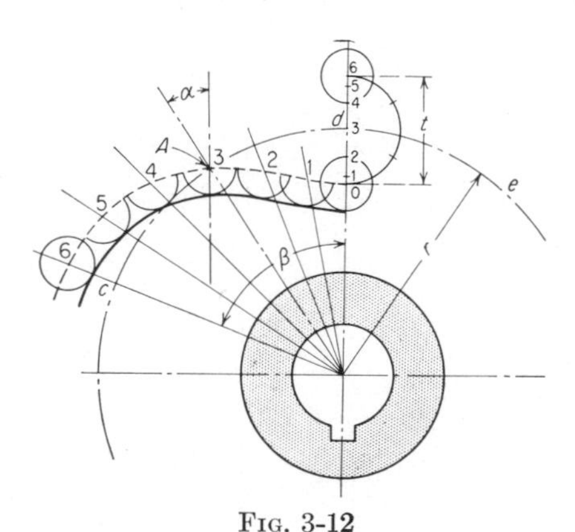

FIG. 3-12

$$\text{Arc } cd = r\beta \qquad \text{in.} \tag{3-1}$$

A cam factor f is defined as

$$f = \frac{l}{t} \tag{3-2}$$

This ratio is dimensionless. Its magnitude depends upon the type of motion and the maximum pressure angle reached during the motion, as can be seen by studying Fig. 3-11.

From Fig. 3-11,

$$CD = l = f\,t = cd \qquad \text{in Fig. 3-12}$$

Therefore, from Eq. (3-1),

$$r = \frac{f\,t}{\beta} \tag{3-3}$$

Because t and β are the same for the motions under study, the radius is directly proportional to the cam factor f. But f $(= l/t)$ obviously is less for harmonic motion than for parabolic motion (see Fig. 3-11). It is evident, therefore, that the cam with the harmonic base curve is the smaller cam, for a given total rise t in a given angular motion β of the cam.

Table 3-1 lists values of the cam factor f which will be found useful in the layout of cams of the type discussed in this article. The method of determining these factors can be understood by careful study of Fig. 3-11. For any one maximum pressure angle and type of motion, l is fixed when t is determined.

TABLE 3-1. CAM FACTOR f FOR DIFFERENT VALUES OF THE MAXIMUM PRESSURE ANGLE α

Type of base curve	Maximum pressure angle α					
	20°	25°	30°	35°	40°	45°
Modified straight line (Fig. 3-5)	3.10	2.59	2.27	2.03	1.89	1.83
Harmonic (Fig. 3-6)	4.31	3.37	2.72	2.24	1.87	1.57
Parabolic (Fig. 3-7)	5.50	4.29	3.46	2.86	2.38	2.00

The pitch circle serves no purpose other than that of obtaining a cam of minimum size for a given pressure angle in cases like the preceding one

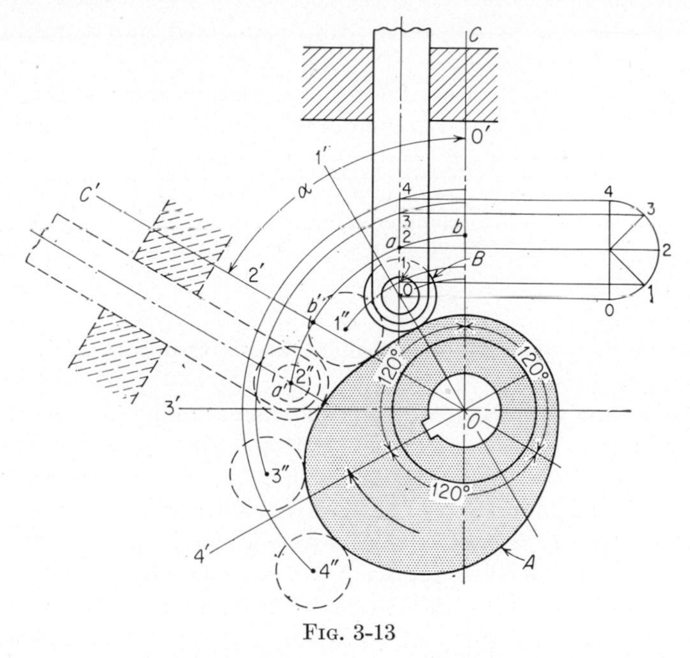

FIG. 3-13

where the *follower is a roller moving in a radial path.* For other motions of the roller follower, and for flat-faced followers, the cam is generally independent of a pitch circle based on pressure angle, and hence no reference will be made to these terms in the discussion of other types of cams.

3-6. Disk Cam with Offset Roller Follower. In Fig. 3-13 the roller follower is required to move in the straight-line path from 0 to 4 with harmonic motion while the cam turns clockwise with constant angular velocity through 120 deg. Since the straight-line path of the roller (extended) does not pass through the axis of the camshaft, this is known as an offset roller follower. Consideration of only the upward motion of the follower as indicated in the figure will be sufficient for the present purpose. The working surface of the cam is determined by the general method outlined in Art. 3-2. One important difference, however, should be carefully noted when comparing this solution with that of Fig. 3-1. Assuming the vertical line through the axis of the cam as the zero or initial line for the layout, the angle of 120 deg is laid off in a counterclockwise sense from this line and subdivided, as required, by the equally spaced radial lines 0′, 1′, 2′, 3′, and 4′. The constructed positions of the roller follower are not located along the radial lines, but are offset from corresponding radial lines the same distance as the actual positions of the roller from the initial radial line *OC*. For example, in position 2″, arc *a′b′* must equal arc *ab* so that when line *OC′* on the cam rotates into initial position *OC* the axis of the roller will be in its required position at 2 along its line of travel.

It should be noted here that the construction shown in Fig. 3-13 is only one of several possible methods that could be used in determining the constructed positions of the roller follower such as shown at 2″. The initial or zero line *OC*, for example, could have been assumed as a line drawn from *O* through the initial, or zero, position of the roller center, in which case the offset such as *a′b′* would vary. Another method is to let the path of the roller center represent the initial or zero line. In this case the four division lines such as 0′, 1′, 2′, 3′, 4′ can be drawn tangent to a circle through *O*, in which case no offsets such as *a′b′* will be required.

3-7. Disk Cam with Reciprocating Flat-faced Follower. In this problem it is required to determine the working surface of a disk cam (Fig. 3-14*a*) to impart motion to a reciprocating flat-faced follower in accordance with the displacement diagram shown in Fig. 3-14*b*. The cam is to displace the follower a total of 1¼ in. during one revolution while turning clockwise at a constant angular velocity. The time for one revolution of the cam has been divided into 16 equal time intervals. The follower is to dwell in its initial position during three time intervals, rise during eight time intervals, dwell in its extreme position during one time interval, and return to its initial position in four time intervals. The displacement diagram given does not represent any definite law of motion for the follower. However, it was drawn so as to give a gradual motion at the beginning and end of the stroke.

The initial, or lowest, position of the follower was taken as 1⅜ in. from the axis of the cam. If the displacement diagram is placed in line with the displacement of the follower, the displacement scale shown in the figure is easily constructed by projection. The most convenient point from which to start the displacement scale is the point *D* where a perpendicular through O, the axis of the cam, intersects the contact surface of the follower. As the cam rotates clockwise, point D in the contact surface of the follower rises to positions 1, 2, 3, 4, etc.; or, following the usual scheme of assuming an inversion of the mechanism while the construction is in progress, as the follower is rotated counterclockwise with the cam stationary, point D moves to the radial positions 1″, 2″, 3″, 4″,

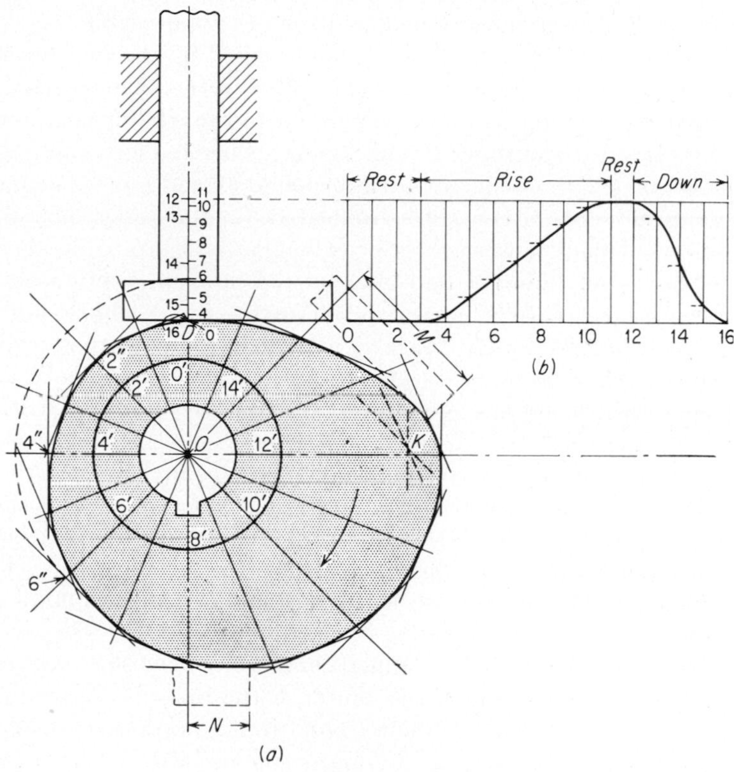

FIG. 3-14

etc. Hence, for the sixth position, e.g., the intersection of an arc of radius $O6$ with radial line $O6'$ of the cam determines the position 6″ of the point D of the contact surface of the follower. Since this contact surface is perpendicular to the center line of the stem of the follower, a line through 6″ perpendicular to $O6'$ determines one line (follower position) to which the working surface of the cam must be tangent. Lines representing the contact surface of the follower having been similarly located for the remaining positions, the working surface of the cam is obtained by drawing a smooth curve tangent to them as shown.

In the layout of cams with flat-faced followers, limitations may be found as to minimum size, because below a certain size it may not be possible to draw the outline of the cam tangent to all the constructed positions of the follower. For example, had the minimum radius OD of the cam been taken as $1\frac{1}{16}$ in., the lines for positions 12, 13, and 14 would have intersected in a point as shown at K, thus requiring the outline of the cam to come to a sharp point. Had the minimum radius OD been taken still smaller, the line for position 13 would fall outside the intersection of the lines for positions 12 and 14, in which case a solution in accordance with the specifications given would be impossible.

The length of face of the follower is determined by the maximum distances of the line of contact to the right or left of D. It is a maximum to the right for position 14 and a maximum to the left for position 8 as shown by the dotted outlines in the figure. For obvious practical reasons the actual length of contact surface should extend somewhat beyond the theoretical points of contact as indicated by dimensions M and N.

If the stem of the follower is cylindrical and its axis located as shown in Fig. 3-14 but offset from the mid-plane of the cam, the friction between the cam and follower face will cause the follower to rotate while reciprocating, thus requiring the follower face to be in the form of a circular disk of radius M. Such a follower is known as a *mushroom follower*. This arrangement has the advantage of assisting lubrication and distributing wear.

For the reciprocating flat-faced follower, the pressure between the cam and follower, friction neglected, is at right angles to the face of the follower. The pressure angle is constant, therefore, and equal to 90 deg minus the angle between the face of the follower and the stem. In the present instance the angle between the face and the stem is 90 deg and hence the pressure angle is zero.

3-8. Disk Cam with Pivoted Roller Follower. Another case of the roller follower is that in which the center of the roller is constrained to move in a circular arc. Let it be required to lay out a disk cam which will give harmonic motion to a pivoted roller follower in the following manner:

Up 15 deg while the cam turns through 120 deg.
Rest while the cam turns through 60 deg.
Down 15 deg while the cam turns through 90 deg.
Rest while the cam turns through 90 deg.

The cam is to have a clockwise rotation at constant velocity. The location of the follower pivot relative to the axis of the cam, and the lowest position of the roller, are shown in Fig. 3-15.

The first step is to subdivide the motion of the follower harmonically. If the chord of the arc through which the roller travels is taken as the diameter of the circle that is used in laying out the harmonic motion, the construction will, in general, be sufficiently accurate, although not exact. In a more exact construction, successive displacements taken from a plotted displacement diagram would be measured *along the line of motion* of the center of the roller. In the construction of Fig. 3-15 it is assumed that the chord differs negligibly from the arc. The numbered positions of the follower correspond to the equal subdivisions (similarly numbered)

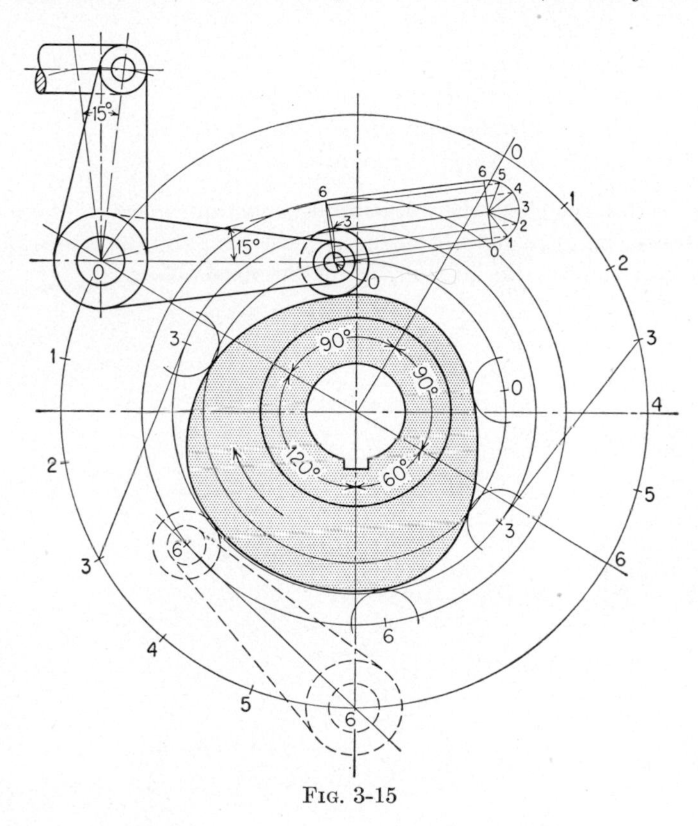

Fig. 3-15

of the angles through which the cam turns while the total rise or fall of the follower is taking place. The next step is to consider the cam mechanism "inverted" and to locate the follower (its working profile is sufficient) in positions corresponding to the numbered positions in the original mechanism. This has been accomplished by a simple geometrical construction as indicated at position 6. Other geometrical constructions that may occur to the student may be used if desired. The working profile of the cam during the periods of rise and fall may now be drawn tangent to the various positions of the working profile of the follower. For the sake of clarity, only positions 0, 3, and 6 have been shown in the figure. In the actual layout all the positions from 0 to 6 would be used. The parts of the profile of the cam that make contact during the periods of rest are arcs of circles having their centers at the axis of the camshaft.

3-9. Disk Cam with Pivoted Flat-faced Follower. As a further example of a cam layout, let it be required to lay out a disk cam which will give harmonic motion to a pivoted flat-faced follower in the following manner:

Up 15 deg while the cam turns through 120 deg.
Rest while the cam turns through 90 deg.
Down 15 deg while the cam turns through 90 deg.
Rest while the cam turns through 60 deg.

The cam is to have clockwise rotation at constant speed. The location of the follower pivot relative to the axis of the cam and the lowest position of the face of the follower are shown in Fig. 3-16.

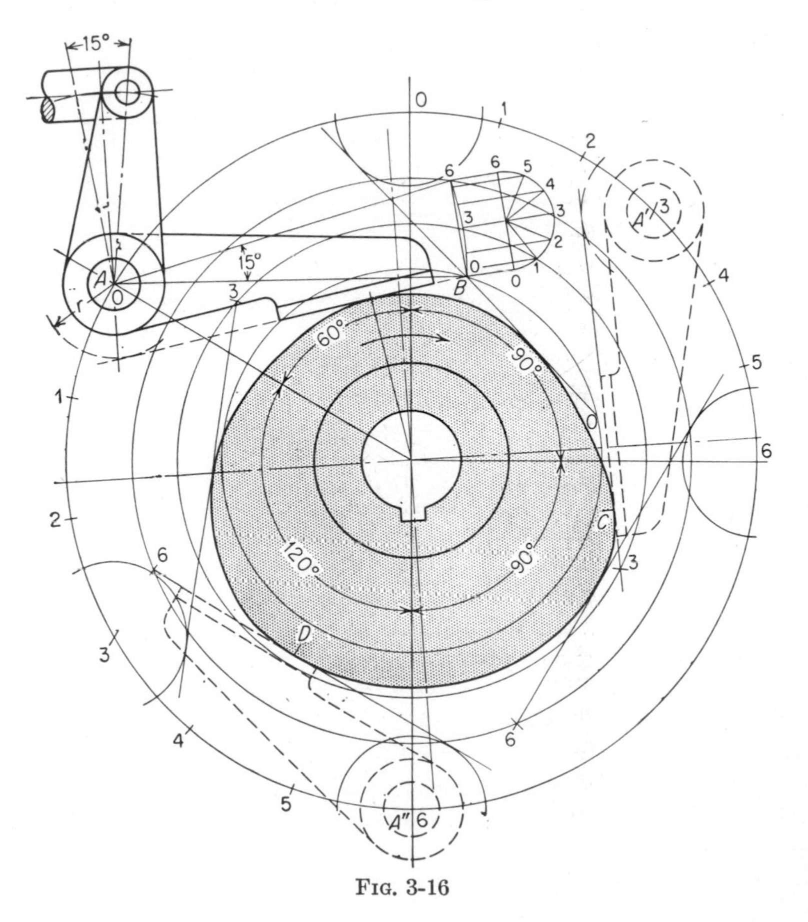

Fig. 3-16

The first step in the layout is to subdivide the angular displacement of the follower in accordance with harmonic motion. A little forethought here will materially reduce the labor of laying out the cam later on. The working profile of the follower is a straight line, and it will eventually be necessary to locate this straight line for the various positions of the follower in the "inverted" mechanism. The one point of the follower that has no motion relative to the fixed link is the center A about which the follower swings. The straight-line working profile of the follower will always be tangent to a circle of radius r drawn from the successive positions of this center. The position of one other point B on the straight line will be sufficient to locate it. If now an arc with A as a center is drawn through any point B and 15 deg laid off on this arc, the total angular motion of the follower will be completely represented. This arc may now be subdivided in accordance with the required motion as shown in the figure. Using the chord instead of the arc for this construction is an approximation but is sufficiently accurate for most conditions. The location of the follower in the various positions in the "inverted" mechanism is now accomplished without difficulty, and the working profile of the cam is drawn in the usual manner, a smooth curve tangent to the follower face in all its locations. For the sake of clarity, only the constructions for positions 0, 3, and 6 have been shown in Fig. 3-16.

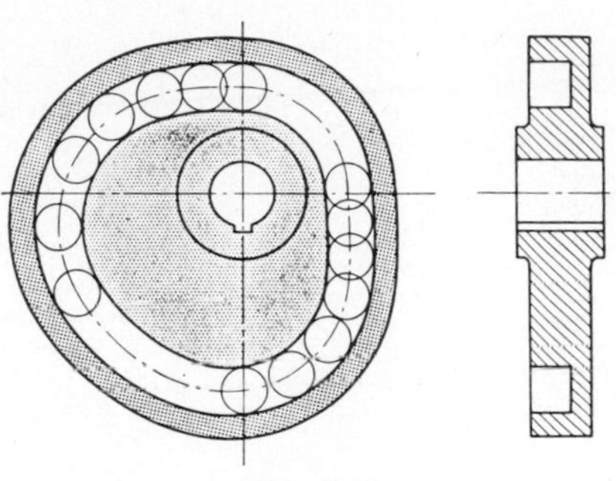

FIG. 3-17

The length and location of that part of the face of the follower that comes into contact with the cam during a complete revolution may be determined by measuring the longest and shortest distances from the follower pivot A to the points of tangency of cam follower. In the figure these distances are $A'C$ and $A''D$, respectively, and, when laid off on the follower in the original mechanism, they completely determine the working face of the follower. For obvious practical reasons a small amount should be added to the theoretical length.

3-10. Other Types of Cams. As previously indicated, the return motion of the follower in the types of cams that have just been discussed must be brought about by some force external to the cam. This is usually accomplished by the use of weights or springs. Where a positive action is desired, the usual method is to add a second working profile of the cam on the outside of the roller in the manner shown in Fig. 3-17.

In designing positive-motion cams, the radius of the follower roller must be considered. This radius may be equal to, but in general should be less than, the shortest radius of curvature of the pitch surface, when measured on the working-surface side. If the radius of the roller is not so taken, the follower will not have the motion for which it was designed.

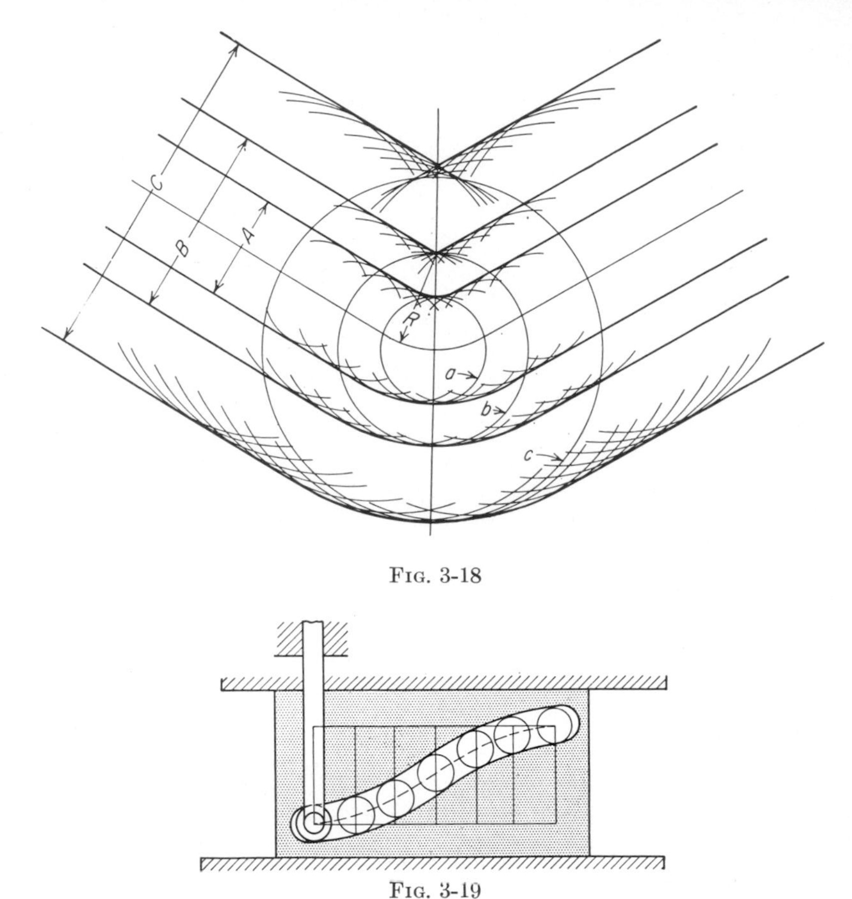

Fig. 3-18

Fig. 3-19

Thus, in Fig. 3-18 let a, b, and c represent three rollers moving in grooves of widths A, B, and C, respectively. The radius of curvature of the pitch surface common to the three grooves at the point under consideration is indicated by R. Roller a has a radius smaller than R, roller b a radius equal to R, and roller c a radius greater than R. It is evident from the figure that, when the radius of the roller is greater than R, as in the case of roller c, the width of groove at that point becomes greater than the diameter of the roller, and positive motion cannot be assured.

A *plate cam*, or *sliding cam*, such as shown in Fig. 3-19, has a reciprocating motion between fixed guides. It carries a curved face or slot that may impart reciprocating motion to a roller follower or oscillating motion to a pivoted-roller-follower arm.

An inversion of the plate cam in which the roller becomes the driver and the plate the follower is sometimes used in light mechanisms such as sewing machines. An example of this type of cam, known as an *inverse cam,* is given in Fig. 3-20.

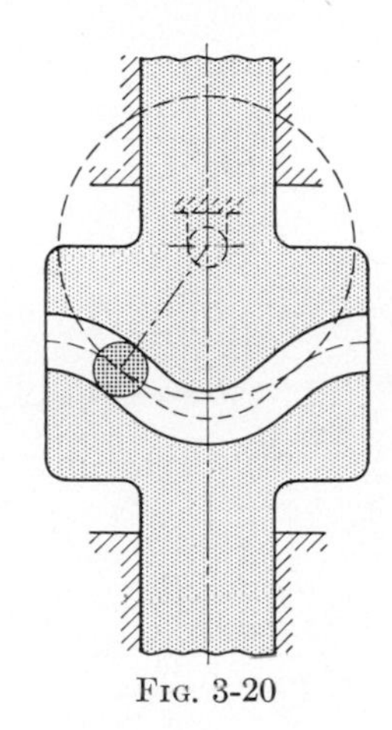

Fig. 3-20

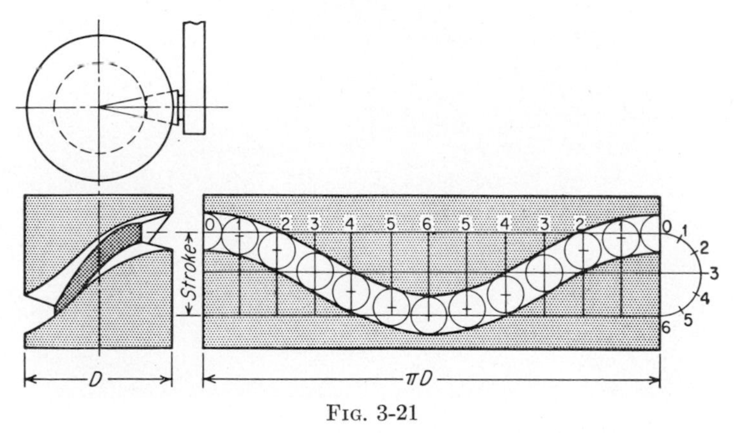

Fig. 3-21

A *cylindrical cam* may be used to give the same motion as a sliding cam. A cam of this form may be looked upon as a sliding cam bent to a cylindrical shape. As the cylinder is rotated on its axis, the follower is given precisely the same motion as when the sliding cam is given a motion of translation. The general method of laying out a cylindrical cam with a pivoted roller follower is shown in Fig. 3-21. The first step is to lay out a displacement diagram in which the height of the diagram is equal to the stroke of the follower and in which the length of the diagram is equal to the circumference of the cylinder. The diagram is next wrapped around the cylinder and the required cam outline transferred to the surface of the cylinder. The groove in the cylinder may now be milled by a cutter of the same size and shape as the roller, the cutter being

FIG. 3-22. *(Courtesy of Brown and Sharpe Manufacturing Co.)*

guided so as to conform to the curve marked on the cylindrical surface. It should be noted that the roller follower is usually the frustum of a cone whose apex is on the axis of the cam. Figure 3-22 illustrates the cutting of a groove in a cylindrical cam.

Another form of cylindrical cam is obtained by fastening adjustable strips on the surface of a cylinder. This form makes it easy to change from one cam profile to another, and is very much used on automatic screw machines.

Disk cams such as illustrated in Fig. 3-23, with roller or mushroom followers, are commonly employed to operate the valves of engines. One application is shown in Fig. 1-1. A simple design with the outline made up of straight lines and circular arcs is indicated in Fig. 3-23.

In Fig. 3-24 is shown a positive-return disk cam whose outline is made up of circular arcs of two different radii. In order that the cam may be

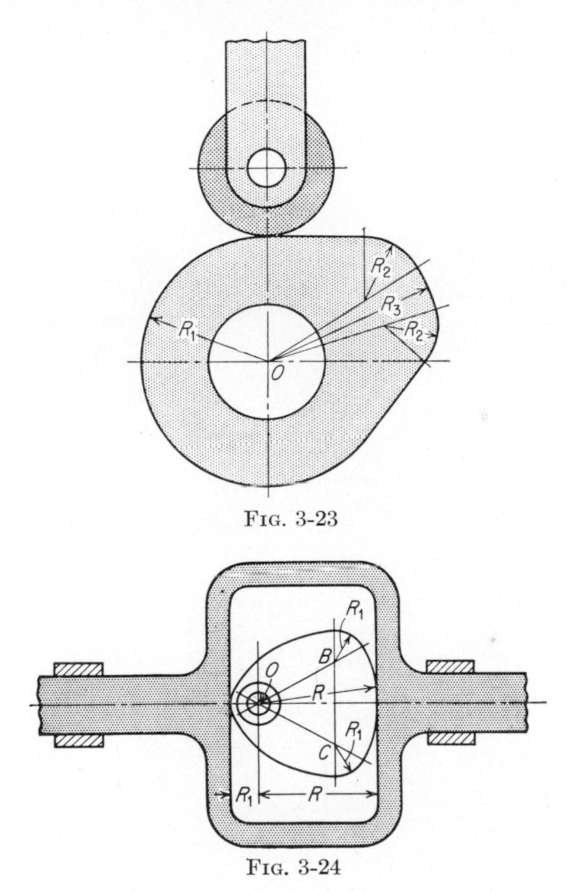

FIG. 3-23

FIG. 3-24

of constant breadth, the centers O, B, and C must be at the apexes of an equilateral triangle. It can be seen that there is a dwell at each end of the stroke while the cam turns through 60 deg. The total displacement $R - R_1$ takes place while the cam turns through 120 deg. Such cams are designed for a given total displacement of the follower. The law of displacement, therefore, must depend upon the assigned outline of the cam.

In Fig. 3-25 is shown an oscillating disk cam that imparts reciprocating motion to a flat-faced follower. Because of its form and action this combination is often spoken of as a *toe-and-wiper cam.* It is seen in the valve mechanism of marine engines.

A well-known form of cam used in stamp mills for pulverizing crushed ore is the *stamp-mill cam* shown in Fig. 3-26. It consists of a reciprocating flat-faced follower actuated by a rotating disk in the form of a double involute cam. When the cam rotates at a uniform rate, the involute

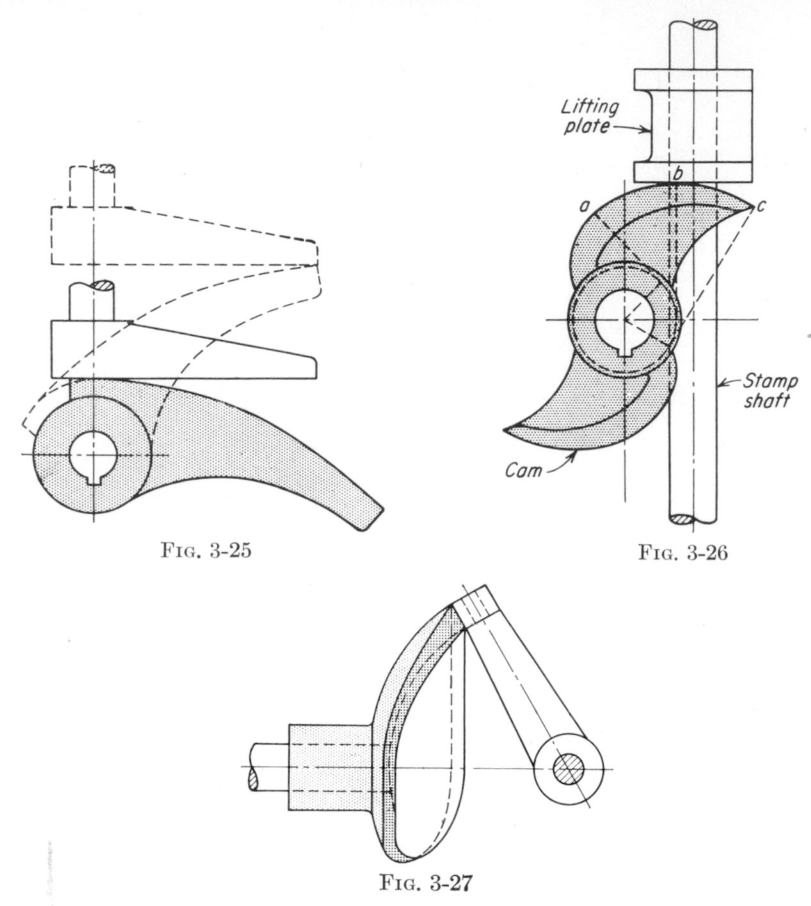

Fig. 3-25

Fig. 3-26

Fig. 3-27

outline of the cam imparts uniform velocity to the follower. The construction of the involute outline is indicated in the figure by the dotted construction at the three points a, b, and c. This construction follows the thought that an involute curve is traced by a knot in a string which is kept taut and unwound from a disk or drum. The stamp shaft in Fig. 3-26 is raised and dropped twice during each revolution of the cam.

A spherical form of cam is shown in Fig. 3-27. There are many other interesting types of cams that enter into the construction of modern machinery in a great variety of forms, but space does not permit their discussion here. The fundamental principles involved in laying out any of these cams are the same as those which have been applied in the examples given in this chapter.

CHAPTER 4

TOOTHED GEARING—SPUR GEARS

The study of gear-tooth action is fundamentally a study of the motion transmitted by a pair of members having curved surfaces in direct contact. Preliminary to the study of gearing, therefore, the general case of a pair of curved members in driving contact will be considered. Figure 4-1*a* shows two such members, with the driving member 2 pivoted at O_2 and the driven member 3 pivoted at O_3. Except for obvious practical limitations the contact surfaces may be of any form, and they may move upon each other with a pure rolling action, a pure sliding action, or with a combined rolling and sliding action. The last condition is the most general.

4-1. Nature of the Motion Transmitted by Curves in Direct Contact. In *pure rolling* action no one point of either member comes in contact with two successive points of the other, for example, two circles rolling upon each other without slipping. If a point on one member comes in contact with *all* successive points of the acting surface of the other, within the limits of the motion, the action is *pure sliding.* This action occurs when one circle (in the above example) is held stationary and the other made to rotate while in contact with it. *Combined rolling and sliding* exists when a pair of circles roll upon each other with some slipping permitted. The precise nature of these actions will now be considered.

In Fig. 4-1, it is evident that all points in 2 must rotate about O_2, and that all points in 3 must rotate about O_3. Consequently, the velocity of any point in either 2 or 3 is represented for the instant by a line through that point perpendicular to the radius connecting it with its center O_2 or O_3. The point of contact P between the members may be considered as a pair of coincident points. Considered as a point on 2, P will be designated as P_2, and as a point on 3 it will be designated as P_3. Then P_2M_2 and P_3M_3 will represent the velocities of P_2 and P_3, respectively.

Motion by direct contact can be transmitted only by normal force between the surfaces. The force between 2 and 3 is, therefore, transmitted in the direction NN, the common normal to the two surfaces at the point of contact. Also, whatever the actual velocities of P_2 and P_3 the components Pn of these two motions along the line of the normal NN

must be equal. If this were not the case, the surfaces would have either to separate or to cut into each other. The tangential components of the velocities of the coincident contact points P_2 and P_3 must be in the direction of the common tangent TT, but they may have different magnitudes,

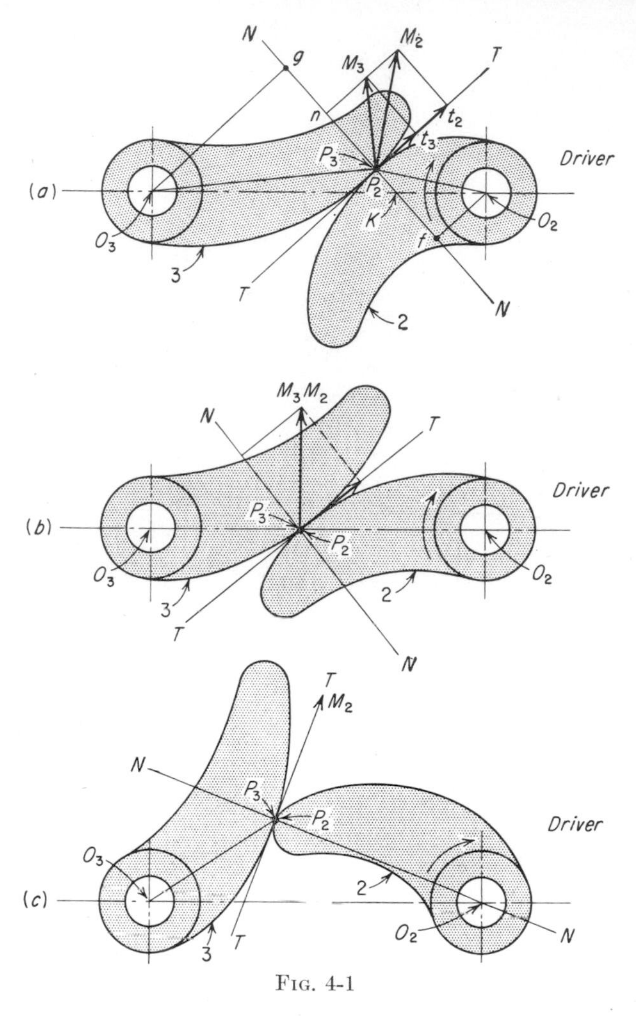

FIG. 4-1

in either the same or opposite directions. In Fig. 4-1a it is seen that these tangential components P_2t_2 and P_3t_3 are in the same direction, but of different magnitudes. It is evident that the difference between the magnitudes of the tangential components represents the rate of the sliding action for this particular phase of the motion. When the tangential components are in the opposite directions, the rate of sliding is evidently their sum. Thus *the rate of sliding is the algebraic difference of the tangential components.*

When the tangential components are equal, the nature of the action is pure rolling. In Fig. 4-1b the members of Fig. 4-1a have been rotated into positions to illustrate this condition, and the velocities P_2M_2 and P_3M_3 are identical. But P_2, as a point on 2, moves for the instant at right angles to O_2P_2, and P_3 as a coincident point on 3 moves at right angles to O_3P_3. Therefore, when P_2M_2 and P_3M_3 are coincident, O_2P_2 and O_3P_3 are both perpendiculars to the same line at the same point, and they must therefore lie in one straight line. In order that this may occur, the contact point P must lie on the line of centers. Thus *the condition of pure rolling is that the point of contact shall always lie on the line of centers.*

When the common normal NN passes through the center about which the driver rotates, as in Fig. 4-1c, the radius O_2P_2 coincides with it and the velocity P_2M_2 of the contact point of the driver lies in the direction of the common tangent TT; hence the normal component of this motion is zero, and P_3 has no motion. It is seen, therefore, that for this particular phase of the motion, the action is for the instant pure sliding.

4-2. Angular-velocity Ratio for a Pair of Curves in Direct Contact. In the general case of a pair of members with curved surfaces in contact, as in Fig. 4-1a, the magnitude of the angular-velocity ratio, hereafter usually abbreviated to simply angular-velocity ratio, of driver to follower is continually varying. This ratio for any instant may be determined by the elementary principles of mechanics. Since the *ratio* of the angular velocities is independent of the *actual* angular velocities, the angular velocity of one member may be assumed, and this affords a means of determining the corresponding angular velocity of the other member at that instant.

If in Fig. 4-1a the angular velocity of link 2 is ω_2, the magnitude of the linear velocity P_2M_2 of P as a point on 2 can be found from the relation

$$P_2M_2 = (O_2P_2)\omega_2 \tag{4-1}$$

in which ω_2 is expressed in *radians* per unit time. The direction of the velocity of P_3 as a point on 3 (perpendicular to O_3P_3) is known, and it has been shown that its normal component must be equal to that of P_2, or Pn. The magnitude of the velocity of $P_3(P_3M_3)$ can now be found, and hence the angular velocity of 3 is

$$\omega_3 = \frac{P_3M_3}{O_3P_3} \tag{4-2}$$

The ratio of the angular velocities is therefore

$$\frac{\omega_3}{\omega_2} = \frac{P_3M_3}{O_3P_3}\frac{O_2P_2}{P_2M_2} \tag{4-3}$$

The above expression may be simplified in the following manner: Drop the perpendiculars O_2f and O_3g from O_2 and O_3 upon the common normal NN. P_2M_2n and O_2P_2f are similar triangles, and hence

$$\frac{P_2M_2}{O_2P_2} = \frac{P_2n}{O_2f}$$

Similarly, P_3M_3n and O_3P_3g are similar triangles and

$$\frac{P_3M_3}{O_3P_3} = \frac{P_3n}{O_3g}$$

Therefore

$$\frac{\omega_3}{\omega_2} = \frac{P_3n}{O_3g}\frac{O_2f}{P_2n} = \frac{O_2f}{O_3g} \tag{4-4}$$

The common normal NN intersects the line of centers O_2O_3 at K, O_2Kf and O_3Kg are similar triangles, and

$$\frac{\omega_3}{\omega_2} = \frac{O_2f}{O_3g} = \frac{O_2K}{O_3K} \tag{4-5}$$

It follows, therefore, that, for a pair of links with curved surfaces in direct contact, *the magnitudes of the angular velocities are inversely as the segments into which the line of centers is divided by the common normal through the point of contact.*

From the above analysis it follows, also, that *for constant angular-velocity ratio the common normal through the point of contact must divide the line of centers in a fixed ratio.* (If the positions of the centers are *fixed,* then the common normal must intersect the line of centers at a fixed *point.*) The propositions just stated are of fundamental importance in the theory of toothed gearing, as will appear later in this chapter.

4-3. Curves That Give Pure Rolling Action. There are many pairs of curves that will satisfy the condition of either constant-velocity ratio or pure rolling. Several cases of the former will be taken up later in connection with the study of gear teeth. Among those curves that will give a pure rolling action, the following may be mentioned: a pair of circles rotating about their centers; a pair of equal ellipses each rotating about one of its foci, with a distance between centers equal to the common major axis; a pair of similar logarithmic spirals rotating about their foci; a pair of equal parabolas; and a pair of equal hyperbolas.

There is but one class of curves, however, i.e., circular arcs rotating about their centers, that can have *both constant angular-velocity ratio and pure rolling action.* It has been shown that for constant-velocity ratio, the common normal must intersect the line of centers in a fixed point (when center positions are fixed); and that for pure rolling the point of contact must lie on the line of centers. If both these conditions are satisfied, the point of contact must be at a fixed place in the line of centers. As a result, the contact radii must be constant and, therefore, the curves are circular arcs.

4-4. Motion Transmitted by Members in Pure Rolling Contact. It has been stated that there are several classes of curves that may be so

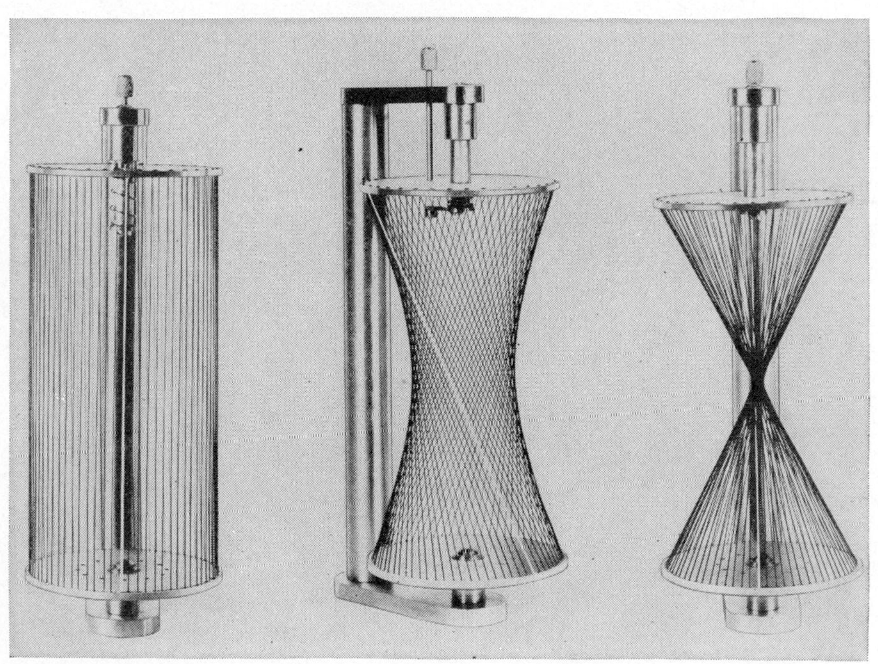

FIG. 4-2. (*Courtesy of Chevrolet Motor Division, General Motors Corporation.*)

paired as to give pure rolling action. In the present article, however, only those cases that have common application in the study of the theory of toothed gearing will be considered, namely, rolling cylinders, rolling cones, and rolling hyperboloids. Figure 4-2 shows the three types of surfaces involved. The stretched strings forming each surface demonstrate that the surface elements are straight lines and that each surface is generated by revolving a properly oriented line about an axis.

1. *Rolling Cylinders.* If two cylinders in contact rotate on each other about fixed parallel axes without slipping, the ratio of the angular velocities of the two cylinders will at all times be constant. Referring to a and b in Fig. 4-3, if there is no slipping at the point of contact P, P_2 and P_3 will have the same velocity at any instant. The angular velocities of the cylinders may be expressed in terms of the linear velocity of the point of contact, as follows:

$$\omega_2 = \frac{V_P}{O_2P} \qquad \text{and} \qquad \omega_3 = \frac{V_P}{O_3P}$$

Therefore

$$\frac{\omega_2}{\omega_3} = \frac{V_P}{O_2P}\frac{O_3P}{V_P} = \frac{O_3P}{O_2P} \tag{4-6}$$

That is, the angular velocities of the cylinders are inversely proportional to their radii. If one cylinder drives the other, the ratio of their angular velocities is designated in a definite manner, being the ratio obtained by

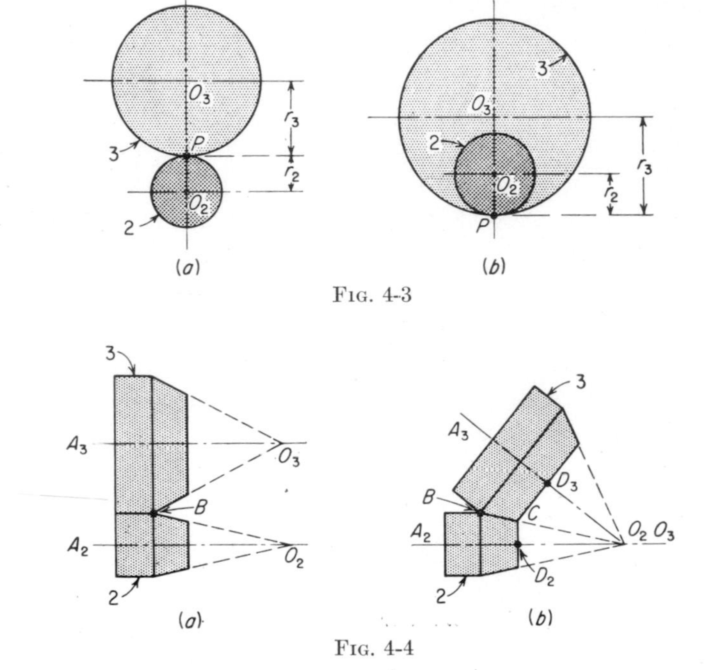

Fig. 4-3

Fig. 4-4

dividing the angular velocity of the driver by the angular velocity of the follower or driven member. This ratio is called the *velocity ratio* of the pair of cylinders. This method of designating velocity ratio is in accordance with universal practice in discussions on gearing.

It should be noted that at a in Fig. 4-3 the cylinders rotate in opposite directions, and that at b they rotate in the same direction.

2. *Rolling Cones.* The derivation of rolling cones from rolling cylinders is illustrated by the use of two right cylinders combined with two right cones as shown in Fig. 4-4. Each cylinder has one base in common with that of one of the cones; hence the axes of this cylinder and cone must coincide. Although the bases of the cones need not be equal, the slant heights must be the same. The bases of the two cones have a common tangent in their plane that passes through B (Fig. 4-4a) and that is

perpendicular to the plane of the paper. If the axis O_3A_3 is caused to rotate about this tangent until the apex O_3 coincides with the apex O_2, the bases of the cones will still have the common tangent at B and will be in contact along element O_2B (Fig. 4-4b). Therefore, the cones will roll upon each other in their new positions, the contact points at their bases having equal velocities along the common tangent, as in the original positions where these bases were in common with those of the two rolling cylinders. Any other corresponding transverse sections of the

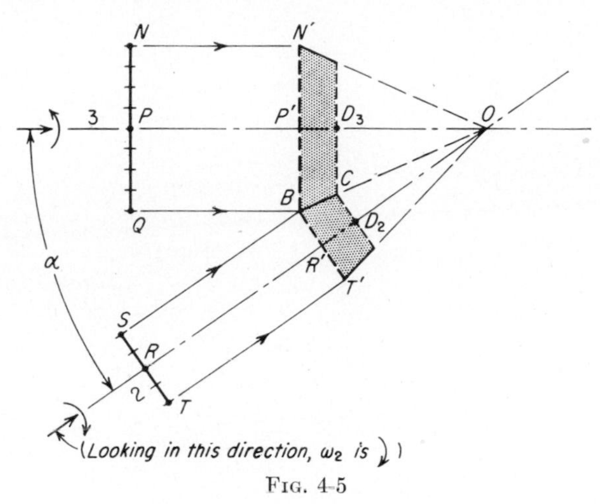

FIG. 4-5

cones equidistant from the common apex O_2, as at C, will roll together with the same velocity ratio as their bases. Therefore,

$$\frac{\omega_2}{\omega_3} = \frac{CD_3}{CD_2} = \frac{r_3}{r_2} \tag{4-7}$$

where r_2 and r_3 are the radii of the bases of the cones 2 and 3.

A pair of rolling cones may be constructed to transmit a given velocity ratio between two given shafts that intersect at some angle α, in the following manner:

On each of the two axes shown in Fig. 4-5, erect a perpendicular line at any point such as at P and at R. Mark off divisions of equal size on the two lines to correspond to the desired velocity ratio. That is, $\omega_3/\omega_2 = RS/PQ = 2/4$ in the figure. Next, consider the two lines NQ and ST to be revolving disks, and slide them to the right along their axes until they touch at B. The two revolving disks, with rolling contact, now have the required velocity ratio. The rolling cones are constructed by drawing the lines BO, $N'O$, and $T'O$.

3. *Rolling Hyperboloids.* A hyperboloid is a surface of revolution which is generated by rotating about an axis a straight line which neither

intersects nor parallels the axis (note Fig. 4-2). Figure 4-6 represents a *pair* of rolling hyperboloids of revolution tangent to each other along a common element AA. Since the elements of the surfaces are straight lines, any pair of elements can be placed in a tangent position. If the axes are fixed in positions corresponding to such tangency, the two surfaces will continue to be tangent along a common element as the two figures rotate about these axes. It is characteristic of the action of these two members as they roll together that there is a certain amount of endwise sliding along the common contact element. In a sense, therefore, there is a departure from pure rolling action. This does not, however, prohibit the use of these forms as pitch surfaces for toothed gears.

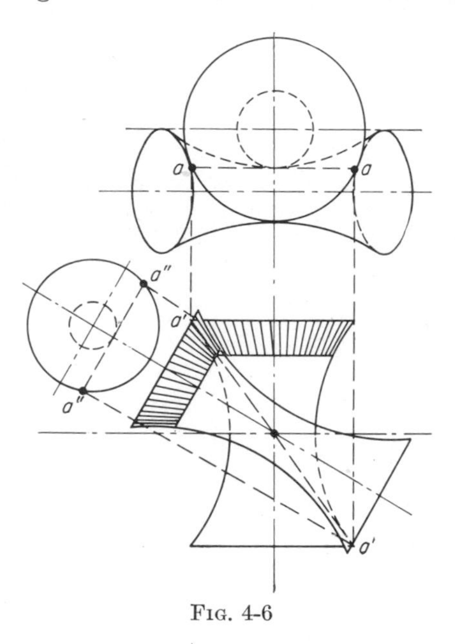

Fig. 4-6

The three surfaces discussed form the basic shapes upon which teeth are placed to form gears. *Spur* gears have parallel axes, as do the corresponding rolling cylinders upon which this type of gear design is based. *Bevel* gears have intersecting axes, as do the corresponding rolling cones. *Hyperboloidal* gears have nonparallel nonintersecting axes, as do the corresponding rolling hyperboloids, and often are considered a special form of bevel gear (see Table 4-1).

4-5. Friction Gearing. When two plain wheels of circular cross section are held in contact with each other by properly constructed bearings, the frictional force between their surfaces will be sufficient to transmit rotary motion from one axis to the other. Such an arrangement constitutes what is known as *friction gearing*, and it is chiefly limited to rolling cylinders, rolling cones, and disk wheels, though other rolling surfaces are possible. Some slipping between the contact surfaces is inevitable; therefore, a constant-velocity ratio cannot be assured. In many instances it is not essential to maintain this definite relation between driver and driven member, and in certain applications the likelihood of slipping when subjected to the excessive loads is desirable, since it serves to protect the driven mechanism against excessive stresses.

Figure 4-7*a* illustrates the principle of a friction drive between parallel shafts. The action is purely frictional as the pair rotate upon each other, and the velocity ratio remains constant unless the resistance of motion exceeds the frictional resistance. Friction gearing of the bevel type for transmitting power between intersecting shafts is illustrated in Fig. 4-7*b*. Figure 4-7*c* shows friction gearing of the disk-wheel form, which has been used extensively in record changers, for varying feed movements in machine tools, and in many other devices transmitting only small power. The small *brush wheel A* (usually the driving member) bears against the face of a disk *B*, and the relative speeds are varied by changing the radial position of wheel *A*.

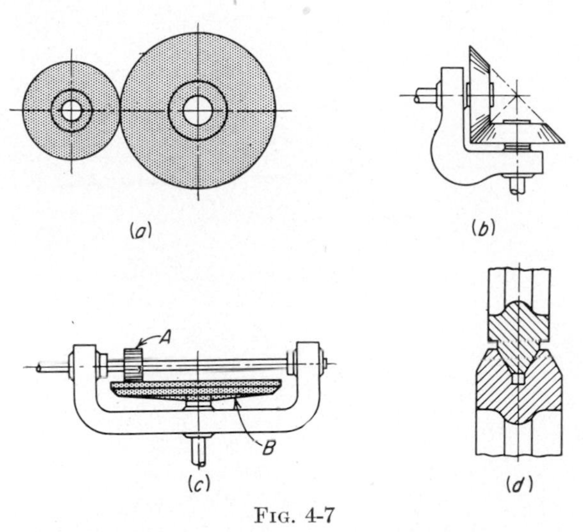

FIG. 4-7

The power that can be transmitted through friction gearing depends upon the physical characteristics of the materials forming the surfaces in contact and also upon the normal pressure between the surfaces. The driving-wheel face is usually made of a material that has a high coefficient of friction, as wood, compressed paper, leather, or rubber; while the driven wheel is usually made of metal. In order to secure greater frictional resistance between the wheels than is possible with the plain surfaces in Fig. 4-7*a*, without at the same time increasing the pressure on the bearings, the wheels are sometimes made with V-shaped grooves and ridges, which mate with a wedging action, as shown in Fig. 4-7*d*.

Although limited in use to the transmission of relatively small powers, friction gearing finds considerable application in speed-controlling devices, light power hoists, coal screens, friction-board drop hammers, etc.

4-6. Toothed Gearing. In the design of toothed gearing the contact surfaces of mating teeth are so shaped that the motion transmitted is the same as that of friction gearing when no slipping occurs. Kinematically, therefore, friction gearing running without slip and toothed gearing are identical, but as pointed out, friction wheels are of limited service in practice, and toothed wheels are employed where considerable power must be transmitted and where a constant-velocity ratio must be maintained at every phase in the cycle of motions. The surfaces of the rolling bodies which are kinematically identical with the toothed wheels which replace them are called *pitch surfaces* of the gears; and right sections of these surfaces are called *pitch lines* or *pitch curves*.

Toothed gearing may be classified according to relation of the axes and pitch surfaces, as in Table 4-1.

TABLE 4-1. CLASSIFICATION OF TOOTHED GEARING

Name	Kind	Relation of axes	Pitch surfaces
Spur gears		Parallel	Cylinders
Bevel gears	Straight	Intersecting	Cones
	Spiral	Intersecting	Cones
	Skew	Not in one plane	Hyperboloids
	Hypoid	Not in one plane	Cones
Helical gears	Parallel	Parallel	Cylinders
	Crossed	Not in one plane	Cylinders
Worm and wheel		Not in one plane	Cylinders*

* Except in case of the hourglass worm (see text).

The characteristics of the various types of gears in the above classification will be discussed in some detail in the articles that follow, after the fundamental principles that apply to toothed gearing in general have been considered.

4-7. Definitions. In order to facilitate the discussion of toothed gearing, it is necessary to define certain terms that are in common use. Those terms that are most frequently mentioned are defined below. It will be noted that the definitions pertaining to gear-tooth parts may, in general, be represented on a right section of the gear. Figures 4-8 and 4-9 will, therefore, be of assistance in obtaining a clear understanding of these definitions.

1. The *pitch surface* is the surface of the imaginary rolling cylinder (cone, etc.) that the toothed gear may be considered to replace.

2. The *pitch circle* is a right section of the pitch surface.

3. The *addendum circle* is the circle bounding the ends of the teeth, in a right section of the gear.

4. The *root* (or *dedendum*) *circle* is the circle bounding the spaces between the teeth, in a right section of the gear.

5. The *addendum* is the radial distance between the pitch circle and the addendum circle.

6. The *dedendum* is the radial distance between the pitch circle and the root circle.

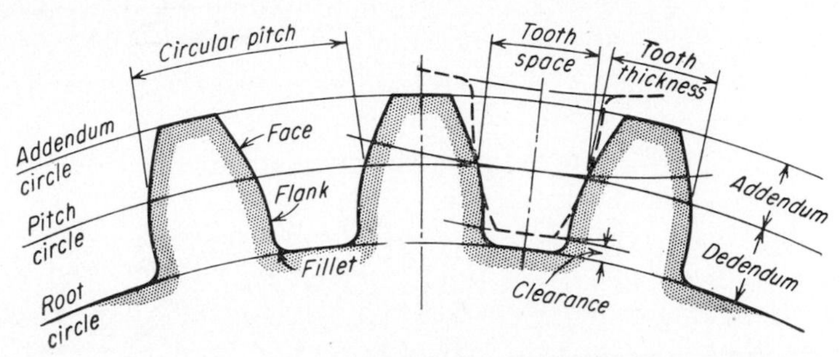

FIG. 4-8

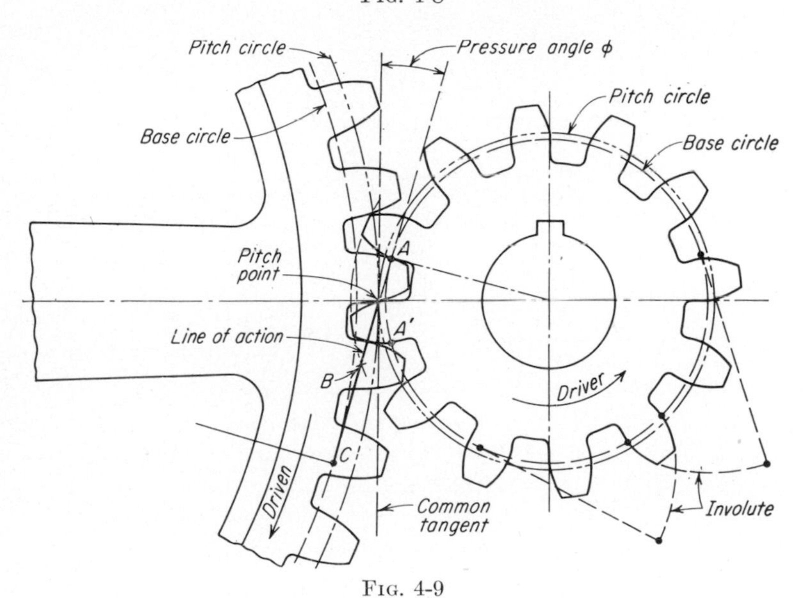

FIG. 4-9

7. The *clearance* is the difference between the dedundum of one gear and the addendum of the mating gear.

8. The *face* of a tooth is that part of the tooth surface lying outside the pitch surface.

9. The *flank* of a tooth is that part of the tooth surface lying inside the pitch surface.

10. The *circular thickness* (also called the *tooth thickness*) is the thickness of the tooth measured on the pitch circle. It is the length of an arc and not the length of a straight line.

11. The *tooth space* is the distance between adjacent teeth measured on the pitch circle.

12. The *backlash* is the difference between the circular thickness of one gear and the tooth space of the mating gear.

13. The *circular pitch* p is the width of a tooth and a space, measured on the pitch circle.

14. The *diametral pitch* P is the number of teeth of a gear per inch of its pitch diameter.

A definite relation exists between the circular pitch and the diametral pitch, which may be determined as follows:

A toothed gear must have a whole number of teeth. The circular pitch, therefore, is equal to the pitch circumference divided by the number of teeth. The diametral pitch is, by definition, the number of teeth divided by the pitch diameter. That is,

$$p = \frac{\pi D}{N} \qquad \text{and} \qquad P = \frac{N}{D}$$

Hence

$$pP = \pi \tag{4-8}$$

where p = circular pitch
P = diametral pitch
N = number of teeth
D = pitch diameter

That is, the product of the diametral pitch and the circular pitch $= \pi$.

15. The *fillet* is the small radius that connects the profile of a tooth to the root circle.

16. A *pinion* is always designated as the smaller of any pair of mating gears. The larger of the pair is called simply the *gear*.

17. The magnitude of the *velocity ratio* of a pair of mating gears is always the ratio of the number of revolutions of the driving (or input) gear to the number of revolutions of the driven (or output) gear, in a unit of time.

18. The *pitch point* is the point of tangency of the pitch circles of a pair of mating gears.

19. The *common tangent* is the line tangent to the pitch circles at the pitch point.

20. The *line of action* is a line normal to a pair of mating tooth profiles at their point of contact.

21. The *path of contact* is the path traced by the contact point of a pair of tooth profiles.

22. The *pressure angle* ϕ is the angle between the common normal at the point of tooth contact and the common tangent to the pitch circles. It is also the angle between the line of action and the common tangent.

23. The *base circle* is an imaginary circle used in involute gearing to generate the involutes that form the tooth profiles. It is the circle drawn from the center of each of a pair of mating gears tangent to the line of action.

4-8. Fundamental Law of Gear-tooth Action. In order to have a pair of toothed gears transmit the same motion as the rolling members (usually cylinders or cones) that they replace, it is necessary for the tooth profiles to have definite forms. There is a fundamental kinematic requirement to which the tooth profiles of any pair of gears must conform if the gears are to transmit the correct motion referred to above. This requirement will now be discussed.

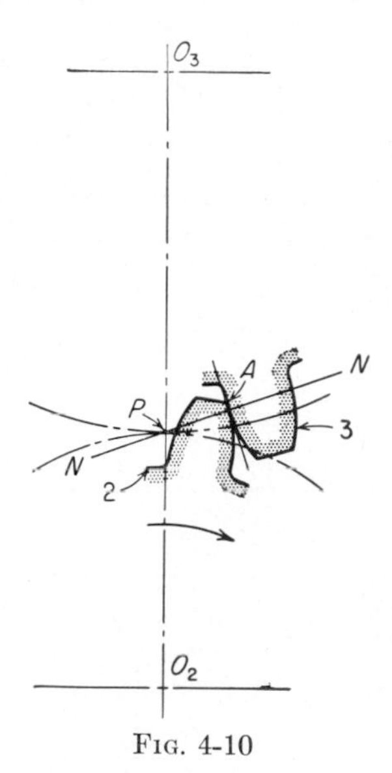

Fig. 4-10

Figure 4-10 shows a pair of gear teeth in contact. All of each gear has been removed except one tooth the resulting mechanism being merely a pair of members with curved surfaces in contact (see Art. 4-1). Now let it be assumed that the pitch circles of the original gears are tangent to each other at the point P on the line of centers. If the tooth profiles in contact are to satisfy the requirement of constant-velocity ratio, the common normal NN at the point of contact A must, at any instant, divide the line of centers in a fixed ratio (see Art. 4-2). For gear centers at fixed positions (the usual case) the common normal NN must intersect the line of centers at a fixed point P; and this fixed point must be at the point of tangency of the pitch circles (called the *pitch point*) if the velocity ratio is to be the same as that of the rolling pitch circles. In fact, this fixed point *establishes* the radii and point of tangency of the rolling pitch circles. The fundamental law of gear-tooth action may now be stated as follows (for gears with fixed center distance): *The common normal to the tooth profiles at the point of contact must always pass through a fixed point (the pitch point) on the line of centers.*

4-9. Usual Forms of Gear-tooth Profiles. A number of tooth forms will satisfy the fundamental requirement of correct gear-tooth action, as outlined in the foregoing articles, but only two have been widely used, namely, the cycloidal form and the involute form, so named because of the curves on which they are based. Up to about 80 years ago the cycloidal tooth form predominated, but now the involute form has replaced the cycloidal almost entirely. Cycloidal outlines are still used in some special cases for both cast teeth and cut teeth. The cycloidal principle is still applied also in the design of pin gearing for watches, clocks, and instruments.

The cycloidal tooth form is discussed briefly later in this chapter largely because of its historical and kinematic interest, and also because it is believed that the student will gain a broader viewpoint of the fundamental theory of gear-tooth action than is possible with his attention confined merely to the one commonly used tooth form, namely, the involute.

INVOLUTE TOOTH PROFILES

The curve most commonly used for gear-tooth profiles is the involute of a circle. This curve is the path traced by a point on a line as the line rolls without slipping on the circumference of a circle. It may also be defined as the path traced by the end of a string as the string is unwound from a circle. The circle from which an involute is derived is called its *base circle.*

In Fig. 4-11, let line MN roll without slipping in a counterclockwise sense on the circumference of the circle whose radius is OA. When the line has reached the position $M'N'$, its original point of tangency A has reached the position A', having traced the involute curve AA' during the motion. As the motion continues, the point A will trace the involute curve $AA'B$. The involute possesses several interesting and useful properties. In the first place let it be noted that, since the line MN rolls without slipping on the circle, the distance aA' is equal to the arc aA. For any instant the instantaneous center of the motion of the line is its point of tangency with the circle. Therefore, the motion of the point that is tracing the involute is perpendicular to the line at any instant, and hence the curve traced will also be perpendicular to the line: i.e., *the normal at any point of an involute is tangent to the base circle.*

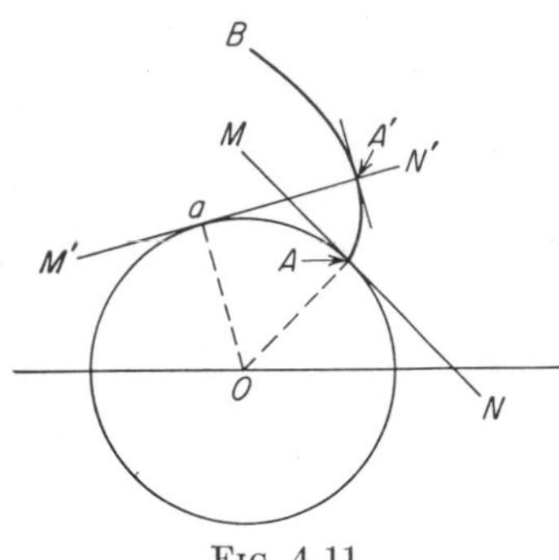

FIG. 4-11

4-10. Construction of the Involute Curve. A method of constructing the involute curve is shown in Fig. 4-12. The arc of the circle to the left of 0 is divided into equal arcs of convenient length, 01, 12, 23, etc.; and the straight line tangent to the circle at 0 is divided, to the left of 0, into equal parts 01′, 1′2′, 2′3′, etc., of the same length as the arcs. Hence when the straight line is rolled on the circle, the point 1′ will come into coincidence with 1, 2′ with 2, 3′ with 3, etc., and the describing point 0 on the line will have moved to 1″, 2″, 3″, etc., corresponding to the dotted positions of the generating line. It is obvious from the method employed in plotting the involute curve that points 1′ and 1″, 2′ and 2″, 3′ and 3″, etc., must be on concentric arcs from O as a center.

A construction that is sufficiently accurate for ordinary purposes is shown in Fig. 4-13. Step off a number of divisions 01, 12, 23, etc., on the circumference of the circle and draw tangents to the circle at these points. From point 1 with a radius equal to chord 01 strike an arc extending from 0 about halfway to tangent 22′. From 2, in turn, with a radius great enough to connect with the preceding arc, strike a second arc extending about halfway to tangent 33′, and so on as indicated in the figure.

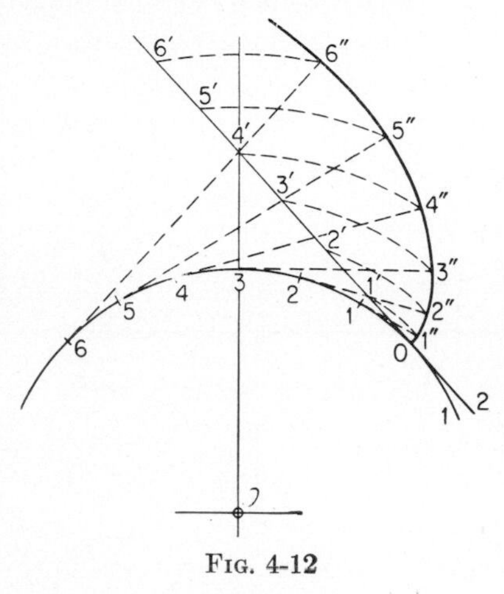

Fig. 4-12

Where, for illustrative purposes, it is desired to draw the teeth of a gear, a template for the purpose may be made from the involute curve that has been constructed in accordance with one of the above methods. A more convenient method, and one that is sufficiently accurate for ordinary purposes, is to find, by trial, arcs that will closely approximate portions of the involute and draw in the tooth outlines with the compass. In most cases it will be found that the locus for the centers of these arcs can be taken on the base circle (see Grant's method, Appendix).

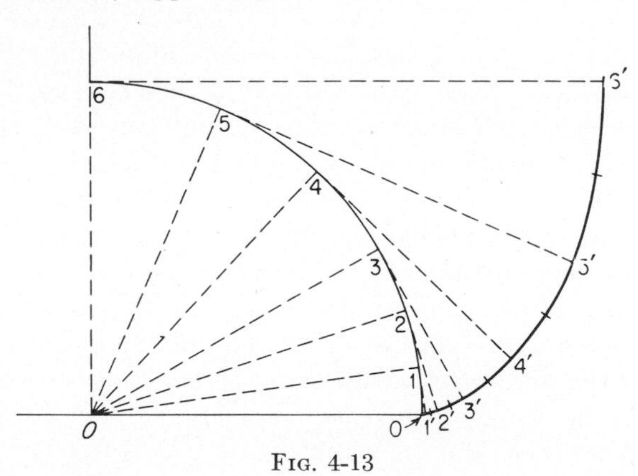

Fig. 4-13

4-11. Involute Teeth Satisfy the Fundamental Law of Gear-tooth Action. In a of Fig. 4-14, the tracing point (it may be thought of as a knot in the cord which is wrapped around the base circle and over the pulley) traces an involute profile on a card attached to the revolving base circle. The cord is always *tangent* to the base circle and *normal* to the involute.

In *b*, as the lower circle revolves, the cord that leaves it is assumed to be wound *onto* the revolving upper circle, the cord being kept taut at all times. The tracing point (a knot in the cord at the position of the contact point of the profiles shown) simultaneously traces the *two* involute profiles, one on the card attached to the lower circle as before, and one

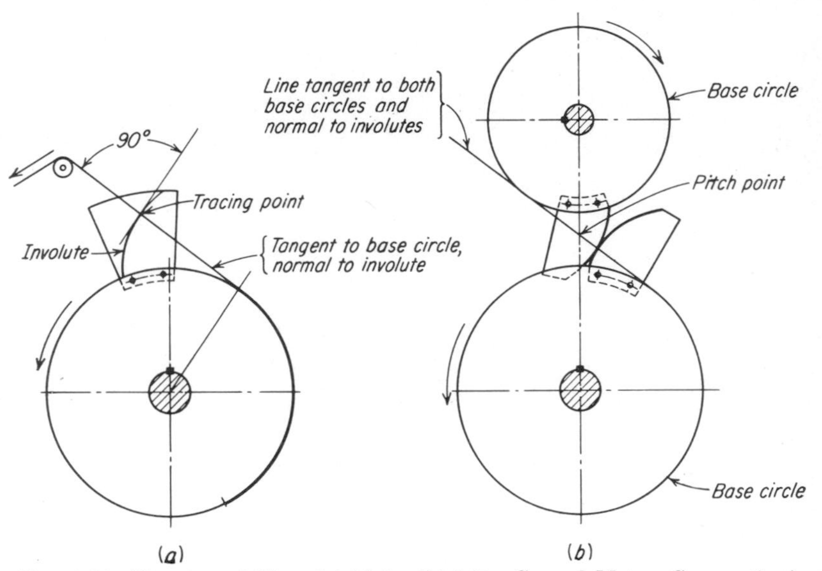

FIG. 4-14. (*Courtesy of Chevrolet Motor Division, General Motors Corporation.*)

on the card attached to the upper circle. Thus the cord, which is tangent to both base circles, is the common normal to the involute profiles at the point of contact, because a tangent to the base circle is always normal to the corresponding involute. This cord (the common normal) cuts the line of centers at a fixed point (the pitch point), and so the fundamental law of gearing is satisfied for meshing involute profiles.

Further, the knot in the cord always is at the contact point because it is generating the two mating profiles simultaneously. Therefore the contact moves along the path of the string, and the line tangent to the base circles is the line of action. It is also the line along which force is transmitted from one tooth profile to the other (disregarding friction) because it is the common normal through the contact point.

It should be mentioned that there can be no involute inside the base circle, and so any portion of a tooth profile below the base circle cannot be an involute form. This will be discussed more fully later in this chapter.

4-12. Characteristics of Involute Tooth Action. As already explained, the common normal to a pair of involute gear teeth in contact is always a line tangent to the two base circles, and this line is called the *line of action.* The point of contact is always situated on this line. The part of the line of action along which contact takes place is determined as follows.

In Fig. 4-15 let 2 and 3 represent the pitch circles of a pair of mating gears, 4 and 5 the corresponding addendum circles, and 6 and 7 the base circles. Let gear 2 be the driver, and assume that it rotates in the direction indicated by the arrow. The shaded lines at A and B represent a pair of mating tooth profiles drawn in at their first and last points of contact. It is evident that these profiles must begin contact at A where the addendum circle of gear 3 cuts the line of action (the locus of all points of involute contact) B_2B_3 tangent to the base circles; and that as the gears rotate, the contact point will move along the line of action B_2B_3 to the point B where this line is cut by the addendum circle of gear 2. With any further motion the profiles will separate, and thus B marks the end of contact. The heavy line APB is called the *path of contact.* Thus, contact along B_2B_3 is limited by the intersection of the two addendum circles and the line of action B_2B_3.

Again in Fig. 4-15, and considering the driving gear 2, it can be seen that the point C where the tooth profile at the beginning of contact cuts the pitch circle will move through the arc CP of the pitch circle as the tooth profile approaches the line of centers, and will move through the arc PC' of the pitch circle as the tooth profile recedes from the line of centers to the point where contact ceases. Arc CP is called the *arc of approach,* and arc PC' the *arc of recess.* The angle α_2 subtended by the arc of approach is called the *angle of approach* and is evidently the angle through which the tooth moves from the position where contact begins to the position where contact is on the line of centers. The angle β_2 subtended by the arc of recess is called the *angle of recess* and is evidently the angle through which the tooth moves from its contact position on the line of centers to the position where contact ceases.

The *arc of action* is the sum of the arcs of approach and recess and is therefore the arc measured on the pitch circle from the position of the tooth at which contact begins to the position of the tooth at which contact ends. The *angle of action* is the sum of the angles of approach and recess and is therefore the angle through which a tooth moves from the position where contact begins to the position where contact ceases.

The methods of determining the arcs of approach and recess DP and PD' and the angles of approach and recess α_3 and β_3 of the driven gear 3 are the same as those just described for the driving gear 2. The *arcs* of approach, recess, and action must be the same for the two gears in contact, because the pitch circles roll together without slipping. The corresponding *angles* will be different, however, because the gears have different radii.

Contact ratio is defined as the ratio of the arc of action to the circular pitch. It can be thought of as the average number of pairs of teeth in contact. It is obvious that this ratio must exceed 1.0 if a succeeding

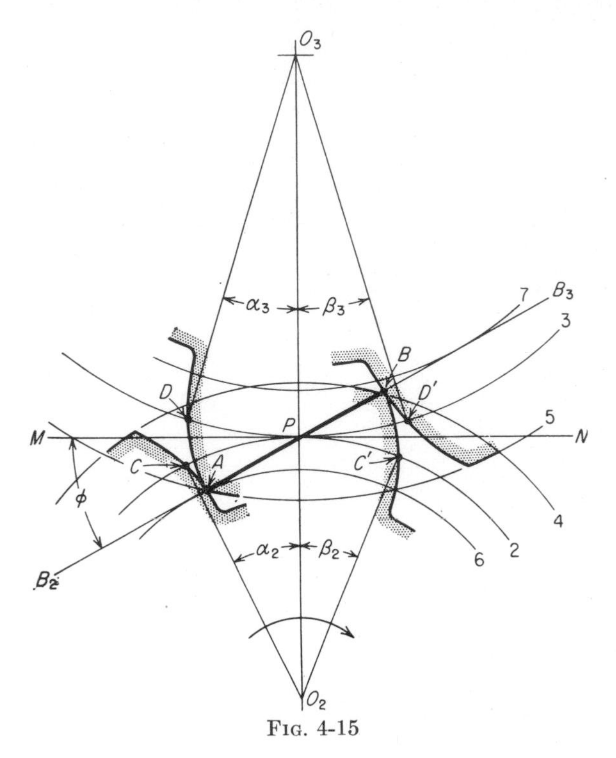

FIG. 4-15

pair of teeth is to come into contact before the preceding pair breaks contact.

The direction of the common normal for involute gear profiles is the same for all positions of the teeth. The line MN (Fig. 4-15) drawn through the pitch point P perpendicular to the line of centers O_2O_3 is tangent to both pitch circles at that point. The angle ϕ which the common normal AB makes with the line MN is called the *pressure angle* (or angle of obliquity) and is evidently a constant for all positions of the mating teeth.

A definite pressure angle actually is not a real property of a gear, as will be shown. The involute profile *is* a definite property, obviously, and therefore so is the base circle. In Fig. 4-15, consider the gear center O_3 to be raised slightly, keeping O_2 at the same location. The base circle of gear 3 is raised by this action, and the common tangent B_2B_3 to the base circles is rotated counterclockwise, cutting the line of centers at a new position P somewhat above the position shown. It is obvious that the pressure angle has been changed by changing the center distance of the gears. Also, because the point P has moved, the pitch circles now have different radii than formerly, although the *ratio* of their radii (and therefore the velocity ratio) remains unchanged. This can be shown as follows.

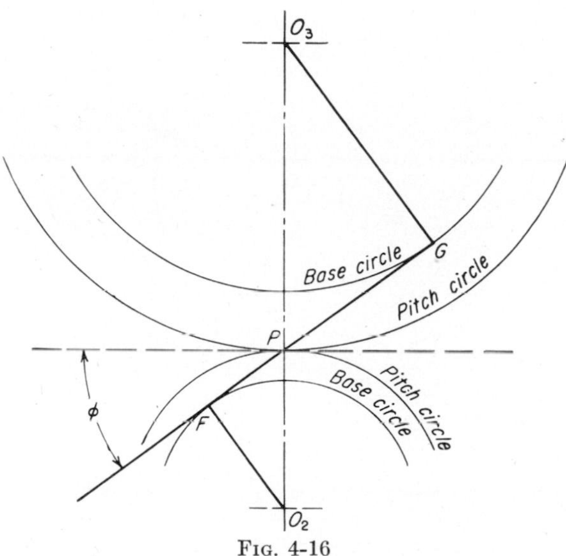

FIG. 4-16

The velocity ratio of a pair of gears has been shown to be equal to the inverse ratio of the radii of the rolling *pitch* circles. That is, as in the representative instance shown in Fig. 4-16, $\omega_3/\omega_2 = O_2P/O_3P$. But from the similar triangles shown, $O_2P/O_3P = O_2F/O_3G$, or the velocity ratio is equal to the inverse ratio of the radii of the *base* circles. As the base circles are properties of the gears, they do not change in size when gear center distance is changed. Therefore, the velocity ratio remains unchanged and the *pitch* circles *must* change in size (although the *ratio* of their radii does *not* change) as center distance is varied. Of course, backlash changes also.

Thus, for involute gears, a change in center distance from a design value does *not* cause a change in velocity ratio—a useful property. In practice, some slight inaccuracies in center distance can occur because of manufacturing tolerances, deformations caused by loads, etc.

4-13. Interference in Involute Gears. Limits of Addendum. Figure 4-17 represents the layout of a pair of involute gears in which the line of action is tangent to the base circles at A_2 and A_3. It will now be shown that for correct tooth action the path of contact must lie entirely within these tangent points, A_2 and A_3. Pinion 2 is the driver, and its addendum circle is the circle passing through the point B_3. The addendum circle of gear 3 passes through B_2.

Consider that contact exists at B_3, and "back off" the teeth to the left, noting especially the travel of the point of contact *downward* over the right face of the pinion 2. At A_2, the contact point is at the base

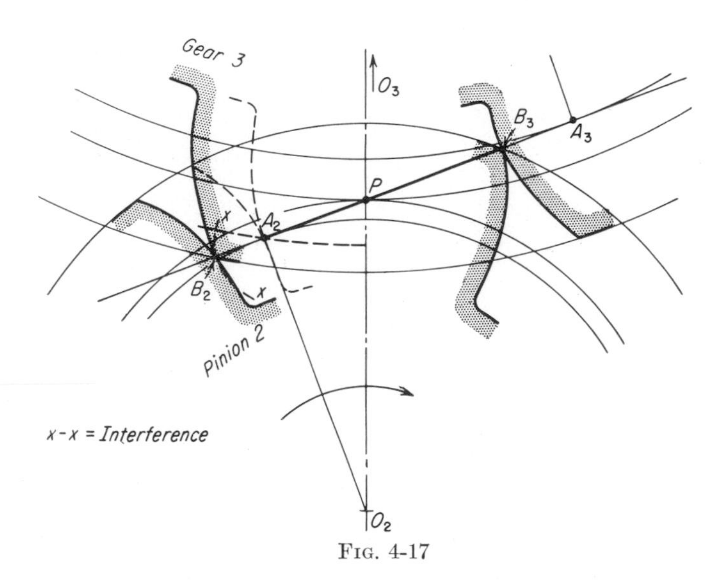

FIG. 4-17

circle for 2. Continued motion of the gears to the left causes contact to occur *inside* the base circle of 2, where there can be no involute. Thus the involute tip of gear 3 is contacting a noninvolute portion of the flank of 2. This occurrence is called *interference.*

It is customary to consider that the flank inside the base circle is a radial line, and when this is true, interference results in a gouging action of the tip against the radial flank (as shown by the dotted *xx* region in Fig. 4-17) and binding of the gears. The common normal at the point of contact is perpendicular to the radial portion of the flank and no longer passes through the original pitch point. Thus the velocity ratio tends to be changed when interfering contact occurs, although other pairs of teeth in involute-to-involute contact may simultaneously be driving at the original velocity ratio. It is obvious that interference, which is one of the chief disadvantages of involute teeth, cannot be tolerated.

The method of checking whether interference will exist in a given pair of gears can be stated as follows: *Interference exists whenever tooth contact is indicated outside the point of tangency of the line of action and the base circle.* Thus in Fig. 4-17 the indicated first point of contact (intersection of tip of tooth of driven gear with line of action) occurs at B_2, which is outside the tangent point A_2. Interference will occur while the teeth move to the right until contact occurs at A_2. Continued motion to the right involves no interference, for involute-to-involute contact occurs *inside* the tangent point. *Last* contact occurs at B_3, still within the tangent point for gear 3 at A_3. Thus there is interference as this particular pair of teeth come *into* contact, but not as they separate.

Several things can be done to eliminate interference. One is to allow no contact outside the tangent point by cutting off the end of the tooth that interferes, forming a stub tooth. Thus in Fig. 4-17, if the tooth on gear 3 is shortened as shown by the dotted line through A_2, there will be no interference. Other possibilities are to leave the tooth the same length but to ease off the tip slightly, or to hollow out the radial flank (called undercutting). Also, increasing the pressure angle can move the point of tangency (such as A_2 of Fig. 4-17) until it falls at the addendum circle (of gear 3 in Fig. 4-17). Then no contact can occur *outside* the tangent point, and interference is eliminated.

All the foregoing methods have been used, singly and in combinations, to eliminate interference. With modern methods of gear manufacture, interference can be eliminated automatically because the design of the cutting tool and the proper relative motion between the tool and the blank result in the cutting away of any interfering portions of the cut teeth. This will be discussed more fully later, when gear systems and production methods are discussed.

In a full-depth 14½-deg involute system the smallest number of teeth in a gear that will mesh with a rack without interference is 32; but in the 14½-deg composite system the cutters have been so designed that the teeth are slightly *eased off* on the end, so that, practically, the minimum number of teeth is 12 (refer to Art 4-16 and Fig. 4-19).

4-14. Layout of a Pair of Involute Gears. The general method of procedure in laying out a pair of involute gears will be illustrated by a numerical example. Let it be required to lay out a pair of involute spur gears in accordance with the following specifications:

The tooth proportions are to conform to the 14½-deg composite system (see later discussion in this chapter).

Distance between shaft centers = 12 in.

Driving shaft turns 400 rpm, clockwise.

Driven shaft turns 200 rpm.

Diametral pitch = 4.

Backlash = 0.

The following computations may now be made:

Pitch radius of the pinion $R_2 = 200/(200 + 400) \times 12 = 4$ in.
Pitch radius of the gear $R_3 = 400/(200 + 400) \times 12 = 8$ in.
Number of teeth in pinion $= PD_P = 4 \times 8 = 32$.
Number of teeth in gear $= PD_G = 4 \times 16 = 64$.
Circular pitch $p = \pi/4 = 0.7854$ in.

From Art. 4-26:

Addendum $= 1/P = \frac{1}{4}$ in.
Dedendum $= 1/P + 0.05p = \frac{1}{4}$ in. $+ 0.05 \times 0.7854$ in. $= 0.289$ in.
Radius of addendum circle (pinion) $= 4 + 0.25 = 4.25$ in.
Radius of addendum circle (gear) $= 8 + 0.25 = 8.25$ in.
Radius of root circle (pinion) $= 4 - 0.289 = 3.711$ in.
Radius of root circle (gear) $= 8 - 0.289 = 7.711$ in.

Lay out the pitch circles, addendum circles, and root circles as shown in Fig. 4-18. Through the pitch point P, draw the line of action E_2E_3

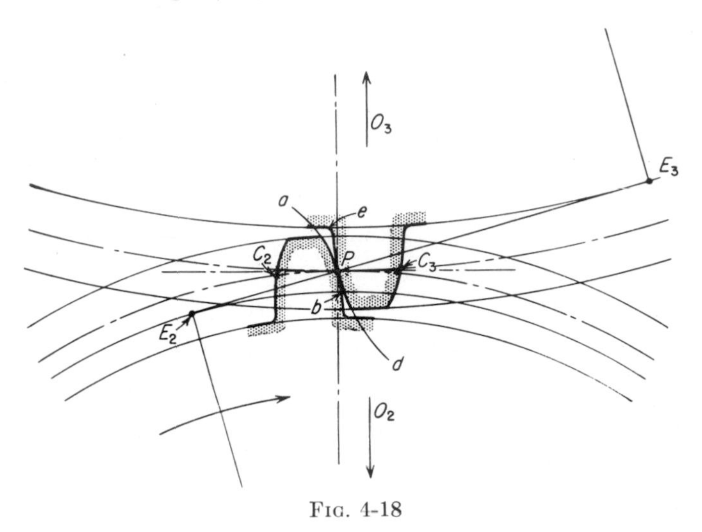

FIG. 4-18

at an angle of 14½ deg with the line tangent to the pitch circles at the pitch point. The base circles are tangent to this line at points E_2 and E_3. It is merely a coincidence here that the base circle through E_3 coincides with the root circle of the gear. The direction of the line of action is governed by the direction of rotation of the driver, which, in this case, is clockwise as indicated by the arrow.

Construct the involute profile bPa of the *pinion* tooth by rolling the line PE_2 on the base circle of the *pinion* and construct the involute profile ePd of the *gear* tooth by rolling the line PE_3 on the base circle of the *gear*. Draw radial lines to complete the parts of the tooth flanks that lie inside the base circles and join the flanks to the root circles by small fillets. Space off the circular thicknesses PC_2 and PC_3 on the pitch circles and through C_2 and C_3 draw the tooth profiles (reversed) to form the opposite sides of the teeth. The tops of the teeth and bottoms of the spaces between adjacent teeth are completed by the bounding addendum and root circles. The complete gears, or as many teeth as desired, can be represented by stepping off the circular thickness on the pitch circles and drawing the profiles through the division points. These profiles may be drawn with the compass, in the manner suggested in Art. 4-10.

The arcs of approach and recess may now be found, if desired, by the methods described in the preceding articles. In Art. 4-12 there was discussed an important relation between the circular pitch and the arc of action that must be considered in the design of any pair of gears. In properly designed gears there must be continuous tooth action; that is, one pair of teeth must not cease contact before the adjacent pair which follows begins contact. It is evident that, when the arc of action and the circular pitch are equal, one pair of teeth is ceasing contact just as the adjacent pair is beginning contact. It follows, therefore, that the circular pitch must not be greater than the arc of action, for in that case continuous contact becomes impossible.

4-15. Interchangeable Involute Gears. A set of gears is said to be interchangeable when any two of the set will run together properly. The conditions of interchangeability for involute gears are that all gears of the set have the *same pitch*, the *same pressure angle*, the *same addendum*, and the *same dedendum*. The standard systems which are in use are discussed later in the chapter.

4-16. Involute Rack and Pinion. When the radius of the pitch circle becomes infinite, as in the case of a rack, the radius of the base circle also becomes infinite, and the involute will become a straight line. Hence *in an involute rack the tooth profiles become straight lines*, and these lines are perpendicular to the line of action. The general layout of an involute rack and pinion is shown in Fig. 4-19. No interference can ever occur between the face of the pinion tooth and the flank of the rack tooth, because this flank is entirely involute in form, but between the face of the rack tooth and flank of the pinion tooth, conditions are such as to give the maximum of interference. Figure 4-19 shows the interference of an uncorrected rack in mesh with a 12-tooth gear having a 14½-deg pressure angle. Slightly less interference is also indicated when two 12-tooth gears mesh with each other. It is evident that the straight-sided rack teeth will mesh properly with any involute gear of the same pitch and the same pressure angle. This fact is of great importance in its bearing on the manufacture of gears, for it makes possible the generating of the teeth of any gear of an interchangeable set with a cutting tool of the simplest form, namely, one having a straight side. This principle will be discussed more fully in Chap. 5.

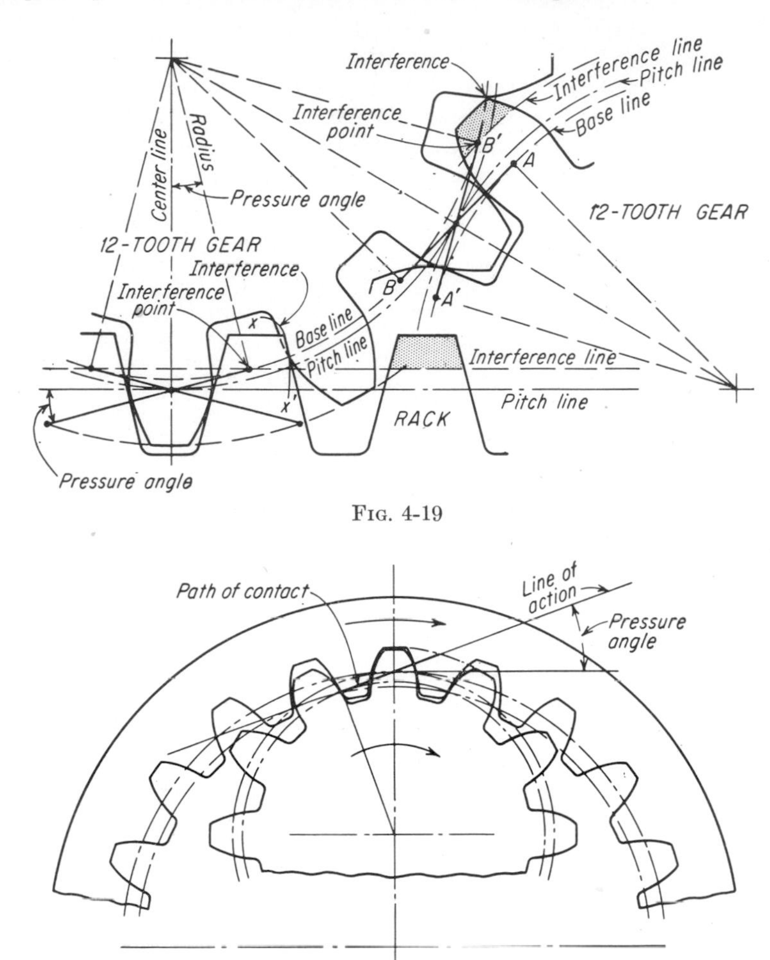

FIG. 4-19

FIG. 4-20

4-17. Involute Internal (or Annular) Gears. Figure 4-20 shows the general layout of an involute internal (or annular) gear and pinion. It will be observed that in the internal gear the tooth profiles are concave instead of convex, as in the case of the external gear (spur gear). Because of this shape an interference condition can exist that is not found in the case of either the external gear or the rack; that is, under certain circumstances a *secondary* interference will occur between the teeth after they have ceased contact along the path of contact. Obviously, the rack with its straight-sided teeth is the limiting case of both the external and the internal gear.

CYCLOIDAL TOOTH PROFILES

A brief undetailed discussion of the cycloidal tooth form is presented in the following pages.

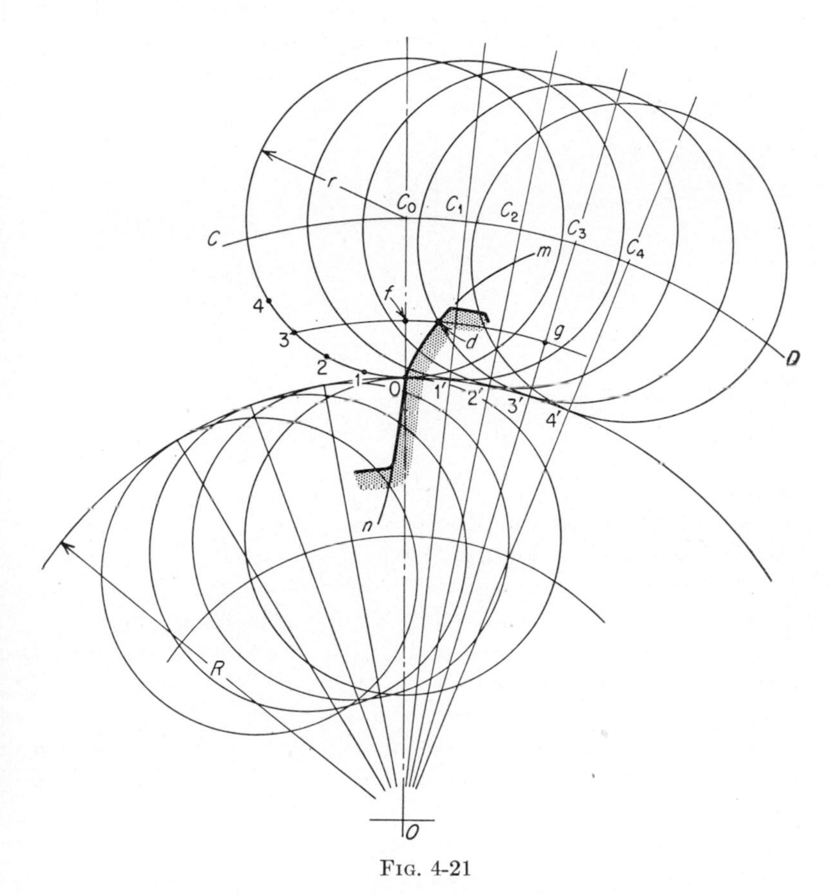

Fig. 4-21

4-18. Construction of the Cycloidal Form. A *cycloid* is the curve described by a point on the circumference of a circle as the circle rolls on a straight line. If the circle rolls on the outside of another circle, the curve described is called an *epicycloid;* when it rolls on the inside of another circle, the curve described is called a *hypocycloid.* Figure 4-21 illustrates a cycloidal tooth profile. The describing circle has been rolled on the *outside* of the pitch circle of the gear to form the face of the tooth, and on the *inside* to form the flank. Figure 4-22 illustrates the effect on the flank produced by using different sizes of describing circles. At *a* the diameter of the describing circle that forms the flank of the tooth is *less* than the radius of the pitch circle, which gives a spreading flank and, therefore, a strong tooth shape. At *b* the diameter of the describing circle is *equal* to the radius of the pitch circle. The hypocycloid in this case becomes a straight line coincident with the diameter of the pitch circle. The result is a *radial* tooth flank, which results in a tooth narrower at the base than at the pitch circle and, therefore, of weak form.

It should be recalled that the part of the flank of an involute gear profile which lies below the base circle is customarily a radial line. Therefore, if desired, this radial line can be considered to be a cycloidal curve

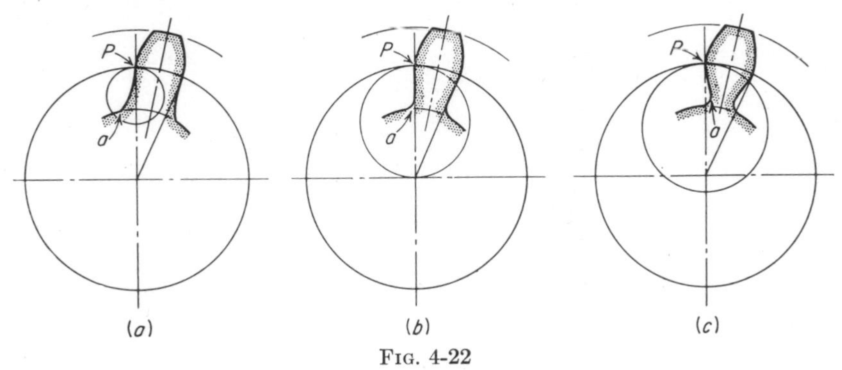

FIG. 4-22

generated by a particular size of describing circle. This knowledge has application in elimination of interference of involute teeth by the modification of an interfering involute tooth tip to a cycloidal form in order to mesh properly with the radial flank of the mating gear.

At *c* the diameter of the describing circle is *greater* than the radius of the pitch circle, and the flank of the tooth is *undercut* as shown; i.e., the hypocycloid *Pa* falls inside the radial line, and a very weak form of tooth results. It follows, therefore, from the standpoint of strength of the teeth, that the diameter of the describing circle should be less than the radius of the pitch circle.

The faces of the teeth of one gear act only on the flanks of the teeth of the mating gear. Hence, the form of the tooth face on one gear is independent of that of its own flank. There is no fixed relation, therefore, as to the size of the describing circles which could be used for a pair of gears, as far as proper tooth action is concerned. There are, however, practical conditions that govern the size of describing circles to be used in any given case.

4-19. Characteristics of Cycloidal Tooth Action. Some of the important characteristics of the action of mating cycloidal teeth will be listed here, without proof.

1. Properly mated cycloidal teeth satisfy the fundamental law of gearing.

2. The path of contact follows the describing circles, as illustrated in Fig. 4-23 by the heavy curve APB (in involute gears, this path of contact is a straight line).

3. The pressure angle is not constant, but varies continuously during tooth contact. In Fig. 4-23, the pressure angle varies from ϕ when contacting teeth are in the positions shown to zero when the teeth are in

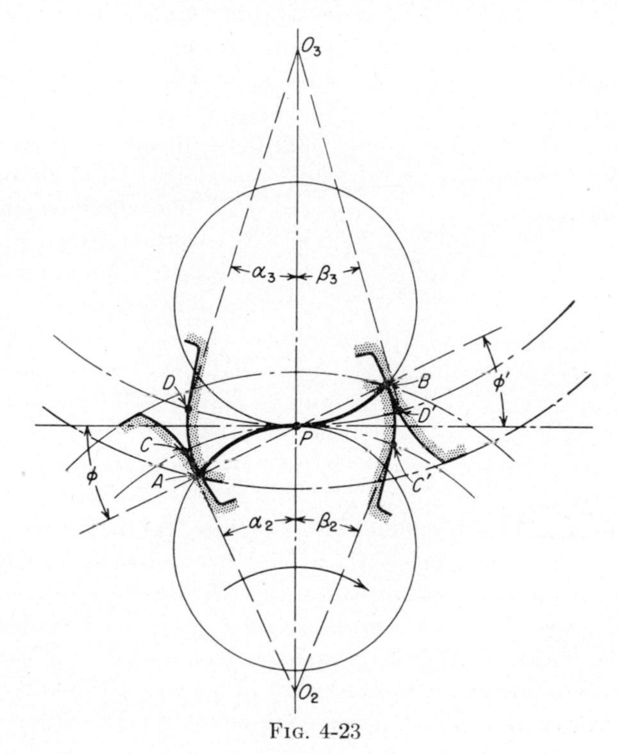

FIG. 4-23

contact at the pitch point P (in involute gears, the pressure angle is constant).

4. The cycloidal rack has curved sides.

5. The conditions necessary for interchangeability in a set of cycloidal gears are that *all the gears of the set have the same pitch, the same addendum and dedendum, and that all the teeth have outlines generated by the same describing circle.*

4-20. Pin Gearing. An interesting form of cycloidal gearing called *pin gearing* is illustrated in Fig. 4-24. Although now confined mainly to clockwork, this is an old type of gearing much used in early times when teeth were commonly made of wood. The kinematic requirement in this class of gearing is that the describing circle shall be equal to one of the pitch circles. The hypocycloid in this case becomes a point, and this point acted upon by an epicycloid described on the other gear by this same describing circle will transmit motion identical with the rolling of the two pitch circles.

The primary tooth forms are, thus, a point P on the pitch circle of the pinion 2 (Fig. 4-24) and the dotted epicycloid which this point describes on the mating gear 3. In actual use point P serves as the center of a pin of sensible diameter, the actual tooth profiles being *parallel* to the primary curves, and at distances from these curves equal to the radius of the pin.

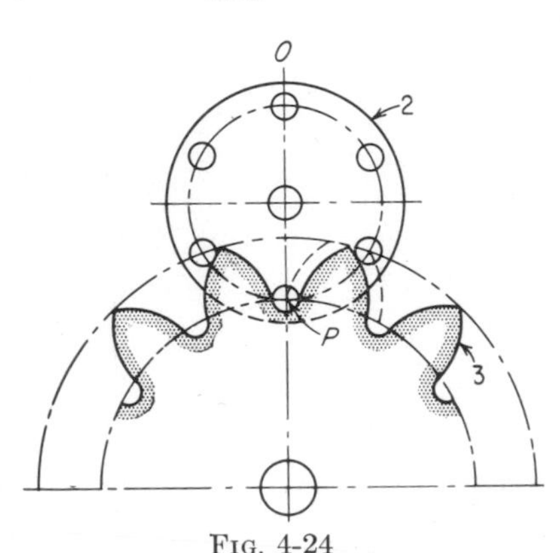

FIG. 4-24

With the primary curves (the point and epicycloid) the angle of action is entirely on one side of the line of centers. In toothed gearing the tooth action is always smoother during recess than during approach. In order to have the action take place during recess and thus secure smooth action in pin gearing, the pin wheel should be the follower. It can be seen, however, that with a pin of sensible diameter there must always be a small angle of approach.

4-21. Comparison of Cycloidal and Involute Tooth Forms. The fact that the involute form of tooth has almost entirely superseded the cycloidal form is in itself an indication that the involute form is superior. In cycloidal gears there is no interference (*if* the exact designed center distance is maintained), the teeth in general have spreading flanks, and a convex surface is always in contact with a concave surface. But these points in favor of the cycloidal form are of little importance now when compared with such properties of the involute form as the simplicity of the single curve, the straight-line rack profile as the fundamental form for cutting tools, the constancy of the pressure angle, and the possibility of varying the center distance without affecting the velocity ratio.

The possibility of changing the center distance of a pair of involute gears without affecting the velocity ratio is one of the very desirable attributes of the involute form. This has been discussed in Art. 4-12. Thus the presence of a center distance different from the design value results in a change in pressure angle, but not in velocity ratio. In an interesting application based upon this principle of change of center distance, two or more gears of slightly different numbers of teeth turning about one axis can be made to mesh correctly with another gear. Thus differential movements can be obtained that are not possible with tooth outlines of any other form.

4-22. Gear-tooth Systems and Proportions. The first toothed wheels were made of wood, and the first metallic gears were cast. Cast gears are still used to some extent in certain machines and appliances where the requirements are not exacting and in places where they are exposed to the elements and operate only occasionally. The teeth of cast gears are made from forms carefully shaped by the patternmaker to secure as close an approximation of conjugate gear-tooth action as possible. The proportions that have been used differ from those for cut teeth only in the provision for greater clearance and greater backlash.

Although the use of toothed gears extends back into ancient times, it was only about 260 years ago that the theory of correct tooth shapes was enunciated, and only about 95 years ago that a means of producing accurately cut teeth was made commercially available. At that time the formed milling cutter of the type shown in Fig. 5-22 was introduced by the Brown and Sharpe Manufacturing Company. This development marked the first real advance in gear manufacture. At this time the cycloidal form of tooth prevailed, but it was not long thereafter that it began to give way rapidly to the involute form.

Since the introduction of the involute form of tooth, a variety of gear-tooth proportions have been devised to meet the changing requirements of industry. The different systems that have come into use have been developed primarily around the method employed to produce them. It is only recently, however, that attempts at standardization have generally been successful.

Before entering into a discussion of these relatively recently adopted gear standards that have gained wide acceptance, a brief review will be given of those systems for cut teeth that have been developed in the past and are still employed to a considerable extent. These systems will generally fall within the following classification:

1. The Brown and Sharpe 14½-deg system.
2. The stub-tooth systems.
3. The unequal-addendum systems.

4-23. The Brown and Sharpe 14½-deg Involute System. When this system was introduced by the Brown and Sharpe Manufacturing Company, it was given the same tooth proportions as the cycloidal system that it has almost completely superseded. This system, like its predecessor, the cycloidal system, was developed primarily for formed milling cutters, and it has long been the recognized standard for this method of cutting gear teeth. This system has been adopted as one of the American Standards and it is now known as the 14½-deg composite system. Table 4-2 of Art. 4-26 gives details of the tooth proportions.

It is noted that in the given proportions the backlash is zero. This is sometimes desirable, but it is more common practice to provide backlash, and when standard cutters are used, this is accomplished by cutting the teeth slightly deeper than standard. In any case, the backlash is only a few thousandths of an inch for ordinary cut gears and does not enter into the kinematic study of gear-tooth action.

The profiles in the Brown and Sharpe $14\frac{1}{2}$-deg involute system are not true involutes, the curves produced by the standard cutters deviating slightly from the true involute in order to avoid interference.

4-24. The Stub-tooth Systems. While the Brown and Sharpe $14\frac{1}{2}$-deg involute system was long the recognized standard for most ordinary purposes, unusual conditions arose where the standard tooth form was not suitable. A demand arose for a stronger tooth shape and one that would minimize interference. Consequently, teeth of greater pressure angle and less depth than the preceding standard have become quite common. Gears with this special form of tooth are called *stub-tooth* gears and have been used successfully in automobile drives, machine tools, hoisting machinery, etc.

Stub teeth are of the involute form, usually having a 20-deg pressure angle, and have a shorter addendum and dedendum than that of the $14\frac{1}{2}$-deg standard. Of the several stub-tooth systems that have been proposed, that adopted by the Fellows Gear Shaper Company (Fig. 4-25) has become the most widely used. In the *Fellows stub-tooth system* the pitch is expressed as a combination of two standard diametral pitches, as for example, $6/8$ pitch which is read *six-eight* pitch and not as a fraction *six-eighths.* Thus, a $6/8$-pitch tooth has a thickness on the pitch line equal to that of a standard 6-pitch tooth, while the addendum is equal to that of a standard 8-pitch tooth. It is thus evident that the first number of the expression for the combined pitch governs the thickness of the tooth, and the second number governs the height of the tooth. Other proportions are shown in Table 4-2 on page 124.

The pitches that have been adopted and widely used are as follows: $4/5$, $5/7$, $6/8$, $7/9$, $8/10$, $9/11$, $10/12$, and $12/14$.

4-25. The Unequal-addendum Systems. The exacting requirements of high-speed gearing led to the adoption by certain gear manufacturers of other special tooth forms that aimed at eliminating the defects of the standard involute system and at securing the best possible quietness, strength, and durability. Important features of these tooth forms, all of which have the true involute outline, are a long addendum for the driving pinion and a short addendum for the driven gear, accompanied by a varying pressure angle; i.e., for a pair of gears of a given velocity ratio the addenda and the pressure angle are chosen so as to give the best running conditions for this particular ratio. Such gears are called *long-and-short-addendum* gears or *unequal-addendum* gears. The latter term will usually be used in this book.

Typical of the form of tooth coming under the above description is the system of gearing originally developed by Max Maag of Zurich, Switzerland, and known as the *Maag system* (Fig. 4-25). The Maag system has been successfully applied in many classes of machinery where high-speed gearing of the spur or helical type is required.

Another tooth form of the unequal-addendum type, known as the *Gleason system for bevel gears*, is in common use in the rear-axle drives of

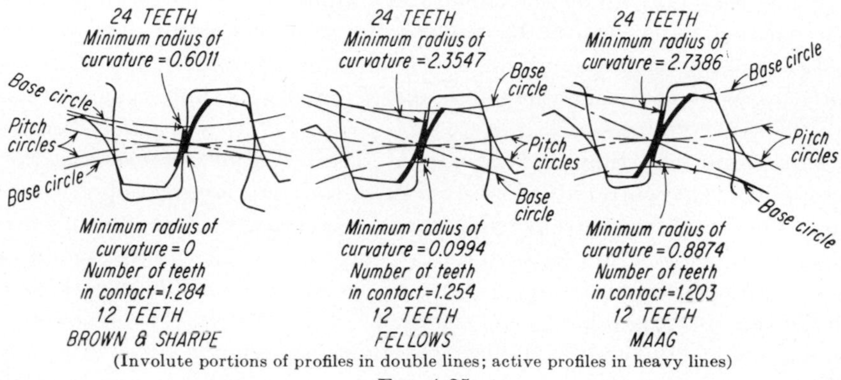

(Involute portions of profiles in double lines; active profiles in heavy lines)

FIG. 4-25

automobiles, and in many other places where high-speed bevel gearing is used.

Some of the advantages claimed for the tooth forms of the unequal-addendum type are as follows:

1. The elimination of interference accompanied by a tooth shape of greater strength. The most advantageous part of the involute curve is used, so that the flanks of the teeth are true involutes, and pinions with a small number of teeth can be used successfully. One of the defects of an equal-addendum system is the interference and undercutting in pinions with small numbers of teeth.

2. The arc of recess is greater than the arc of approach, and there is a maximum of near-rolling type of contact and a minimum of sliding contact. These factors result in smoother and quieter running and less wear than in the case of equal-addendum teeth.

3. The teeth of the pinion and gear can be proportioned so as to be of equal strength, whereas with equal-addendum teeth this is not possible without resorting to the use of different materials for the pinion and gear.

In the case of the unequal-addendum tooth forms such as referred to above, the tooth proportions vary with the velocity ratio, and hence the proportions cannot be given by simple formulas as in the case of equal-addendum teeth, where the proportions are constant. For example, the addenda and pressure angle for a pair of gears of 2:1 ratio would be different from the addenda and pressure angle of a pair of gears of the same pitch having a 4:1 ratio. Elaborate tables are required to cover the data for all the ratios in common use. The tables for the Gleason system for straight-tooth bevel gears are given in Chap. 5. Since the tooth profiles of all such systems are involutes, the fundamental theory of the gear-tooth action is the same for all.

Figure 4-25 is given for the purpose of comparing the tooth proportions in a particular case, for the Brown and Sharpe 14½-deg involute system, a stub-tooth system, and an unequal-addendum system. In each of the three cases represented, the pitch, pitch diameters, and velocity ratio are the same.

4-26. American Standard Spur-gear Tooth Forms. Since any two spur gears that will mesh properly with a rack will mesh properly with each other, rack-tooth proportions may be used as the basis for an *interchangeable system* of spur gears. Of the various spur-gear tooth standards that have been proposed for an interchangeable system, the best known and most widely used have been the Brown and Sharpe 14½-deg involute system and the Fellows 20-deg stub-tooth system. Of the unequal-addendum systems for *noninterchangeable* spur gears, that proposed by Maag is typical and is perhaps the best known. It was probably the first used to produce to any extent noninterchangeable spur gears on a commercial basis.

In June, 1921, the American Standards Association, with the American Gear Manufacturer's Association and the American Society of Mechanical Engineers as joint sponsors, began a survey of the various standards on gears in use in both the United States and Europe with the view of developing standards to meet the requirements of industry. With the findings of this survey as a background, the American Standards Association in November, 1932, approved and designated as American Standards, four spur-gear tooth forms that are listed as follows:

14½-deg composite system.

14½-deg full-depth involute system.

20-deg full-depth involute system.

20-deg stub involute system.

The proportions approved for the basic rack of the 14½-deg composite system are shown in Fig. 4-26. In this system the 12-tooth pinion has been adopted as the smallest gear of the set, and the tooth profile of the basic rack is a combination of involute and cycloidal curves. Reference to Fig. 4-27 will indicate how the proportions given in Fig. 4-26 were established. The line bPc is drawn at an angle of 14½ deg with the pitch line 3 of the rack, circle 2 being the pitch circle of a 12-tooth pinion with its center at O_2. The corresponding base circle 2′ is tangent to line bPc at c. Since there can be no involute action beyond c, the distance ce is the maximum addendum of a 14½-deg involute rack tooth. Since

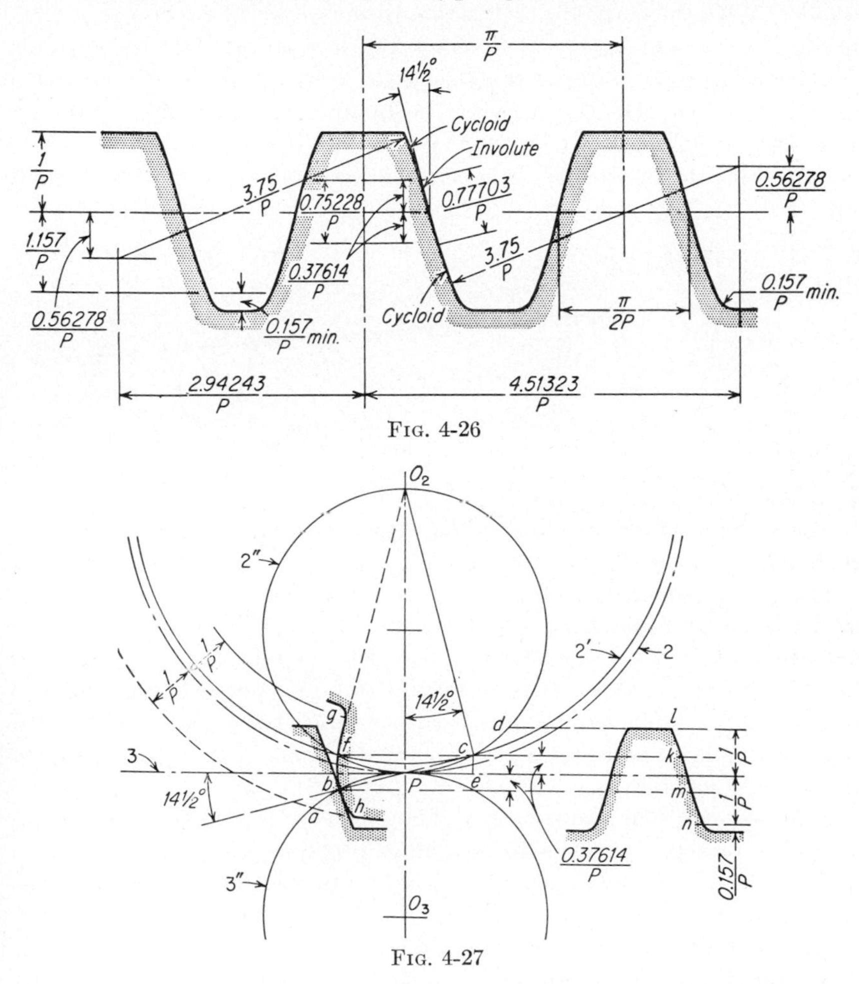

Fig. 4-26

Fig. 4-27

PcO_2 is evidently a right triangle, point c lies on a circle 2″ having a diameter equal to the radius of the 12-tooth pinion. Hence, if kl is described by rolling 2″ on pitch line 3 and fg is described by rolling 2″ on the inside of 2, the cycloidal face kl of the rack tooth will mesh properly with the hypocycloidal flank fg of the pinion tooth. Beyond c the path of contact cd would therefore follow the describing circle 2″. The flank mn of the rack tooth will mate properly with the face bh of the pinion if the flank mn is made a cycloid described by rolling 3″ on pitch line 3, and the face bh is made an epicycloid by rolling 3″ on the outside of 2. The rack-tooth profile is therefore cycloidal from n to m, involute from m to k, and cycloidal from k to l; whereas the pinion-tooth profile is hypocycloidal from g to f, involute from f to b, and epicycloidal from b to h. With the diameter of 2″ equal to the radius of 2, the hypocycloid fg becomes a radial line. If the rack tooth is to be basic for an interchangeable system of gears, the generating circle 3″ must be made equal to 2″, in which case the path of the point of contact $abPcd$ becomes symmetrical with respect to the center line O_2P. It was in the above manner that the basic rack of the 14½-deg composite system as shown in Fig. 4-26 was derived. In this figure are given the radii and location

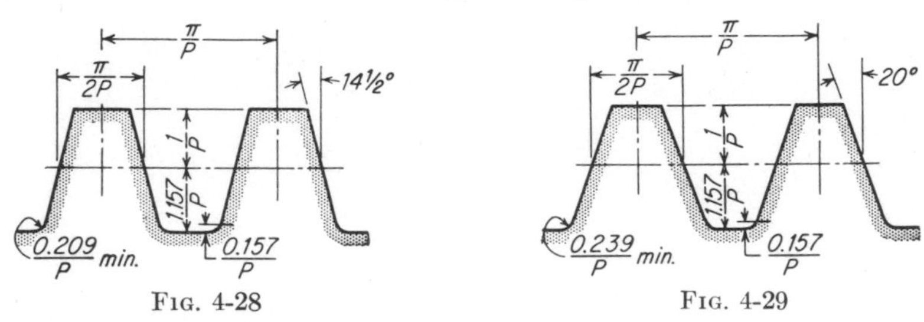

Fig. 4-28

Fig. 4-29

of the centers of circular arcs that very closely approximate the true cycloidal faces and flanks of the rack.

The tooth proportions of the 14½-deg composite system are identical with those of the old Brown and Sharpe 14½-deg involute system (see Art. 4-23), and the basic racks of the two systems have substantially the same tooth profiles. It is important to note, therefore, that the new system was established by practically adopting the old system as a definite standard, and that gears cut in accordance with the two systems are interchangeable.

The 14½-deg composite system finds its chief application where the teeth are produced by the form-milling process, particularly by those concerns that have been using the Brown and Sharpe cutters for many years in jobbing and repair work.

The tooth proportions of the 14½-deg full-depth involute system (Fig. 4-28) and the 20-deg full-depth involute system (Fig. 4-29) are identical with those of the 14½-deg composite system, but the teeth of the basic racks have straight sides throughout the full working depth. These two systems are employed when the teeth are cut by the generating methods (see Chap. 5), and their purpose is to obtain full involute action. However, to obtain full involute tooth action on pinions of 31 teeth and smaller in the 14½-deg full-depth involute system and on pinions of 17 teeth and smaller in the 20-deg full-depth involute system, the outside diameter of the pinion must be increased. To maintain standard center distance where a gear is substituted for the rack, the outside diameter of the gear must be decreased the same amount. Tabulations of the amounts of increase and decrease in diameter and the corresponding circular tooth thicknesses on the pitch circle of both gear and pinion are included in tables published by the American Standards Association.

For gears with a relatively large number of teeth, say 40 or more, the 14½-deg full-depth involute system is customarily employed, but for gears with fewer teeth, the 20-deg full-depth involute system is given preference.

The proportions for the basic rack of the 20-deg stub involute system are given in Fig. 4-30. This system, also, applies to gearing where the generating method is employed. A pinion having 14 teeth is the smallest that will mesh with the rack without interference. It may be observed, also, that these proportions approach very closely those of the Fellows stub-tooth system. The 20-deg stub-involute system is particularly applicable where the tooth numbers are relatively small in both pinion and gear, as in automobile-transmission assemblies. It is also used to some extent for heavy mill gears with large tooth numbers, where the increased strength of the shorter teeth is a distinct asset.

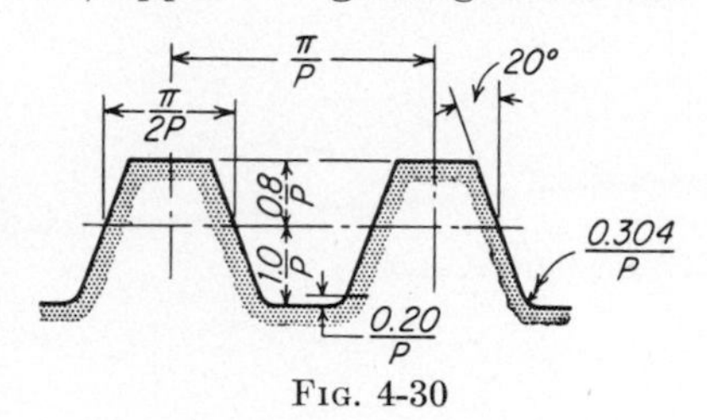

FIG. 4-30

Table 4-2 is given for convenience in making a comparison of the standard spur-gear tooth forms that have been discussed in this and in previous articles.

Table 4-3 is abstracted from the 1950 American Standard 20-degree Fine Pitch System for Spur and Helical Gears. Provision is included in this system for using unequal-addendum modification if desired. The *fine-pitch* series includes gears of 20 diametral pitch and finer having a 20-deg pressure angle. Gears in instruments, computers, servomechanisms, etc., often fall in the fine-pitch category.

4-27. Other Tooth Forms. As new gearing problems and applications arise, the industries concerned are faced with the necessity of selecting and producing gear forms and proportions which will solve their specific problems. Thus the art and science of gearing are continuously changing. The 1932 American Standards Association specification of gear-tooth forms and proportions is no longer adequate to represent fully current practice (although the involute form remains the accepted shape). Industries which use large quantities of gears, such as the automotive and aircraft industries, are in positions to set their own practices, and many of these practices will be and have been incorporated into standards of the American Gear Manufacturer's Association and the American Standards Association.

In the automotive industry, pressure angles for spur gears in the range

TABLE 4-2. COMPARISON OF STANDARD INTERCHANGEABLE INVOLUTE SYSTEMS

	14½-deg Brown and Sharpe system	14½-deg composite system	14½-deg full-depth involute system	20-deg full-depth involute system	Fellows stub-tooth system‡	20-deg stub-involute system
Pressure angle, deg	14½	14½	14½	20	20	20
Addendum	$\frac{1}{P}$	$\frac{1}{P}$	$\frac{1}{P}$	$\frac{1}{P}$	$\frac{1}{P_d}$	$\frac{0.80}{P}$
Working depth	$\frac{2}{P}$	$\frac{2}{P}$	$\frac{2}{P}$	$\frac{2}{P}$	$\frac{2}{P_d}$	$\frac{1.60}{P}$
Clearance	$\frac{0.157}{P}$	$\frac{0.157}{P}$	$\frac{0.157}{P}$	$\frac{0.157}{P}$	$\frac{0.25}{P_d}$	$\frac{0.20}{P}$
Dedendum	$\frac{1.157}{P}$	$\frac{1.157}{P}$	$\frac{1.157}{P}$	$\frac{1.157}{P}$	$\frac{1.25}{P_d}$	$\frac{1}{P}$
Whole depth	$\frac{2.157}{P}$	$\frac{2.157}{P}$	$\frac{2.157}{P}$	$\frac{2.157}{P}$	$\frac{2.25}{P_d}$	$\frac{1.80}{P}$
Circular thickness*	$\frac{p}{2}$	$\frac{p}{2}$	$\frac{p}{2}$	$\frac{p}{2}$	$\frac{\pi}{2P_n}$	$\frac{p}{2}$
Tooth space*	$\frac{p}{2}$	$\frac{p}{2}$	$\frac{p}{2}$	$\frac{p}{2}$	$\frac{\pi}{2P_n}$	$\frac{p}{2}$
Fillet radius†	$\frac{0.157}{P}$	$\frac{0.157}{P}$	$\frac{0.209}{P}$	$\frac{0.239}{P}$	$\frac{0.25}{P_d}$	$\frac{0.304}{P}$

The symbols P and p denote diametral and circular pitch, respectively.

All dimensions are in inches.

* In these proportions backlash has not been taken into account.

† These may be assumed as minimum values. The fillet radius has not been standardized.

‡ For this fractional form of pitch, the symbol P_n represents the figure in the numerator and P_d the figure in the denominator. Example: $P_n/P_d = 4/5$.

from 17 to 22½ deg are common. Many gear sets use systems of unequal addendums. When large quantities are involved, each pair of meshing gears can properly be designed to best suit a specific application. In many gears, use is made of a full-rounded root such as that shown in Fig. 4-31 in place of the more usual root form because of the increased resistance to breakage under heavy loads.

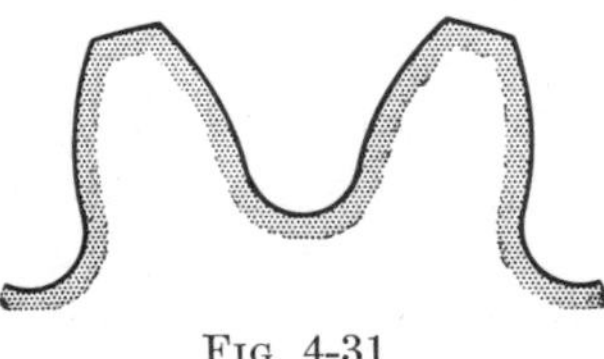

FIG. 4-31

Tooth shapes are often tapered (or crowned), with the center portion slightly thicker than the edge, as shown in Fig. 4-32. This modification keeps the load-carrying portion near the central part of the tooth, allowing some tolerance in misalignment of shafts and deflection which otherwise would cause the load to be carried on the edge of the tooth. Such a

TABLE 4-3. 20-DEG INVOLUTE FINE-PITCH SYSTEM STANDARD TOOTH PROPORTIONS

Addendum	$\frac{1.000}{P}$
Dedendum	$\frac{1.200}{P} + 0.0002$
Working depth	$\frac{2.000}{P}$
Whole depth	$\frac{2.200}{P} + 0.002$
Clearance	$\frac{0.200}{P} + 0.002$
Tooth thickness on pitch diameter	$\frac{1.5708}{P}$
Pitch diameter	$\frac{N}{P}; \frac{n}{P}$
Outside diameter	$\frac{N+2}{P}; \frac{n+2}{P}$
Center distance	$\frac{N+n}{2P}$

N = number of teeth in gear.
n = number of teeth in pinion.
P = diametral pitch.
All dimensions are given in inches.

modification also can compensate the profile for deflections at the contacting position causing (locally) a noninvolute shape of an original involute. Figure 5-8 shows a pair of Zerol *bevel* gears which have been modified to give local tooth bearing. The gears were painted with marking compound and then run for a few seconds to show the local bearing areas which appear dark in the photograph.

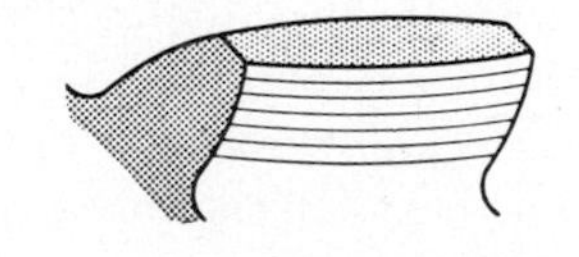

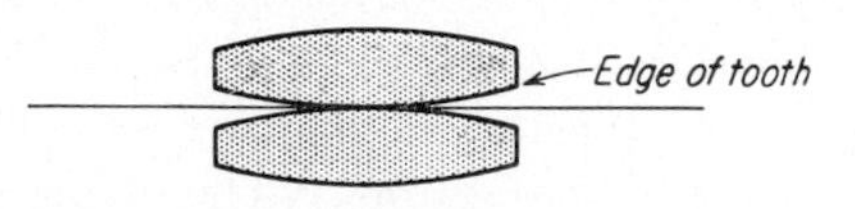

Crowned shape is exaggerated for easy visualization

FIG. 4-32

Obviously, no comprehensive coverage of the specialty of gearing is possible within the scope of this book. Rather, basic consideration has been given to the motions of a pair of meshing gears and the reasons for the accepted use of the involute tooth profile. For details concerning the multitude of facets of the field of gearing, complete books and the aforementioned standards (which are continuously being studied and revised) should be consulted.

The following chapter (Chap. 5) surveys the types of gears other than spur gears (which have been discussed in this chapter), namely, bevel, helical, and worm gears (note again Table 4-1). In addition, Chap. 5 contains information on methods of gear manufacture.

CHAPTER 5

GEAR TYPES AND MANUFACTURING METHODS

The previous chapter was concerned primarily with spur gears. The present chapter discusses the other types of gears which are in common usage, and also indicates briefly some of the methods of manufacturing gears. It might be well at this point for the reader to review Table 4-1 and Fig. 4-2.

In Fig. 5-1 are shown examples of many of the types of gears which are in use today. They can all be classified under the general headings indicated in Table 4-1; namely, spur gears, bevel gears, helical gears, or worm and wheel combinations.

The various types (except spur gears) will be discussed in the following articles.

5-1. Classification of Bevel Gears. The form of gearing used to connect shafts that intersect[1] is called bevel gearing. In Art. 4-4 the contact surfaces of a pair of bevel friction wheels were shown to be surfaces of the frustums of a pair of cones having a common apex. If teeth are formed on these surfaces in a manner analogous to the methods used in spur gearing, the cones become the pitch surfaces of a pair of bevel gears. In the great majority of bevel-gear drives the shafts are at right angles, but the angle between the shafts may be either greater or less than 90 deg. In such cases the gears are called *angular bevel gears.* When the angle between the shafts is 90 deg and the two gears of a pair are equal, the gears are called *miter gears.* When the pitch angle (see Fig. 5-3) of a bevel gear is 90 deg, it is called a *crown gear.*

Bevel gears are nearly always made to work together in pairs and are not interchangeable. The involute form of tooth is used, and the Brown and Sharpe 14½-deg involute tooth proportions have been widely used in the past for ordinary work; but for the better class of bevel gearing and especially for high-speed gearing, the teeth conform to standards that have been developed for bevel gears (see Art. 5-15). Bevel gears may be divided into four classes (see also Table 4-1):

1. Straight-tooth bevel gears, in which the elements of the tooth surfaces are straight lines, converging at the apex of the pitch cone. The pitch surfaces are cones.

[1] The shafts are nonintersecting in the case of skew bevel gears.

2. Spiral bevel gears, in which the teeth are curved. The pitch surfaces are cones.

3. Skew bevel gears, which may be used to connect nonparallel nonintersecting shafts. The pitch surfaces are hyperboloids (see Art. 4-4) and the teeth are straight.

FIG. 5-1. (*Courtesy of Chevrolet Motor Division, General Motors Corporation.*)

4. Hypoid gears, which connect nonparallel nonintersecting shafts. The pitch surfaces are cones and the teeth are curved.

5-2. Form of Bevel-gear Teeth. The motion of a pair of spur gears comes under the head of plane motion, and the relative motions can be represented by a right section. Thus, an involute tooth profile is described by a point in a line that rolls on a base circle. In reality this is simply a representation, in a plane, of a tooth surface that is swept up by a line in a plane as the plane containing the line rolls on a base cylinder. This simple treatment cannot be applied to bevel gears. Although each bevel gear considered separately has plane motion (rotation) about its own axis, the relative motion, considering the two gears rolling together, is spherical motion.

1. *Theoretically Correct Form.* A simple illustration will show that the method of forming the surface of an involute bevel-gear tooth is fundamentally the same as that for forming the surface of an involute spur-gear tooth. In Fig. 5-2 let the cone OHI represent the base cone of the bevel

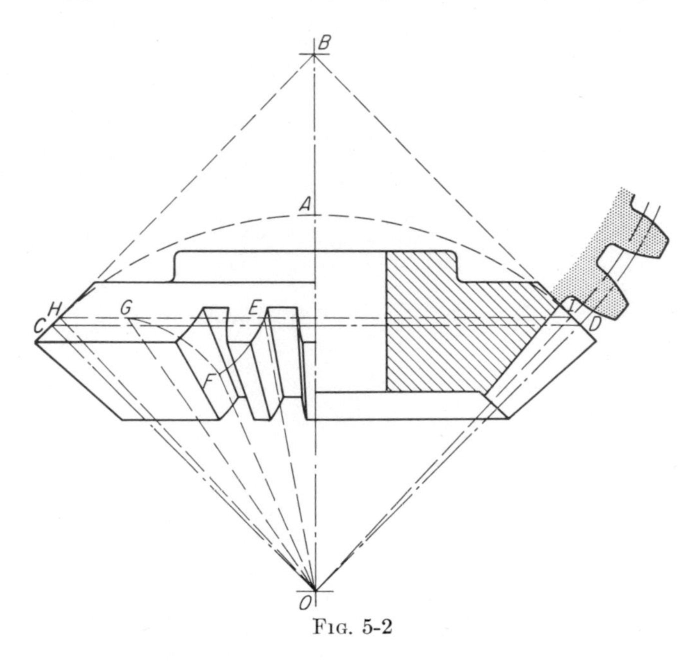

FIG. 5-2

gear shown, on which the involute tooth surfaces are to be formed. Imagine the base cone to be enclosed in a thin flexible covering that is cut along the line OE. Now take hold of the edge OE and, keeping the covering taut, unwrap it. The surface OEF swept up by the edge or element OE is the desired involute surface. The point E is constrained to remain a constant distance from the cone center O, equal to OE, and consequently moves in the surface of a sphere HAI. For this reason the curve EF is called a *spherical involute.*

2. *Tredgold's Approximation.* The difficulty of laying out a tooth profile that is a curve lying in the surface of a sphere is apparent, since the surface of a sphere cannot be developed on a plane surface. No appreciable error will be introduced if the conical surface CBD is substituted for the spherical surface CAD. The cone CBD, called the *back*

cone, is tangent to the sphere at the circle CD, this circle representing the common base of the pitch cone and back cone. The back cone is thus *normal* to the pitch cone; i.e., the elements of the back cone are perpendicular to corresponding elements of the pitch cone. It can be seen in the figure that the comparatively short distance necessary to include the entire tooth profile practically coincides with the spherical surface. Within the limits of practice, therefore, such profiles agree quite closely with the theoretically correct forms.

The above approximation was first published by Tredgold and is known as *Tredgold's approximation.* In the application of the method to the layout of bevel-gear teeth, the teeth are laid out on the developed surface of the back cone as shown at the right of Fig. 5-2, and the back-cone surface is then wrapped back to its original position. The profiles thus constructed on the back cone will form the ends of the bevel-gear teeth, and straight lines drawn from the pitch-cone center O and following the profiles will sweep up the tooth surfaces, all elements of which are straight lines converging at O.

The teeth are laid out on the developed back cone by exactly the same methods as used for spur gears. In Fig. 5-2 the slant height BD of the back cone is the pitch radius of the equivalent spur gear and BI is the radius of the base circle.

5-3. Bevel-gear Nomenclature. As far as the layout of the teeth and their action are concerned, the definitions and nomenclature for bevel gears are in every respect the same as those given for spur gears in Arts. 4-7 and 4-12. Bevel gears, however, require the additional nomenclature shown in Fig. 5-3, which is self-explanatory.

In bevel gearing *formative number of teeth* is the number of teeth of the given pitch that would be contained in a complete spur gear having a pitch radius equal to the back-cone distance of the bevel gear. This number of teeth is used in laying out the teeth, in the kinematic analysis of the tooth action, in calculating the strength of the teeth, in designing and selecting the proper cutting tools, etc.

5-4. Layout of a Pair of Straight Bevel Gears. Figure 5-4 shows a pair of straight-tooth bevel gears laid out in accordance with the following specifications:

14½-deg composite system (Arts. 4-23 and 4-26)
Diametral pitch = 2½
Velocity ratio = 2:1
Angle between shafts = 69 deg
Number of teeth in pinion = 15
Number of teeth in gear = 30
Length of face of the gears = 2¾ in.

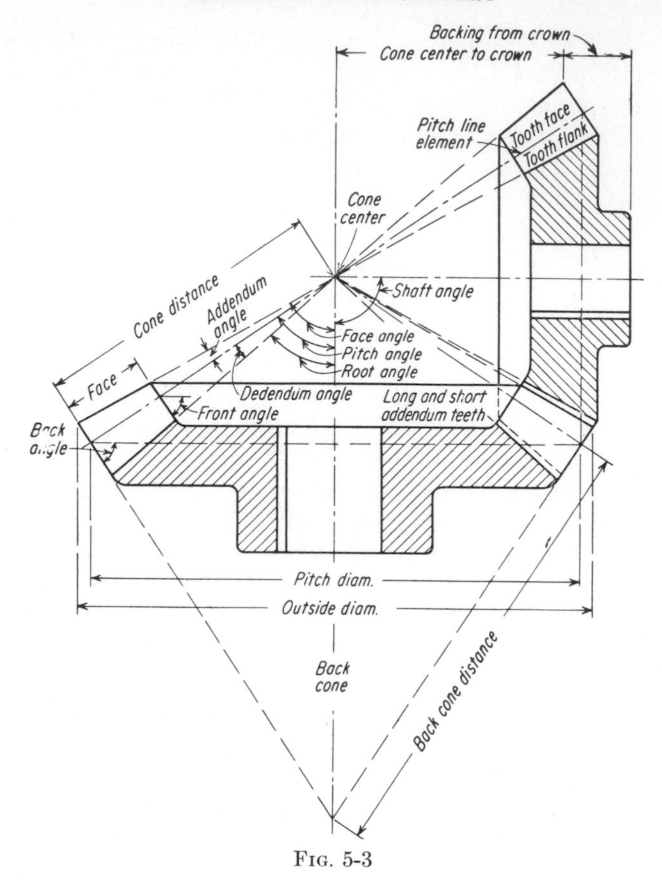

Fig. 5-3

The following items may be calculated before starting the layout (all dimensions given in inches).

	Pinion	Gear
Pitch diameter	6.000	12.000
Back-cone distance	3.228	8.853
Addendum	0.400	0.400
Dedendum	0.463	0.463
Formative number of teeth	16.14	44.26

Lay out the pitch cones in accordance with the method described in Art. 4-4, and then draw the back cones *HDB* and *HEC*. The gears will be shown in sectional view, the usual practice for working drawings. At *H* lay off on the back cones the addendum and dedendum distances for the pair of teeth in contact. Next lay off the face length *HK* on the contact element of the pitch cones and draw the lines that converge at O, representing the tops and bottoms of the pair of teeth. Repeat the above tooth layout at E and D and complete the sectional view of the gears by drawing in suitable hub, rim, and web proportions, as shown in the figure.

Although the above layout is sufficiently complete for ordinary shop purposes, it is frequently necessary to show the true shapes of the tooth profiles. To do this, develop parts of the back cones as shown. The

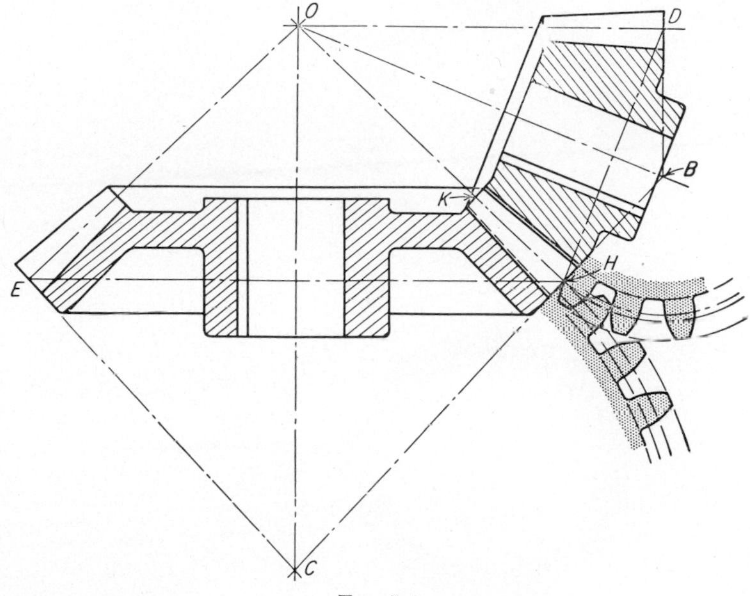

FIG. 5-4

teeth will then be laid out as the teeth of a pair of involute spur gears having pitch circles whose radii are BH and CH and whose pitch is the same as that of the bevel gears. The formative number of teeth for the pinion $= 2BH \times P = 2 \times 3.228 \times 2.5 = 16.14$, and the formative number of teeth for the gear $= 2 \times CH \times P = 2 \times 8.853 \times 2.5 = 44.26$. That is, the teeth are laid out as the teeth of a pair of equivalent spur gears having 16.14 teeth in the pinion and 44.26 teeth in the gear, whose center distance is BC and whose pitch is the same as that of the bevel gears.

Before generating machines were developed, the teeth of bevel gears were proportioned as in Fig. 5-4 and then cut on the milling machine by the use of formed cutters of the Brown and Sharpe type. To a very limited extent bevel gears are still cut by this method, which is inherently incorrect because the space between teeth should change as the cutter proceeds from one end of the teeth to the other. When the requirements are at all exacting, the teeth of bevel gears are proportioned in accordance with the Gleason system for bevel gears (Art. 5-15) and cut to a high degree of accuracy on generating machines. It should be observed that if teeth having the same diametral pitch and velocity ratio as those in Fig. 5-4 had been laid out in accordance with the Gleason system they would appear as in Fig. 5-3.

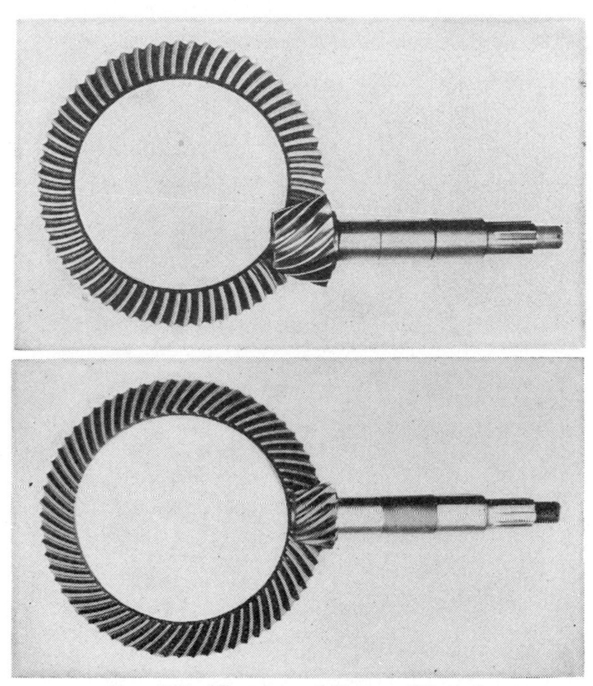

Fig. 5-5. (*Courtesy of Gleason Works.*)

5-5. Spiral Bevel Gears. Spiral bevel gears offer one of the best combinations of quietness, strength, and durability available in gearsets. In the past, they have been almost universally used for the rear-axle drive of the automobile, but at the present time hypoid gears have replaced them to a considerable extent because hypoids allow the drive shaft from the engine to be lowered, as indicated in Fig. 5-5. Spiral bevels are also used in machine tools, motion-picture machinery, sewing machines, and many other classes of machinery where quiet and smooth operation at high speeds is essential, or where the quantity is such that the cost is less than for straight-tooth bevel gears.

Theoretically, the teeth of spiral bevels are curved on a spiral, but in order to facilitate manufacture they are actually curved on the arc of a circle that very closely approximates the true spiral within the limits of the length of the tooth. The use of a circular arc instead of a spiral not only makes rapid production possible but results in an adjustability valuable from the standpoint of assembly and operation.

The outstanding advantage of spiral bevel gears as compared with straight-tooth bevel gears is the greater smoothness and quietness of operation of spiral bevel gears, particularly at high speeds.

The reasons for the smooth and quiet action of spiral bevel gears are:

1. More teeth are in contact than in the corresponding straight-tooth bevel gears.

2. The teeth of spiral bevel gears engage with one another gradually, the contact beginning at one end and gradually working over to the other end, whereas in straight-tooth bevel gears the contact takes place along the entire face of the tooth at the same instant. As illustrated in Fig. 5-6, there is no lengthwise sliding of contacting spiral-bevel-gear teeth.

A disadvantage is that spiral bevels produce a radial load perpendicular to their axes and a thrust load along their axes, and these loads must be considered in the design of the bearings and mountings.

The points of superiority of spiral bevel gears over straight-tooth bevel gears are identical in almost every respect with those of helical gears over spur gears (see Art. 5-11).

FIG. 5-6. By cutting the teeth of a straight bevel gear in an infinite number of sections and arranging them in continuously advancing steps, a gear with curved oblique teeth is evolved, in which each section rolls with its mating section and there is no lengthwise sliding. (*Courtesy of Gleason Works.*)

The fundamental theory of toothed gearing as applied to straight-tooth bevel gearing applies also to spiral bevel gearing.

The tooth proportions of spiral bevel gearing conform to what is called the *Gleason system for spiral bevel gears* (see Art. 5-15).

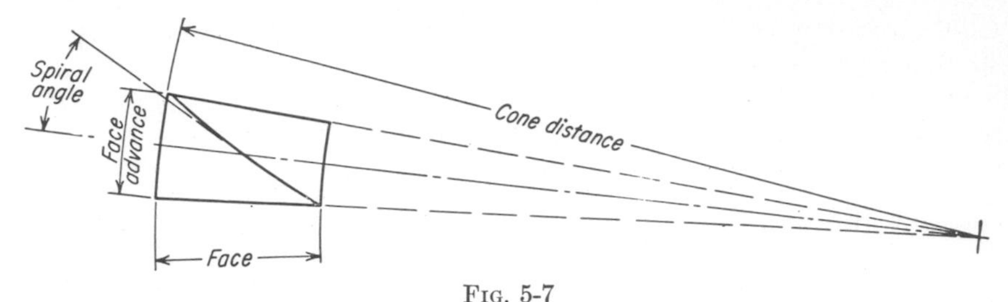

FIG. 5-7

The diagram in Fig. 5-7 shows how the angle of spiral (analogous to helix angle in helical gears) is designated.

Zerol gears (illustrated in Fig. 5-8) are spiral bevel gears with zero spiral angle. These gears have the low axial thrust characteristics of *straight* bevel gears, and the teeth can be cut and ground on the same

FIG. 5-8. Zerol bevel gears. (*Courtesy of Gleason Works.*)

machines used for manufacturing other spiral bevels. An example of the application of Zerol gears is found in aircraft engine-propeller gearing.

5-6. Skew Bevel Gears. Skew bevel gears may be used to connect shafts whose axes do not intersect, as illustrated in Fig. 5-9. The teeth are constructed on rolling hyperboloids as pitch surfaces (see Art. 4-4). The tooth elements are straight lines which, however, do not converge to a common point as in the case of ordinary straight-tooth bevel gears. The teeth are in contact along straight lines, and there is sliding action along the tooth elements as well as at right angles to the tooth elements. Skew bevel gears are rarely used in machine construction because of the difficulties involved in producing correct tooth forms. This type of gearing, however, has certain well-recognized advantages, and numerous efforts have been made to overcome the difficulties in its design and production.

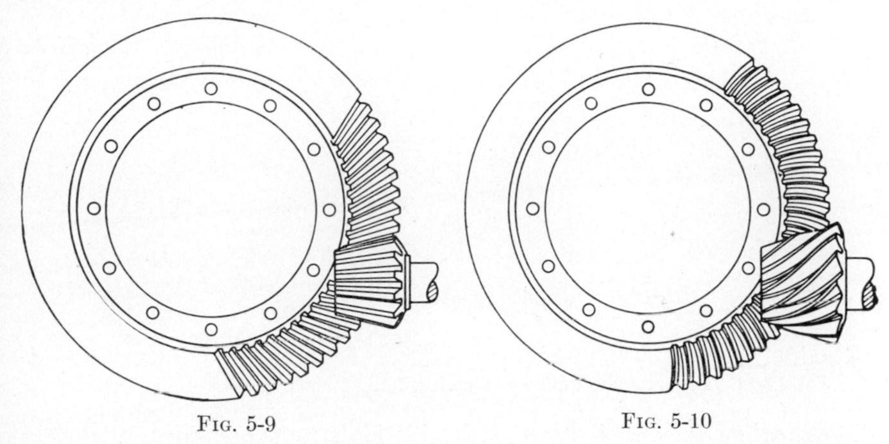

FIG. 5-9 FIG. 5-10

5-7. Hypoid Gears. The outcome of a development in the attempt to secure the advantages possessed by skew bevel gears has been a type of bevel gear called the *hypoid gear* (illustrated in Figs. 5-5 and 5-10), which was developed by the Gleason Works, Rochester, N.Y. The term *hypoid* is a contraction of the term hyperboloid and is believed to have been first used by G. B. Grant in his "Treatise on Gearing." Although the correct pitch surfaces of these gears are hyperboloids, they are approximated by sections of cones in actual practice. The error is small over the relatively short length of a tooth (review Fig. 4-2). Hypoid gears possess the main characteristics of skew bevel gears, namely, offset axes and a certain amount of sliding action in the direction of the tooth elements. But hypoid gears have curved teeth, whereas true skew bevel gears have straight teeth constructed on rolling hyperboloids of revolution as pitch surfaces.

The hypoid gear was developed in 1925 and was first used in the rear-axle drive of the Packard automobile. Its chief application has been in the automobile, where it has become the standard rear-axle drive for passenger cars (Figs. 5-1, 5-5, and 5-10). It is also having gradually increasing use in the industrial field. The use of hypoid gears in the automobile permits lowering of the drive shaft and is thus advantageous in the design of cars with low bodies.

Some important advantages result from the application of hypoid gears to industrial purposes. The pinion may be mounted on a continuous shaft, bearings may be placed on both sides of the gear and pinion, and in some cases, the raised or lowered position of the pinion allows arrangements of machinery that would not otherwise be possible. In the application shown in Fig. 5-11, hypoid gears are used in a wire-drawing mill which draws the wire at progressively increasing speeds. Owing to the offset axes, the pinion of a pair of hypoid gears is larger and stronger than that of the equivalent pair of spiral bevel gears (note Fig. 5-5). Also, the teeth of the hypoid gears have a certain amount of endwise sliding. Although this sliding action results in quieter operation, the

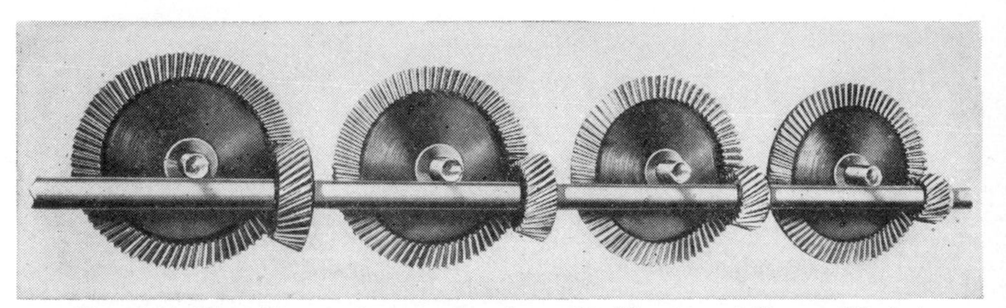

FIG. 5-11. Hypoid gears for wire-drawing mill. (*Courtesy of Gleason Works.*)

sliding, when in combination with the high unit pressures, makes it necessary to give special attention to the matter of lubrication.

5-8. Helical Gearing. Helical gearing is a term applied to all classes of gears in which the teeth are of helical form. Helical gearing is used to connect parallel shafts as well as nonparallel nonintersecting shafts. The pitch surfaces are cylindrical as in spur gearing, but the teeth, instead of being parallel to the axes, wind around the cylinders helically like screw threads.

The distinction between helical gearing and worm gearing may be stated as follows: If the number of threads, or teeth, on the pitch cylinder is such that no one thread makes a complete turn, the gear is called a helical gear. If, on the other hand, a thread (from one to four is the usual number) makes a complete turn, the result is a *worm* and the mating gear is called a *worm wheel.*

Helical gearing is used mainly between parallel shafts, but it is also employed to some extent to connect nonparallel nonintersecting shafts, usually at 90 deg.

5-9. Helical Gears for Parallel Shafts. Helical gears connecting parallel shafts are illustrated in Fig. 5-12*a*. The conception of a helical gear is simplified by considering it as a spur gear with the teeth twisted. Assume a spur gear to be cut by a series of equidistant planes perpendicular to the axis and let the resulting slices or laminations be twisted relative to each other (as one would fan a deck of cards) as shown in Fig. 5-13. The result is a *stepped gear* in which there is a series of narrow teeth derived from the original gear coming into contact successively with similarly constructed mating teeth, instead of a pair of wide teeth coming into contact at the same instant along the entire length, as in the original pair of gears. If the slices become thin enough, lines joining corresponding points of the stepped teeth become helices, and the gear is,

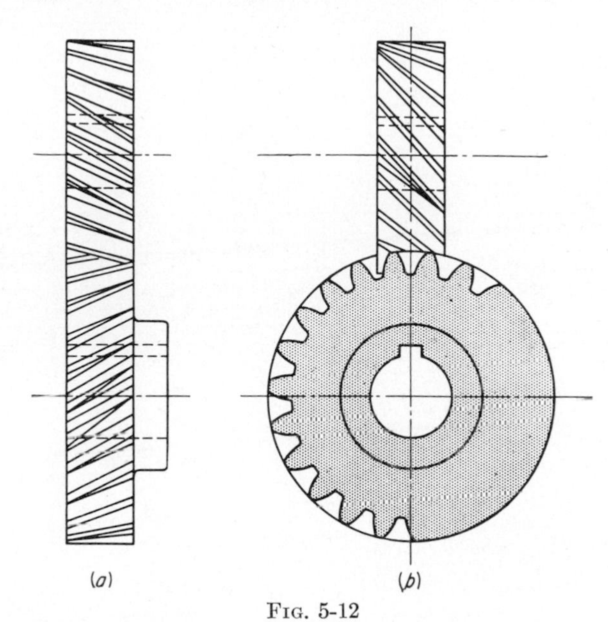

Fig. 5-12

therefore, called a helical gear. The teeth of helical gears with parallel axes have line contact, just as do spur gears. An instantaneous line of contact, however, is not parallel to the axes but runs diagonally across the tooth faces, so that in successive sections all conjugate points of a pair of profiles are in contact at the same time. The teeth of helical gears are of the involute form, usually of a stub type, with a pressure angle greater than that of the standard 14½-deg systems.

In Fig. 5-13 helices formed by the intersection of the pitch surface and two adjacent tooth surfaces are shown. The distance between the helices measured around the pitch surface in a plane at right angles to the axis is the *transverse circular pitch*. The distance measured in a similar manner at right angles to the helices is called the *normal circular pitch*. The *helix angle* is always the angle between the tangent to the helix and the axis of the gear, as shown in the figure. The velocity ratio follows exactly the same law as for spur gears.

It is readily visualized that the direction of the force of one tooth against the mating tooth, which is along the common normal, produces an axial thrust in a single pair of helical gears. This objection is overcome by using *double-helical* or *herringbone* gears (Fig. 5-14), which are simply two single-helical gears with teeth of the opposite hand combined.

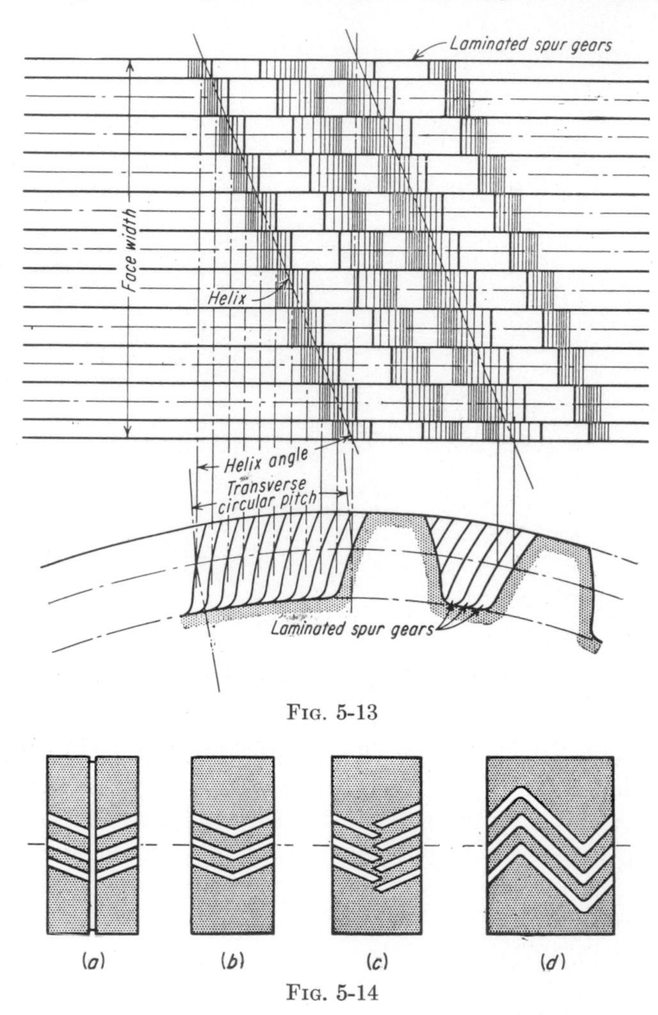

Fig. 5-13

Fig. 5-14

Double-helical gears, as manufactured in this country, are of two general types, as follows:

1. The type in which the teeth meet at a common apex at the center of the face as shown in Fig. 5-14*b*. A modification of this type in which the central part has been removed with the object of providing tool clearance is shown in Fig. 5-14*a*.

2. The type known as the *Wuest* gear in which the teeth are staggered

as shown in Fig. 5-14*c*. The central portion is cut away somewhat by the tools but the blank is not grooved as in Fig. 5-14*a*.

Figure 5-14*d* illustrates an interesting type of helical gear for connecting parallel shafts called the *Citroën gear*, manufactured by the Citroën Gear Company, Ltd., Paris, France. The teeth may be cut to any required shape and may be either double helical or double herringbone.

FIG. 5-15. (*Courtesy of Westinghouse Electric Corporation.*)

Since the teeth of Citroën gears are produced with an end-milling cutter, the teeth can be formed readily on bevel gears also, in either the double-helical or double-herringbone form.

Figure 5-15 is a view of a 17,500-hp articulated marine-propulsion gear unit using herringbone gears. Two drive turbines deliver power to the gears near the top of the photograph, and the double-reduction gearset transmits the power to the propeller shaft connected to the large gear near the bottom of the picture.

5-10. Advantages of Helical Gears. The outstanding advantage of helical gears, as compared with corresponding spur gears, is that helical gears run more smoothly and more quietly at high speeds and under other severe service conditions.

The reasons for the smooth and quiet action of helical gears are:

1. There are more teeth in contact than in the corresponding spur gears.

2. The teeth come into engagement gradually, the contact beginning at an end and gradually working over to the other end. In spur gears the contact takes place along the entire face of the tooth at the same instant.

5-11. Helical Gears for Nonparallel Nonintersecting Shafts. While helical gearing for nonparallel nonintersecting shafts may be designed for any angle between the shafts, the shafts are usually at right angles

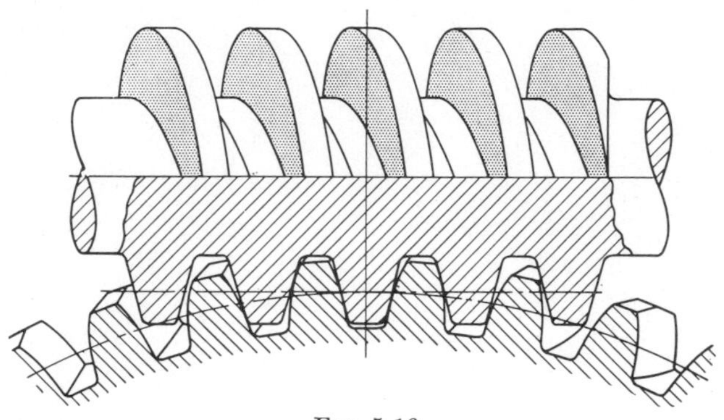

FIG. 5-16

as in Fig. 5-12*b*. The general case of helical gearing may be considered as exemplified by some angular relation of shafts between 0 and 90 deg, and shafts parallel and at right angles might be considered to represent special cases of helical gearing.

It is important to note that, although the design is essentially the same, the tooth action of helical gears for nonparallel nonintersecting axes is quite different from that of helical gears for parallel shafts. The former have merely point contact, while the latter have line contact. In the former case, also, there is a large amount of sliding in the direction of the common tangent to the tooth elements, which is entirely absent in the latter case. Helical gears for nonparallel nonintersecting shafts may consequently be used only for comparatively light service. Such gears are, therefore, of secondary importance.

5-12. Worm Gearing. Worm gearing is commonly employed to obtain higher velocity ratios than can conveniently be obtained from other forms of gearing. In a worm-gear combination, the shafts are nonintersecting and nearly always make an angle of 90 deg with each other. The worm, which is usually the driver, is a special form of helical gear in which the angle of the helix is such that the tooth becomes a thread making one or more complete turns around the pitch cylinder. The worm gear (or wheel), the driven member, becomes a form of helical gear in which the helix angle is the complement of that of the worm.

There are two classes of worm gearing in use: the *straight or parallel type* illustrated in Fig. 5-16, and the *hourglass type* (often known as Hindley worm gearing), shown in Fig. 5-17. Of these two classes the straight type is more frequently used.

5-13. Straight Worm Gearing. The pitch surfaces of straight worms are cylinders, and the teeth are of the involute form. The proportions now regarded as standard are given in Art. 5-16. The contact between a pair of teeth is theoretically line contact.

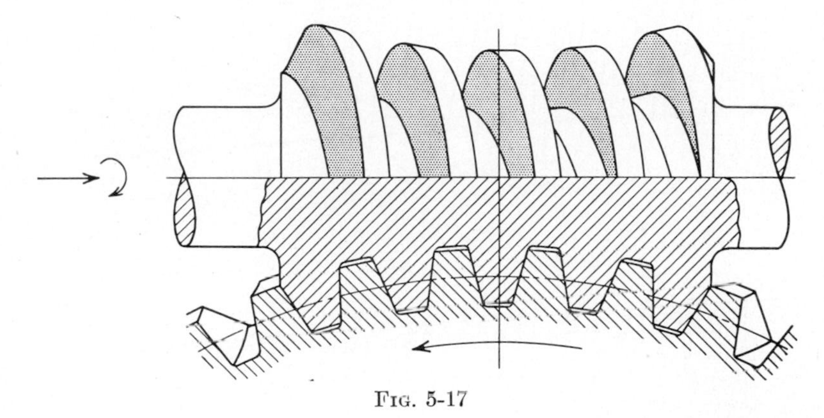

Fig. 5-17

The pitch of the worm is the distance between corresponding points of adjacent threads or teeth measured parallel to the axis of the worm, and is called the *linear pitch* or *axial pitch.* The linear pitch of the *worm* must, of course, be equal to the circular pitch of the worm *gear.* The *lead* of a worm thread is the axial distance traversed by a point moving on the helix in one complete turn of the helix. Worms are designated as right-hand (or left-hand) single-thread worm, right-hand (or left-hand) double-thread worm, etc. A right-hand worm-and-gear combination corresponds to a right-hand bolt and nut. The directions of motion can be determined by regarding the worm to be a bolt, and as it rotates, by visualizing the direction which the nut (represented by the meshing gear) moves. For example, in Fig. 5-17, as the right-hand worm rotates clockwise (viewed from the left) it causes the contacting gear portion shown to move to the left. In a single-thread worm the lead equals the axial pitch; in a double-thread worm the lead equals twice the axial pitch; etc. The distinction between lead and pitch is made clear by examination of Fig. 5-18, in which the difference between a single- and a multiple-thread worm is clearly shown.

The velocity ratio in worm gearing does not depend upon the diameters of the worm and gear but upon the ratio of the number of teeth on the worm gear to the number of threads on the worm. One revolution of a single-thread worm turns off one tooth of the mating gear, one revolution of a double-thread worm turns off two teeth, etc. Thus, for a single-thread worm and a worm gear having 36 teeth, the velocity ratio is 36:1; and for a triple-thread worm and a worm gear with 36 teeth the velocity ratio is 36:3, or 12:1.

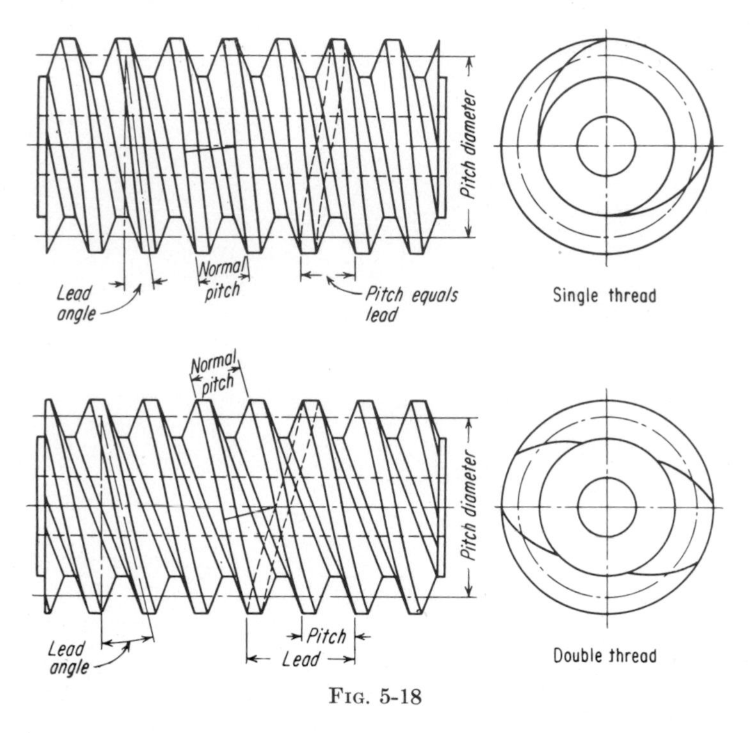

Fig. 5-18

In general, when the pressure angle is $14\frac{1}{2}$ deg, 32 teeth is the smallest number advisable for the worm gear on account of interference, or undercutting (see Art. 4-13). Hence, if the velocity ratio is less than 32:1, the worm must be provided with a multiple thread in order to avoid this condition.

It can be seen in Fig. 5-19 that, in a central plane containing the axis, the straight worm is an involute rack, and the teeth, therefore, have straight sides. In the central plane perpendicular to the axis of the worm gear, the worm gear is an ordinary involute spur gear. The worm and gear are, therefore, equivalent to an involute rack and pinion, and in making a layout of a worm and gear, the method employed is the same as that for the involute rack and pinion.

A characteristic of worm gearing that is not found in other types of gearing discussed in this chapter is its self-locking property; that is, a worm drive, in general, is not reversible. This is obviously a useful property in many applications such as elevators, hoists, automobile steering devices, etc. In rare cases, as in the cream separator and other centrifuges, for example, the worm gear is the driver, in which case the helix angle is very large.

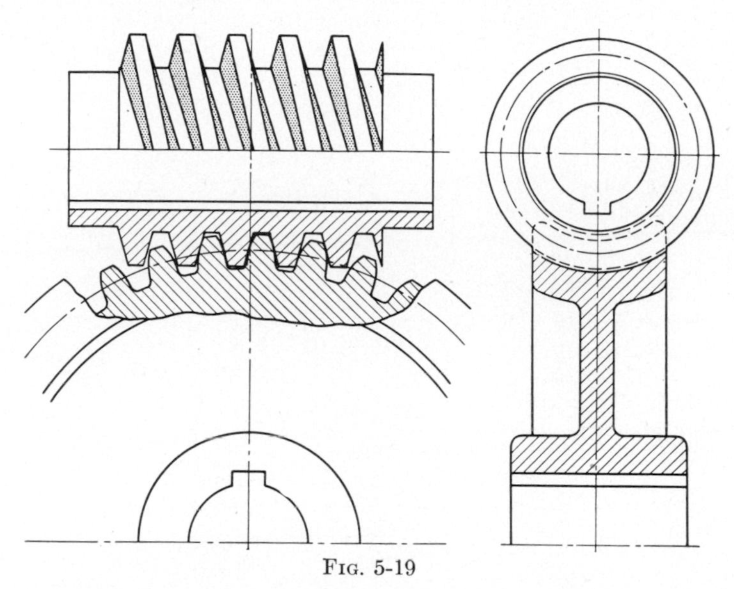

Fig. 5-19

5-14. Hourglass Worm Gearing. This type of gearing derives its name from the hourglass shape of the worm (Fig. 5-17). The objective in making the worm conform to the shape of the gear is to obtain more contact surface between the mating gears than in the straight-worm type. In securing this advantage, however, manufacturing difficulties are greater, and the mounting requirements are more exacting than in straight worm gearing.

5-15. Standard Tooth Proportions for Bevel Gears. Standards have been developed by the Gleason Works, Rochester, N.Y., covering in a very comprehensive manner the tooth forms of straight-tooth bevel gears, spiral bevel gears, and hypoid gears. The teeth are of the unequal-addendum type and the gears are not interchangeable. This system, known as the *Gleason system for bevel gears*, applies to any pair of generated straight-tooth or spiral bevel gears operating at right angles where the pinion is the driver and has 10 or more teeth in the case of straight-tooth bevel gears and 5 or more teeth in the case of spiral bevel gears. The data covering that part of this system relating to straight-tooth bevel gears are given in Tables 5-1 and 5-2. Cutters also have been

TABLE 5-1. GLEASON WORKS SYSTEM FOR GENERATED STRAIGHT-TOOTH BEVEL GEARS*

(For straight-tooth bevel gears only, operating at right angles, where pinion is driver and has 10 or more teeth)

	Gear	Pinion
Addendum	$\dfrac{\text{Addendum for 1 DP (Table 5-2)}}{\text{DP}}$	$\dfrac{2.000 \text{ in.}}{\text{DP}}$ − addendum of gear
Dedendum	$\dfrac{2.188 \text{ in.}}{\text{DP}}$ − addendum of gear	$\dfrac{2.188 \text{ in.}}{\text{DP}}$ − addendum of pinion
Circular thickness:		
Ratios using 14½-deg pressure angle	$\dfrac{1.071 \text{ in.}}{\text{DP}}$ + (0.5 × addendum of gear) − $\dfrac{K \text{ (Table 5-2)}}{\text{DP}}$	$\dfrac{3.142 \text{ in.}}{\text{DP}}$ − circular thickness of gear
Ratios using 17½-deg pressure angle	$\dfrac{0.971 \text{ in.}}{\text{DP}}$ + (0.6 × addendum of gear) − $\dfrac{K}{\text{DP}}$	
Ratios using 20-deg pressure angle	$\dfrac{0.871 \text{ in.}}{\text{DP}}$ + (0.7 × addendum of gear) − $\dfrac{K}{\text{DP}}$	

Ratio	Pressure angle, deg
14 or more teeth in pinion	14½
13:13 to 13:24	17½
13:25 and higher	14½
12:12 and higher	17½
11:11 to 11:14	20
11:15 and higher	17½
10:10 and higher	20

$$\text{Working depth} = \frac{2.000 \text{ in.}}{\text{DP}} \qquad \text{Full depth} = \frac{2.188 \text{ in.}}{\text{DP}}$$

* The description in Art. 5-15 and data in Tables 5-1 and 5-2 represent the system as originally presented to and adopted by AGMA in 1922. Certain revisions were adopted in 1941 that relate principally to the selection of ratio and pressure angle that will eliminate interference, or undercut, in all cases.

made available for pressure angles other than the three shown. Similar tables of data are available for spiral bevel gears and hypoid gears, but space does not permit their inclusion here. In these tables the term *diametral pitch* is designated by the abbreviation DP. When this term is designated by a symbol in this text, the letter P is used (see Art. 4-7).

For *fine-pitch* (20 diametral pitch or finer) straight bevel gears, there is an American Standard similar to that already mentioned in Art. 4-26 for spur and helical hears. The pressure angle is 20 deg, the working depth is $2.000/P$, the whole depth is $2.188/P + 0.002$, and the system is of the unequal-addendum variety.

TABLE 5-2. GLEASON WORKS SYSTEM FOR STRAIGHT-TOOTH BEVEL GEARS
Addendum* for 1 diametral pitch

Ratios		Addendum, in.	Ratios		Addendum, in.
From	To		From	To	
1.00	1.00	1.000	1.42	1.45	0.760
1.00	1.02	0.990	1.45	1.48	0.750
1.02	1.03	0.980	1.48	1.52	0.740
1.03	1.04	0.970	1.52	1.56	0.730
1.04	1.05	0.960	1.56	1.60	0.720
1.05	1.06	0.950	1.60	1.65	0.710
1.06	1.08	0.940	1.65	1.70	0.700
1.08	1.09	0.930	1.70	1.76	0.690
1.09	1.11	0.920	1.76	1.82	0.680
1.11	1.12	0.910	1.82	1.89	0.670
1.12	1.14	0.900	1.89	1.97	0.660
1.14	1.15	0.890	1.97	2.06	0.650
1.15	1.17	0.880	2.06	2.16	0.640
1.17	1.91	0.870	2.16	2.27	0.630
1.19	1.21	0.860	2.27	2.41	0.620
1.21	1.23	0.850	2.41	2.58	0.610
1.23	1.25	0.840	2.58	2.78	0.600
1.25	1.27	0.830	2.78	3.05	0.590
1.27	1.29	0.820	3.05	3.41	0.580
1.29	1.31	0.810	3.41	3.94	0.570
1.31	1.33	0.800	3.94	4.82	0.560
1.33	1.36	0.790	4.82	6.81	0.550
1.36	1.39	0.780	6.81	∞	0.540
1.39	1.42	0.770			

* To obtain addendum select from table value corresponding to ratio given by the formula:

$$\text{Ratio} = \frac{\text{number of teeth in gear}}{\text{number of teeth in pinion}}$$

TABLE 5-2. GLEASON WORKS SYSTEM FOR STRAIGHT-TOOTH BEVEL GEARS (*Continued*)
Values† of K for circular thickness formula

No. of teeth in pinion / Ratios	Values of K for different ratios, in.								
	10	11	12	13	14	15–17	18–21	22–29	30 and up
1.00–1.25	0.025	0.010	0.000	0.000	0.000	0.000	0.000	0.000	0.000
1.25–1.50	0.070	0.015	0.040	0.015	0.015	0.000	0.000	0.000	0.000
1.50–1.75	0.100	0.050	0.070	0.040	0.030	0.010	0.000	0.000	0.000
1.75–2.00	0.120	0.080	0.100	0.045	0.050	0.020	0.000	0.000	0.000
2.00–2.25	0.140	0.105	0.120	0.050	0.065	0.030	0.010	0.010	0.010
2.25–2.50	0.160	0.125	0.140	0.060	0.080	0.045	0.030	0.030	0.025
2.50–2.75	0.175	0.145	0.155	0.070	0.090	0.060	0.045	0.040	0.035
2.75–3.00	0.190	0.160	0.170	0.080	0.100	0.070	0.060	0.050	0.040
3.00–3.25	0.205	0.170	0.180	0.090	0.110	0.080	0.070	0.060	0.045
3.25–3.50	0.215	0.180	0.185	0.100	0.120	0.090	0.080	0.065	0.050
3.50–3.75	0.225	0.190	0.190	0.110	0.125	0.095	0.085	0.070	0.055
3.75–4.00	0.230	0.195	0.195	0.120	0.130	0.100	0.090	0.070	0.060
4.00–4.50	0.240	0.200	0.205	0.135	0.140	0.110	0.095	0.080	0.065
4.50–5.00	0.250	0.210	0.210	0.150	0.150	0.115	0.100	0.085	0.070
5.00 and higher	0.255	0.220	0.215	0.165	0.160	0.120	0.100	0.085	0.070

† Select value corresponding to number of teeth in pinion and ratio given by formula.

For spiral bevel gears (and for helical gears) the pressure angle usually indicated is the *normal* pressure angle, measured in a plane normal to the spiral (or helix), and, for spiral bevels, at the mean cone distance unless otherwise specified. Normal pressure angles of 16, 16½, 17, 17½, 18½, 20, 22½, and 25 deg are in use.

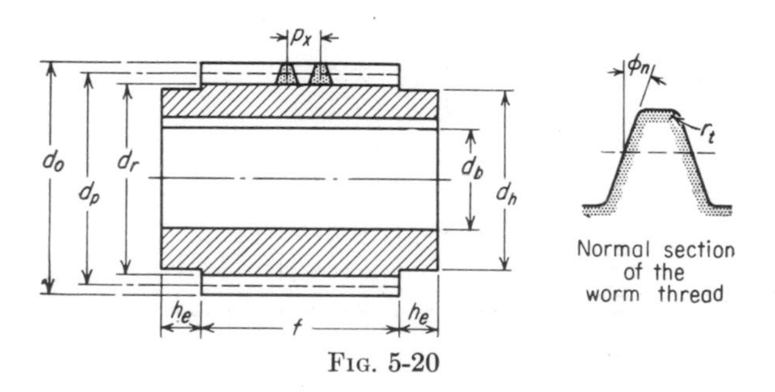

FIG. 5-20

5-16. Standard Proportions for Worm Gearing. The proportions recommended by the American Gear Manufacturer's Association for worm gearing of the type shown in Figs. 5-20 and 5-21 for general com-

TABLE 5-3.* PROPORTIONS OF STANDARD COMMERCIAL WORMS AND WORM WHEELS RECOMMENDED BY THE AMERICAN GEAR MANUFACTURER'S ASSOCIATION (GEAR RATIOS 10:1 TO 100:1)
(Refer to Figs. 5-20 and 5-21)

Terms for worm (Fig. 5-20):	Symbol	Single- and double-thread worms	Triple- and quadruple-thread worms
Axial pitch	p_x		
Pitch diam	d_p	$2.4p_x + 1.1$	$2.4p_x + 1.1$
Outside diam	d_o	$3.036p_x + 1.1$	$2.972p_x + 1.1$
Root diam	d_r	$1.664p_x + 1.1$	$1.726p_x + 1.1$
Hub diam	d_h	$1.664p_x + 1.0$	$1.726p_x + 1.0$
Bore (max.)	d_b	$p_x + 0.625$	$p_x + 0.625$
Face length	f	$p_x\left(4.5 + \frac{N}{50}\right)$	$p_x\left(4.5 + \frac{N}{50}\right)$
Hub extensions	h_e	p_x	p_x
Hub length	...	$f + 2p_x$	$f + 2p_x$
Keyway	...	AGMA Std.	AGMA Std.
Addendum	...	$0.318p_x$	$0.286p_x$
Whole depth	...	$0.686p_x$	$0.623p_x$
Lead angle	λ	$\cot \lambda = \frac{d_p \times 3.1416}{\text{lead}}$	$\cot \lambda = \frac{d_p \times 3.1416}{\text{lead}}$
Nor. tooth thickness	...	$0.5p_x \cos \lambda$	$0.5p_x \cos \lambda$
Top round	r_t	$0.05p_x$	$0.05p_x$
Lead	...	No. of thds. $\times p_x$	No. of thds. $\times p_x$
Nor. pressure angle	ϕ_n	14½ deg	20 deg

Terms for wheel (Fig. 5-21):	Symbol	Single- and double-thread worms	Triple- and quadruple-thread worms
Circular pitch	p		
No. of teeth	N		
Pitch diam	D_G	$N \times 0.3183p$	$N \times 0.3183p$
Throat diam	D_t	$D_G + (0.636p)$	$D_G + (0.572p)$
Outside diam	D_o	$D_t + (0.4775p)$	$D_t + (0.3183p)$
Hub diam	D_h	$1.875 \times$ bore	$1.875 \times$ bore
Keyway	...	AGMA Std.	AGMA Std.
Face width	F	$2.38p + 0.25$	$2.15p + 0.2$
Hub extensions	...	0.25 bore	$0.25 \times$ bore
Hub length	H	$F + 0.5 \times$ bore	$F + 0.5 \times$ bore
Radius of face	R_f	$0.882p + 0.55$	$0.914p + 0.55$
Radius of rim	R_r	$2.2p + 0.55$	$2.1p + 0.55$
Edge round	R_e	$0.25p$	$0.25p$
Center distance	C	$(D_G + d_p) \times 0.5$	$(D_G + d_p) \times 0.5$
Nor. pressure angle	ϕ_n	14½ deg	20 deg

* These proportions do not apply to shaft worms, gearing furnished with bearings or housings, or gearing of units.

mercial use are given in Table 5-3. The range covered by these proportions may be defined as follows:

Axial pitches, in.	$\frac{1}{4}$, $\frac{5}{16}$, $\frac{3}{8}$, $\frac{1}{2}$, $\frac{5}{8}$, $\frac{3}{4}$, 1, $1\frac{1}{4}$, $1\frac{1}{2}$, $1\frac{3}{4}$, 2
Multiplicity of threads	Single to quadruple
Gear ratios	10:1 to 100:1
Relation of axes	90 deg only

The *thread form* regarded as *standard* is the form produced by a straight-sided milling cutter having a diameter not less than the outside diameter of the worm or greater than $1\frac{1}{4}$ times the outside diameter of the worm, the sides of the cutter having a pressure angle of $14\frac{1}{2}$ deg for single- and double-thread worms and pressure angle of 20 deg for triple- and quadruple-thread worms.

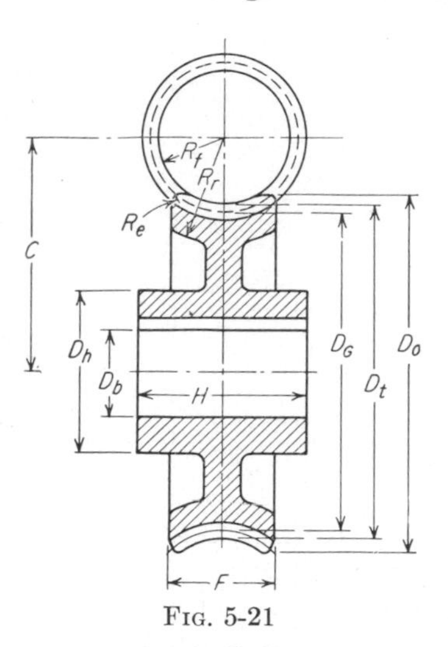

FIG. 5-21

The American Standard Design for Fine-pitch Worm Gearing provides for the small axial pitches of 0.030, 0.040, 0.050, 0.065, 0.080, 0.100, 0.130, and 0.160 in. A normal 20-deg pressure angle is specified for the cutting and grinding tools, and a range of helix angles from 0.5 to 30 deg is recommended. Standard worm pitch diameters lie in the range from 0.250 to 2.000 in.

5-17. Methods of Producing Gear Teeth. Gear teeth are produced commercially by two distinct processes: casting and maching. Formerly all gear teeth were cast,[1] the molds being formed from complete patterns of the gears. At the present time, however, cast teeth are generally produced by a machine-molding process which is superior to the earlier method. Even with the improved foundry methods, however, cast teeth are not suitable for high speeds and are found usually only in the cheaper classes of machinery.

Many small gears made of zinc, tin, aluminum, and copper alloys are die-cast, with resulting good accuracy and surface finish.

In the machining process two general methods are followed: *forming* and *generating*. In the forming method the teeth are cut by a rotary milling cutter or a reciprocating shaper cutter that has been formed to the exact shape required, by a planing tool which forms the outline required by following a previously shaped template, or by broaching.

5-18. Producing Gear Teeth by the Forming Process. The *rotary milling cutter* and its action are illustrated in Fig. 5-22 for a spur gear

[1] In ancient times wooden gears were used.

with straight teeth, and in Fig. 5-23 for a helical gear. The cutter is shaped so that the required tooth outlines are formed by setting the cutter to the proper depth and cutting the spaces between adjacent teeth. The standard rotary cutters are made by the Brown and Sharpe Manufacturing Company and may be obtained in sets to cover a wide range of work. This is one of the oldest and most widely used methods of producing cut teeth. It is extensively employed in the production of many of the gears that are required to meet only ordinary conditions.

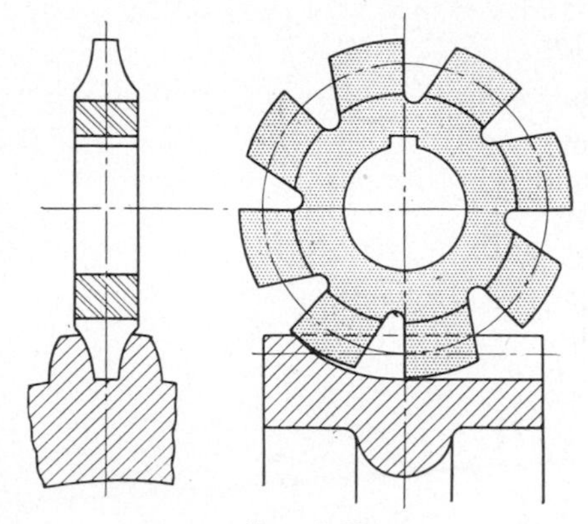

FIG. 5-22

The teeth of Citroën gears illustrated in Fig. 5-14*d* are cut with a rotating end-milling cutter formed to the normal contour of the teeth.

Stamping, which may be considered a process in which all the teeth of the gear are formed by a single stroke of the cutter (die), is used extensively in producing small thin gears such as are used in watches, clocks, instruments, etc.

FIG. 5-23. (*Courtesy of Brown and Sharpe Manufacturing Co.*)

The *template planing method* is shown in Fig. 5-24 as applied in cutting the teeth of bevel gears. The apex of the pitch cone of the bevel gear is located at O. OA is the axis of a roller in contact with a master form or template, called a *former*, whose shape corresponds exactly to an enlarged profile of the tooth to be cut. It can be seen that, as the roller swings about O and remains in contact with the *former*, the line OA will sweep up the required tooth surface. In the actual cutting process the point of the tool reciprocates along the line OA, the cutting action taking place as the tool travels toward the cone center O. In cutting the teeth of spur gears by this method, the point of the tool is constrained to move parallel to the axis of the gear, instead of being directed toward the apex of a cone.

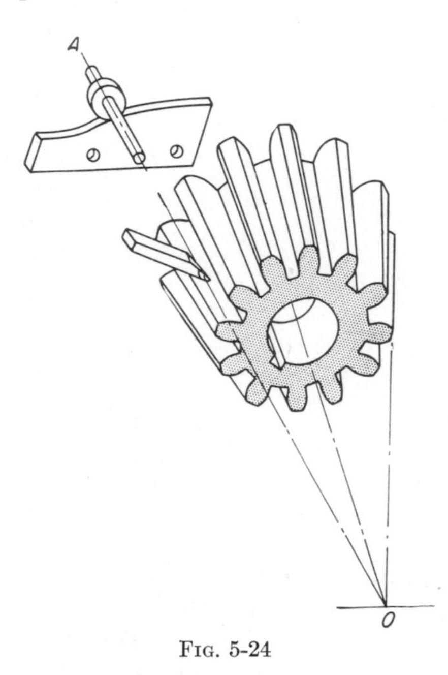

FIG. 5-24

In one type of *shaper* the formed cutters feed in all around the circumference as the gear blank reciprocates up and down.

In the broaching process, the finishing teeth of the broach are shaped so that they conform to the desired shape of the space between gear teeth. Both internal and external gears can be broached.

Gears also are made by extruding a rod and slicing it, and by the techniques of powder metallurgy.

5-19. Methods of Generating Spur- and Helical-gear Teeth. In the *generating method* a machine generates mathematically correct tooth profiles by virtue of the motions given to the cutter and the gear blank. The pitch circle of the cutter (which is shaped like a gear or a rack) is forced to roll with the pitch circle of the gear blank, as though two gears were in mesh, while the cutter also reciprocates or has some other motion which cuts teeth on the blank. The cut teeth, therefore, are *forced* to assume the proper shape. It might be noted that if the conditions of interference exist (see Art. 4-13) the generating cutter chops out the interfering portion of the blank. Thus *undercutting*, with consequent weakening of the tooth near its base, becomes the practical result of indicated interference.

Machines for generating gear teeth are of three distinct types: one employs a rotating worm-shaped cutter called a hob; another employs a reciprocating pinion-shaped cutter; and in the third type, a reciprocating rack-shaped cutter is used.

The *hobbing process* consists in rotating and advancing a worm-shaped cutter through a rotating blank. Figure 5-25 illustrates a hob in position for generating the teeth of a spur gear. The usual hob consists of a straight-sided worm-shaped cutter, its axial section representing the developed form of the generating rack. As the cutter and blank revolve

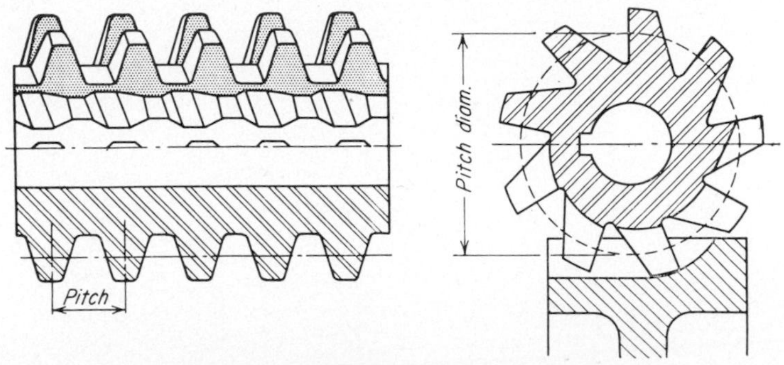

Fig. 5-25

Fig. 5-26. (*Courtesy of Barber-Coleman Company.*)

at the proper relative speeds, the hob presents a profile to the blank which appears to be a *rack* meshing with the desired teeth on the blank. The cutter teeth therefore cut the necessary spaces between teeth of the blank as they move into the blank.

Figure 5-26 is a photograph of the hobbing of a helical gear. Note that the hob is set at an angle to the gear being cut.

The hobbing process has the advantage of requiring only one hob to cut gears with any number of teeth of the same pitch. All motions are continuous and, except for the feeding of the hob through the blank, all motions are rotary. Hobbing is one of the most rapid methods of producing gear teeth.

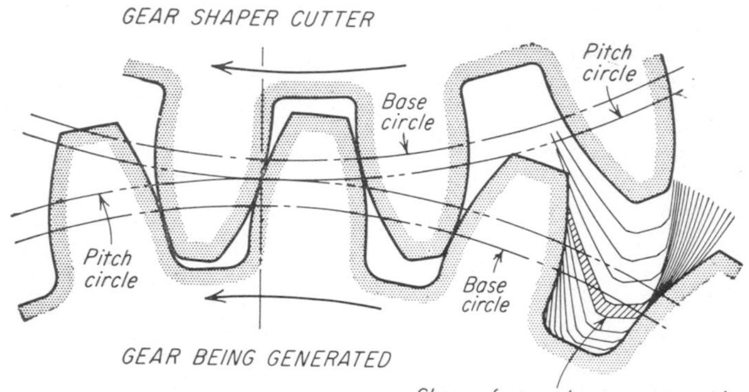

FIG. 5-27. (*Courtesy of The Fellows Gear Shaper Company.*)

FIG. 5-28. (*Courtesy of The Fellows Gear Shaper Company.*)

The Fellows gear shaper is the best-known machine employing the *pinion-shaped cutter*. The cutting process is indicated in Figs. 5-27 and 5-28. Both the cutter and the gear blank are rotated the proper amount between the strokes of the cutter, as though there were rolling of the pitch circles. In this machine the cutter is made exactly like a pinion that would engage with the gear about to be made. The cutter differs in one respect from the pinion in that its addendum is made long enough to provide clearance on the gear to be cut. The cutter is mounted on a vertical ram that has a reciprocating motion; and both cutter and work are made to rotate as if they were two similar gears in engagement. One cutter cuts gears with any number of teeth of the same pitch. This method is extensively used in the automotive industries in producing the well-known Fellows form of stub tooth.

The Maag machine is typical of the generating machines using the *rack-shaped cutter*. The cutting process is similar to that of the Fellows

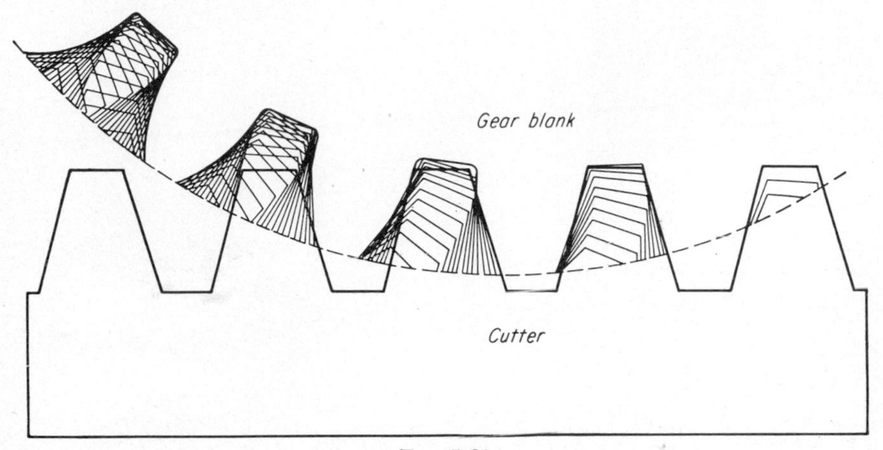

FIG. 5-29

gear shaper, and is shown in Fig. 5-29, as applied in generating the teeth of a spur gear. Between each reciprocating stroke of the cutter the blank is revolved and advanced the proper linear distance along the face of the rack cutter to correspond to rolling of the blank pitch circle on the rack pitch line. When the blank has advanced a distance equal to the circular pitch, it returns to its starting point without any rotary motion, and the cutting process begins again. This method of generating gears has the same advantage as hobbing and cutting with pinion-shaped cutters in that one cutter cuts gears with any number of teeth of the same pitch.

Any of the methods described above may be adapted to the cutting of either spur or helical gearing. No one method has all the advantages. Specific requirements determine which is best to use.

5-20. Methods of Generating Bevel-gear Teeth. The generating of bevel-gear teeth is greatly complicated by the fact that the tooth elements converge toward the apex of the pitch cone. The reciprocating-rack principle still can be employed, but instead of the cutting tool

Fig. 5-30. (*Courtesy of Gleason Works.*)

consisting of several rack teeth, it must be in the form of *one side* only of a rack tooth.

1. *Straight-tooth Bevel Gears.* The cutting processes employed in two types of Gleason bevel-gear generators are shown in Fig. 5-30. In both types, the cutters have motion *perpendicular* to the plane of the paper while simultaneously moving as shown *in* the plane of the paper as though they were gear teeth meshing with the desired teeth on the blank. The cutting edges of the tools represent the sides of the teeth of a crown gear, which is to the bevel-gear system what a rack is to the spur-gear system. As the tools rotate about the axis of this imaginary meshing crown gear they cut in the direction of the tooth elements. The gear to be cut is set so that the apex of its pitch cone coincides with the apex of the crown gear, and so that the pitch surfaces of both are tangent to each other. Means are then provided to rotate tools and gear about their axes, maintaining the same relative speed of rotation that they would have if they were finished gears properly in mesh. Since the tools are taking the place of the crown gear, they cut away the interfering metal on the gear tooth and thus leave the correct tooth profile.

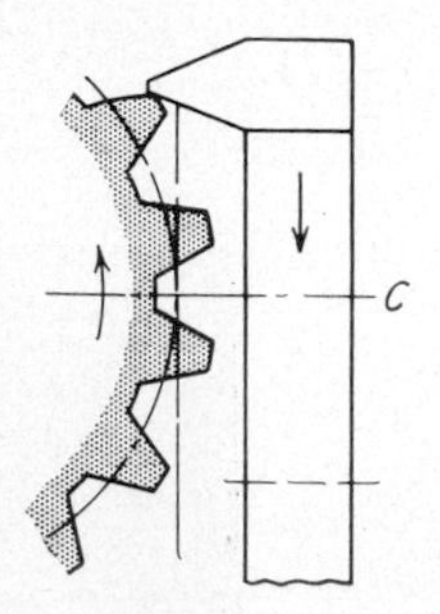

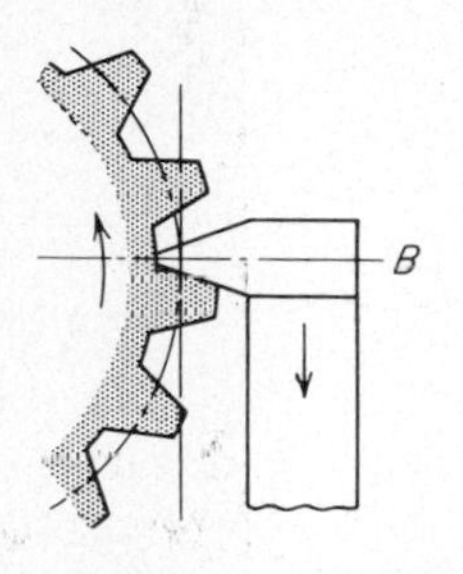

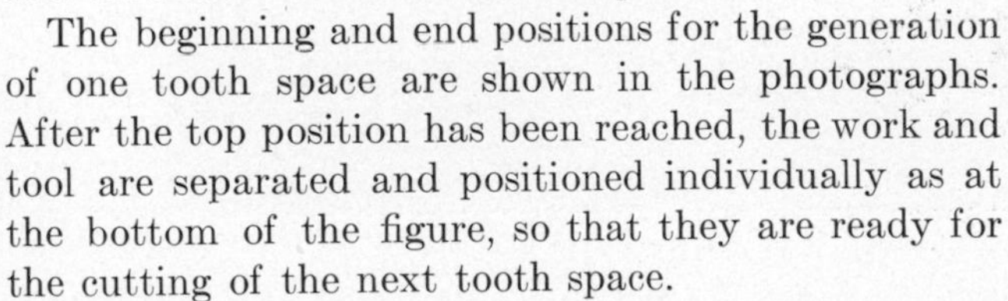

The beginning and end positions for the generation of one tooth space are shown in the photographs. After the top position has been reached, the work and tool are separated and positioned individually as at the bottom of the figure, so that they are ready for the cutting of the next tooth space.

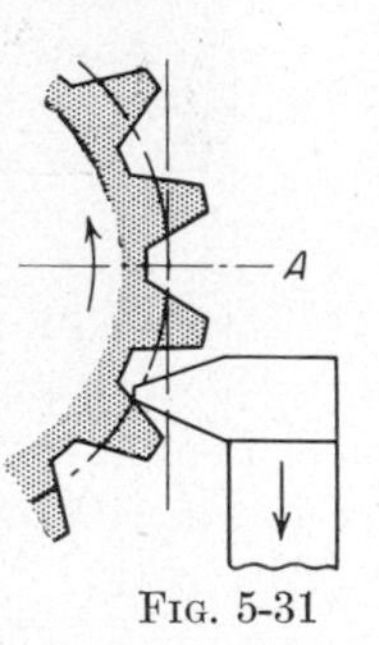

Fig. 5-31

The cutting motion perpendicular to the plane of the paper is a reciprocating one for the two tools on the right of Fig. 5-30. The tools on the left (the completing cutters) *rotate* about an axis somewhat to their right. There are *two* tools here, as there are also for the reciprocating cutters, one to cut each side of a tooth. Two tools are required in either scheme because the tooth space must converge toward the cone apex in bevel gears, as already mentioned.

An additional feature of the Gleason schemes shown is the provision for crowning the tooth automatically during the generation of the tooth. This is done by providing for motion which allows slightly more metal to be cut from the ends of the tooth than from the middle. The reasons for crowning were indicated briefly in Art. 4-27.

In gears of coarse pitch the gears are previously gashed, or roughed out. In gears of fine pitch, however, the teeth may be generated from the solid blank.

2. *Spiral Bevel Gears.* The principle of action of the Gleason spiral-bevel-gear generator is shown in Figs. 5-31 and 5-32. A circular cutter of the face-mill type is used with blades whose cutting edges are straight-sided to correspond to the tooth shape of a crown gear, which is to the

FIG. 5-32. (*Courtesy of Gleason Works.*)

bevel-gear system what a rack is to the spur-gear system. The teeth are generated by rolling the revolving cutter and the gear blank with the proper relative motions to conform to the meshing action of the desired gear tooth with the crown-gear tooth to which the cutter corresponds. The cutter, then, in order to mesh with the gear, must cut out the space between teeth of the gear.

5-21. Grinding, Shaving, Lapping, and Burnishing of Gear Teeth. With the increasing use of hardened gears and the demand for accurate tooth profiles and smooth surface finish with a minimum of imperfections, machines of various types have been developed for performing the finishing operations of grinding, shaving, lapping, and burnishing. Some grinding machines are of the form-grinding type, and others are of the generating-rack type. The machines that operate on the rack principle have the generating features common to the generating machines that employ cutting tools. The chief object in grinding gear teeth is to make corrections for the irregularities resulting from the distortion due to heat-treatment. Figure 5-33 is a diagram showing how the flat side of a revolving grinding wheel generates involute tooth curves as the gear is caused to roll back and forth along an imaginary rack.

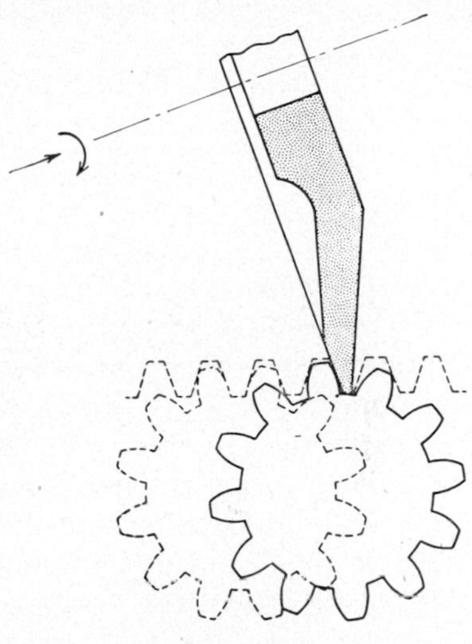

Fig. 5-33

Burnishing and lapping processes are also extensively employed in producing smooth and polished tooth surfaces to meet the exacting requirements of modern industrial gears. *Burnishing* consists of rotating a gear under pressure between hardened burnishing gears which are slightly oversize. This is done to refine the tooth form by plastic flow of the metal before hardening. In the *lapping* process, a lapping gear is run with the gear being finished, with an abrasive compound between them. Also, two gears which will run as a pair may be lapped together. Usually an axial reciprocating motion accompanies the rotation to produce uniformity of the surfaces.

A finishing process known as *gear shaving* is widely employed to correct the inaccuracies of gear teeth before the gear is hardened, thus either reducing to a minimum the stock that must be removed by the final grinding or lapping operation after heat-treatment or making these subsequent finishing operations unnecessary. The shaving tool resembles a gear, but with each tooth sliced by planes perpendicular to the tool axis. The edges of the slices are cutting edges. When run with a gear, the tool cutting edges have a slicing or shaving action across the tooth faces of the gear. Very fine hairlike chips are removed, and the profile is refined to the desired shape.

Gear grinding and shaving processes can incorporate actions which crown the tooth being finished, making the tooth slightly thicker at the center than at the ends. The advantages of such a form have been discussed in Art. 4-27.

CHAPTER 6

INTERMITTENT-MOTION MECHANISMS

In addition to the common agencies for transmitting motion, namely, linkwork, cams, gears, belts, ropes, and chains, there are numerous devices for controlling and regulating the motions of machine parts and adapting such motions to a great variety of purposes. By far the greatest number of these supplementary mechanisms consist of combinations such as ratchets used in feed and stop mechanisms, escapements for controlling clockwork, clutches and brakes for controlling motions between a pair of shafts; or the combinations may be of a more complex nature as in indexing and reversing mechanisms for automatic machines. Such combinations are usually classified as *intermittent-motion mechanisms*, and they will be discussed in this chapter under the following subdivisions:

1. Ratchets, escapements, intermittent gearing, and indexing devices; for producing rotary or rectilinear motion by steps.
2. Reversing mechanisms; for producing reversible running of trains of mechanisms.
3. Clutches and brakes; for stopping and starting parts having rotary motion.

6-1. Intermittent Movements. The action of automatic machinery is based upon intermittent movements so synchronized that one or more members perform a certain function and then remain at rest while other members perform their functions. From various combinations of the two main movements possible in a machine, namely, rotation and translation, all the complicated movements in machinery are composed. For example, in the case of rotary motion, any number of revolutions or any fractional part of a revolution may be followed by a period of rest equivalent to any number of revolutions or fractional part of a revolution. Combinations may be varied in any way, and a series of different motion periods followed by required rest periods.

6-2. Ratchet Mechanisms. A wheel provided with suitably shaped teeth, receiving an intermittent circular motion from an oscillating or reciprocating member, is called a ratchet wheel. A simple form of ratchet mechanism is shown in Fig. 6-1. A is the ratchet wheel, and B

is an oscillating lever carrying the driving pawl C. A supplementary pawl at D prevents backward motion of the wheel. When arm B moves counterclockwise, the pawl C will force the wheel through a fractional part of a revolution dependent upon the motion of B. When the arm moves back (clockwise), the pawl C will slide over the points of the teeth while the wheel remains at rest because of the fixed pawl D, and will be ready to push the wheel on its forward (counterclockwise) motion as before. The amount of backward motion possible varies with the pitch of the teeth. This motion could be reduced by using small teeth, but this procedure would result in weak teeth, and the expedient is sometimes adopted of placing several pawls side by side on the same axis, the pawls being of different lengths. The contact surfaces of wheel and pawl should be so inclined that they will not tend to disengage under pressure.

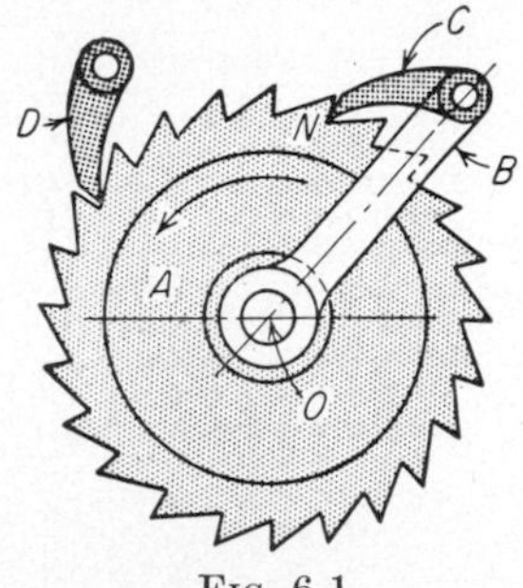

FIG. 6-1

FIG. 6-2. (*Courtesy of Stephens-Adamson Manufacturing Co.*)

As shown in Fig. 6-1, this means that the common normal at N should pass between the pawl and ratchet-wheel centers. If this common normal should pass outside these limits, the pawl would be forced out of contact under load unless held by friction. In many ratchet mechanisms the pawl is held against the wheel during motion by the action of a spring.

The usual form of the teeth of a ratchet wheel is that shown in Fig. 6-1, but in feed mechanisms such as used on many machine tools it is necessary to modify the tooth shape for a reversible pawl so that the drive can be in either direction.

Ratchet mechanisms occur in a great variety of forms and find a wide range of application. They may be seen in ratchet drills, lifting jacks, feed and stop mechanisms in machine tools, capstans, windlasses, typewriters, etc. Figure 6-2 illustrates a ratchet device used on a pan-feeder conveyor.

6-3. Escapements. Under escapements may be grouped a number of self-acting stopping and releasing ratchet mechanisms in which the driven link is alternately released and stopped. The most familiar examples are found in clockwork and timing devices where the driving weight or spring is permitted to move the parts and, therefore, the hands by a definite amount at regular intervals.

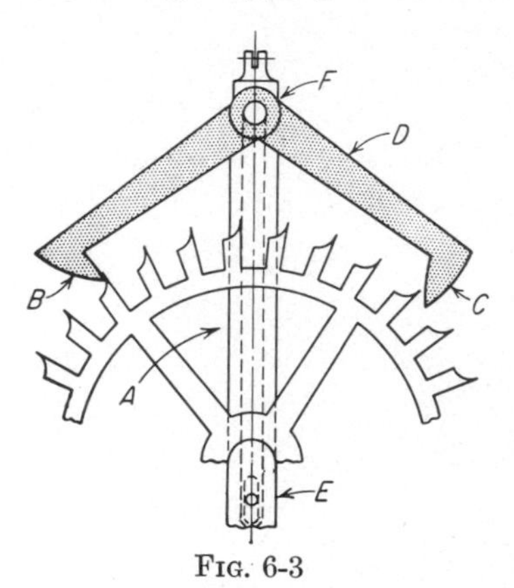

FIG. 6-3

One common form of escapement such as used in clocks is shown in Fig. 6-3. The escape wheel A turns intermittently in the direction indicated by the arrow and is supplied with long pointed teeth acting upon the pawl ends B and C, which are known as pallets. Spring (or other) torque is supplied to wheel A to maintain its motion. The double pawl D oscillates, alternately allowing and stopping the rotation of A at definite intervals. The timing of these intervals is governed by the oscillating pendulum E, which is attached to the double pawl D at F. The pendulum has a normal period of vibration which depends upon its length, and if the pendulum is so heavy that the rotative effort caused by the spring torque on the escape wheel cannot alter this period, the pendulum in swinging will control the motion of the wheel. The force exerted by the spring on the wheel and then by the wheel on the pawl can be made sufficient to overcome the frictional resistance which acts to stop the pendulum, and thus the amplitude of the vibrations can be maintained. Many modifications of the escapement have been devised to meet special requirements. The problem of design is an intricate one requiring the skill of specialists. In watches, chronometers, etc., a balance wheel is used instead of a pendulum to regulate the period of the vibrating member, but all these escapements operate on the same general principle.

6-4. Intermittent Gearing. A pair of rotating members may be so designed that, for continuous rotation of the driver, the follower will alternately roll with the driver and remain stationary. Such an arrangement is known by the general term *intermittent gearing.* This type of gearing occurs in some variety in counting mechanisms, motion-picture machines, feed mechanisms, etc.

1. *The Simplest Case.* The simplest form of intermittent gearing, illustrated in Fig. 6-4, has the same kind of teeth as ordinary gears designed for continuous motion. This example is merely a modification of a pair of 20-tooth gears to meet the requirement that the follower shall advance one-tenth of a turn for each turn of the driver. The interval of action is the two-pitch angle α (indicated on both gears), the single tooth on the driver engaging with each space on the follower to produce the required motion of one-tenth turn of the follower. During the remainder of a driver turn, the follower is locked against rotation in the manner shown in the figure. In order to vary the relative movements of the driver and follower, the meshing teeth may be arranged in various ways, to suit requirements. For example, the driver may have more than one tooth, and the periods of rest of the follower may be uniform or may vary considerably. Counting mechanisms are often equipped with gearing of this general type.

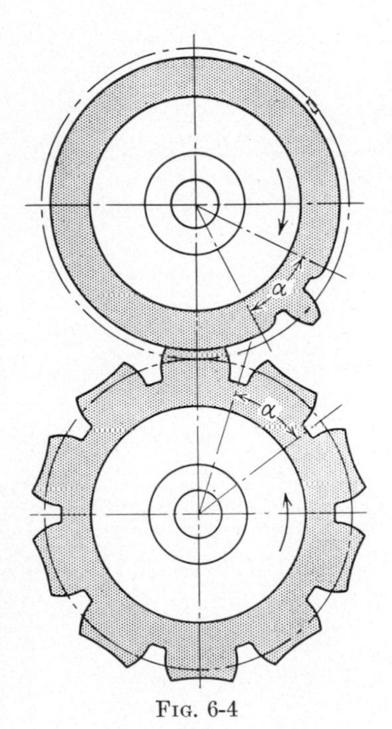

Fig. 6-4

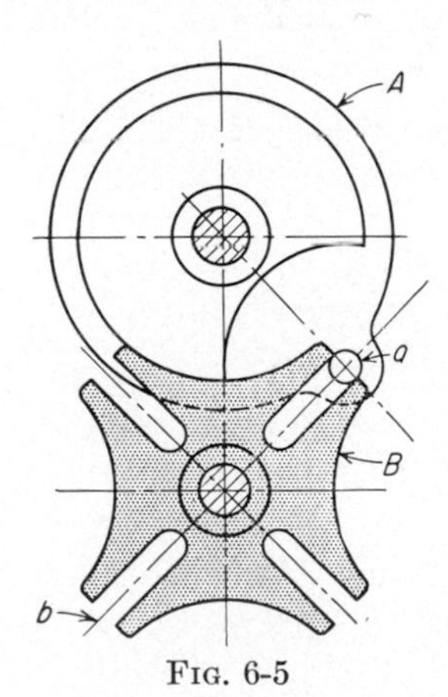

Fig. 6-5

2. *The Geneva Wheel.* An interesting example of intermittent gearing is the *Geneva wheel* combination shown in Fig. 6-5. In this particular case the driven wheel B makes one-fourth of a turn for one turn of the driver A, the pin a working in the slots b causing the motion of B. The circular portion of the driver, coming in contact with corresponding hollow circular parts of the driven wheel, retains it in position when the

pin or tooth a is out of action. The wheel A is cut away near the pin a as shown, in order to provide clearance for wheel B in its motion. If one of the slots is closed, A can make only a little more than $3\frac{1}{2}$ revolutions in either direction before pin a strikes the closed slot and thus stops the motion. The device in this modified form was early used in watches, music boxes, etc., to prevent overwinding, and from this application received the name *Geneva stop*. In the arrangement as a stop, wheel A is secured to the spring shaft, and B turns on the axis of the spring barrel. The number of slots or interval units in B depends upon the number of times it is desired to turn the spring shaft.

As applied in some types of motion-picture machines, the wheel B has four slots, so that it is given a quarter turn for every revolution of wheel A. Wheel B is secured to the sprocket that meshes with the openings in the edge of the film and jerks it down before the lens, picture by picture, as the pin in wheel A pulls B around one-quarter of a turn.

The Geneva wheel is sometimes used on machine tools for indexing or rotating some part through a fractional part of a revolution.

6-5. Indexing Devices. By indexing, as a shop term, is meant the operation of exactly dividing the periphery of a circular piece into a number of equal parts, or what is the equivalent, one turn of a wheel, or 360 deg, into a number of equal angular divisions. Mechanisms in considerable variety have been devised to produce this result, many of them concerned with the production of toothed gearing. Indexing is mostly done on milling machines and gear-cutting machines. The indexing mechanisms on gear-cutting machines are usually more or less complex. Indexing in its simplest form is found on the milling machine where, for general service, the well-known type of index head, also called dividing head, which forms a part of the standard equipment of the machine, is used.

1. *The Index Head.* Index heads vary considerably in design but all embody the essential features of the mechanism of the plain index head outlined in Fig. 6-6. Spindle A is carried in a headstock with which comes a companion tailstock. Work can then be held either between centers or in a chuck attached to the spindle. On the spindle is a 40-tooth worm wheel B driven by a single-threaded right-hand worm C. On the end of the worm shaft and located at the side of the head is the adjustable index crank D with handle E and plunger pin F which can be let into any hole in the index plate G. This plate, which, with sector H, is shown in more detail in Fig. 6-7, carries the essential dividing element of the apparatus in the form of a number of circles of equally spaced holes.

Each index head is provided with several interchangeable index plates in order to cover a wide range of work. The plate shown in Fig. 6-7 has circles of 15, 16, 17, 18, 19, and 20 equally spaced holes, but for the sake of clearness in illustrating the use of the plate, only the circle of 17 holes is shown.

As 40 turns of the crank are required to make one turn of the work, since the work turns with the spindle, in order to find how many turns of the crank are necessary for a certain division of the work, 40 is divided by the number of divisions desired. The rule then may be stated as follows: *Divide* 40 *by the number of divisions desired, and the quotient will be the number of turns, or part of a turn, of the crank which will give each desired division.* Applying the rule, to make 40 divisions of the work the crank would be given one complete turn to obtain each division.

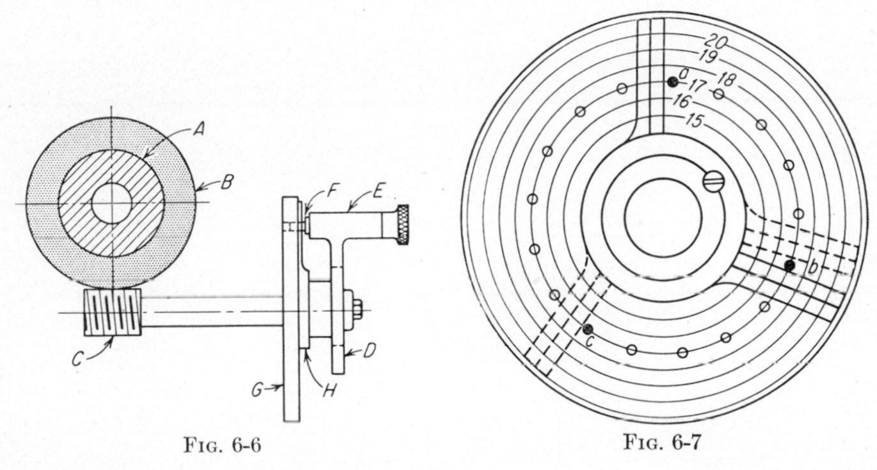

FIG. 6-6 FIG. 6-7

To obtain 20 divisions the crank would be given two complete turns for each division.

2. *Function of the Index Plate.* The index plate becomes necessary when the crank has to be turned only a part of the way around to obtain the necessary divisions. For example, if the work is to be given 120 divisions the crank must be turned one-third the way around, and an index plate should be selected that has a circle with a number of holes divisible by three, as 15 or 18 (Fig. 6-7).

3. *Function of the Index Sector.* The index sector is of service in obviating the necessity of counting the holes at each partial turn of the crank. The sector has two arms that can be set and clamped together at any angle. To illustrate the use of the sector it will be assumed that it is desired to divide the work into 136 parts. Applying the rule, 40 divided by 136 gives $5/17$, showing that the crank must be rotated five-seventeenths of a turn to obtain each of the 136 divisions on the work. An index plate with a circle of 17 holes is selected (Fig. 6-7), and the sector is set to measure off 5 spaces. Starting with the crankpin in hole *a*, for example, a cut would be made in the work and the crank

would then be turned to bring the pin into the hole b and a second cut made in the work. The sector would then be turned to the position shown by dotted lines. After the second cut had been made, the crank would again be turned to bring the pin into hole c, and so on until the operation had been repeated 136 times. The operator of the machine is provided with index tables so that for any given case the number of

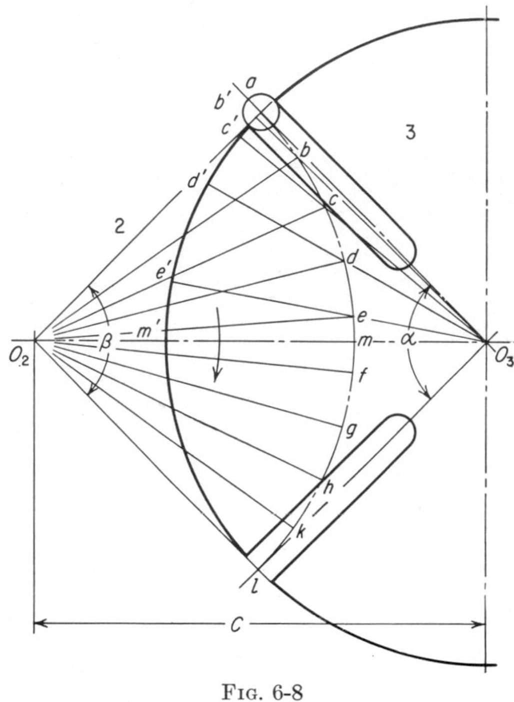

Fig. 6-8

turns of the crank and number of holes in the circle may be readily ascertained.

For rapid indexing in simple numbers, as in milling four- or six-fluted reamers, provision is usually made for indexing directly through a plate on the spindle A itself. This plate has circles commonly of 24, 30, or 36 holes and indexes any number dividing directly into these numbers. To change from one indexing method to the other, a cam or eccentric throw-out is provided for disengaging the worm. With the worm out of mesh with the worm wheel, the spindle is then indexed by hand.

Index heads can be used not only on milling machines, for which they are primarily designed, but also on shapers, planers, and any other machine tool where there is work that must be indexed.

4. *The Geneva Wheel as an Indexing Device.* The process of indexing involves the starting of a mass from rest, moving it through a fixed

distance, bringing it back to rest, and accurately locating it. Regardless of the magnitude of the mass the starting and stopping are accompanied by a shock. In the design of certain kinds of automatic machinery where provision must be made for indexing the work or the tools it is necessary to reduce this shock as much as possible, and good results have been achieved by the use of the Geneva wheel.

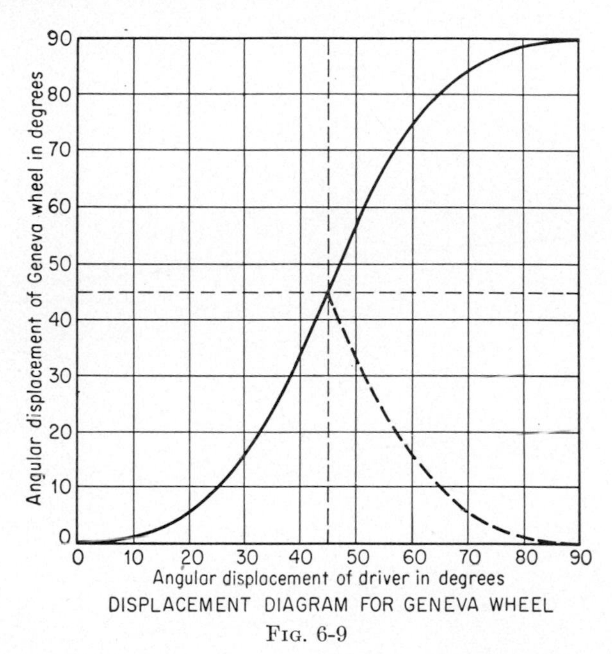

FIG. 6-9

The mechanism consists of a driving and a driven member arranged so that the driving member, rotating uniformly about its axis, carries a roller that engages a slot in the driven member and causes the latter to rotate with a variable angular motion until the roller leaves the slot. The driven member then remains stationary until the driving member has rotated to a position where engagement is made with the next slot.

In Fig. 6-8, the roller shown entering the slot at a moves with uniform angular motion about the center O_2 of the driver in the direction indicated, leaves the slot at l, and continues its motion until point a is again reached where it is then ready to engage the next slot. In the meantime the first slot of the Geneva wheel has moved about its center of rotation O_3 through the angle aO_3l, when it again comes to rest.

It will be seen from the figure that, since the driving arm O_2a is of constant length, it forces the roller moving in the slot of the driven member toward the center O_3 of the driven member as the latter rotates about that center. The effective arm O_3a will therefore decrease to O_3m

as its minimum length. At this point the driven member 3 is rotating at its maximum velocity. The driven member 3 has a gradual motion at the start as shown by the travel of the roller from *a* to *b* while the point *a* of the wheel moves from *a* to *b*′. Acceleration of the wheel 3 is positive (velocity is increased) until the roller is on the center line O_2O_3 at *m*, after which the acceleration is negative until the roller leaves the slot at *l*, where the wheel comes to rest.

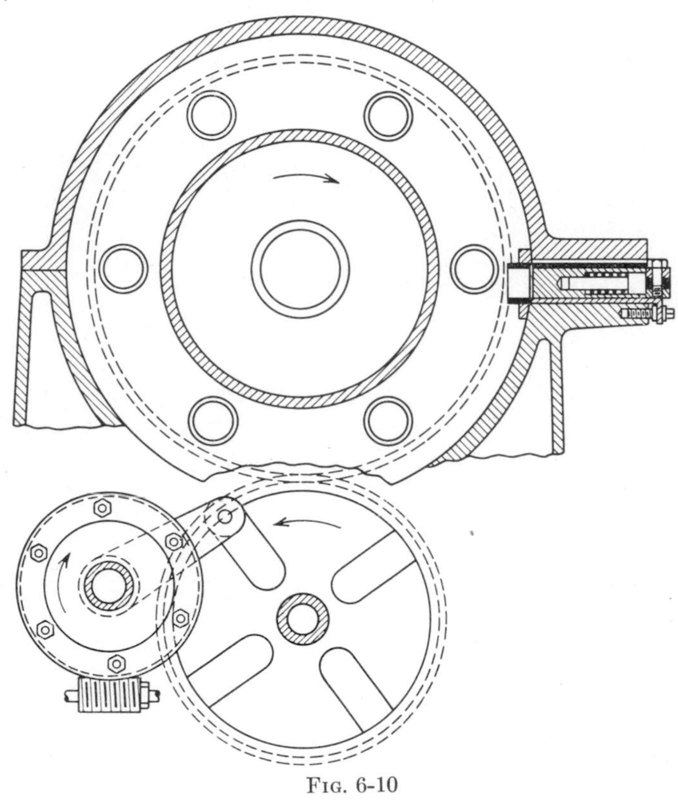

FIG. 6-10

The relative angular motions of the driving and the driven member are clearly shown by the angular-displacement diagram in Fig. 6-9. If the second half of the motion of the driven member is plotted as shown by the dotted line in the figure, then the ordinates of the curves in the lower half of the figure will represent, to a suitable scale, the angular velocities of the driven member corresponding to the constant angular velocity of the driving member at any given positions indicated by the abscissas.

With the mass characteristics of the driven member known, and means for finding its velocity in any phase of its motion, the accelerations and consequently the forces acting may be determined by the methods outlined in Part II of this book.

In laying out a Geneva wheel it is necessary to know the diameter and number of slots. The procedure then is to draw the circle of radius O_3a (Fig. 6-8) and lay off the angle $\alpha = 360/N$, where N is the number of slots. Next the angle α is bisected by the line O_3O_2, and through point a the line aO_2 is drawn tangent to the circle, making an angle $\beta/2$ with the center line O_3O_2. The length of driving arm O_2a and center distance C between driver and follower are thus determined. If greater accuracy is required than can be obtained by the graphical method, these distances can easily be calculated.

In Fig. 6-10 is shown an indexing mechanism in which it was found necessary to introduce a geared connection between the Geneva wheel and the member to be indexed. A locking device, to prevent any rotation of the last driven member while the driving member is moving around to the position of engagement with the next slot, is shown in the figure.

6-6. Reversing Mechanisms. A reversing mechanism is an arrangement of machine parts whose function is to produce reversible motion of sliding or rotating members or entire trains of mechanism. The types of reversing mechanisms in common use vary greatly, both as to principle of operation and as to form. Some are hand controlled, but by far the greater number are operated automatically. Some must be so designed that the reversal of motion occurs at a fixed point in a cycle of carefully timed movements, and with others the point of reversal can be changed by means of adjustable dogs or trips that are attached to a moving part and control the action of the reversing mechanism.

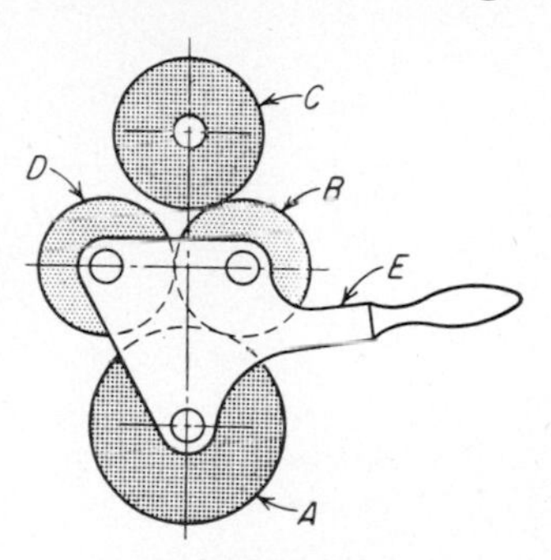

FIG. 6-11

1. *The Idler-gear Arrangement.* One of the simplest methods for obtaining reversal of motion in a gear train provides for the insertion of an additional gear in the train. In Fig. 6-11, the two intermediate gears B and D are carried on a swinging arm E pivoted on the axis of gear A. The arm E can be adjusted for engaging either of the intermediate gears with gear C. In the position shown, the drive is from C through B to A. When the arm is shifted to the other extreme position, the drive is through both intermediate gears, or from C through D and B to A, thus reversing the direction of rotation of A.

2. *The Planetary-gear Arrangement.* Reversal of motion may be obtained by means of a planetary gear train, as illustrated diagrammatically in Fig. 6-12. This figure shows the principle of operation of the transmission gearing used in some designs of automobiles but does not show all the gearing.

The gear a that is keyed to the crankshaft is the driver in each case. At a is shown the combination for obtaining the slow forward speed. The internal gear c is held stationary by the application of a brake band to its periphery. The pinions b carried by the driven member are then forced by the driving gear a to roll around inside the internal gear c, thus transmitting a slow rotating motion to the driven member attached to the pinions.

The reversal of motion is shown by the combination at b. The member carrying the pinions is prevented from rotating by the application of another brake band, so that the pinions are forced to rotate on their

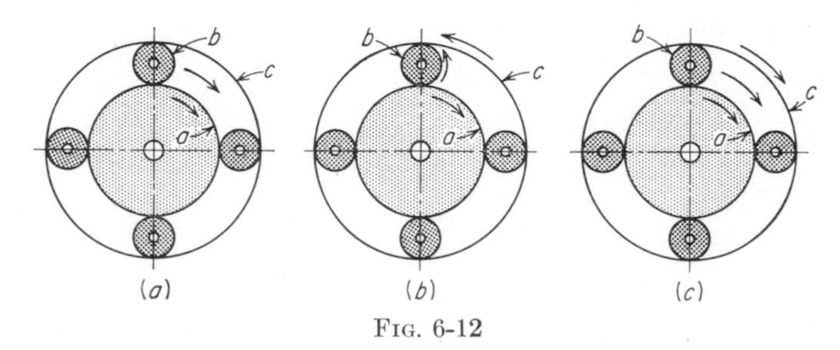

FIG. 6-12

studs and thus produce rotation of the internal gear c. In this case the internal gear that is attached to a driving sprocket becomes the driven member.

For obtaining the direct high-speed drive the gears are all locked together and revolve as a unit with the crankshaft, as shown by the combination at c. The relative motions of the gears of each combination are clearly indicated by the arrows. The analysis for determining the speed ratios in such a system of gears is presented in Chap. 7.

3. *The Bevel-gear Device.* A bevel-gear type of reversing mechanism is illustrated diagrammatically in Fig. 6-13. This arrangement is found in many classes of mechanisms in which the reversing action is automatic. Gear a, which is usually the driver, is constantly in mesh with the pinions b and c. The pinions are loose on the driven shaft and have a toothed clutch d interposed between them. The clutch is free to slide axially on a feather key that forces it to revolve with the shaft. Each pinion is provided with a clutch end so that its engagement with the clutch compels it to revolve with the shaft. Since the pinions rotate in opposite directions, the rotation of the driven shaft is reversed as the clutch is shifted from one pinion to the other. No motion is transmitted to the driven shaft when the clutch is in the neutral position shown in the figure. The clutch is shifted by means of the yoke e, which may receive its motion in various ways.

4. *The Use of Open and Crossed Belts.* Reversal of rotation may be accomplished by the use of open and crossed belts. One familiar device for reversing the motion of the drive pulley of a machine is outlined in Fig. 6-14. This is the older form of drive that has been much used for

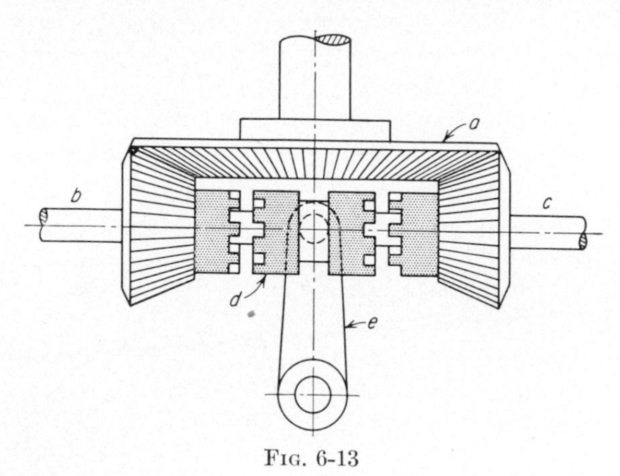

Fig. 6-13

an ordinary engine lathe. Power is taken from a continually running line shaft A by means of open and crossed belts B and C which drive countershaft pulleys D and E in opposite directions, thus reversing the rotation of the machine drive pulley F. Provision is made for starting

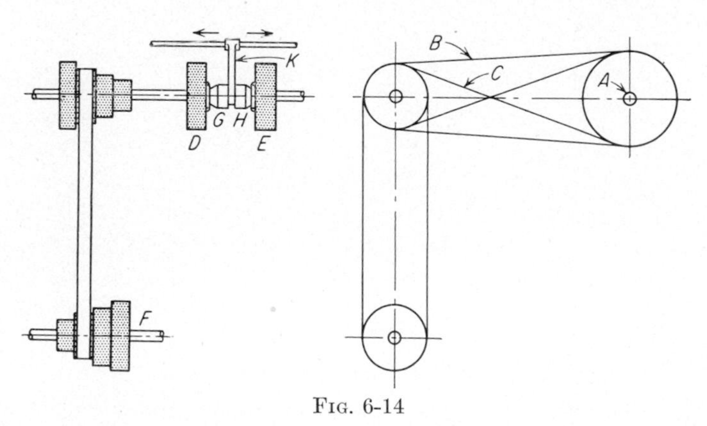

Fig. 6-14

and stopping and for reversing the machine by means of friction clutches G and H connected with each pulley. The shifting yoke K is usually operated by a hand-controlled lever. The older countershafts for many other machine tools requiring a reversal of motion have been commonly arranged in this manner.

CLUTCHES AND BRAKES

6-7. Clutches and Clutch Mechanisms. A clutch is a form of connection between a driving and a driven member on the same axis, so designed that the two members may be engaged or disengaged at will either by a hand-operated device or automatically by the action of some power-driven device. Several distinct types of clutches are in common use, and they are made in a great variety of designs. The common types of clutches may be divided into two general classes, namely, the positive clutch and the friction clutch. The positive clutch is simpler and more powerful and is used when it is not objectionable to start the driven member suddenly and where the resistance to motion is not so great as to cause injurious shock when the clutch engages suddenly. Since a friction clutch can be engaged gradually and will slip while the driven member is being accelerated, it can be used successfully with great differences in speed.

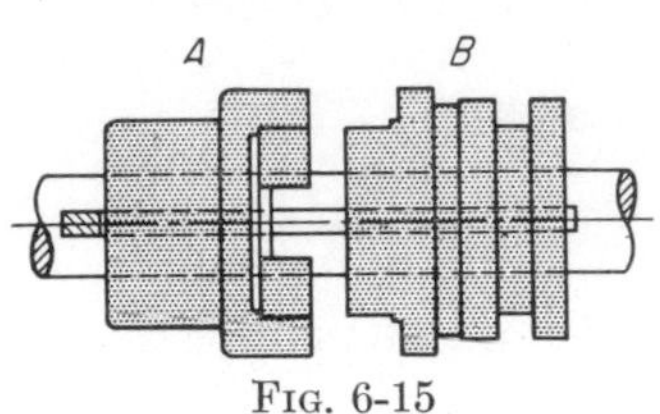

FIG. 6-15

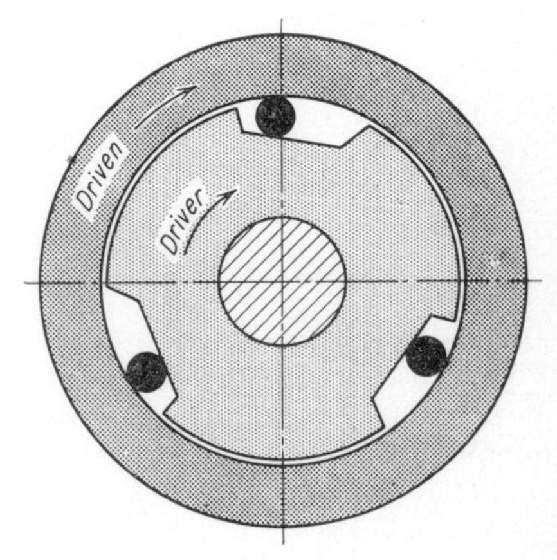

FIG. 6-16

1. *Positive Clutches.* A simple form of *jaw clutch* is shown in Fig. 6-15. Part A is keyed to its shaft, and part B is splined so that it can be moved axially by a shifting device that engages it by means of a groove in the flange as shown. Several forms of this clutch differ mainly in regard to the shapes of teeth or angle of the engaging surfaces.

Figure 6-16 illustrates a type of *overrunning clutch.* This device might also be considered to be a form of ratchet. As the driver delivers torque to the driven member, the rollers or balls are wedged into the tapered recesses, giving positive drive. Should the driven member attempt to drive the driver in the directions shown, the rollers or balls become free and no torque is transmitted.

This type of device has been incorporated in automotive overdrives and "free-wheeling" units.

2. *Friction Clutches.* The different types of friction clutches vary mainly in regard to the form of the friction surfaces and with respect to the kinds of material used to obtain sufficient frictional resistance. The

frictional surfaces may be *conical* or *cylindrical*, or in the form of one or more *flat rings* or *disks*. A simple form of conical clutch is shown in Fig. 6-17. Motion is transmitted from the driving member A to the driven member B by the frictional resistance of the conical surfaces.

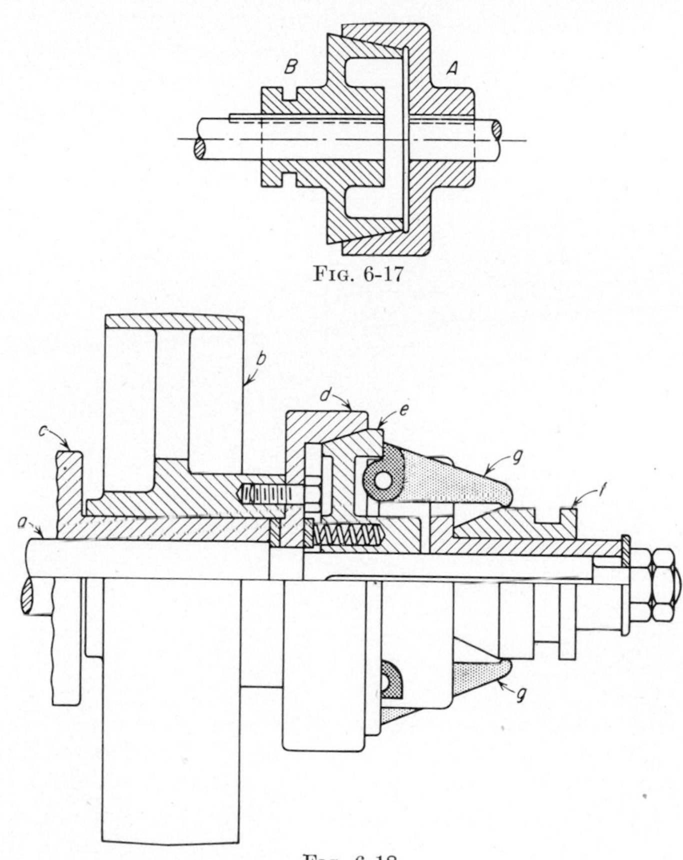

Fig. 6-17

Fig. 6-18

3. *Clutch-operating Mechanisms.* Since friction clutches require the development of heavy pressure just as they close up, the toggle principle (see Chap. 2) is very generally employed in the operating mechanism of the clutch. In some types of friction clutch, the members are maintained in engagement by means of springs which are compressed in order to release the clutch. Where toggle mechanisms are employed, springs are frequently used to separate the friction members and thus eliminate any tendency to drag when the clutch is open.

An example of a conical clutch and its operating mechanism as applied to a machine tool is shown in Fig. 6-18. The driving pulley b runs loose on the hub of the bearing c, and fastened to the pulley is the cast-iron cone d. The sliding cone e, which engages with cone d, is keyed to the driving shaft a by means of a feather key. The cones are engaged by means of the sliding member f and the levers g. Several helical springs, one of which is shown in the figure, are placed in the hub of the cone e for the purpose of disengaging the two cones when the sliding member f is shifted so as to relieve the pressure between the cones.

In the type of clutch shown in Fig. 6-19, the operating mechanism consists of an assemblage of several vise arms, d and e, each double

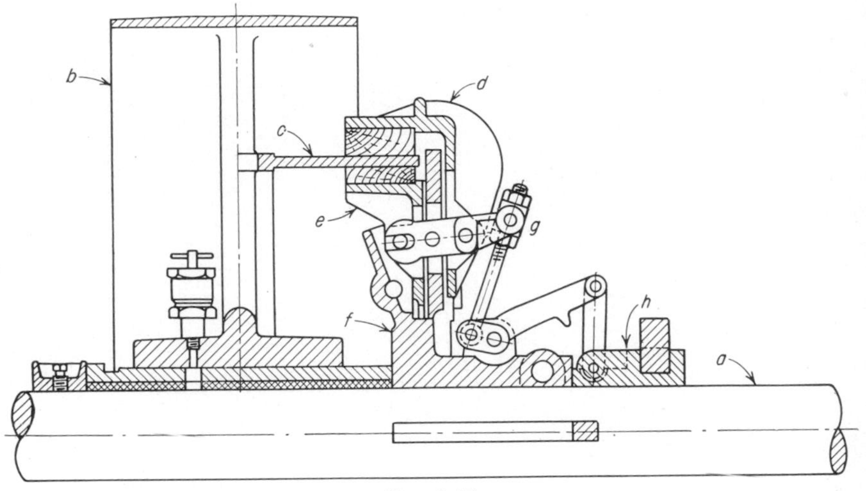

FIG. 6-19

gripping an annular ring c integral with the driving pulley b, which runs loose when not in engagement with the clutch. By means of the toggle mechanism at g, the vise arms d and e, which are lined with wood, asbestos fiber, or other friction material, are forced to grip the annular ring from both sides. The shaft a is driven by means of the member f, which is keyed to it. The clutch is thrown in and out of engagement by means of the sliding member h.

A clutch of the disk type commonly used in automobiles is depicted diagrammatically in Fig. 6-20. As the clutch pedal is depressed, the sleeve and other parts move in the directions indicated by the arrows, freeing the clutch disk. The transmission shaft then can remain motionless while the engine crankshaft continues to revolve. Releasing the clutch pedal allows the compression springs to force the pressure plate to the right, squeezing the clutch disk between the pressure plate and the flywheel disk. The frictional forces at the contact surfaces now transmit torque from the engine crankshaft to the transmission shaft. The clutch disk is faced on both sides by suitable friction material such as asbestos.

6-8. Brakes. In their action, brakes are very closely related to friction clutches, and in some starting and stopping mechanisms the clutch and brake principles are used in combination. The study of clutches and brakes more properly belongs in books on machine design, and, therefore,

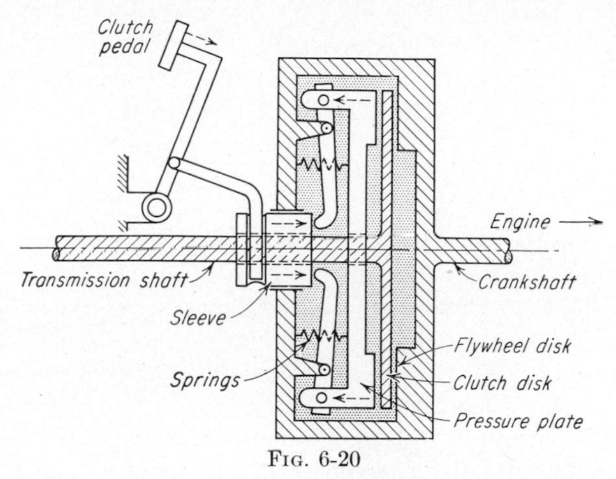

FIG. 6-20

no attempt will be made to go further than the few illustrations already shown, which have been given merely for introductory purposes as examples of intermittent-motion devices.

A vast number of ingenious devices of kinematic interest could properly be included under the heading of this chapter, but the object has been merely to call attention to this important class of mechanisms by a few simple examples rather than to attempt a thorough treatment. An extended discussion of such mechanisms may be found in several well-known treatises on mechanism.

CHAPTER 7

TRAINS OF MECHANISM

Since the primary purpose of a machine is to modify energy, all machines must have as constituent parts a mechanism or a series of mechanisms between the point where the power is received and the point where the useful work is delivered. Similarly, mechanisms, whether they form constituent parts of machines or are employed merely in their primary function of modifying motion, are made up of combinations of units arranged in trains. Such combinations of units are called *trains of mechanism*. These trains may consist of a great variety of constituents: linkwork, cams, gears, chains, belts, ropes, etc., any or all.

When the distance between shafts is comparatively great, rope, belt, or chain connections may be used. When the distance between shafts is comparatively small and positive transmission is essential, gears are used. When this last condition is not a requisite, and the distance between shafts is too small to use belts or chains advantageously, frictional gears are occasionally employed. It is possible with a train of mechanism to secure any desired result as to plane and direction of the ultimate motion, and as to variety of motion, whether rotary, reciprocating, constant, intermittent or irregular, etc. Trains of mechanism are common to all classes of machinery, and in connecting the source of power to the useful work, they must provide separately or in combination mechanical advantage, a definite velocity ratio, flexibility as to speed ratio, and compactness.

7-1. Typical Examples. A typical example of a comparatively simple train of mechanism is found in the ordinary metal-cutting lathe of the older style. The mechanism of this lathe is shown in diagrammatic form in Fig. 7-1. The main train at A drives the spindle S, which carries the workpiece and rotates it against the cutting resistance of the tool. The first secondary train B, an end view of which is also shown, drives the lead screw; this screw engages a split nut on the tool carriage or saddle, so as to give the latter a smooth and steady motion along the *ways* for screw cutting. The belt at C connects L and M, and the gears at D drive the feed rod, which by means of gear trains E and F can give a longitudinal feed to the whole saddle or a cross-feed to the slide resting on the saddle.

The cone pulley *a*, with pinion *b* fastened to it, is the first member of the main train. This combined pulley and gear is driven by a belt from an overhead countershaft and turns freely on the spindle to which the gear *e* is fastened. Through the back-gear train *b-c-d-e*, a slow and powerful movement is given to the spindle. For high-speed work the back gears *c* and *d*, which are fastened to a hollow shaft *G*, are thrown

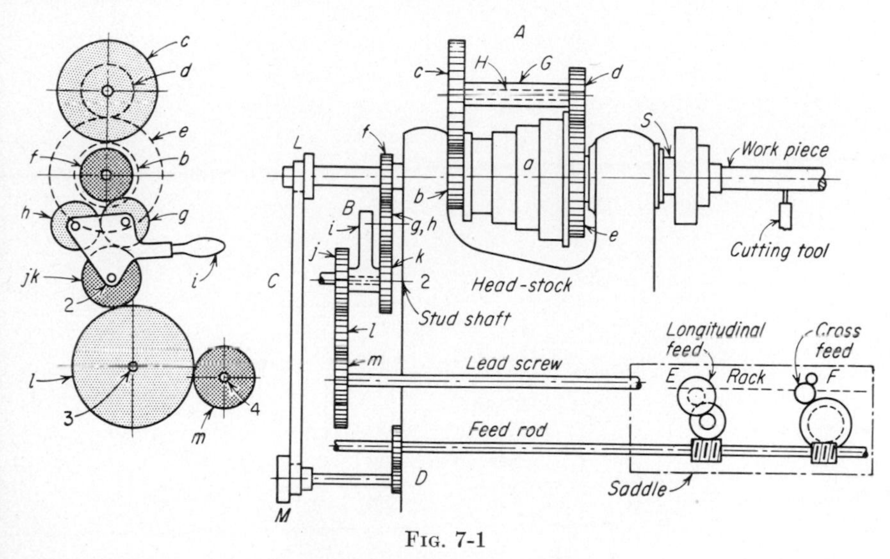

FIG. 7-1

out by means of the shaft *H*, which has eccentric bearings in the head-stock. Gear *e* is then locked to cone *a*, so that the belt drives the spindle directly. As regards the reversing gears, it is only necessary to state that the plate carrying these intermediate gears is pivoted on axis 2 and that arm *i* can be latched in position for forward, neutral, and reverse. Different gears can be slipped on studs 2, 3, and 4 so as to give different speed ratios of lead screw to spindle, stud 3 being adjustable to any position required.

The belt drive for the feed rod is an old-fashioned arrangement, modern lathes having some form of all-gear drive. In the old type of lathe the lead screw could be used for ordinary longitudinal feeding, but its general use would cause unevenly distributed wear, so that it is generally reserved exclusively for the operation of screw cutting.

The tool-carriage feed trains *E* and *F* are carried in the *apron* that extends down from the saddle, and they connect with the fixed rack on the lathe bed for the longitudinal feed and with the screw for the cross-feed, respectively.

An example of a more complex train of mechanism is shown in Fig. 7-2. This figure is a diagram of the driving mechanism that has been used in one type of automatic gear-cutting machine. The constituents common to trains of mechanism occur here in considerable variety, namely, pulleys; spur, bevel, and worm gearing; cams; linkwork; etc. The mechanism as a whole is composed of a number of auxiliary trains

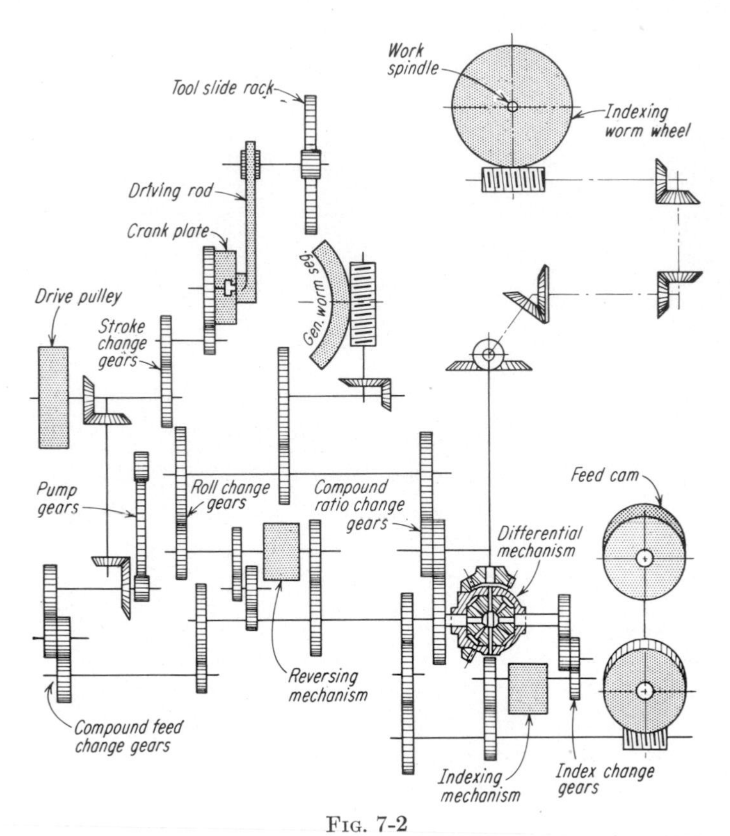

Fig. 7-2

branching off from the main drives to the cutting tools and the work, to actuate the indexing, reversing, and feeding mechanisms, etc. In a machine of this type careful timing is necessary, so that the velocity ratios are generally exact.

7-2. Gear Trains. If motion is transmitted entirely through gearing, the combination of gears is called a gear train. Gear trains may be divided into two classes, *simple* and *compound*. In a simple gear train each shaft of the mechanism carries one gear only, as in Fig. 7-3. In a compound train of gears each shaft except the first and last carries two gears which are fastened rigidly together as in Fig. 7-4. It is kinematically possible to transmit motion between any two shafts by a single pair of gears, but there are practical considerations that often make this arrangement impracticable.

A numerical example will show the advantage of a compound train over a simple train. In Fig. 7-4 let a, b, c, and d have 12, 60, 16, and 64 teeth, respectively. The magnitude of the velocity ratio ω_a/ω_d is

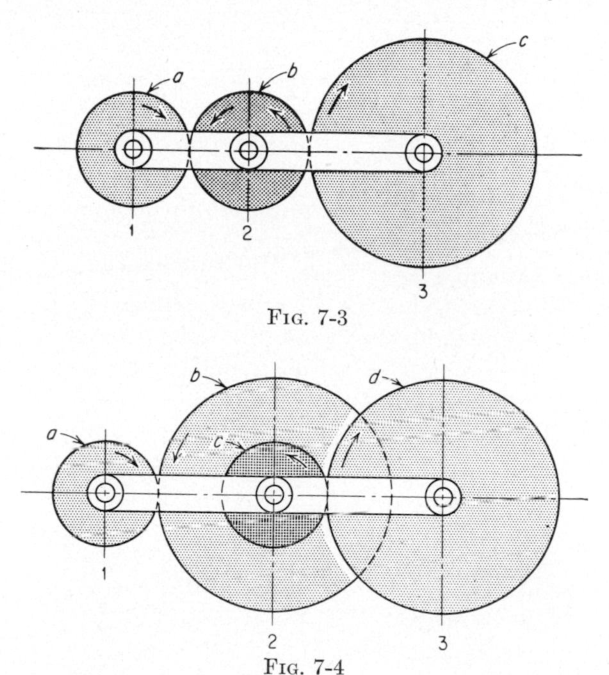

FIG. 7-3

FIG. 7-4

therefore $^{60}\!/_{12} \times {}^{64}\!/_{16} = {}^{20}\!/_{1}$. To get the same velocity ratio by employing a simple train, a 12-tooth pinion meshing with a 240-tooth gear would be required, assuming 12 to be the smallest number of teeth practicable for the pinion. If the diametral pitch of the gears is taken as 4, the diameter of the 240-tooth gear would be 60 in., whereas in the compound train the diameter of the largest gear would be 16 in. Moreover, in case of the single pair, the pinion would be subjected to twenty times the tooth wear of the gear, a very undesirable condition. In addition, the faster pair of gears in the compound train may have a smaller pitch and narrower face than the slower pair, a condition that results in further compactness.

Still further compactness in a compound gear train can be obtained by placing the first and last gears on the same axis, as shown in Fig. 7-5, one of the gears sometimes being attached to a sleeve or hollow shaft. This arrangement is sometimes referred to as a *reverted* gear train. A familiar example of this arrangement is observed in clockwork, where the hour and minute hands are attached to gears on the same axis.

Compound trains of the type shown in Figs. 7-4 and 7-5 find wide application in speed-reducing mechanisms used in connection with high-speed prime movers such as the steam turbine (see Fig. 5-15) and the high-speed electric motor. Such a train is generally enclosed in a housing, forming a unit that connects the prime mover to the driven machine, and is called a *speed reducer*. Descriptions of several types of speed reducers in common use are given in Art. 7-11.

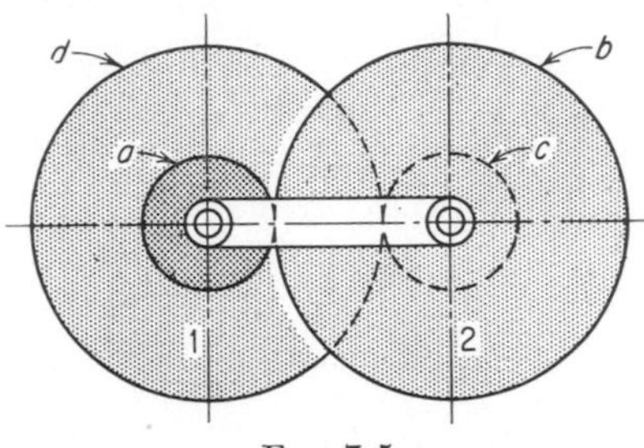

FIG. 7-5

7-3. Velocity Ratio in a Gear Train. In Chap. 4 the magnitude of the velocity ratio of a pair of mating gears was defined as the ratio of the magnitude of the angular velocity of the driving gear to that of the driven gear. A similar relation can be defined as the velocity ratio of a gear *train*.

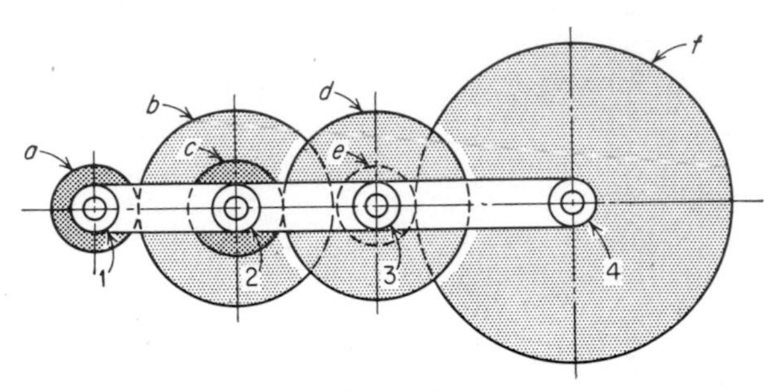

FIG. 7-6

Assume four axes, 1, 2, 3, and 4 (Fig. 7-6), to be arranged as shown and connected by toothed gears represented by their pitch circles a, b, c, etc. Gear a meshes with b, c meshes with d, and e meshes with f. Both the gears b and c are secured to the shaft 2; hence they must rotate as one piece, having the same angular velocity at any instant. Likewise d and e are both secured to shaft 3 and thus have the same angular velocity. Let the angular velocity magnitudes in rpm of the shafts 1, 2, 3, and 4 be represented by n_1, n_2, n_3, and n_4, respectively. Two gears that mesh together must have the same pitch; hence the numbers of teeth are proportional to the circumferences, to the diameters, or to the radii. But their angular velocities are inversely as the radii, and therefore inversely as the numbers of teeth in the gears. If a, b, c, etc., are the numbers of teeth in the gears designated by these letters,[1] it follows that

$$\frac{n_1}{n_2} = \frac{b}{a} \qquad \frac{n_2}{n_3} = \frac{d}{c} \qquad \frac{n_3}{n_4} = \frac{f}{e}$$

Therefore

$$\frac{n_1}{n_4} = \frac{n_1}{n_2}\frac{n_2}{n_3}\frac{n_3}{n_4} = \frac{b}{a}\frac{d}{c}\frac{f}{e} = \frac{bdf}{ace} \tag{7-1}$$

In this train a is the driver and b is the follower in the first pair; c is the driver and d the follower in the second pair; and e is the driver and f the follower in the third pair. The important conclusion expressed by n_1/n_4 in the above equation may be put into the following general statement, good for all such trains:

The magnitude of the angular-velocity ratio of the first driving gear to the last driven gear equals the continued product of the numbers of teeth in the driven gears divided by the continued product of the numbers of teeth in the driving gears.

In finding the velocity ratio of a train, any of the factors n_1/n_2, n_2/n_3, etc., may be expressed in terms of the numbers of teeth, radii, diameters, or revolutions per unit of time of the pair of gears involved. If the last relation is used, however, the ratio is direct and not an inverse ratio as in the case of the other terms. Nor is it necessary that these different factors all be given in the same terms. Thus in Fig. 7-6, if a has 15 teeth and b has 60 teeth; if c is 3 in. in diameter and d is 15 in. in diameter; if e makes 75 rpm and f makes 25 rpm, then

$$\frac{n_1}{n_4} = \frac{60}{15} \times \frac{15}{3} \times \frac{75}{25} = \frac{60}{1}$$

Hence, if shaft 1 makes 120 rpm, shaft 4 will make 2 rpm.

7-4. Value of a Gear Train. In most problems involving gear trains the speed of the first gear is known and that of the last required. It is convenient, therefore, in making calculations to have a factor by which to multiply the first in order to get the last speed. This factor is the reciprocal of the velocity ratio of a train. The relation may be expressed by writing the reciprocal of Eq. (7-1) as follows:

$$\frac{n_4}{n_1} = \frac{ace}{bdf} \qquad \text{or} \qquad n_4 = n_1\frac{ace}{bdf} \tag{7-2}$$

The reciprocal of the velocity ratio, i.e., the continued product of the numbers of teeth in the driving members divided by the continued product of the numbers of teeth in the driven members, is known as the *value of the train* (or *train value*). Referring to Fig. 7-4, if a, b, c, and d

[1] This scheme will be followed throughout the chapter.

have 12, 60, 16, and 64 teeth, respectively, the train value is

$$(a/b)(c/d) = 12/60 \times 16/64 = 1/20$$

and if $n_1 = 100$ rpm, $n_3 = n_1(ac/bd) = 100 \times 1/20 = 5$ rpm.

In making gear-train calculations, it is a matter of choice as to whether the results are obtained by consideration of the over-all velocity ratio or train value. A little practice, however, will show that the use of the train value is generally more convenient.

7-5. Directions of Rotation. If two gears are placed in direct contact, their sense of rotation, except in the case of an internal gear, will be opposite; but if one intermediate gear is placed between them, their sense of rotation will be the same. In Fig. 7-3 let the letters that represent the gears also designate their respective numbers of teeth. Gear a drives b, which in turn drives c. Gear b is, therefore, both a driven and a driving gear. Such a gear is called an *idler.* The magnitude of the velocity ratio is

$$\frac{n_a}{n_c} = \frac{b}{a}\frac{c}{b} = \frac{c}{a}$$

Thus the introduction of an idler in a train serves to change the sense of rotation, but not the magnitude of the velocity ratio or the train value. The chief function of an idler is to change the sense of rotation. Sometimes it is used to bridge the space between driver and follower and so enable a given velocity ratio to be transmitted by use of smaller gears than would otherwise be possible. If a train is made up entirely of spur gears, the adjacent shafts rotate in opposite sense and the alternate shafts in the same sense. If there is an odd number of shafts, the first and last axes will rotate in the same sense; and if there is an even number of shafts, the first and last shafts will rotate in the opposite sense. In general *the safest procedure is to determine the directions of rotation in a train of mechanism by inspection and not attempt to go by rule.*

A familiar example of the case in which an idler provides a simple and convenient reversing device is embodied in the *tumbler gears* common to many mechanisms. A common application of the tumbler-gear arrangement occurs in the headstock of the screw-cutting lathe (Fig. 7-1), where it provides a convenient means of changing the direction of feed, or for cutting either a right- or left-hand thread. Two idlers, g and h, are mounted on a swinging arm i so that they are always in mesh, but at different center distances from axis 2 of the pivot. In the configuration shown, the train runs f-g-k, and f and k rotate in the same direction. By swinging the arm through a small angle, h is brought into mesh with f. The train now runs f-h-g-k, and f and k rotate in opposite directions.

7-6. Gear Trains with Nonparallel Axes. In the preceding articles only gear trains with parallel axes have been described. The methods deduced in Art. 7-3 for the velocity ratio and value of a train are general and apply to all cases when the proper substitutions are made. A train is shown in Fig. 7-7 in which the axes are not all parallel. A pinion *a* on shaft 1 drives the spur gear *b* on shaft 2; a pair of bevel gears *c* and *d*

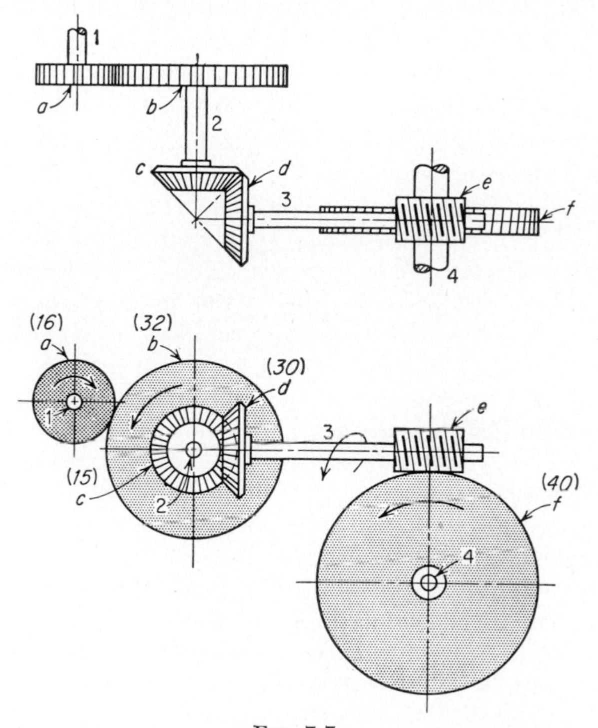

Fig. 7-7

connect shafts 2 and 3, and a worm *e* on shaft 3 drives the worm wheel *f* on shaft 4. If the numbers of teeth on *a*, *b*, *c*, *d*, and *f* are 16, 32, 15, 30, and 40, respectively, if the worm is right-hand and single-thread, and if shaft 1 runs at a speed of 320 rpm,

$$\frac{n_4}{n_1} = \frac{n_4}{n_3}\frac{n_3}{n_2}\frac{n_2}{n_1} = \frac{1}{40} \times \frac{15}{30} \times \frac{16}{32} = \frac{1}{160}$$

Therefore, $n_4 = 320 \times 1/160 = 2$ rpm.

Directions of rotation have been visualized and are shown in Fig. 7-7.

7-7. Selection of Gears for a Train. When no restrictions or limitations are imposed as to the number, size, pitch, etc., of the gears, the variations in the make-up of gear trains that would transmit a given velocity ratio are almost unlimited. No definite rules or formulas can be followed, the process being mainly one of *cut and try* until the desired result is obtained. There are, however, certain limiting conditions in practice that confine the selection to combinations within comparatively narrow limits. Almost every drive is subject to certain special restrictions as to space and arrangement of the gears. The power transmitted determines the pitch of the gears in any given case, and since it is common practice to limit the number of teeth in a pinion to 12, the minimum size of the pinion is fixed. When the function of the train as a whole is to reduce the speed, smoothness of action requires that there should be no *gearing up* at any point in the train; i.e., the driving gear should be smaller than the driven gear in every case.

1. *The Hunting Tooth.* Although it is a debated question, it is considered by many that greater evenness of wear on the teeth will result when a given pair of teeth on two gears comes into contact the least number of times. To illustrate this point, assume a pair of mating gears to have 30 teeth each; then a given tooth of one gear will come into contact with a given tooth of the other gear at each revolution of each gear. Now if the number of teeth in one gear is increased to 31, the velocity ratio is nearly the same as before, and yet a given pair of teeth will come into contact only after 30 revolutions of one of the gears and 31 revolutions of the other. The tooth inserted is called a *hunting tooth,* because a pair of teeth, after once being in contact, gradually separate and then approach by one tooth in each turn, and thus appear to hunt each other as they go around. A hunting-tooth arrangement is frequently found in automobile rear-axle drives, where, for example, a 13-tooth pinion driving a 53-tooth ring gear is just as satisfactory, as far as velocity ratio is concerned, as the exact-multiple ratio of 13 to 52. But there are a great many cases where an exact-multiple ratio is required, thereby rendering the use of a hunting tooth impossible.

2. *Illustrative Examples.* It is evident from the above discussion that in selecting gears for a train there are some general lines of attack which may be followed, and which may best be understood by studying certain typical problems.

a. As an example let it be required to select the gears for a train arranged as shown in Fig. 7-4. The train value is to be $\frac{1}{20}$ and the gears selected are to be subject to the limitation that no gear is to have fewer than 12 teeth. The first step in the solution of the problem is to write the equation

$$\frac{\omega_d}{\omega_a} = \frac{a}{b}\frac{c}{d} = \frac{1}{20}$$

It is desirable in general to have the two factors a/b and c/d as nearly alike as possible. The square root of $\frac{1}{20}$ is between $\frac{1}{4}$ and $\frac{1}{5}$; so a trial pair of factors may be taken as $\frac{1}{4}$ and $\frac{1}{5}$. Then,

$$\frac{a}{b}\frac{c}{d} = \frac{1}{4} \times \frac{1}{5} = \frac{13}{52} \times \frac{12}{60} = \frac{1}{20}$$

and the equation is satisfied if a, b, c, and d have 13, 52, 12, and 60 teeth, respectively. By factoring differently or by using other numbers of teeth for the pinion, other successful combinations of gears may be selected.

The above arrangement has what is considered by many the disadvantage of having the same tooth of a pinion always making contact with the same tooth in a gear. This condition may be avoided and the hunting-tooth condition obtained by the following arrangement:

$$\frac{a}{b}\frac{c}{d} = \frac{12}{52} \times \frac{13}{60} = \frac{1}{20}$$

b. As a further example let it be required to select the gears for a train with the axis of the last gear coincident with the axis of the first gear, as in Fig. 7-5. The train value is to be $\frac{1}{20}$, no gear is to have fewer than 12 teeth, and all gears are to be of the same pitch. As in the preceding example, a trial pair of factors may be taken as $\frac{1}{4}$ and $\frac{1}{5}$ so that

$$\frac{a}{b}\frac{c}{d} = \frac{1}{4} \times \frac{1}{5} = \frac{1}{20}$$

Now *since the pitches are all the same,*

$$a + b = c + d$$

Let x represent this sum. Then a value must be chosen for x such that it may be broken up into two parts whose ratio is 1:4 and also two parts whose ratio is 1:5. If x is made equal to the least common multiple of $(1 + 4)$ and $(1 + 5)$, the conditions will be satisfied. This least common multiple is 30. Then $a/b = \frac{6}{24}$ and $c/d = \frac{5}{25}$. But the values in the numerators are too small for the numbers of teeth required in the pinions, so that numerator and denominator of each fraction must be multiplied by some number such that no number will be less than the number of teeth allowed in the smallest gear. In this case multiplying $\frac{6}{24}$ by $\frac{3}{3}$ and $\frac{5}{25}$ by $\frac{3}{3}$ gives

$$\frac{a}{b}\frac{c}{d} = \frac{18}{72} \times \frac{15}{75}$$

Therefore the conditions of the problem are satisfied when a, b, c, and d have 18, 72, 15, and 75 teeth, respectively. When the pitches of the two pairs of gears are not the same, a slight modification of the method is necessary. The solution in such a case is left to the student.

7-8. Speed-changing Gear Trains. Change Gears. The gears of a train may all be interposed permanently between the driving and driven members, or certain gears in the train may be changed from time to time to vary the velocity ratio of the train. Gears thus subject to change are called *change gears*. Speed-changing gear trains are mostly used in machine tools and motor cars. The oldest and simplest arrangement of change gears is that in which certain gears may be removed from their shafts or studs and others substituted in their place.

The most convenient and desirable arrangement, and the one that represents the modern tendency in design, is that in which the gears are not removed from the machine, but are enclosed in a box or housing called a *gearbox* or transmission, so designed that various combinations of gears may be selected automatically and the speed changes quickly obtained by manipulating a lever that slides or shifts the gears into desired positions.

The mechanism of the ordinary engine lathe, particularly in respect to its screw-cutting provision, illustrates a typical application of a gear train in which speed changes are required. In lathes of the older type the change gears are removable, whereas most modern lathes are of the quick-change type with an elaborate train of gears in which the speed changes are made in a manner similar to the gearshift of automobiles.

1. *Speed-changing Mechanism of the Engine Lathe.* The general arrangement of gearing in a simple lathe of the older type, equipped for screw cutting, is shown in Fig. 7-1 and was briefly described in Art. 7-1. On the spindle S are the gears b, e, and f and the stepped cone a. The cone is connected by belt to the countershaft, which supplies the power. The gear e is keyed to the spindle, and the stepped cone is free to rotate on the spindle except when it is locked to gear e by means of a pin. The gear b is fastened to the stepped cone and rotates with it; c and d are the *back gears* and are keyed to a shaft that is parallel to the spindle. This shaft has eccentric bearings so that the back gears may be thrown out of mesh with b and e.

When the back gears are thrown out, e and the stepped cone are locked together, and as many speeds can be obtained as there are steps on the cone, whereas with the back gears thrown in, e and the stepped cone are not locked, and the number of speeds of the spindle is therefore doubled.

The purpose of the spindle gear f is to drive the thread-cutting train. The tumbler gears g and h are for the purpose of changing the direction of rotation of the lead screw. Gear g meshes with k, which is the inside stud gear, and the arrangement is such that f and g may be thrown out of mesh and h be made to mesh directly with f. f and k are usually made the same size, but, when they are not the same, k is usually made twice the size of f, so that k makes one-half as many revolutions as the spindle. On the same shaft with k is the outside stud gear j. Meshing with j is an intermediate gear l, which meshes with the lead-screw gear m.

In thread cutting it is necessary to change the stud gear j and the lead-screw gear m. The intermediate gear l is held on a slotted bracket called a *quadrant*, not shown in the figure. This allows l to be adjusted to accommodate different-sized gears j and m.

In order to cut a greater range of threads without adding more change gears, the intermediate gear l may be *compounded*, that is, replaced by two gears of different sizes, and fastened together, one of them meshing with the stud gear and the other with the lead-screw gear. When compounded, the intermediate gears are subject to change and therefore become change gears along with the stud gear and lead-screw gear.

One of the inherent advantages of the involute system, namely, the possibility of varying the center distance without affecting the velocity ratio, is manifest in change-gear sets, because no provision is made for adjusting these gears to the theoretical center distance. To do so would greatly complicate the design.

The cutting tool is held in a toolholder mounted on a carriage that is moved back and forth along the work by means of a split nut engaging the lead screw. If there were eight threads per inch on the lead screw, then eight revolutions of the lead screw would advance the cutting tool 1 in., and if the work made four revolutions to eight of the lead screw, four threads per inch would be cut on the work.

2. *Selecting Change Gears for the Screw-cutting Train.* An illustrative example will show the method of selecting the change gears to do a given piece of work. A certain lathe of 16-in. swing has a lead screw of four threads per inch, and a set of change gears of the following numbers of teeth: 24, 30, 36, 42, 48, 54, 60, 66, 69, 72, 78, 84. Let it be required to find the numbers of teeth in the stud and lead-screw gears j and m, when f and k are the same size.

Since the gears g, h, and l are idlers, they need not enter into the calculations, and since the spindle and stud shafts make the same number of revolutions, it is only necessary to find the number of revolutions of the stud shaft.

Let n = number of threads per inch to be cut on the work.
N = number of threads per inch on lead screw.
t = number of teeth on lead-screw gear.
T = number of teeth on stud gear.

Then

$$\frac{N}{n} = \frac{T}{t} \tag{7-3}$$

Let it be required to select change gears for cutting six threads per inch on the work.

$$\frac{T}{t} = \frac{N}{n} = \frac{4}{6} = \frac{2}{3}$$

Any two gears of the set may be taken that have numbers of teeth in the ratio of T to t. In this particular case a 24-tooth gear would be selected for the stud and a 36-tooth gear for the lead screw.

If the change gears are compounded, the pair of intermediate gears, unlike the idler l, which they replace, affect the velocity ratio between the spindle and the lead screw because of their difference in diameters.

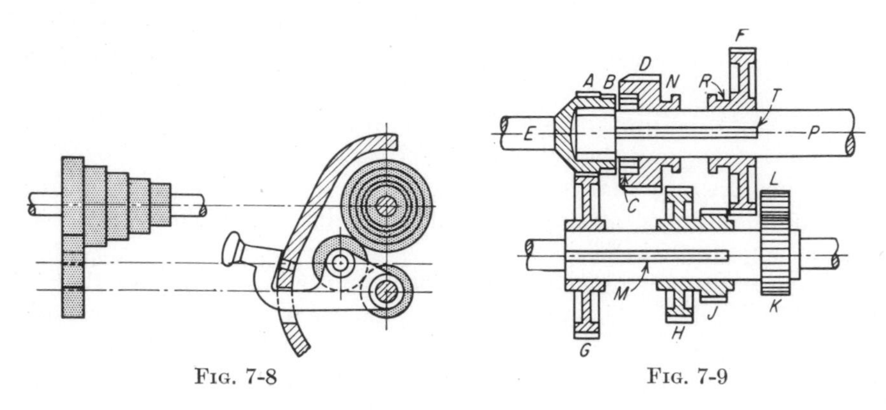

Fig. 7-8 Fig. 7-9

The velocity ratio as found by the preceding method must be multiplied by the ratio of the two intermediate gears. Thus if 35 threads are to be cut on the same lathe, it will be necessary to use the compound intermediate gears. Take the ratio $TT/tt = 4/35$ and split it into the obvious factors $2/5$ and $2/7$, which can be realized with gears of 24, 60, 20, and 70 teeth.

The primary unit of nearly all quick-change lead-screw trains is a cone-and-tumbler pair (Fig. 7-8) based on the arrangement shown by gears j, l, and m (Fig. 7-1). The secondary train, which will not be discussed here, multiplies and divides the cone ratio by 2. Of the two gears j and m in Fig. 7-1, one will serve as a pinion (constant), and the other will become a cone (in effect, a gear of varying diameter). Whether f or j is expanded into a cone depends upon the system of screw threads.

The above screw-cutting trains are given as examples of the old and the modern arrangements; but among lathes of various makes there are many modifications in detail to be found. All ordinary screw-cutting lathes, however, have mechanisms that are fundamentally those given above.

3. *The Automobile Transmission.* Figure 7-9 is a drawing of a variable-speed transmission showing how the speed-changing principle is applied in the automobile gearshift drive. The standard arrangement gives three forward speeds to the drive shaft (one without using the gears) and one reverse. The power from the engine is delivered through a friction clutch (not shown) to the shaft E and to this shaft is secured a gear A having also a part of a jaw clutch B on its right-hand side. Gear A meshes with another gear G on the countershaft M, which carries also the gears H, J, and K keyed to it, and, whenever the engine shaft E operates, the gears G, H, J, and K are running. On the right is shown the power shaft P, which connects with the rear or driving axle; this shaft is in line with E and carries the gears D and F and also the inner part C of the jaw clutch, which engages with the outer part B. The gears D and F are forced to rotate with P by means of keys at T, but both are sliding gears and may be moved separately along the splined shaft by means of a shifting lever that connects with the collars at N and R, respectively. In addition to the gears already mentioned there is another, an idler gear L, which meshes with K and runs on a bearing behind the gear K.

When it is desired to operate the car at maximum speed (third or high), the operator throws F into the position shown and pushes D to the left so that the clutch piece C engages with B, in which case P runs at the same speed as the engine shaft E. The second highest speed (second or intermediate) is obtained by slipping D to the right until it comes into contact with H, the ratio of the gears (train value) then being $(A/G)(H/D)$; F remains as shown. For the lowest speed (first or low) D is placed as shown in the figure and F moved into contact with J, the ratio of the gears in this case being $(A/G)(J/F)$. The shaft P and the car are reversed by moving F to the right until it meshes with L, the gear ratio being $(A/G)(K/L)(L/F)$. The shifting arrangement is such that it is not possible to have more than one set of gears in operation at one time.

7-9. Planetary Gear Trains. In all the gear trains so far discussed, the axes of the gears have remained in fixed positions relative to each other. The distinguishing feature of a *planetary* gear train is that some of the gear axes *change* position relative to others. For example, in Fig. 7-10 gear b rolls around the outside of stationary gear a as the arm c revolves. Thus the axis of gear b changes its position. The motion of the planets around the sun is planetary motion. So is the motion of the *planet* gears (such as b) around the *sun* gear (such as a).

1. *Simple Planetary (Epicyclic) Gear Train.* In Fig. 7-10 let a and b be two gears mounted on an arm c, so that, if c were fixed, a and b would form a simple gear train. Now let it be assumed that a is the fixed

member of the train. Then c can rotate about O, carrying b with it, b itself rotating relative to c around its axis at O'. It is required to find the number of revolutions that b will make relative to the fixed member a for every revolution of c around O.

First, let a be disconnected from the fixed frame, so that a, b, and c (locked together) can make one revolution as one piece, in the direction indicated, around O. That b will make one revolution can be seen by noting the positions of any line on b, as PO', during different phases of the revolution, as shown. Next if arm c is held stationary and fixed gear a is rotated backward one revolution, a is brought back to its original position, and thus its resultant motion is zero. At the same time b must receive a/b turns forward. Then the total number of revolutions that b will make for one revolution of arm c around O (with no net rotation of a) is $1 + a/b$, and its direction of rotation will be the same as that of c. It is evident that c can be rotated in either direction, and the result obtained will still hold.

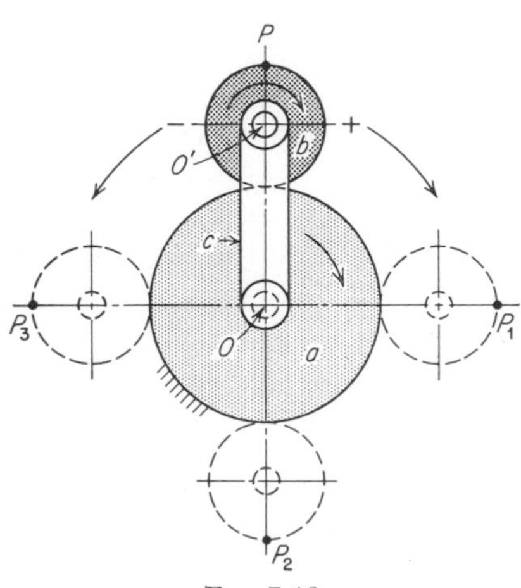

Fig. 7-10

The analysis for the total motion of each link may be tabulated in the following manner: Assume clockwise rotation as +, and counterclockwise rotation as −. Then considering the whole mechanism locked to the arm c and given one revolution about O in the positive direction, each member will rotate +1 revolution. Next consider that the arm c is fixed (then there is only an ordinary gear train) and that gear a is given one revolution in the negative direction. The total motions in this case are: −1 revolution for a; $+a/b$ revolutions for b; and 0 revolutions for c. The total motion of each link is then given by the following table:

	a	b	c
Motions with arm c	+1	+1	+1
Motions relative to arm c	−1	$+\frac{a}{b}$	+0
Total motions	0	$1+\frac{a}{b}$	+1

The basic idea can be stated as follows: the total angular motion of any gear is equal to the angular motion of the arm c plus the angular motion of the gear *relative to* the arm. In the table, these are stated as *motions with arm c* and *motions relative to arm c*.

It is important to note that the *arm* mentioned above is always the member attached to and carrying the *moving gear* axes. It is noted that the procedure employed is an analysis for the motions in a *planetary train*, and yet each step of the procedure involves no planetary action at all.

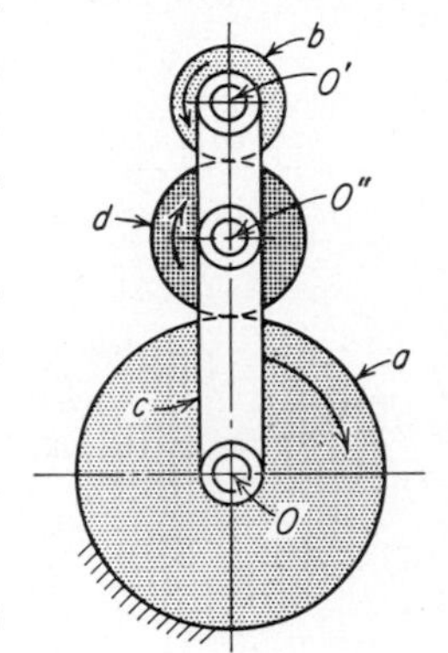

FIG. 7-11

From the table, the angular-velocity ratio between the gear b and the arm c may be expressed as follows:

$$\frac{n_b}{n_c} = \frac{1 + a/b}{1}$$

and the signs indicate that both rotate in the same direction.

If an idler is placed between a and b, as in Fig. 7-11, or if a is an internal gear (Fig. 7-12), the direction of motion of b is reversed. Figures 7-10, 7-11, and 7-12 are not practical gearsets as they appear but have been used to introduce simply the method of analysis and the types of motion in planetary gear trains. The following paragraphs and articles will present (among other things) examples of actual planetary trains.

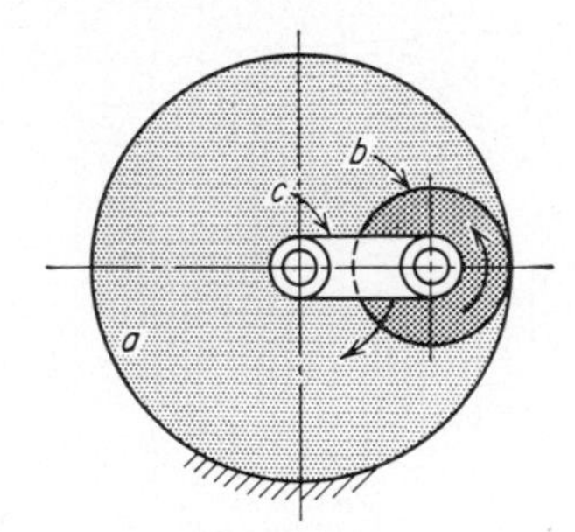

FIG. 7-12

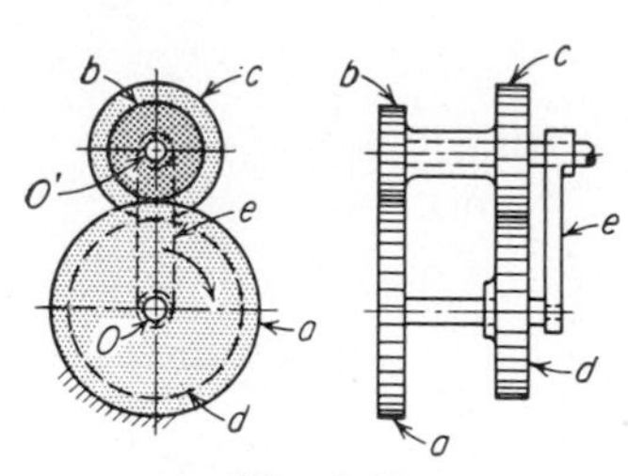

FIG. 7-13

2. *Compound Planetary Gear Train.* The tabular scheme may be followed if other gears such as c and d are added as in Fig. 7-13, forming a compound reverted train. If b is fastened to c, and d is free to revolve about O, the motions are as shown in the table on page 190. It follows that

$$\frac{n_d}{n_e} = \frac{1 - ac/bd}{1} \qquad \text{or} \qquad \frac{n_e}{n_d} = \frac{1}{1 - ac/bd}$$

	a	b	c	d	e
Motions with e	+1	+1	+1	+1	+1
Motions relative to e	−1	$+\frac{a}{b}$	$+\frac{a}{b}$	$-\frac{a}{b}\frac{c}{d}$	0
Total motions	0	$1+\frac{a}{b}$	$1+\frac{a}{b}$	$1-\frac{ac}{bd}$	+1

This form of planetary train is used extensively for obtaining large velocity ratios between the arm e and the last gear d.

As an extreme example: Let a, b, c, and d have 99, 100, 101, and 100 teeth, respectively. Then

$$\frac{n_e}{n_d} = \frac{1}{1 - ac/bd} = \frac{1}{[1 - (99 \times 101)/(100 \times 100)]} = \frac{10{,}000}{1}$$

Since these gears are mounted on parallel shafts, the distance between centers must be the same for each pair. *Theoretically, therefore, the sum of the numbers of teeth for each pair should be the same.* Since this is not the case in the example given, it becomes necessary to deviate slightly

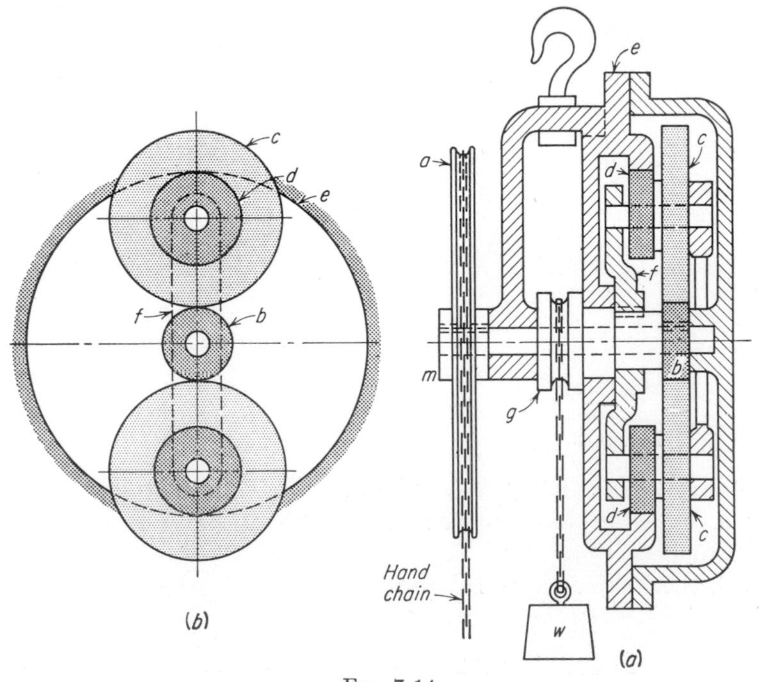

FIG. 7-14

from the theoretically correct center distances, making suitable adjustments as to backlash, etc. With involute gears this would not affect the velocity ratio.

The application of planetary gearing to hoisting devices will be obvious from the above. Planetary gears are also used for feed mechanisms on large boring bars, in machines for making wire rope, in speed reducers, etc.

3. *The Triplex Hoist.* In the triplex hoist, shown in diagrammatic form in Fig. 7-14*a*, is found an interesting practical application of the planetary gear train. The frame *e* is provided with bearings that carry the hoisting sprocket *g*, and in the casting *f* (the *arm*), which is keyed to the hoisting sprocket, are studs, each carrying a pair of compound gears *cd*, the smaller one (*d*) meshing with an internal gear integral with the frame *e*, and the larger one (*c*) of the pair meshing with a pinion *b* on the end of the shaft *m* to which the hand sprocket wheel *a* is attached. When an operator pulls on the hand sprocket chain, the hand sprocket wheel *a* is rotated, and with it the pinion *b*, which in turn sets the compound gears *cd* in motion. As one of these gears meshes with the fixed annular gear on the frame *e*, the only motion possible is for the spider and the studs carrying the compound gears to revolve, and thus carry with them the hoisting sprocket wheel *g*.

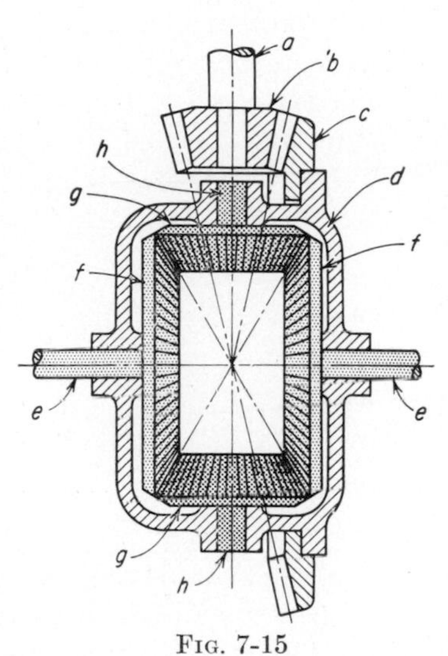

Fig. 7-15

The velocity ratio between hand sprocket wheel *a* and hoisting sprocket wheel *g* is evidently the same as that between gear *b* and rotating casting *f*. In finding the velocity ratio for any specific case, it is convenient to draw the skeleton outline of the gear train as shown in Fig. 7-14*b*. The two pairs of gears *c* and *d* are for the purpose of distributing the load. In finding the velocity ratio it is only necessary to consider one-half of the symmetrical arrangement, i.e., the gears that are lettered in Fig. 7-14*b*.

4. *Differential Gear Trains. The Automobile Rear-axle Drive.* Figure 7-15 shows the arrangement of bevel gears in the automobile rear axle or differential. Shaft *a* is driven from the engine and has keyed to it the bevel pinion *b* meshing with ring gear *c*, which is fastened to the frame (or spider) *d*. The frame turns freely on the axles *e*, to which gears *f* and the rear wheels are keyed. The frame *d*, to which *c* is fastened, carries the pinions *g*, which are free to turn on the studs *h*. There are

usually two or three of these pinions, in order to distribute the load. When the automobile is going straight ahead, a drives b, and all the other gears revolve as a unit with c, without any relative motion. As soon, however, as the car starts to turn a corner, say, toward the right, the left-hand wheel will have to travel faster than the right-hand wheel, and therefore the gears begin to move relative to each other, the action being that of a planetary train.

Let it be assumed that the left-hand wheel is jacked up so that it may turn freely while the right-hand wheel is left on the floor, thus holding right-hand gear f from turning. Consider fixed gear f as the first gear of the train and the frame d as the arm carrying gears g. The velocity ratio of free gear f relative to fixed gear f may be found by use of the tabular method already discussed.

	f(fixed)	d (or c)	g	f(free)
Motion with d	+1	+1	...	+1
Motion relative to d	−1	0	...	$+\frac{f}{g}\frac{g}{f}$
Total motion	0	+1	...	$1 + \frac{fg}{gf} = 2$

The table shows that the left-hand wheel (f) turns twice as fast as gear c, and in the same direction (both signs +).

7-10. Planetary Gear Trains with Two Inputs. In the previous analyses of planetary trains there was always one gear which was stationary.

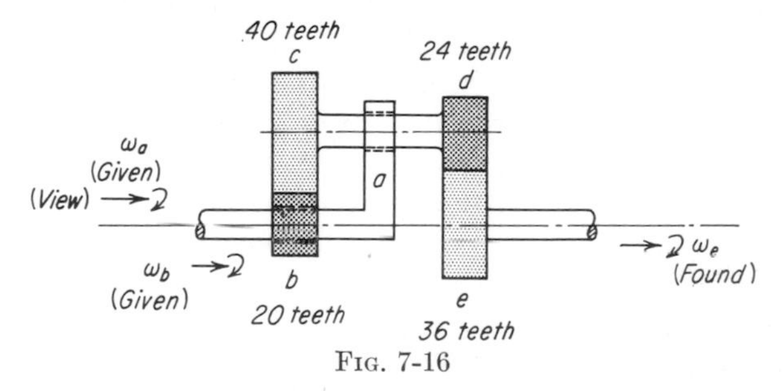

Fig. 7-16

If now that gear is given a definite amount of rotation, thus constituting the second input to the gear system, the rotation of the output shaft can be determined by application of essentially the same ideas which have already been discussed.

For example, in Fig. 7-16, ω_a and ω_b are input shaft angular velocities, and the problem is to find ω_e, the angular velocity of the output shaft.

The principle of superposition is applicable: i.e., the total rotation of e is the algebraic sum of the rotation of e caused by b (with a fixed) and the rotation of e caused by a (with b fixed). Thus the analysis consists of two parts:

1. Find ω_e/ω_b with a fixed. Call this value k_1. Then $\omega_{e1} = \omega_b(k_1)$. Since the arm a is considered fixed, the analysis involves an ordinary gear train (no planetary action).

2. Find ω_e/ω_a with b fixed. Call this value k_2. Then $\omega_{e2} = \omega_a(k_2)$. The analysis is for a planetary gear train with one fixed gear, as discussed in the previous articles.

The *total* $\omega_e = \omega_{e1} + \omega_{e2} = \omega_b(k_1) + \omega_a(k_2)$, which is merely the sum of the contributions of a and b. For the gear system of Fig. 7-16,

$$\omega_e = \tfrac{1}{3}\omega_b + \tfrac{2}{3}\omega_a \tag{7-4}$$

The verification is left to the reader. Algebraic signs for numerical values inserted for ω_b and ω_a must be the same if the directions are the same, and opposite if the directions are opposite. It is observed that, if either ω_b or ω_a is zero, the analysis becomes exactly that of the previous articles.

The automobile rear-axle differential (Fig. 7-15) acts as a double-input planetary train when the car travels around a curve. An analysis shows that the ring gear c rotates at one-half the algebraic sum of the two wheel speeds.

Other applications of double-input planetaries include servomechanisms and computing devices. A planetary system can deliver continuously to the output shaft an algebraic average of two input angular motions (the differential does this); or, more generally, it can multiply each input by a constant, add the results algebraically, and deliver the answer to the output shaft.

7-11. Speed Reducers. The turbine with its high speed is more efficient than the low-speed engine; and the small high-speed motor is more efficient, just as powerful, and less costly than the larger one of slow speed. The use of the high-speed prime mover, however, calls for a compact and efficient device that will decrease or step down the speed to a value suitable for practical conditions, i.e., to a speed or range of speed at which the driven member operates economically and efficiently. Belt, chain, and rope drives, open gearing, or any combination of these methods of speed reduction may be found in practically every industry. Although these methods of speed reduction possess certain individual merits, they are also accompanied by many objectionable features. Since the maximum practical ratio for a belt drive is about 6:1, and that of a chain or spur-gear drive 8:1, it is evident that great reduction is possible only by the use of a series of reducing units. It is evident then that transmission requiring considerable speed reduction would necessitate the use of a great deal of space.

Open drives such as the above are also difficult to lubricate properly and are subject to wear and deterioration, particularly where exposed to dirt, dust, grit, moisture, and acid and alkaline fumes. Another very important consideration is that of safety. If not carefully guarded, open drives are a constant source of danger because of the rapidly moving exposed parts.

The advent of the efficient high-speed prime movers and the disadvantages of the prevalent methods of speed reduction resulted in the development of units known generally as *speed reducers.* These units vary in design, but they all consist essentially of a train of gears totally enclosed in a metal housing. Speed reductions in ratios over 1,000:1 are accomplished in standard constructions, and special construction makes ratios of almost any value possible. Speed reducers can be classified in four general categories: (1) worm gear, (2) spur gear, (3) internal gear, and (4) planetary gear.

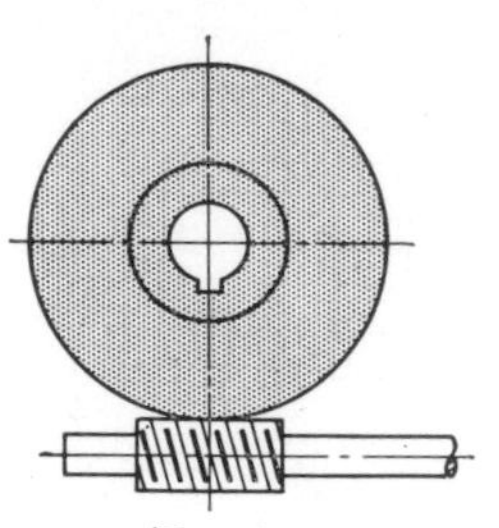

FIG. 7-17

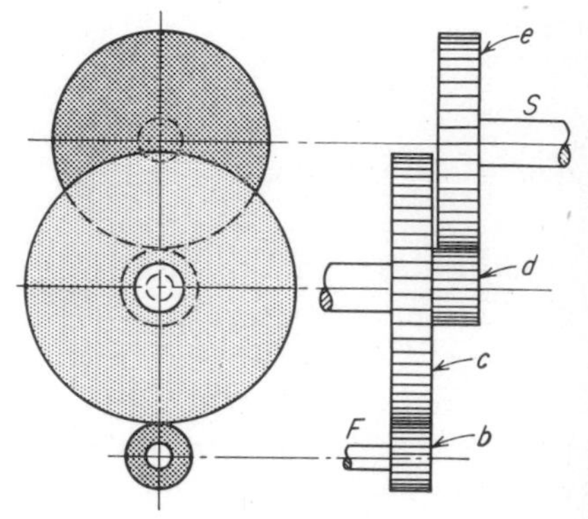

FIG. 7-18

1. *The Worm-gear Reducer.* The worm type of reducer (Fig. 7-17) is commonly used to secure high-speed reductions between shafts that are at right angles to each other. Combined bevel- and spur-gear reducers are also used for angle drives.

2. *Spur-gear Reducers.* The usual reducer of this type consists of a train of gears arranged as in Fig. 7-18 or 7-19, each having its peculiar advantages. The arrangement in Fig. 7-18 is commonly found in large steam-turbine drives in which double-helical gears are used (see Fig. 5-15 for a 17,500-hp reducer). The most used arrangement is that of Fig. 7-19, the outstanding advantages being compactness and reduction in a straight line. The low-speed shaft S is to all outward appearances a continuation of the high-speed shaft F. The high-speed pinion b meshes usually with three gears c making contact at 120-deg points of the pinion. This three-point application of the load is for the purpose of keeping down

the size of a single-tooth load. Keyed to the same shaft of each of these gears is another smaller gear *d*, which in turn drives a larger gear *e* making contact at 120-deg points of the larger gear. This large gear is keyed to the slow-speed shaft *S*. Any reduction required may be attained by proper gear ratios, or by making series combinations of the units just described.

3. *The Internal-gear Reducer*. In the internal-gear type of reducer (Fig. 7-20), the small high-speed pinion *b* makes three-point contact

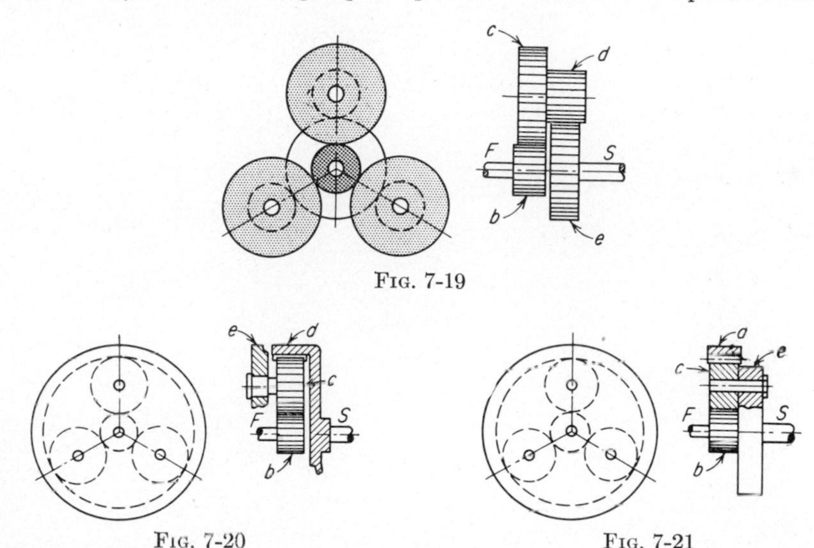

Fig. 7-19

Fig. 7-20

Fig. 7-21

with three intermediate gears *c*, which in turn make three-point contact with an internal gear *d*. The intermediate gears *c* are *journaled* in the frame *e*, and the internal gear *d*, which is keyed to the low-speed shaft *S*, is free to rotate. Greater speed reduction is obtained by making series combinations of this single stage.

4. *The Planetary-gear Reducer*. One type of planetary speed reducer is shown in Fig. 7-21. Here, as in the previous cases, the high-speed pinion *b* meshes with the three intermediate gears *c*. The gears *c* are *journaled* on studlike shafts that protrude from a plate *e*, which is keyed to the low-speed shaft *S*. As the three intermediate gears *c* rotate about their centers, they mesh with a stationary internal gear *d*. This causes the entire unit, consisting of the three intermediate gears, the plate in which they are journaled, and the low-speed shaft, to rotate. Greater speed reduction is obtained by making series combinations of this single stage.

Figure 7-22 is a schematic diagram of the well-known Hydra-Matic automobile transmission. Two planetary sets in series are used singly or in combination to give three forward speeds plus a direct straight-through drive (high gear). A third planetary set furnishes a reverse. The forward planetary sets are activated and connected by means of the clamping bands and clutches shown.

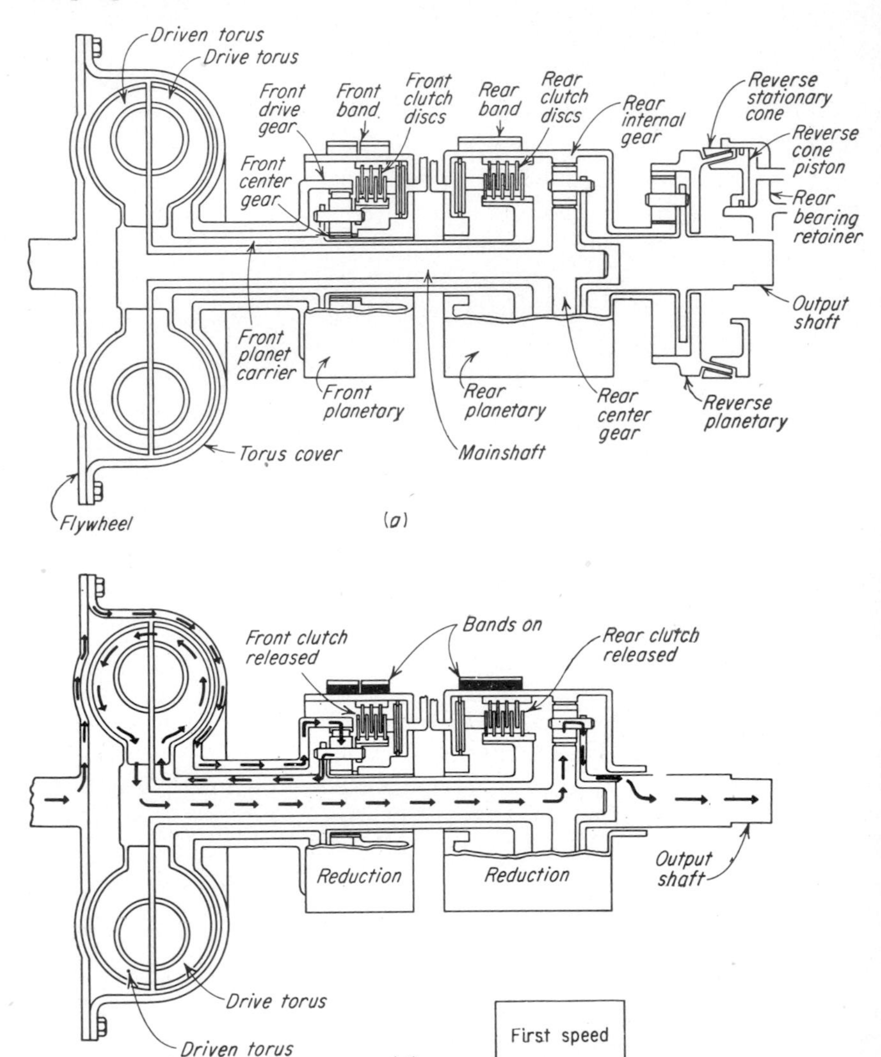

FIG. 7-22. (*Courtesy of Detroit Transmission Division, General Motors Corporation.*)

In *b* of Fig. 7-22 is shown the lowest-speed gear setting of the four possible ones. There is speed reduction through both planetary sets in series. The front band stops rotation of the front center gear, causing the front planet carrier (the arm) to revolve as the engine or flywheel shaft revolves. The motion of the planet carrier causes the drive torus (part of the fluid coupling) to rotate. As the driven torus revolves, it turns the rear center gear. The rear bands are clamped and do not allow the rear internal gear to turn. Therefore the rear planet carrier (the output shaft) must revolve. This shaft is connected to the drive shaft to the rear axle.

To by-pass the front planetary set and thereby obtain only the speed reduction of the rear set, the front *clutch disks* are engaged and the front set of *bands* is released. This locks the front planetary so that it revolves as a unit, transmitting flywheel rotational speed directly to the drive torus of the fluid coupling. The rear planetary action is the same as before.

The general method of operation has been indicated, and the combinations used to obtain other speed ratios will not be discussed.

PART II

KINEMATICS AND DYNAMICS OF MACHINERY

CHAPTER 8

VELOCITIES IN MACHINES

Kinematics and dynamics of machinery is the study of motions and forces in machines. On the motion side this subject differs from mechanism (Part I of this book), in that it is concerned primarily with velocities and accelerations rather than with displacements.

Three ends are to be gained from a study of the kinematics and dynamics of machinery, namely:

1. An ability to visualize the motions and forces existing in machines.
2. A knowledge of the fundamentals of mechanics as applied to machines.
3. A knowledge of the details of some of the methods employed in machine analysis.

The first two ends are the most important and they embrace, in fact, the only part of the subject which the engineer can hope to retain in large measure as a permanent part of his mental equipment. The third objective, however, should not be slighted. For, although the engineer is forced to store many such details of his profession in his notebook or library rather than in his mind, his ability to make ready use of such information in later years often depends to a large extent upon the practice he gets in using it as a student.

In the analysis of machines it is often advantageous to employ graphical methods more extensively than is usual in the study of analytical mechanics, and in the following chapters considerable emphasis is placed upon the use of such methods.

As stated in Chap. 1, a majority of machine members are constrained to have plane motion; i.e., their points move in parallel planes. In such cases it is possible to study the motion of a machine member by means of its projection in a plane parallel to the plane of motion; that is, by means of a working drawing of the machine of which the member is a part. This is due to the fact that all the points of the member situated on any one line perpendicular to the plane of motion have exactly the same motion. In this book major emphasis is placed upon the analysis of machines in which the parts move with plane motion.

8-1. Methods of Analysis for Velocities. In this chapter two general methods will be developed, and a third will be introduced briefly, for determining the velocity of any point in a machine having constrained motion when the velocity of any other point is known. The two main methods will be called the *method of relative velocities* and the *method of instantaneous centers.* The third method is the *method of orthogonal components.* These methods may be applied either graphically or analytically but are more often applied graphically. In many problems, however, a solution is more easily effected either analytically or by a combination of graphical and analytical work. Nevertheless, it is believed that the visualization inherently required in a graphical approach promotes understanding as well as providing a working technique of rather broad applicability, and so considerable emphasis is placed on graphical methods.

The reasons for making a velocity analysis vary somewhat with the type of machine analyzed. In high-speed machinery, it is important to have a knowledge of the inertia forces set up in a given machine. In order to determine these inertia forces, the accelerations of certain points of the machine must be determined, and this usually requires that a complete velocity analysis be made first. In many instances some idea of the accelerations may be obtained from a velocity analysis alone. In quick-return mechanisms used on shapers, slotters, and other machine tools, a velocity analysis will show the cutting and return speeds of the tool. It is frequently more convenient to determine the mechanical advantage of a given mechanism by means of a velocity analysis than by means of a force analysis. Finally, the sliding velocity between two links in contact is a factor in the problem of lubrication.

8-2. Vector Quantities. The fundamental quantities of mechanics may be divided into two groups, since they are either vector or scalar quantities. A vector quantity has both magnitude and direction (including *sense* of direction), whereas a scalar quantity has magnitude only. A vector is a straight line, the length of which is proportional to the magnitude of the quantity which it represents, and the direction (including sense) of which (indicated by the angular position of the line and an arrowhead on the line) is the same as that of the quantity. In this book the single word *direction* will usually be used to denote all the directional properties of a vector, including the sense. The initial point on the straight line which is a vector is called the origin or tail of the vector, and the final point at the tip of the arrowhead is called the extremity or head.

Examples of vector quantities are displacement, velocity, acceleration, force, and momentum. Examples of scalar quantities are distance (without regard for direction), speed, and energy. It is shown in textbooks on vector analysis that it is possible to manipulate vectors in much the same manner as algebraic quantities. That is, a set of operating rules may be formulated by means of which operations very similar to those commonly performed upon algebraic quantities (such as addition, multiplication, differentiation, etc.) may also be performed upon vectors. The operations suited to the purposes of this text, however, are primarily vector addition and vector subtraction in a plane.

The addition and subtraction of two vectors may be understood by referring to Fig. 8-1. The vectors shown at *a* are added at *b*, the vector V_3 being their sum. A vector equation may be written to symbolize this operation, as shown in the figure. For either side of the equation,

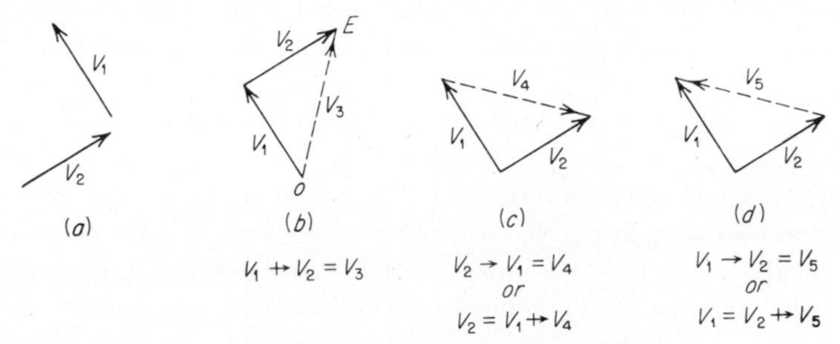

FIG. 8-1. Addition and subtraction of vectors.

the figure shows an identical result—a beginning at O and an ending at E. The origin of the vector representing the sum is coincident with the origin of the first vector of the sum, and its extremity is coincident with the extremity of the last vector. At *c* and *d* of Fig. 8-1 are shown examples of vector subtraction. Transposing in the first equation of *c* gives the second one, and it is evident that the vector diagram at *c* checks both equations. Any vector equation involving subtraction can be rearranged so that only addition is involved by transposing each subtractive term to the opposite side of the equation. By so doing, it may sometimes be easier to visualize and check the correspondence of the vector equation and the vector polygon which represents it. The signs $\nrightarrow$ and $\rightarrow$ should be noted. They indicate that addition and subtraction are vectorial and not algebraic.

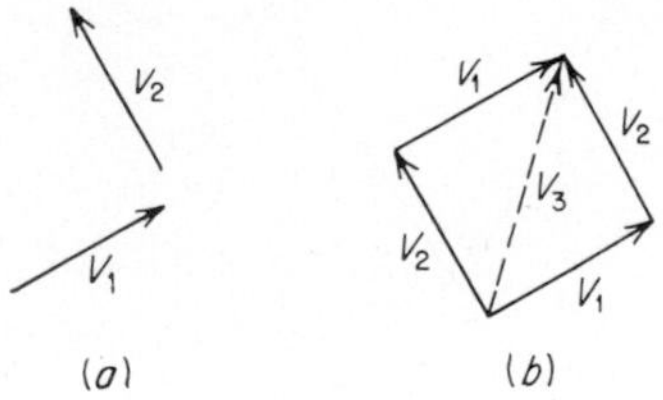

$$V_1 \nrightarrow V_2 = V_3 = V_2 \nrightarrow V_1$$

FIG. 8-2. The order of adding vectors does not affect their sum.

The fact that the order in which two vectors are added has no effect upon their sum is brought out in Fig. 8-2. The same statement applies to any number of vectors.

8-3. Motion. In a strict sense, we cannot know *absolute motion*, because we know of no location in the universe which we can say is absolutely stationary. Thus all motion is relative motion, and we can only say that a point is stationary or moves relative to an observer. When we say that an automobile is traveling relative to a *fixed* observer at 50 mph, we really mean that 50 mph is the speed or velocity relative to an observer on the earth over which the automobile is passing. This is the essential viewpoint used in the study of motion in machines and mechanisms. Because we are not conscious of our own motion if we, as observers, stand at a fixed spot on the surface of the earth, we adopt the earth's surface as our frame of reference. In discussions of motion in this book, terms such as velocity or acceleration (sometimes referred to as "absolute" velocity or "absolute" acceleration) mean velocity or acceleration relative to the surface of the earth near the point or body in question.

The term "relative" velocity (or acceleration) usually implies that the observer is not stationary on the earth's surface. For example, if car A moves past car B, the velocity of A relative to B is the velocity of A as seen by an observer in car B. Stated in another way, the velocity of A relative to B is merely the additional velocity which A has over and beyond the velocity of B. The "relative" velocity is therefore only the (vector) *difference* between the velocities of A and B. This will be discussed more fully later.

8-4. Motion of a Point. The motion of a rigid body is defined in terms of the motions of its points. In order, therefore, to study machine motions it is first necessary to consider the motion of a single point. As mentioned previously, it will be assumed that the motions dealt with hereafter are *plane motions* unless otherwise stated.

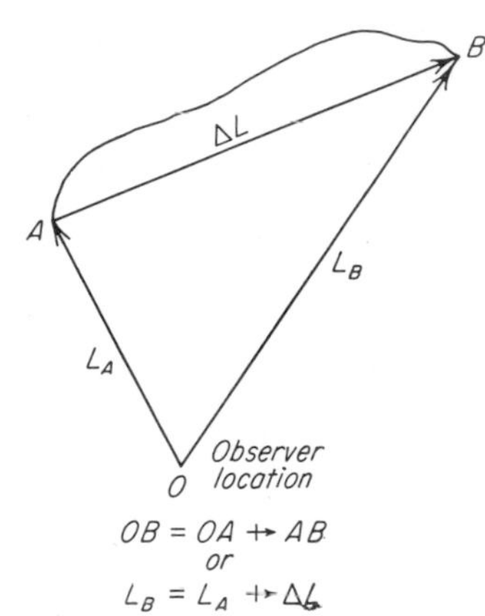

Fig. 8-3. Position and displacement vectors for a moving point.

1. *Position of a Point.* Motion of a point involves change of position of the point, and hence the first requirement in the study of motion is a means of defining the position of a point at any instant. The position of a point in a plane may be defined:

(1) In terms of its distance from a pair of fixed rectangular axes.

(2) In terms of the length and direction of a line drawn to it from a single fixed point in the plane, the direction of the line being measured by means of the positive angle that it makes with some fixed line in the plane.

In graphical representation the position of a point in a plane is usually referred to the paper on which the representation is made and is indicated by means of a letter or some other suitable character.

In Fig. 8-3, the position vectors OA and OB specify the positions of points at A and B relative to an observer at O. (An observer might conceivably use a transit and range finder to indicate the directions and magnitudes of such vectors.)

2. *Displacement of a Point.* Referring to Fig. 8-3, if a point moves from A to B along any path whatsoever (as indicated by the irregular curve), its change of position is measured by the straight line AB; i.e., it is the distance AB measured along the straight line in the direction AB. This change of position is called the *displacement* of the point.

In terms of the position vectors L_A and L_B, a vector equation can be written.

$$L_B = L_A \nrightarrow \Delta L \tag{8-1}$$

which says that the new position B is the same as the old position A except for the difference or change of position ΔL. This *vector change of position* ΔL (or AB) is the displacement of the point (a vector quantity),

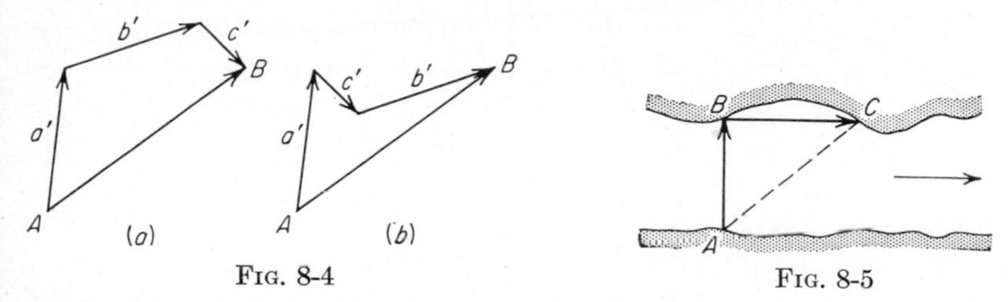

FIG. 8-4

FIG. 8-5

as before. The displacement obviously does not depend upon the particular position of the observer. It should be noted that, although the total distance traveled by the point is the length of the irregular curve from A to B, the change of position of the point is the straight line AB.

If a point is given a number of *successive* displacements, the total displacement of the point will be equal to the vector sum of the several displacements. This is indicated in Fig. 8-4, where it is also shown that the total displacement is the same regardless of the order in which the intermediate displacements are taken.

Displacements of a point which occur *simultaneously* can be illustrated by the example depicted in Fig. 8-5. A boat moving at right angles to the current in a river (along AB) also is carried downstream by the current. The total displacement AC of the boat will be the sum of the displacements AB, given it by the man rowing it, and BC, given it by the current. Actually, the two component displacements occur *simultaneously* so that the path of the boat is along the diagonal AC, with the same final displacement AC as if the displacements AB and BC had taken place separately.

3. *Velocity and Speed of a Point.* The concept of velocity is introduced by considering the time required for a given displacement. In Fig. 8-6 let a point be moving along the plane curve AB. It is at position 1 at

time t and at position 2 at time $(t + \Delta t)$. The displacement of the point during the interval is ΔL. Its *average velocity* during the interval is a vector quantity having the same direction as ΔL and a magnitude of $\Delta L/\Delta t$. The average *speed* of the point during the same interval is $\Delta s/\Delta t$.

The average velocity of a moving point during an infinitesimal time interval is called its *instantaneous velocity*. In this case the quantities Δs and ΔL become the infinitesimals ds and dL, respectively. Since the latter are practically equal in length, the instantaneous speed ds/dt is the same as the magnitude of the instantaneous velocity, and, since the direction of dL is practically the same as that of the tangent, the direction of the instantaneous velocity is also the same as that of the tangent.

The infinitesimal quantity ds is equal to the product $R(d\theta)$, where R is the radius of curvature and $d\theta$ is the angle between the two radii drawn from the center of curvature to the ends of the infinitesimal segment ds. The magnitude of the instantaneous velocity is therefore equal to $R(d\theta/dt)$.

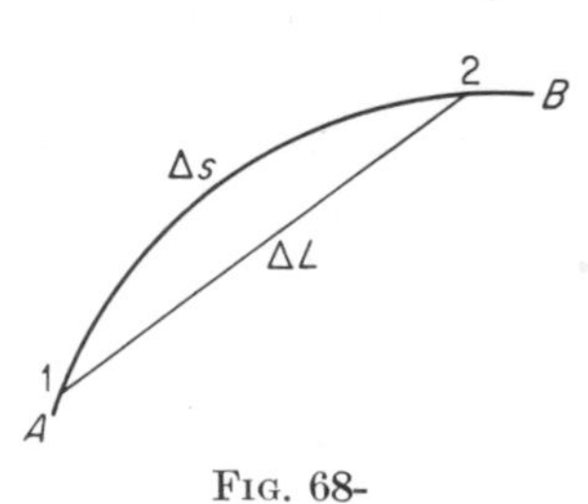

Fig. 68-

$$V = R \frac{d\theta}{dt} \tag{8-2}$$

If the radius of curvature is thought of as moving with the point, the quantity $d\theta/dt$ may be considered its angular velocity.

The term *velocity* is usually intended to mean *instantaneous* velocity and will be so understood hereafter.

It is apparent that a point which is given simultaneous displacements also is given simultaneous velocities when the time involved is considered. For the example (Fig. 8-5) of a man rowing a boat across a river,

$$(\text{Vel. of boat}) = (\text{vel. given it by man}) \nrightarrow (\text{vel. of stream}) \tag{8-3}$$

The total velocity is equal to the vector sum of the component velocities, and the velocity of a point in general may be considered the combined effect of several component velocities. The velocity polygon may be thought of as a large-scale drawing of the infinitesimally small displacement diagram as an aid to visualization.

8-5. Angular Motion. The *angular position* of a line in a plane is defined in terms of the angle that the line makes with some arbitrarily chosen reference line in the plane. The *angular displacement* of a line is the change of its angular position. If, for example, the line AB in Fig. 8-7 moves to a new position as indicated by the letters $A'B'$ it is said to have undergone an angular displacement of $\Delta\theta$ units of angle, counterclockwise.

The angular displacement of a line does not completely define its change of position. For example, the line AB might have been moved to some other position as indicated by the letters $A''B''$ and still have been given the same angular displacement $\Delta\theta$. Its final *angular position*, however, would be the same regardless of whether it were moved to $A'B'$ or $A''B''$.

If the angular displacement of a line takes place during the time interval Δt, then the *average angular velocity* of the line during the interval

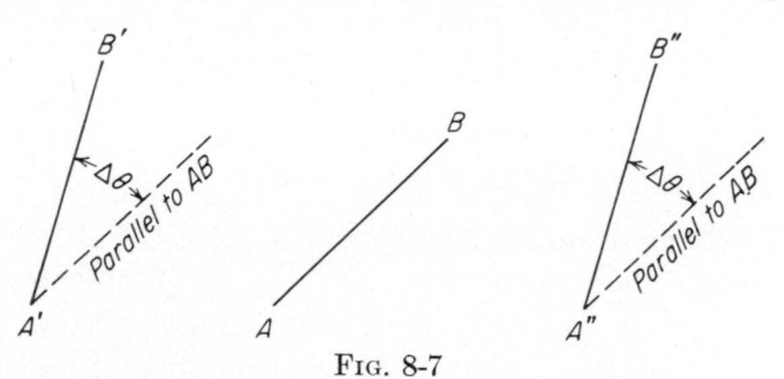

FIG. 8-7

is given by the ratio $\Delta\theta/\Delta t$. If an infinitesimal time interval is considered, the ratio $d\theta/dt$ gives the *instantaneous angular velocity* (or simply *angular velocity*).

$$\text{Angular velocity} = \omega = \frac{d\theta}{dt} \tag{8-4}$$

To find the angular velocity of a *rigid body* in plane motion, it is only necessary to find the angular velocity of any line on the body in the plane of motion. It should be obvious that all lines of a rigid body must have the same angular motion.

8-6. Relative Motion. 1. *Relative Displacement of Two Points.* The essential idea of relative motion has been stated in the previous articles of this chapter. For two moving points, for example, their relative velocity is merely the *vector difference* of their "absolute" velocities. Similar statements apply for the relative displacement and relative acceleration of the two points. The use of vector equations and corresponding vector diagrams will help to clarify the matter.

It was shown in Art. 8-4 that the displacement of a moving point is the vector change of its position relative to some fixed point. Instead of referring the position of a moving point to some fixed point, it may, at any given instant, be referred to the position occupied by some other moving point at the same instant. The position of one moving point relative to another is defined in exactly the same terms as those given in Art. 8-4 for the position of a moving point relative to some fixed point, and it is, therefore, a vector quantity of the same kind. Referring to Fig. 8-8a, let the two moving points A and B be considered. The position of the point B relative to the point A is given by the line AB measured in the direction AB. If the points are given the displacements ΔL_A and ΔL_B to the positions A' and B', the position of B relative to A will then be given by the line $A'B'$. Now let BB'' be drawn parallel and equal to ΔL_A. Since by construction $A'B''$ is equal to the relative position vector AB, it follows that $B''B'$ $(= \Delta L_{BA})$ is the vector change of relative position. By analogy this vector is called the displacement of point B relative to the point A. *The displacement of one point relative to another, therefore, is defined as the vector change of the relative position of the two points.* It can be seen from the figure that

$$\Delta L_B = \Delta L_A \mathrel{+\!\!\!\!\rightarrow} \Delta L_{BA}$$

or

$$\Delta L_{BA} = \Delta L_B \rightarrow \Delta L_A \tag{8-5}$$

It should be evident that the displacement ΔL_{AB} of A relative to B is equal and opposite the displacement ΔL_{BA} of B relative to A.

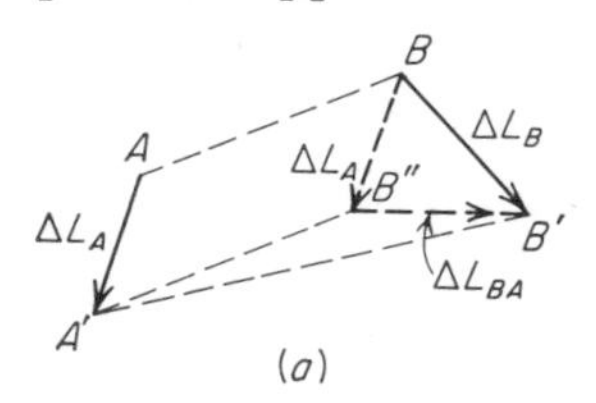

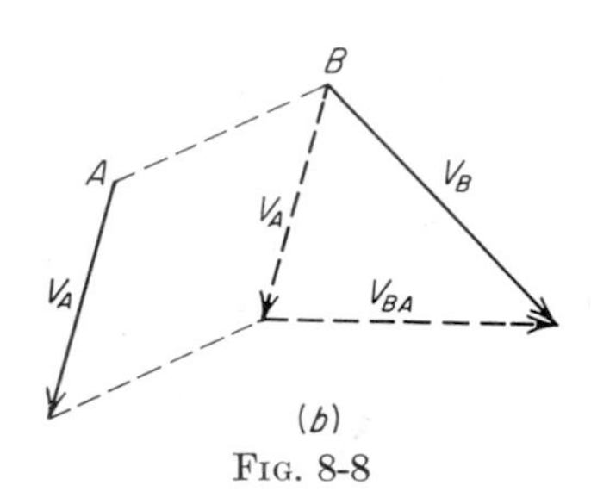

Fig. 8-8

2. *Relative Velocity.* Now let the displacement of two points A and B, during an infinitesimal interval of time dt, be considered. The infinitesimal displacements themselves are too small to be shown in a figure, but if the velocities of A and B are known, they may be represented by vectors as shown at b in Fig. 8-8, and these vectors will be proportional to the actual infinitesimal displacements. It follows that the vector V_{BA} is proportional to the infinitesimal displacement of B relative to A. It is evident from the figure that

$$V_B = V_A \mathrel{+\!\!\!\!\rightarrow} V_{BA}$$

or

$$V_{BA} = V_B \rightarrow V_A \tag{8-6}$$

The foregoing is a fundamental principle of extreme importance and should be thoroughly understood before proceeding any further. The expedient of considering the velocity vectors as large-scale representations of the actual infinitesimal displacements is of assistance in visualizing relative motion. Thus the "displacement" V_B may be considered the net effect of two simultaneous "displacements" V_A and V_{BA}. The final displacement would be the same if the two simultaneous displacements were taken separately, and the point B may be thought of as moving first with A, i.e., parallel to V_A to the end of "displacement" V_A, and thence on alone to its final position at the ends of "displacements" V_B and V_{BA}. During the first part of the journey, its position relative to A remains unchanged, and therefore all the change of relative position must take place during the second part. It follows that V_{BA} must be the "displacement" of B relative to A, and if the vectors are again taken to represent velocities, the vector V_{BA} must represent the relative velocity. Thus, *the velocity of one point relative to another is defined as the time rate of the vector change of the relative position of the two points.*

3. *Relative Angular Velocity of Two Rigid Bodies.* In Fig. 8-9 are shown reference lines drawn on two rigid bodies having angular velocities ω_2 and ω_3. The angular velocity of 3 relative to 2 is defined as the difference between ω_3 and ω_2. In equation form,

$$\omega_{32} = \omega_3 - \omega_2 \tag{8-7}$$

Also

$$\omega_{23} = \omega_2 - \omega_3 = -\omega_{32} \tag{8-8}$$

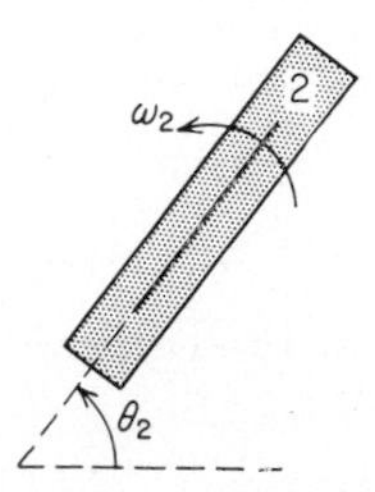

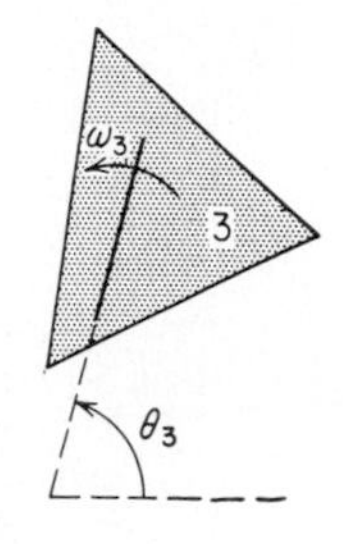

FIG. 8-9

Careful attention must be given to consistency of algebraic signs of numerical values used in the above equations. It might be noted that, if body 2 were stationary, ω_{32} would be identical with ω_3, the "absolute" angular velocity of 3. The equations show that the angular velocity of 2 relative to 3 is equal to and opposite the angular velocity of 3 relative to 2 (and the equations and conclusions are true for *any* two bodies whatsoever that might be numbered 2 and 3).

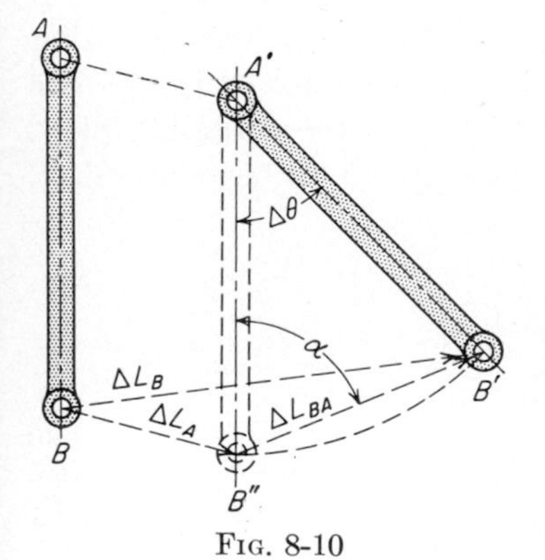

FIG. 8-10

METHOD OF RELATIVE VELOCITIES

8-7. Relative Velocity of Two Points on the Same Rigid Body. It was shown in Art. 8-4 that any displacement of a point may, for the purpose of analysis, be assumed to be made up of two or more simultaneous displacements, the vector sum of which is equal to the actual displacement. Referring to Fig. 8-10, let it be assumed that the body AB has been displaced to the position $A'B'$ during the time interval Δt, and let it be further assumed that the displacement of any point of the body such as B is made up of the two simultaneous displacements BB'', equal to the displacement AA' of some other point A, and $B''B'$, the displacement of B relative to A.

If the displacements BB'' and $B''B'$ are taken separately, the net result will be the same; i.e., if A and B move to A' and B'', and then if

B moves from B'' to B' along the arc as shown, the final positions of the two points will be the same as though the displacements occurred simultaneously. Since two points determine the position of a body having plane motion, it is evidently quite possible for the body to move as a whole along with the two points A and B in the manner just described, and it follows that if the body *is* moved in this manner the displacement of any third point will be made up of two displacements similar to those of B.

The vector equation for the displacement of the point B is

$$\overline{BB'} = \overline{BB''} \nrightarrow \overline{B''B'} = \overline{AA'} \nrightarrow \overline{B''B'}$$

or

$$\Delta L_B = \Delta L_A \nrightarrow \Delta L_{BA}$$

Dividing both sides of the equation by Δt,

$$\frac{\Delta L_B}{\Delta t} = \frac{\Delta L_A}{\Delta t} \nrightarrow \frac{\Delta L_{BA}}{\Delta t} \qquad \text{or} \qquad V_{B\text{avg}} = V_{A\text{avg}} \nrightarrow V_{BA\text{avg}}$$

The magnitude of $V_{BA\text{avg}}$ is given by

$$V_{BA\text{avg}} = \frac{\Delta L_{BA}}{\Delta T} = \overline{AB}\,\frac{2 \sin \Delta\theta/2}{\Delta T}$$

where $\Delta\theta$ is the angular displacement of the body. The direction $V_{BA\text{avg}}$ makes an angle α with both the initial and final positions of the line AB. This angle is obviously the same for any other point of the body.

Now let it be assumed that the displacements all take place during the infinitesimal interval of time dt. The average velocities become instantaneous velocities; the angle α becomes, except for negligible quantities, equal to 90 deg; and the quantity $(2 \sin \Delta\theta/2)/\Delta t$ becomes $d\theta/dt$, the angular velocity of the body. Therefore

$$V_B = V_A \nrightarrow V_{BA} \tag{8-9}$$

$$V_{BA} = \overline{AB}\,\frac{d\theta}{dt} = \overline{AB}\,\omega \tag{8-10}$$

In summary,

1. The velocity of any point of a rigid body is given by the vector sum of the velocity of some other point and the velocity of the first point relative to the second.

2. The velocity of any point of a rigid body relative to any other point of the body is a vector quantity having a magnitude equal to the product of the angular velocity of the body and the distance between the points, and a direction at right angles to the line connecting the points.

Let it be noted that the first statement is merely a special case of the statement of the relation that exists between the velocities of two independent points and their relative velocity.

8-8. Applications of the Method of Relative Velocities. Points on the Same Link. The simplest and most fundamental application of the method of relative velocities is that shown in Fig. 8-11. The velocity of point A of the rigid body is assumed to be completely known, and it is also assumed that the direction of the velocity of the point B is along the line XX. The velocity of the point B is

$$V_B = V_A \nrightarrow V_{BA}$$

The direction of V_{BA} is perpendicular to the line AB. If, therefore, the vector V_A is laid off from B and a line drawn through its extremity at right angles to AB, the intersection of this line with the line XX will determine the velocity vectors V_B and V_{BA} as shown. The vector diagram must be consistent with the vector equation which it represents.

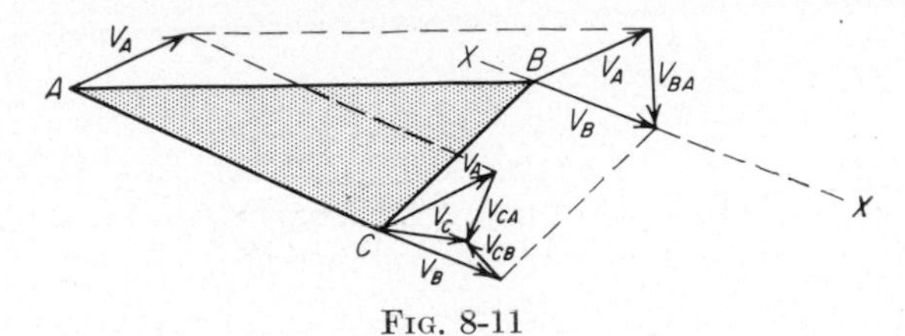

FIG. 8-11

Now let it be required to determine the velocity of any third point C. First, let the two following vector equations be written:

$$V_C = V_A \nrightarrow V_{CA}$$
$$V_C = V_B \nrightarrow V_{CB}$$

The simultaneous solution of these two vector equations may be obtained as follows: Let the vectors V_A and V_B be laid off from the point C, and let lines be drawn through their extremities perpendicular to AC and BC, respectively. The intersection of these two lines locates the extremity of the vector V_C. If now the construction at C is compared with the preceding equations, it will be found to satisfy both of them, and the vector V_C must, therefore, represent the velocity of the point C.

The angular velocity of the rigid body is

$$\omega = \frac{V_{BA}}{AB} = \frac{V_{CA}}{AC} = \frac{V_{CB}}{BC} \qquad \text{(clockwise)}$$

8-9. Application to a Simple Mechanism. The Velocity Polygon. In Fig. 8-12*a* is shown a four-link mechanism in which V_A, the velocity of point A on link 2, is assumed to be known and V_B, the velocity of a point on link 4, is required. The velocity of A has a magnitude determined from the angular velocity and length of link 2, ($V_A = \overline{O_2A}\omega_2$). It has a direction perpendicular to O_2A. The point A, being the center of a turning joint, is actually the position of a pair of coincident points, one on link 2 and one on link 3. These points must have the same velocity, as otherwise links 2 and 3 would separate at A. The velocity of one point on link 3 is therefore known as soon as V_A is known. Likewise B represents a pair of coincident points on links 3 and 4. Hence, since A and B may be considered as points on link 3, the analysis of the preceding article applies. Therefore $V_B = V_A \nrightarrow V_{BA}$. In the vector addition representing this equation, the sides of the triangle are:

V_A, known in magnitude and direction.

V_B, known to be perpendicular to O_4B.

V_{BA}, known to be perpendicular to BA.

The vector polygon is shown in Fig. 8-12*b*. First the known vector V_A is laid off, to some scale. Next a line *mm* is drawn from the extremity

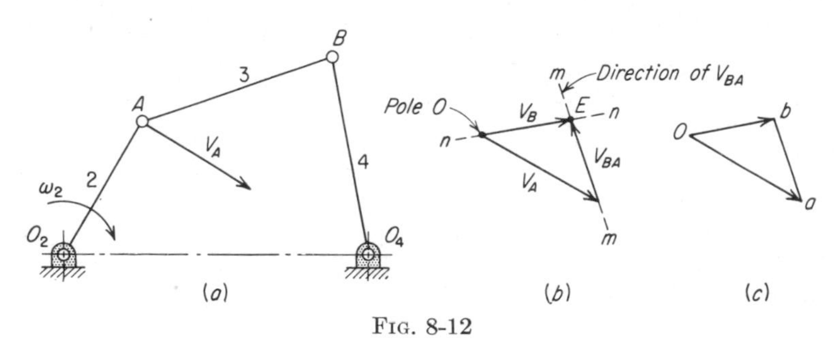

FIG. 8-12

of V_A perpendicular to BA. From the origin of vector V_A a line *nn* is drawn perpendicular to O_4B. The intersection of these lines determines the vectors V_B and V_{BA} completely. *The proper placement of the arrowheads on the intersecting lines is determined by ensuring that the diagram maintains the equality expressed by the vector equation. Both sides of the equation are started from the pole shown and both add outward to end at point E.*

Making the vector addition on the drawing of the mechanism at the point whose velocity is required, as shown in Fig. 8-11, has certain advantages but results in confusion of lines in complex mechanisms. Hence, in applying the method of relative velocities, such constructions will generally be made off the figure, as in Fig. 8-12*b*.

In Fig. 8-12*c* is shown the same diagram with a simplified notation. In this diagram it is observed that the absolute-velocity vectors radiate from a pole *o*, and the relative-velocity vector is represented by the line joining the extremities of the absolute-velocity vectors. The small letters at the extremities of the vectors correspond with the capital letters on the mechanism. Thus *oa* represents V_A, *ob* represents V_B, and *ab* and *ba*, in the order given, represent V_{BA} and V_{AB}, respectively. This diagram, Fig. 8-12c, is known as the *velocity polygon*. In drawing velocity polygons it is important to keep in mind that all absolute-velocity vectors are lines radiating from the pole, whereas all relative-velocity vectors are lines joining the ends of absolute-velocity vectors. In what follows arrows representing the sense will be placed on absolute-velocity vectors but usually will be omitted from relative-velocity vectors for convenience in reading these vectors in either sense as desired.

In Fig. 8-13*a* the four-link mechanism of Fig. 8-12*a* has been redrawn with link 3 extended to include a third point *C*, the velocity of which is to be determined. First, a pole *o* is chosen (Fig. 8-13*b*), and, with *oa* (the velocity of *A*) known, *ob* (the velocity of *B*) is determined, as in Fig. 8-12*c*.

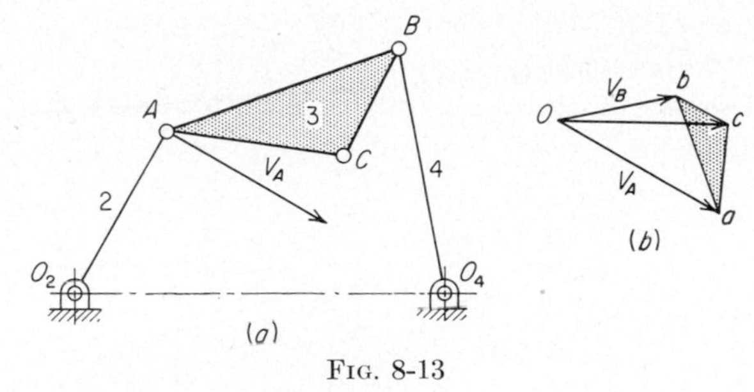

FIG. 8-13

To determine the velocity of point *C*, consider the following equation:

$$V_C = V_A \nrightarrow V_{CA}$$

In this equation, V_A is known in both magnitude and direction, and V_{CA} is known to be perpendicular to the line *CA*. If from the extremity of vector *oa* in the velocity polygon (Fig. 8-13*b*) a line is drawn perpendicular to *CA*, it is known from the equation that the extremity of vector *ac* will lie on this line. The complete determination of V_C, however, requires the use of another equation

$$V_C = V_B \nrightarrow V_{CB}$$

In this equation V_B is known in both magnitude and direction, and V_{CB} is known to be perpendicular to the line *CB*. If, therefore, from the extremity of vector *ob* in the velocity polygon, a line is drawn perpendicular to *CB*, it is known from the equation that the extremity of the vector *bc* will lie on this line. The intersection *c* of the two lines perpendicular respectively to *AC* and *BC* determines the magnitudes of the relative velocity vectors *ac* and *bc* and at the same time the direction and magnitude of vector *oc*, which represents V_C, the velocity of *C*. This method of determining the velocity of *C* is essentially the same as that employed in connection with Fig. 8-11; it has been described in detail to bring out the manner in which the velocity polygon may be used to reduce graphical work.

In the velocity polygon (Fig. 8-13*b*), the line *ab* representing V_{BA} was drawn perpendicular to the line AB. The lines *ac* and *bc* were drawn perpendicular respectively to AC and BC. Thus, the three sides of the shaded triangle *abc* of the velocity polygon are perpendicular respectively to the sides of the link ABC, and it follows that the triangle *abc* in the polygon is similar to the triangle ABC in the mechanism. For this reason, the triangle *abc* in the velocity polygon is known as the *velocity image* of the link ABC. Each line or link in the mechanism has its image in the velocity polygon. Lines *ab*, *bc*, and *ac* are images of lines AB, BC, and AC, respectively. Line *oa* is the image of O_2A (link 2), *ob* is the image of O_4B (link 4), etc. The image of link 1 (the fixed link) is the pole *o*, the point of zero velocity, since link 1 has no motion.

8-10. Application to Sliding Members Such as Cams and Gears. In the cam mechanism (Fig. 8-14), which can also be taken for the purposes

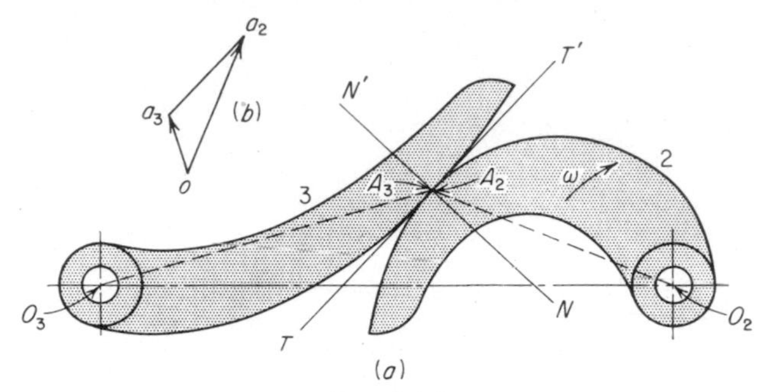

FIG. 8-14

of analysis as representing a pair of gear teeth, link 3 is driven by link 2. At the instant under consideration, point A_2 of the driver is in contact with point A_3 of the follower. From the angular velocity of the driver (link 2), the linear velocity V_{A_2} of point A_2 can be found. That is, $V_{A_2} = (O_2A_2)\omega_2$. The problem is to find the velocity V_{A_3} of point A_3 on the follower. It is known that

$$V_{A_3} = V_{A_2} \nrightarrow V_{A_3A_2}$$

Velocities V_{A_2} and V_{A_3} are known to be perpendicular respectively to O_2A_2 and O_3A_3. To draw the velocity polygon, the direction of $V_{A_3A_2}$ must be at least partially known. Since 2 and 3 are rigid links, there can be no relative motion in the direction of the normal NN'. Therefore A_2 and A_3 can move relative to each other only along the common tangent TT'. The polygon is shown in Fig. 8-14*b*. Vector oa_2 is laid off equal to the known velocity V_{A_2}. A line is drawn from *o* in the known direction of V_{A_3} and a line from a_2 in the known direction of $V_{A_3A_2}$ (parallel to the tangent TT'). The intersection of these two lines at a_3 gives V_{A_3} and $V_{A_3A_2}$. With the velocity of one point on 3 known, the velocity of any other point on the link, or the angular velocity of the link, can easily be determined.

8-11. Application to a Complex Mechanism. In Fig. 8-15 is shown the skeleton outline of a toggle press. Link 2, the driving link, has a constant angular velocity. Link 8 is the last driven link. If crank O_2A has a length of 10 in. and an angular velocity of 1 radian/sec, the point A will have a velocity of 50 fpm. Knowing V_A, the problem is to determine V_F (the velocity of the ram, link 8) by the method of relative velocities. The mechanism is shown at about the beginning of its stroke, where V_F is large compared with its value during the pressure portion of the stroke.

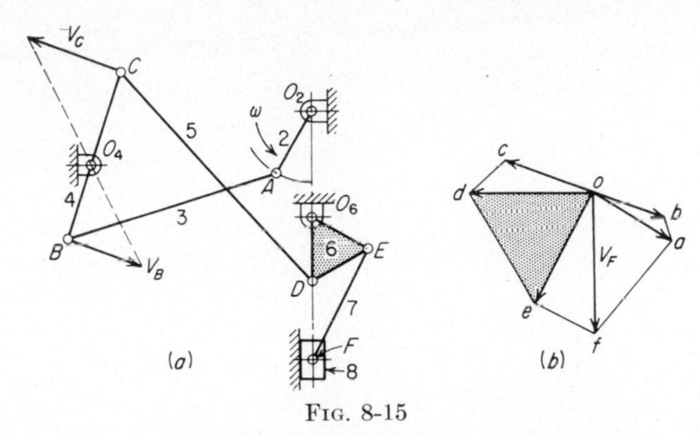

FIG. 8-15

Figure 8-15*b* is the velocity polygon, which was drawn beginning with V_A (equal to *oa*) and progressing through the mechanism to V_F (equal to *of*). A velocity scale of 1 in. = 50 fpm was used, and V_F on being scaled off was found to be equal to 73 fpm for the configuration of the mechanism shown. Note that V_{FA} (= *af*) is *not* necessarily perpendicular to a line drawn from F to A, because the length of such a line can change; therefore V_{FA} can have a component *along* the FA direction as well as perpendicular to it.

It is left to the reader to write the necessary vector equations and relate them to the vector diagram of Fig. 8-15*b*.

8-12. Methods of Attack. The solution of a velocity problem requires in general the exercise of a certain amount of ingenuity, as almost every problem is more or less a special case. The best method of obtaining a solution depends largely upon the conditions of the particular problem at hand. By way of illustration let the mechanism shown in Fig. 8-16 be considered. The velocity of the point A is known, and the velocities of the points B and C are required. The velocities V_B and V_{BA} are determined by the methods used in the preceding articles. In the determination for point C, however, these methods break down because of the impossibility of securing intersections. In order to determine V_C, let it first be noted that

$$V_{BA} = (AB)\omega_3 \qquad \text{and} \qquad V_{CA} = (AC)\omega_3$$

Therefore,

$$\frac{V_{CA}}{V_{BA}} = \frac{AC}{AB}$$

The vector V_{CA}, being proportional to V_{BA} in the ratio AC/AB, may be determined by means of the ordinary similar triangle construction shown at b in the figure. This construction may, if desired, be made on the drawing of the mechanism itself as shown by the dashed lines. The velocity of C is now found as the vector sum of V_A and V_{CA}. The vector diagram has been drawn at point C in the figure.

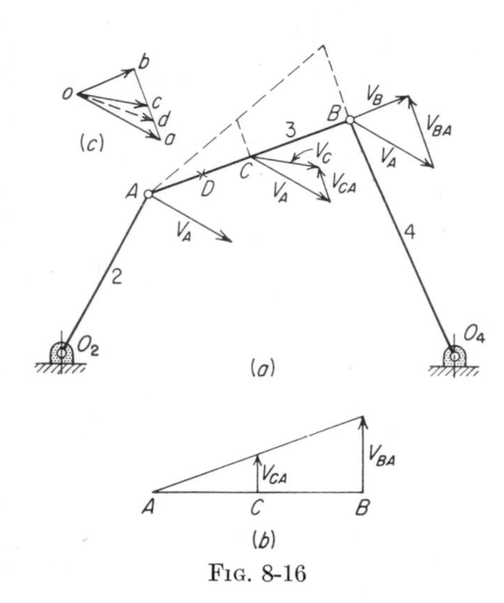

FIG. 8-16

The velocity image of ACB is at c in Fig. 8-16. As an illustration of a use of the image, the velocity of a point D halfway between A and C has been found by locating the position d in the image halfway between a and c. The velocity of D is the vector od.

8-13. Practical Considerations. It is seldom the case that the velocities are required for a single position only of a mechanism. Usually the velocities must be determined for a sufficient number of positions to show how the motions of the machine members vary during one complete cycle. Time and trouble can be saved by giving a little consideration to the method of attack before actually starting the problem; and, as previously stated, practically every problem may profitably be considered as a special case. The problem given in the following article is intended as an illustration of the foregoing statements; it is not in any sense intended to illustrate a *specific* method of procedure applicable to any large number of different problems.

8-14. Velocity Analysis of a Crank Shaper. Let it be required to make a velocity analysis of the crank shaper shown in Fig. 8-17. In order to vary the stroke of the ram the length of the crank arm may be varied by the bevel gears and screw shown in the figure. To avoid extended discussion, the analysis will be made here for one length of crank arm only. This involves the determination of the velocities of the several members of the machine for all positions during one complete cycle; i.e., during one revolution of the crank. The general method of procedure is to make velocity determinations for a number of positions during the cycle and to approximate the velocities corresponding to the intermediate positions from the data thus obtained, usually from a plotted curve.

The crank gear of the shaper under consideration is assumed to turn at a constant angular velocity of 10½ rpm. The length of the crank arm is 6 in. It is required to determine the velocity of the ram for all

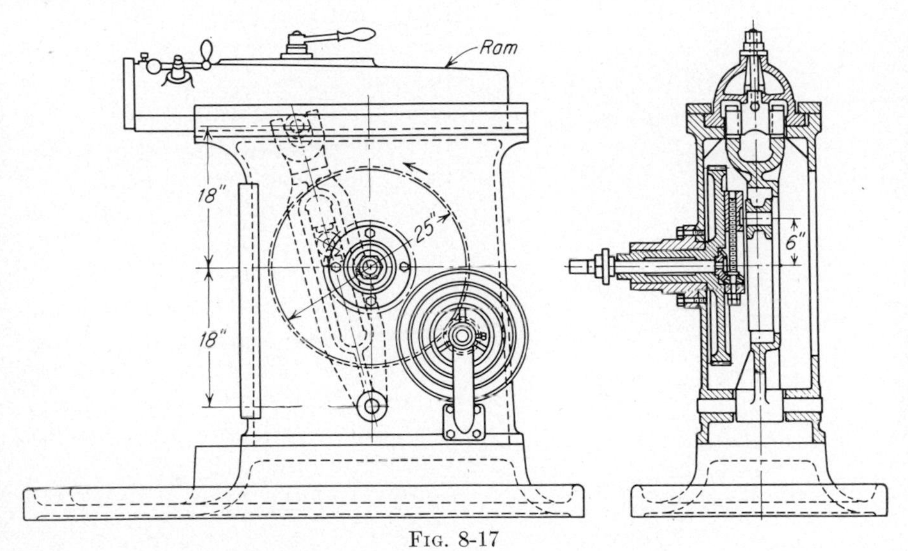

Fig. 8-17

positions of the crank, the mechanical advantage of the mechanism, and any other items of interest that may appear as the analysis progresses.

1. *Preliminary Analysis.* The first steps in the analysis are to choose a suitable scale (a scale of 1½ in./ft was chosen for the original drawing) with which to draw the skeleton outline of the mechanism in any convenient position, as shown in Fig. 8-18*a*, and to calculate the velocity of the crankpin as follows:

$$V_{A_2} = \frac{6(2\pi)}{12} 10\tfrac{1}{2} = 33 \text{ fpm}$$

The requirement now is to determine V_{B_5}, the velocity of the cutting tool. The analysis will proceed from link to link, starting with known velocity on link 2. After choosing a suitable scale (a scale of 1 in. = 30 fpm was chosen for the original drawing), the velocity V_{A_2} (which is also V_{A_3}) of the center of the crankpin may be laid off as the vector oa_2 in the polygon at b in the figure. Now let the velocity of the point A_4 be considered. The vector equation relating the velocity of the point on link 4 to a point on link 3 is

$$V_{A_4} = V_{A_4A_3} \nrightarrow V_{A_3}$$

A_4 is a point on link 4 coincident with the point A_3 on link 3 and may be thought of as being situated on link 4 directly beneath the point A_3.

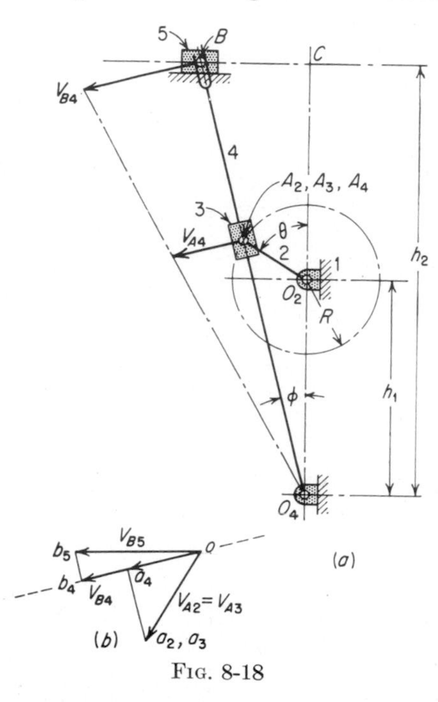

FIG. 8-18

Since link 4 rotates about the fixed center O_4, the velocity V_{A_4} will have a direction perpendicular to the line O_4A_4, and this direction may therefore be laid off along oa_4 in the polygon. The relative motion of A_3 and A_4 is along the center line of the slide, which in this case is the line O_4A_4. If, therefore, a line is drawn from the extremity of vector oa_3 and parallel to the line O_4A_4, the intersection of this line with the line oa_4 will determine the point a_4 and therefore the magnitudes of the velocities V_{A_4} and $V_{A_4A_3}$. The point of link 4 situated at B has a velocity of which the direction is perpendicular to the line O_4B. Since link 4 rotates about a fixed axis, the velocity V_{B_4} is proportional to V_{A_4} in the ratio O_4B/O_4A, and may be determined by the similar-triangle method shown on the figure and drawn as ob_4 in the polygon. The ram, link 5, is constrained to move in a horizontal direction, i.e., it has a motion of horizontal translation, and the point B_5 will, therefore, also move in a horizontal direction. The relative motion of B_4 and B_5 is along the center line of the fork, i.e., along the line O_4B. The vector equation is $V_{B_5} = V_{B_5B_4} \nrightarrow V_{B_4}$. The velocities $V_{B_5B_4}$ and V_{B_5} may be determined by drawing a line through the extremity of ob_4, in the polygon parallel to the line O_4B, to intersect the horizontal line ob_5 drawn from the pole o. The vector ob_5 represents the velocity of the cutting tool.

2. *Analysis for All Positions of the Crank.* The velocity determination having been made for a single position of the crank, ways and means may now be considered for reducing the labor involved in dealing with the other positions. As the next step in the solution, let the skeleton outline be redrawn as shown in Fig. 8-19 and the crankpin circle divided

into 24 equal parts, starting at the extreme right where the ram is at the beginning of the cutting stroke. Now let the velocity polygon of Fig. 8-18 (which is the polygon for crank position 11) be redrawn for the present purpose, as shown at *b* in Fig. 8-19. If this polygon is turned through 90 deg, as shown at *c* in Fig. 8-19, it will be observed that oa_2 is parallel to the crank O_2A, that ob_4 is parallel to the rocker arm O_4B, and that ob_5 is in the direction of the vertical center line of the mechanism. This suggests the possibility of using a modified drawing of the

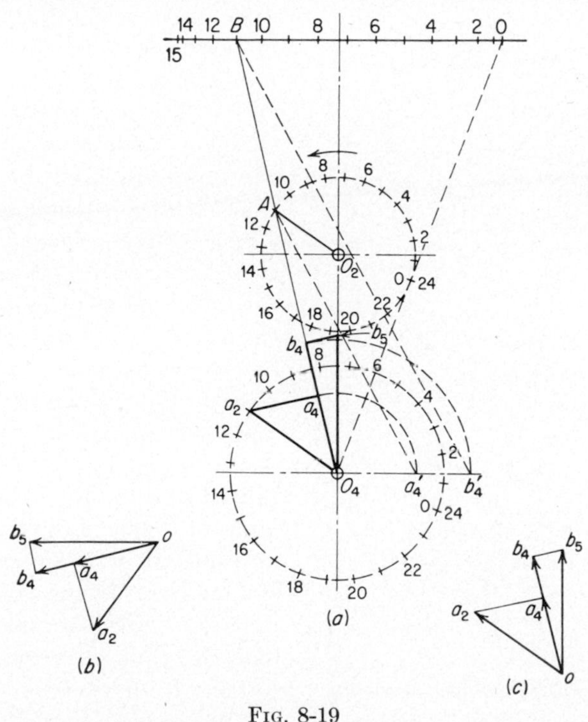

FIG. 8-19

mechanism itself as the velocity polygon. (A *revolved velocity polygon* such as shown in Fig. 8-19*c* could be drawn for any mechanism, and its use is recommended by some authors because of the convenience in drawing the vectors in the polygon parallel instead of perpendicular to the center lines of the links of the mechanism.)

The next step in the solution is to construct the revolved polygon of *c* in Fig. 8-19 on the mechanism, as shown by the heavy lines, with the pole *o* placed at the center O_4. The O_4a_2 line represents the velocity of point A_2 (and A_3). Point a_4 is established at the location where a line from a_2 drawn perpendicular to the rocker arm O_4B meets the rocker arm. The parallel dotted lines drawn from *A* and *B* show how the proportion $oa_4/ob_4 = O_4A_4/O_4B_4$, necessary to locate point b_4, can be conveniently obtained. From b_4 a line is drawn perpendicular to the rocker arm O_4B, and point b_5 is located at the intersection of this line with the vertical center line O_4O_2. The line O_4b_5 now represents the velocity of the ram (but revolved 90 deg from the true direction). Now let the circle be drawn with O_4a_2 (the velocity V_{A_2}) as a radius, with numbered divisions corresponding to those of the crankpin circle. The velocity polygon for any other position of the crank, and hence the desired velocity V_{B_5} of the ram, can now be easily determined by lightly drawing in the new configuration of the mechanism and drawing the lines of the revolved polygon, from O_4 as the pole, parallel to or coincident with the links of the mechanism as illustrated for crank position 11 in the foregoing discussion.

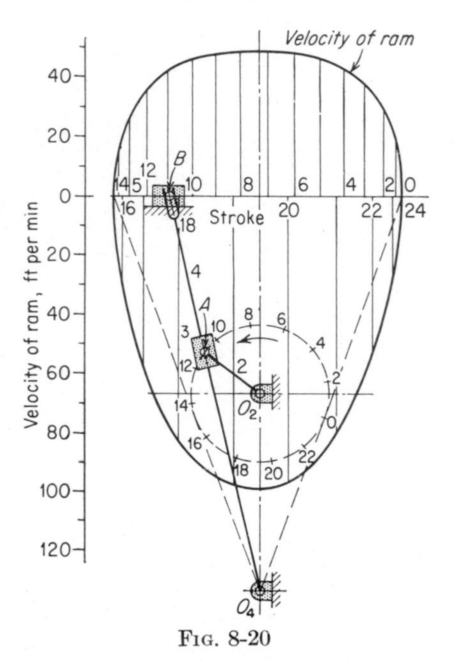

FIG. 8-20

Since it is the velocity vector ob_5 (which represents the velocity of the ram) that is required for each of the 24 crank positions, it will be observed as the work progresses that merely the location of the intermediate points will suffice, and many of the lines in the constructions radiating from the pole O_4 can be omitted. The motion of the ram on each side of the mid-position is symmetrical; so the velocity determinations need be made for one-half the crank circle only.

3. *The Velocity-Space Diagram.* Starting with the ram in the zero position, at the extreme right (Fig. 8-20) and continuing through the complete cycle, the velocity of the ram V_{B_5} (vector ob_5 from Fig. 8-19) has been plotted on the stroke of the ram as a base line. The ordinates of the curve above the base line represent the speed of the cutting tool during the working stroke (right to left), and the ordinates below the base line represent the speed of the tool during the idle, or return, stroke. Not only is this a useful diagram in the study of an existing mechanism of this kind, but it is also useful in the original design, where various arrangements and proportions of links may be tried out and the diagrams compared in order to secure the most desirable arrangement. In some designs modifications of the standard or conventional linkage are made in order to secure a flatter curve on the cutting stroke, i.e., a more uniform cutting speed.

4. *The Velocity-Time Diagram.* Since the angular velocity of the crank is assumed to be constant, the 24 equal divisions on the crank circle represent equal intervals of time. If, therefore, 24 equal spaces of any convenient length are laid off on a horizontal line, as shown in Fig. 8-21, and if the velocities of the ram are laid off as ordinates perpendicular to this line, the velocity-time diagram for the motion of the ram will be obtained. This diagram also brings out very clearly the relation between the cutting speed and the return speed of the tool.

The mechanical advantage during the cutting stroke may be determined from the velocity analysis. The mechanical advantage may be

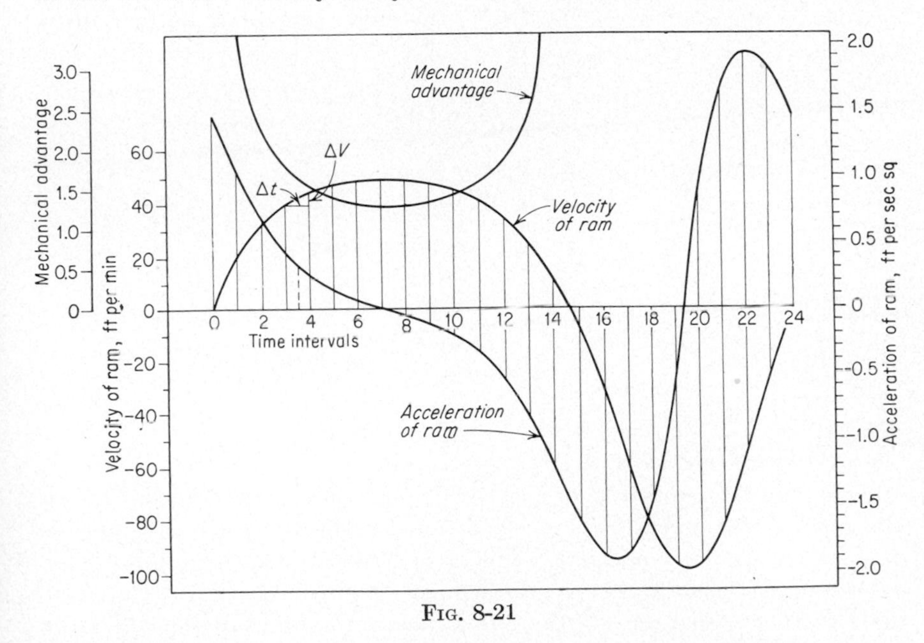

FIG. 8-21

defined as the ratio of the resistance encountered by the tool to the tangential force exerted upon the crank gear at the pitch line, on the assumption of 100 per cent efficiency. It may be shown that this ratio is equal to the ratio of the pitch-line velocity of the crank gear to the velocity of the cutting tool. The pitch-line velocity of the crank gear, whose pitch diameter is 25 in., may be calculated as follows:

$$V = R\omega = \tfrac{1}{2} \times \tfrac{25}{12} \times 2\pi \times 10.5 = 68.71 \text{ fpm}$$

If, therefore, 68.71 is divided by the velocity of the ram for any position in the cutting stroke, the mechanical advantage for that position will be obtained. These values have been calculated and plotted in Fig. 8-21. The curve shows very clearly that near the end of the stroke the mechanical advantage becomes very large, so that if the tool meets any unusual obstruction when the ram is in this position, very large forces are set up.

5. *Acceleration and Inertia Force of Ram.* Although the discussion of acceleration is presented in Chap. 9, the usual reader of this book probably has sufficient background to follow the rather simple discussion which follows. He needs only to recognize that acceleration is the time rate of change of velocity, and that forces are required to produce accelerations. The velocity-time diagram furnishes data for an approximate determination of the inertia forces of the ram. In Fig. 8-21, a length equal to 24 position spaces represents the time required for one revolution of the crank. One space, therefore, represents an increment of time

$$\Delta t = \frac{60}{10.5 \times 24} = 0.238 \text{ sec}$$

Between numbered positions 3 and 4, for example, the increment of velocity V_{B_5} is

$$\Delta V_{B_5} = 0.740 - 0.660 = 0.080 \text{ fps}$$

The average acceleration is

$$A_{B_5} = \frac{\Delta V_{B_5}}{\Delta t} = \frac{0.080}{0.238} = 0.336 \text{ ft/sec}^2$$

After a suitable scale has been chosen (in this case 1 in. = 0.6 ft/sec^2) the average acceleration is laid off midway between positions 3 and 4. This having been done for all such positions, and a curve drawn through the points, the approximate accelerations of the ram corresponding to the numbered positions may be obtained by scaling the ordinates at these points. The weight of the ram of this machine is estimated as 300 lb. The inertia force of the ram at position 4, for example, where the scaled value of the acceleration is 0.258 ft/sec^2, is

$$F = \frac{W}{g} A_{B_4} = \frac{300}{32.2} \times 0.258 = 2.42 \text{ lb}$$

The maximum inertia force (near positions 16 and 22) is about 18 lb.

METHOD OF INSTANTANEOUS CENTERS

8-15. Instantaneous Center of Rotation of a Rigid Body. From the figure and discussion of Art. 8-7 it is apparent that the displacement of a body having plane motion may, for the purpose of analysis, be considered as made up of a translation *with* one of its points and a rotation *about* that point. Upon this concept the method of relative velocities was based. The method of instantaneous centers is based upon a similar concept, namely, that any displacement of a body having plane motion may be considered as a pure rotation about some center.

Referring to Fig. 8-22, let rigid body AB be displaced to any new position $A'B'$. Regardless of how this is done, it is evident that the change of position *could* have been accomplished by pure rotation of the triangle OAB about the center O, which is located at the intersection of the perpendicular bisectors of AA' and BB'.

The displacements of A and B can be written as:

$$\text{Displacement of } A = (OA)2 \sin \frac{\Delta\theta}{2}$$

$$\text{Displacement of } B = (OB)2 \sin \frac{\Delta\theta}{2}$$

For a very small displacement, the angle $\Delta\theta$ becomes infinitesimal, and the body can be considered to be rotating instantaneously about the point O as an instantaneous center.

Dividing both sides of the above equations by Δt, and allowing $\Delta\theta$, Δt, and the displacements to become infinitesimal, yields for the magnitudes of the instantaneous velocities of points A and B

$$V_A = (OA)\omega$$
$$V_B = (OB)\omega$$

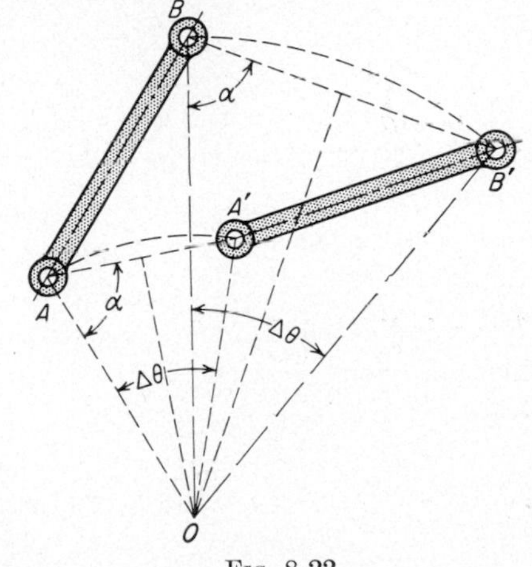

FIG. 8-22

Here, ω is the angular velocity $d\theta/dt$ of OA and OB, the radii drawn from instantaneous center O to points A and B. The directions of the velocities of A and B are the same as those of the infinitesimal displacements of A and B, which are obviously perpendicular to the radii OA and OB.

Because A and B are representative of any two points of any rigid body in plane motion, generalization is possible.

$$V = R\omega \tag{8-11}$$

in which ω must be expressed in radians per unit time, and the linear units of V and R must correspond. In this equation R is the radius from the instantaneous center of rotation for the body to the point having the velocity V, the entire body having the angular velocity ω.

The practical significance of the instantaneous-center concept can best be brought out by means of a simple example. Referring to Fig. 8-23a, let it be assumed that the link AB has plane motion, that V_A is the known velocity of point A at a given instant, and that the direction of V_B at the same instant is also known. The instantaneous radii of the points A and B are, as seen above, perpendicular to the directions of the velocities V_A and V_B respectively, and hence may be drawn from A and

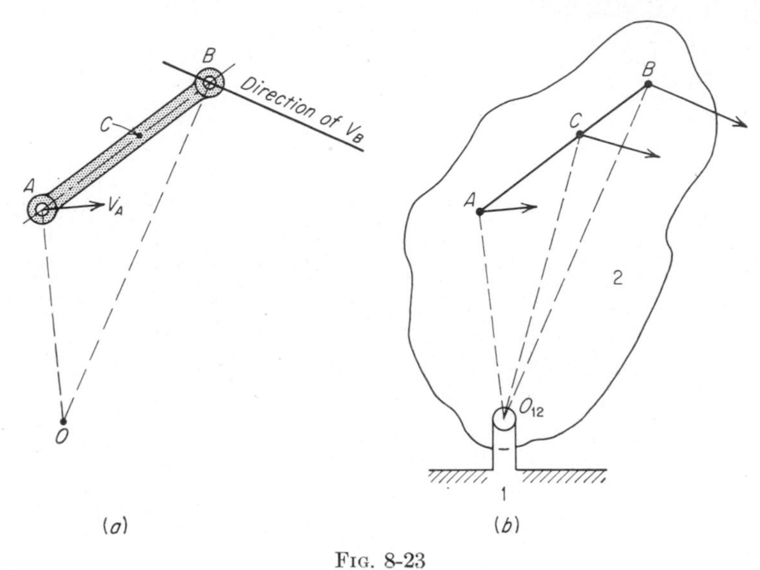

FIG. 8-23

B as indicated by the dashed lines. Their intersection locates the instantaneous center O. The magnitude of V_B can then be determined from the relations:

$$V_A = (OA)\omega \qquad V_B = (OB)\omega$$

i.e.,

$$\frac{V_B}{V_A} = \frac{OB}{OA} \tag{8-12}$$

the velocities being, therefore, proportional to their instantaneous radii. Similarly, any third point C has a velocity perpendicular and proportional to its instantaneous radius OC (not shown).

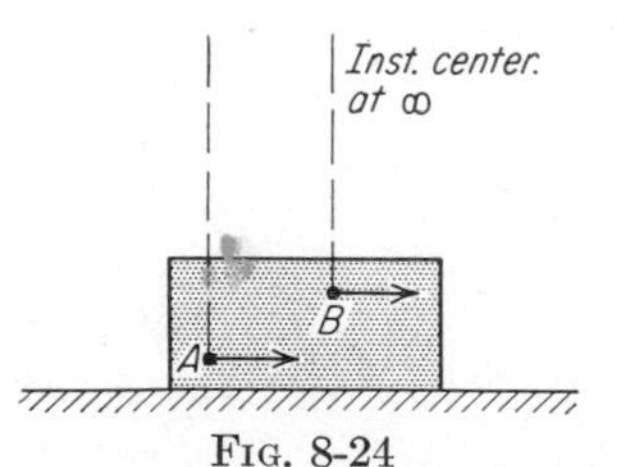

FIG. 8-24

It may be assumed, as an aid in visualization, that the body is extended and pivoted at O as shown in b of Fig. 8-23, but *only for the instant under consideration.*

It is important to note that the link AB is not *necessarily* constrained to *continue* to rotate about the point O and will not in general do so. In velocity analyses, however, we are concerned only with the conditions that exist at each specified instant during the motion of a body, and it is convenient, therefore, to think of the motion of the link AB as being a rotation about the instantaneous center O at each instant under consideration.

A special case of the instantaneous center is illustrated in Fig. 8-24. Here the link has a motion of straight-line translation, and hence the instantaneous radii of any two of its points are parallel and intersect at infinity.

8-16. Instantaneous Center of a Pair of Links. The previous section dealt with the concept of an instantaneous center of rotation for a single body or link. Such a center is, in effect, a location at which the *moving* link may be considered to be pivoted (connected by a hinge joint) to a *stationary* link *for an instant for the purpose of velocity analysis at that instant.* Figure 8-23*b* illustrates this idea.

The instantaneous center of a *pair* of links is merely an extension of this idea, and is the point about which one of the links may be considered to be rotating with respect to the other link at a given instant, for the purpose of velocity analysis, without regard to whether or not one of the links is stationary. Referring again to Fig. 8-23, it is evident that the instantaneous center O_{12} is the common position of the *only* pair of coincident points of links 1 and 2 that have the same velocity, namely, zero. Now let it be assumed that link 1 is set in motion but that the *relative* motion of the two links remains unchanged for the instant; the last requirement is met if the relative velocity of every pair of coincident points is the same as before. The relative velocity of the pair at O_{12} was zero, since the absolute velocity of each of the two points was zero. With link 1 in motion, and with the relative motion of the two links remaining unchanged, the relative velocity of the pair of coincident points at O_{12} must remain the same as before, namely, zero; hence the absolute velocities of these two points must be identical. This leads to the following very useful definition. The instantaneous center of *any* pair of links is at the common position of that pair of coincident points, one on each link, that have the *same* absolute velocity. An equivalent definition is that it is at the common position of the pair of coincident points having zero relative velocity. At such a position the two links can be visualized as being connected by a hinge joint for an instant, for the two links have identical instantaneous velocities (magnitudes and directions) at this point.

Methods for finding and using these centers are discussed in more detail in following sections, but first a few examples will be given to assist understanding. In Fig. 8-25, the numbers 12 (one two), 13 (one three), 23 (two three), etc., are used to indicate the positions of the instantaneous centers of the pairs of links designated by the pairs of numbers. At *a* is shown a four-bar linkage. The locations of instantaneous centers 12, 23, 34, and 14 are obvious. When two links are connected by a turning joint, then certainly at the center of that joint the two links are constrained to have the same velocity. Center 13 is located at the intersection of the two lines drawn, as shown, perpendicular to the velocities of points A and B of link 3. This center is the instantaneous center of rotation for link 3, just as 12 and 14 are the centers of rotation at this instant for links 2 and 4, respectively.

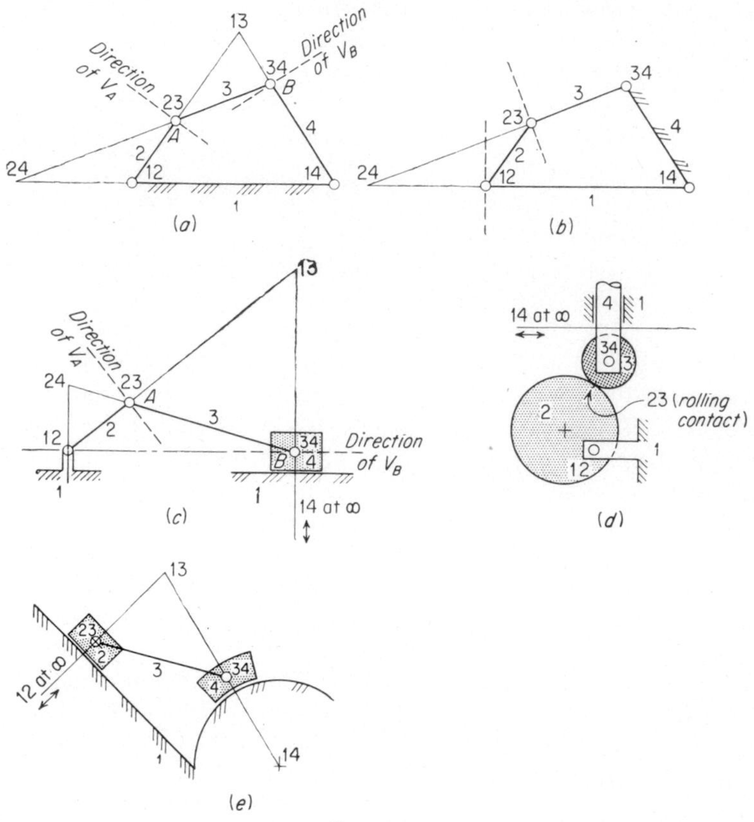

FIG. 8-25

In a four-link mechanism, there are six possible combinations of two links, and so there are six possible instantaneous centers. Five of these have been determined. The location of the sixth one, 24, depends upon the fact that, when a mechanism is *inverted*, the *relative motions* among the parts remain unchanged. Therefore, the position of the center for relative motion between any two links remains unchanged. To apply this idea to Fig. 8-25*a*, consider the inversion of the mechanism shown at *b* in which link 4 is stationary rather than link 1. The location of instantaneous center 24 then follows exactly the same thought and procedure used in locating 13 in the original mechanism at *a*. The center 24 then remains at the same position when the linkage is reinverted to become the original mechanism.

It should be noted that 23, 34, and 24 are all similar types of instantaneous centers. None is a center of rotation as are 12, 13, and 14, because none of the links involved is a fixed link. Position 24 is the location of a point which has the same velocity whether that point is regarded as a part of link 2 (extended) or as a part of link 4 (extended). Points 23 and 34 have this same characteristic for the pertinent links. Such points might be termed *transfer points*, as contrasted to *centers of rotation*, because they are points which allow an analysis to proceed from one link to another through transfer of velocity from one link to another at a point of common velocity.

In the additional mechanisms shown in Fig. 8-25, the instantaneous centers which can be visualized readily are marked. Some centers not shown are not so easily visualized or determined by means so far discussed. The following section in this chapter presents a simplifying aid. The reader is urged to study the situations of Fig. 8-25 carefully. In particular, he should notice the location of 23 in *d*. When one link rolls on another the points of contact *must* have the same velocity, since neither point is slipping or sliding past the other. Thus the instantaneous center for rolling contact between two bodies is always at the contact point.

It is obvious that many *instantaneous* centers are, in addition, fixed or permanent centers as well.

8-17. The Law of Three Centers. The method of locating the instantaneous centers of a four-link mechanism, based on the application of simple fundamental principles, was demonstrated in the preceding article. It is obvious that this method would be quite complicated and laborious for mechanisms of six, seven, eight, or more links. This article suggests a simpler and more practical method of locating instantaneous centers, namely, the use of what is known as the *law of three centers*. This law may be stated as follows: *When any three bodies have plane motion, their three instantaneous centers (or centers of relative rotation) lie on the same straight line.*

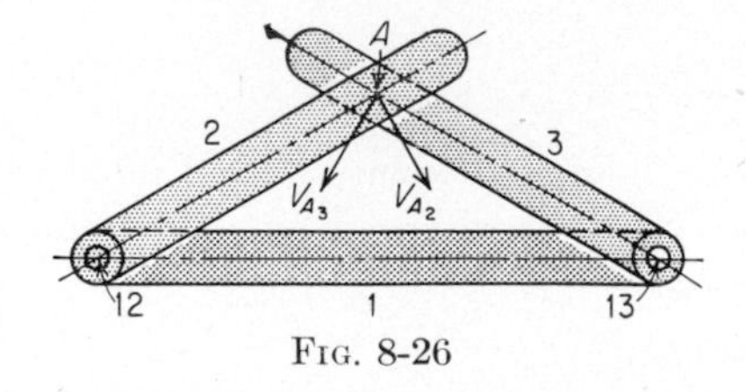

FIG. 8-26

In Fig. 8-26 are shown any three bodies having plane relative motion. Assuming that the locations of instantaneous centers 12 and 13 are known as shown, the problem is to show that the third possible instantaneous center, 23, must lie on the straight line connecting 12 and 13. First, this will be shown to be true when link 1 is stationary. Let it be assumed that center 23 might have a position such as that of A in Fig. 8-26. As a point on 2 the direction of motion of A is indicated by the vector V_{A_2} at A perpendicular to radius $12A$. As a point on 3 the direction of motion of A is indicated by the vector V_{A_3}, at A perpendicular to radius $13A$. It is evident that, even though these vectors may be of equal length, they do not coincide as is necessary if the two coincident points are to have zero relative velocity. This condition is possible only when A, that is, 23, lies on the line joining 12 and 13. Therefore, the three centers 12, 23, and 13 must lie on the same straight line, as stated in the law of three centers.

If link 1 is *not* stationary, it is only necessary to recognize that inversion of a mechanism does not change the *relative* motion among the links. Thus the position of 23 (which is a center of relative rotation of links 2 and 3) is not changed merely because link 1 is a moving link. The law of three centers is valid, therefore, for any three links having plane relative motion.

8-18. Procedure for Locating Instantaneous Centers. Before instantaneous centers may be *used* in analyzing for velocities in a mechanism, their locations must be determined. It may not always be necessary to know the positions of *all* the instantaneous centers of a particular mechanism in order to solve for the desired velocities. However, the following procedure is general enough to fit any situation, and it will be assumed that all centers are to be located.

1. Find the number of centers. Because there is one center for each pair of links, the total number N is the possible number of combinations of the n links of the mechanism taken two at a time.

$$N = \frac{n(n-1)}{2} \qquad (8\text{-}13)$$

2. Write down a list of all the centers.

3. Locate by inspection as many as possible, and check them on the list.

4. Locate the remaining centers by use of the law of three centers.

As an example, consider Fig. 8-27. The number of centers is found to be six, and these are listed at b in the figure. Centers 12, 23, 34, and 14 are found by inspection and checked off in b as shown. Centers 13 and 24 remain unknown. To locate 13, consider a combination of links 1, 3, and 2. The three instantaneous centers are 13, 12, and 23, and they must all lie on one line, according to the law of three centers. This line is drawn through centers 12 and 23. By considering a combination of links 1, 3, and 4, it is apparent that center 13 must lie also on the line drawn through centers 14 and 34. Because 13 must lie on each of two lines, it can lie only at the intersection of the two, and is located as shown in the figure.

The location of center 24 follows a similar procedure, with 24 lying at the intersection of the line through 23 and 34 and the line through 12 and 14.

A simple bookkeeping technique will be explained which is helpful in using the law of three centers, particularly when a mechanism has many links. For each *link*, a *point* is placed on the paper and marked, as shown in Fig. 8-27*e*. For each pair of links one line can be drawn in this bookkeeping diagram, and so there are as many lines possible as there are instantaneous centers. Next, as many centers as possible should be visualized in the mechanism and recorded by drawing lines in the bookkeeping diagram. In Fig. 8-27, centers 12, 23, 34, and 14 are easily

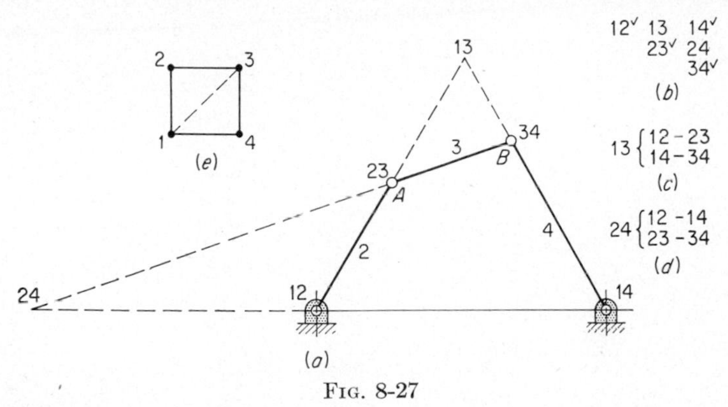

FIG. 8-27

visualized and so the lines connecting 1 and 2, 2 and 3, 3 and 4, and 1 and 4 are drawn. The next step is to draw a line representing one of the unknown centers. If a line (such as 13) can be drawn which completes two triangles at the same time, then the instantaneous center represented by that line can be found. For example, the line 13 indicates that center 13 lies on a straight line with 14 and 34, and also on a straight line with 12 and 23, according to the law of three centers. Center 13 therefore lies at the intersection of the two lines 14-34 and 12-23, as shown. In the bookkeeping diagram, the line from 1 to 3 is then drawn in as a solid line to indicate that this center is now known. It is left to the reader to apply the same technique to locate center 24.

It is obvious that at times the intersection of two lines, and therefore the locations of some instantaneous centers, may fall off the sheet containing the drawing of a mechanism. Sometimes it is possible to solve the problem without use of the particular center involved, but not always. Special constructions or techniques or a different method of analysis, such as the relative-velocity method, might be used. It is wise for the analyst to remain flexible in his approach and to have at his command more than one method of velocity analysis.

Figures 8-28 and 8-29 present as examples for study several mechanisms in which the instantaneous centers have been located. In the bookkeeping diagrams, the solid lines represent centers which were first visualized, and the dotted lines represent centers which were then found by applying the law of three centers.

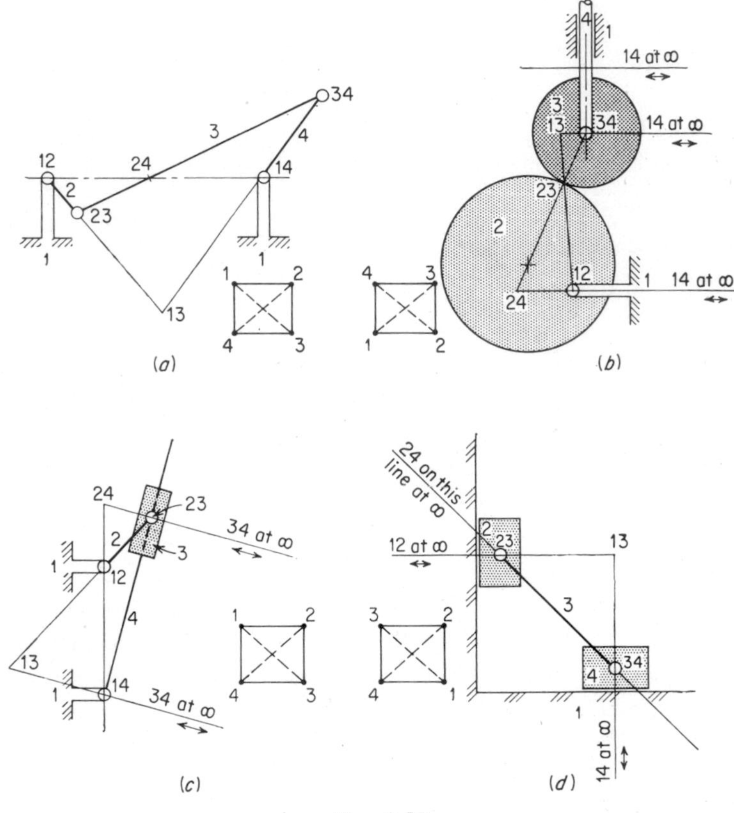

FIG. 8-28

Figure 8-29a perhaps requires explanation. The figure indicates that *when two links are in sliding contact, their instantaneous center must lie somewhere on the common normal through the contact point,* and this is true in general. The point of contact P is in reality two coincident points, P_2 on link 2 and P_3 on link 3. While the cams 2 and 3 are in driving contact, these two points cannot have relative motion along the direction of the common normal NN'. It is evident that the instantaneous relative motion of points P_2 and P_3 must be of a sliding nature along the common tangent TT'. The instantaneous center for the relative motion, then, must lie along a line at right angles to the common tangent, i.e., along the common normal through the contact point. In Fig. 8-29*a* the intersection of this common normal with the line connecting 12 and 13 locates center 23, the third of the three instantaneous centers of this particular mechanism of three links.

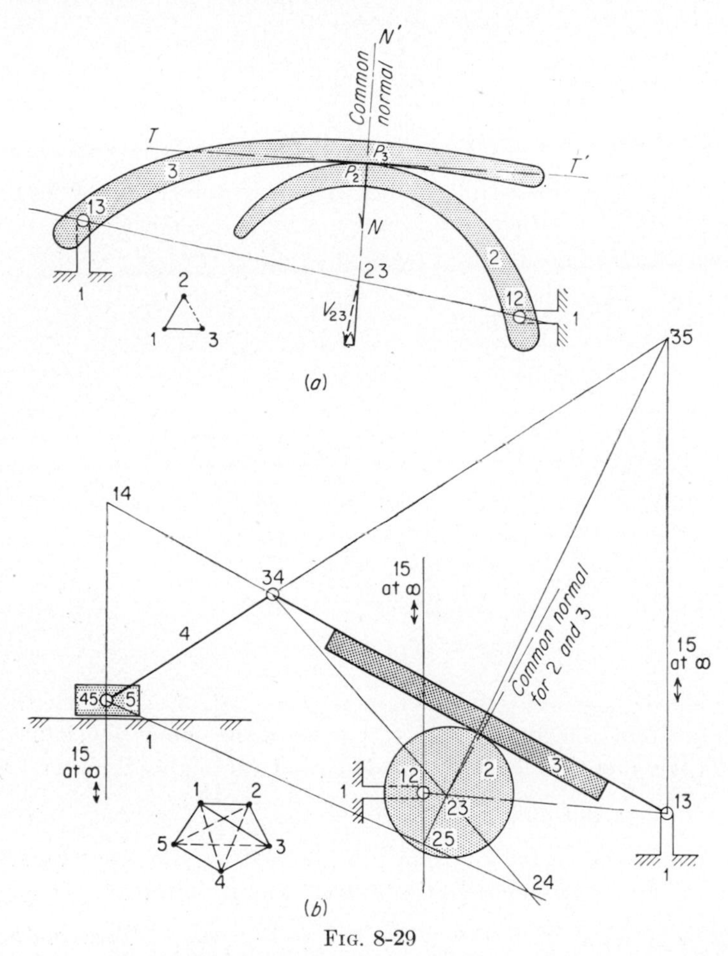

FIG. 8-29

8-19. Application of Instantaneous Centers to the Solution of Velocity Problems. This article and following articles will show how instantaneous centers are used in the solution of velocity problems. Figure 8-30 is Fig. 8-27 redrawn showing all the instantaneous centers. Assuming that V_A, the velocity of A, is known, let it be required to find V_B, the velocity of B. According to the instantaneous-center concept, links 3 and 1 may be considered to be connected by a hinged joint at 13 at the instant shown, because 1 and 3 have the same velocity at that location.

In effect, then, 13 is a center of rotation for link 3, for purposes of velocity analysis. Point A, which is a point on both link 2 and link 3, is moving with a velocity as shown by vector V_A. Since A and B are two points on link 3, the velocities V_A and V_B are directly proportional to the distances 13-23 and 13-34, respectively, of these points from center 13. The equations demonstrating this are

$$V_A = (13\text{-}23)\omega_3 \qquad V_B = (13\text{-}34)\omega_3$$

Therefore,
$$\frac{V_A}{13\text{-}23} = \frac{V_B}{13\text{-}34}$$

This equality is expressed in the dotted-line graphical construction for finding V_B shown in the figure. When a vector is placed in a position other than its true position, for construction purposes, its symbol will be primed as shown in the case of vectors $V_{A'}$ and $V_{B'}$.

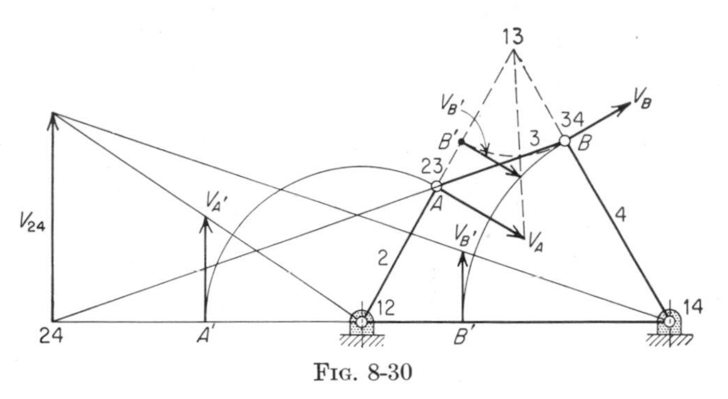

FIG. 8-30

The velocity V_B may also be found by the use of center 24, the two coincident points on links 2 and 4 that have the same velocity. Since center 24 is a point on link 2 (extended) that turns about center 12, it is moving with an absolute velocity $V_{24} = V_A \dfrac{12\text{-}24}{12\text{-}23}$, i.e., with a velocity represented by the vector V_{24} shown in Fig. 8-30. Now, center 24 is also a point on link 4 (extended) that turns about center 14. Therefore, point B is moving with a velocity $V_B = V_{24} \dfrac{14\text{-}34}{14\text{-}24}$. This method of determining V_B is shown graphically by the solid-line construction in Fig. 8-30.

It is sometimes necessary to determine the ratio of the angular velocities of two members of a mechanism with respect to a third member. Such determinations by the use of instantaneous centers may be shown as follows:

In Fig. 8-30 let it be recalled that center 23 is the point of common velocity for links 2 and 3. From the fundamental relation $\omega = V/R$, the angular velocity of link 3 relative to link 1 is equal to the velocity of point 23 divided by the distance 13-23, i.e., expressed as an equation;

$$\omega_{31} = \frac{V_{23}}{13\text{-}23}$$

likewise

$$\omega_{21} = \frac{V_{23}}{12\text{-}23}$$

therefore

$$\frac{\omega_{31}}{\omega_{21}} = \frac{12\text{-}23}{13\text{-}23}$$

similarly

$$\frac{\omega_{41}}{\omega_{21}} = \frac{12\text{-}24}{14\text{-}24} \tag{8-14}$$

Thus, the angular-velocity relations between links of a mechanism may be expressed in terms of distances between instantaneous centers without involving linear velocities in any manner whatever.

8-20. Application to Slider-crank Mechanism. In Fig. 8-31 is shown the skeleton outline of a slider-crank mechanism and the locations of the

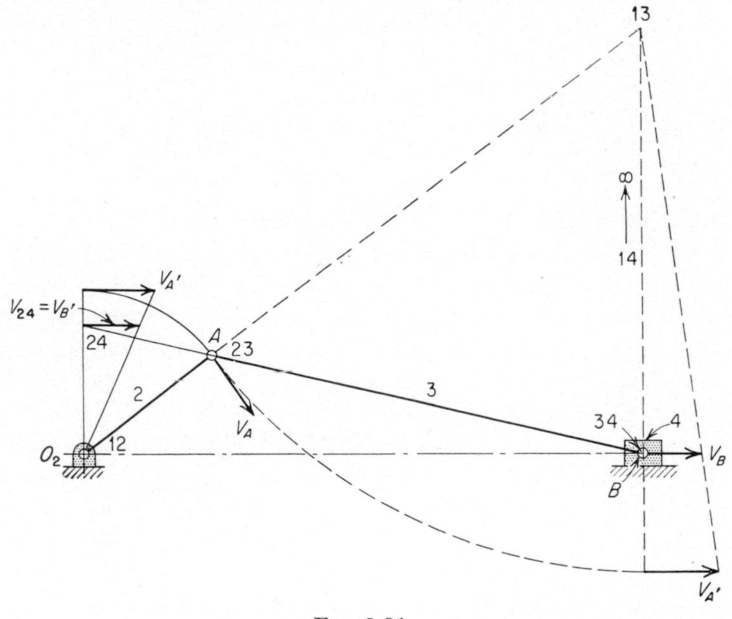

FIG. 8-31

instantaneous centers. It is assumed that the velocity V_A of the crank-pin center A is known. The velocity V_B of the piston can be found by using the center 13 as an instantaneous center of rotation for link 3, the dotted-line construction indicating the method. Alternatively, the center 24 may be used. The velocity at this point is *determined* as though the point were part of link 2, and then *used* as though it belonged with link 4. Thus velocity is transferred to link 4 from link 2 at a point of common velocity, point 24. This construction is shown in solid lines.

8-21. Application Involving Cams or Gears. In Fig. 8-32 is shown a three-link cam mechanism and the locations of the three instantaneous centers. A similar situation exists for a pair of gear teeth in contact. Link 2, turning in the sense indicated by the arrow, drives link 3.

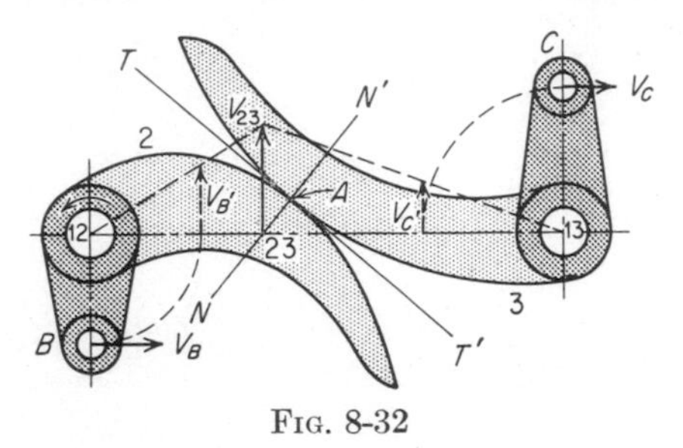

FIG. 8-32

Now let it be assumed that the velocity of any point B on the driver is known and that the velocity of any point C on the follower is required. Since center 23 is the point of common velocity in the two links, it becomes, as in the preceding examples, the transfer point in passing from velocities on 2 to velocities on 3. The solution is shown graphically on the figure.

It is evident (Art. 8-19) that the ratio of the angular velocity of the follower to that of the driver is given by the expression

$$\frac{\omega_{31}}{\omega_{21}} = \frac{12\text{-}23}{13\text{-}23}$$

8-22. Application to a Complex Mechanism. The solution of a problem involving a mechanism of six links is shown in Fig. 8-33. Since there are six links and therefore 15 centers the problem of locating all the centers is somewhat more involved than in the preceding illustrations. Although it is seldom that all the centers are required, the complete determination will be made here because of its value as an illustrative problem. The procedure in locating the centers follows the methods already discussed. In the bookkeeping diagram at b the solid lines represent the instantaneous centers initially visualized. The dotted lines represent those which were then found by using the law of three centers. In a complex mechanism it is helpful to recognize four-member linkages, since the unknown centers in such a linkage lie at the intersections of opposite sides of the linkage (see Figs. 8-30 and 8-31).

After finding all the instantaneous centers, let it be assumed that the velocity of point A on link 2 is known and that the velocity of point D on link 6 is required. Link 6 has a motion of translation, and hence the velocities of all its points are identical. The velocity V_{26} of the point on 2 at the center 26 is easily determined, as shown. The point on 6 at this center has the same velocity. Hence V_D, the velocity of D, is now known. The figure shows how V_D may be checked, using center 25.

The solution by means of 25 is possible because the velocity of a point on 2 is given and the velocity of a point on 5 is required.

Other solutions not shown are possible, also. For example, starting with link 2, velocity might be transferred to link 3 at center 23, then to link 5 at center 35 (after making use of the center of rotation 13 for link 3), and finally to link 6 at center 56 by making use of the center of rotation 15 for link 5.

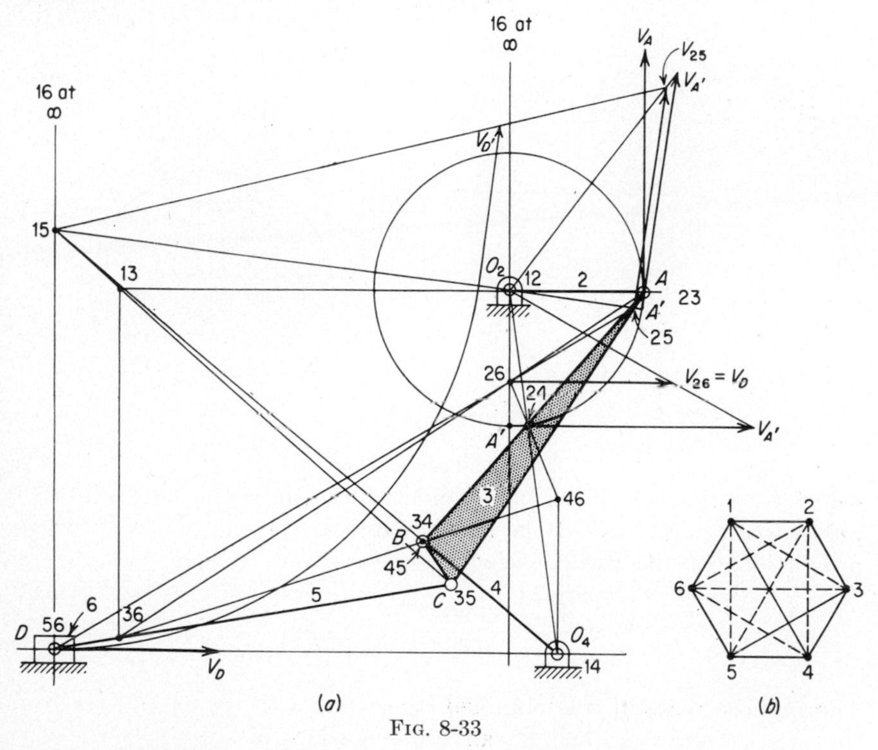

FIG. 8-33

8-23. Application to Planetary-gear Trains. The solution of two planetary-gear-train problems will be presented as examples of the usefulness of instantaneous centers in solving gear problems. Most gear problems can also be solved by other means, but the use of more than one method allows a check of results.

In the planetary-gear train shown in Fig. 8-34, the letters a, b, and c represent the numbers of teeth of the gears. The pitch of the teeth is 10, and the numbers of teeth are $a = 75$, $b = 30$, and $c = 15$. Applying the tabular method of Chap. 7 it will be found that the ratio of the number of turns of the last gear c of the train to the number of turns of the arm d is

$$\frac{n_c}{n_d} = \frac{1 + a/c}{1} = \frac{1 + 75/15}{1} = \frac{6}{1}$$

Now as a check of this solution, find the instantaneous centers, which in this case are designated by letters, and then lay off a vector of any convenient length at *bd*, the end of arm *d*. Since center *bd* is the point of common velocity for links *b* and *d*, its velocity is the same whether considered to rotate about center *ad* or center *ab*. It is evident from the construction shown in the figure that vector $\overline{mo}$ represents the velocity of a point on the pitch circle of gear *c*, and that vector $\overline{mn}$ represents the

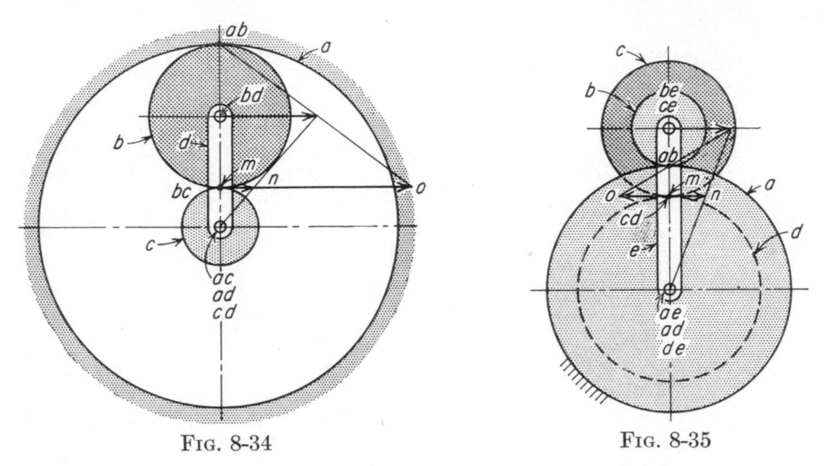

FIG. 8-34

FIG. 8-35

velocity of the coincident point on arm *d*. Since the two coincident points are rotating about the same center *ac* (or *ad*), $\overline{mo}$, and $\overline{mn}$ are proportional to the angular velocities of *c* and *d*. That is

$$\frac{n_c}{n_d} = \frac{\overline{mo}}{\overline{mn}} = \frac{3.96}{0.68} = \frac{6}{1} \qquad \text{approx}$$

The numerical result was obtained by scaling vectors $\overline{mo}$ and $\overline{mn}$ from the original drawing, and it can be seen that this result checks very closely with the exact result obtained above by the tabular method.

Next consider Fig. 8-35. The pitch of the gears is 8, and the numbers of teeth are $a = 60$, $b = 18$, $c = 33$, and $d = 45$. Applying the tabular method it is found that

$$\frac{n_d}{n_e} = \frac{1 - ac/bd}{1} = \frac{1 - 60 \times 33/18 \times 45}{1} = -\frac{1.444}{1}$$

Applying the graphical construction in exactly the same manner as in Fig. 8-34, it is found that the ratio of the number of turns of the last gear *d* in the train to the number of turns of arm *e* is

$$\frac{n_d}{n_e} = \frac{\overline{mo}}{\overline{mn}} = \frac{1.08}{0.75} = -\frac{1.44}{1} \qquad \text{approx}$$

Again it is observed that this result is in very close agreement with the exact result obtained by the tabular method.

It is obvious that the graphical work described above could be duplicated step by step analytically. The details will not be worked out here other than to call attention to the fact that the vector $\overline{mo}$ in Fig. 8-34 is obviously twice the length of the vector at point *bd*; also that the vector $\overline{mn}$ is one-third the length of the vector at *bd*. Therefore, $\overline{mo}$ is *exactly* six times as long as $\overline{mn}$.

8-24. Centrodes. The relative motions of any two links may be considered as a series of rotations about successive positions of their instantaneous center of relative rotation. A pair of curves in contact, one attached to and moving with each link and always containing this center,

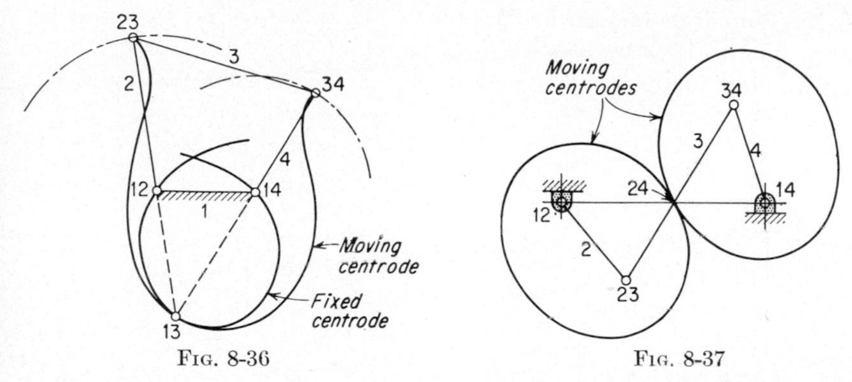

FIG. 8-36

FIG. 8-37

constitute what are known as the *centrodes* of relative motion of these links. It follows, therefore, that there is a pair of centrodes for each pair of links in a mechanism.

For example, if any link, such as link 3 (Fig. 8-36) has motion relative to the fixed link 1, then this motion may be regarded as a series of rotations about successive positions of the instantaneous center 13. If a curve is drawn connecting all the positions of the instantaneous center 13, this curve is called the *fixed centrode*. This locus of center 13 may be regarded as a curve attached to link 1. Similarly, if a plane is attached so as to move with link 3, a second curve that always contains the center 13 can be drawn on this plane. The second curve is called the *moving centrode*. If these curves were given material backing and constrained to roll upon each other without slip they could replace the actual connecting links 2 and 4 and continue to give the same relative motion of the links 1 and 3. Centrodes such as those of Fig. 8-36 are mainly of theoretical interest. Centrodes of symmetrical form such as circles and ellipses may, on the other hand, be of more practical interest. For example, if a wheel rolls along a rail the instantaneous center of the wheel is always the point of contact with the rail. The top of the rail is therefore the locus of this center, i.e., the fixed centrode. Also, the rim of the wheel always contains the instantaneous center and is, therefore, the moving centrode.

In a pair of gears the instantaneous center of relative motion is always the point of contact of the pitch circles. These pitch circles, therefore, constitute a pair of centrodes of relative motion and are both moving centrodes. In the case of the crossed four-bar linkage Fig. 8-37, the centrodes of relative motion are a pair of ellipses. In general, it may be said that, except in the case of symmetrical curves such as those mentioned above, centrodes are of little practical significance.

ORTHOGONAL COMPONENTS

8-25. Fundamentals of the Method. The fundamental idea underlying this method can be stated simply as follows:

If two points move so that the distance between them does not change,

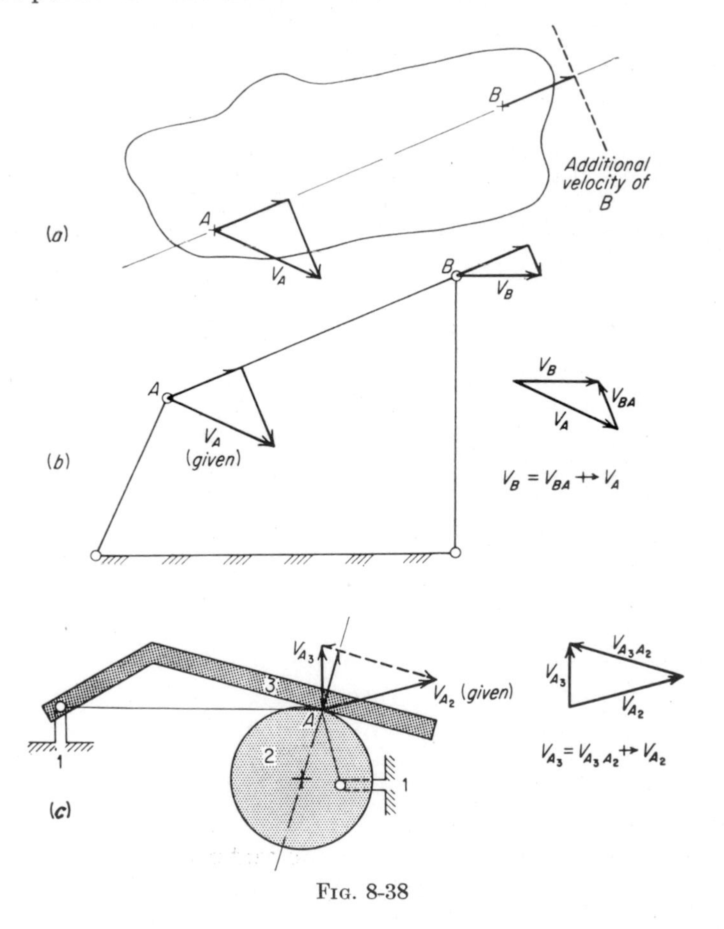

FIG. 8-38

then each point must have the same component of velocity along the line connecting the points.

This statement is obvious for two points a fixed distance apart. In Fig. 8-38*a*, V_A is given. Point *B* is another point on the same link at a fixed distance from *A*. A line is drawn from *A* to *B*, and V_A is broken into orthogonal (right-angled) components along this line and perpendicular to it as shown. The component toward *B* is the velocity which *A* has directly toward *B*, and if the distance *AB* is not to change, *B* must move with an identical velocity component along the *AB* direction, as shown. This component is not the entire velocity of point *B*, but all additional velocity of *B* must be *perpendicular* to *AB*. Therefore, the component shown plus (vectorially) an additional perpendicular component gives the total velocity of point *B*. If the direction of V_B is known, the solution is complete. Examples are shown at *b* and *c* of Fig. 8-38.

The method is simple and quick to use in many instances. However, it is no more convenient, in general, then the relative-velocity method and, in fact, the two methods are quite similar. With Fig. 8-38*b* and *c* are shown velocity vector diagrams, from which it is seen that the relative velocity in each case is merely broken into two pieces in the orthogonal-component analysis. It is more meaningful and useful to have the complete relative velocities available, and these are obtained directly in the relative-velocity analysis without additional effort. The method of orthogonal components is therefore a less valuable technique than the method of relative velocities.

SPECIAL METHODS AND COMBINATIONS OF METHODS

8-26. Example Situations. It occasionally happens that a velocity problem is encountered that cannot be solved directly by the general methods thus far described. The general fundamental principles which have been discussed still apply, of course, but some ingenuity and flexibility are required in determining the best way to apply these principles to any one specific problem. No general rule can be laid down for the solution of all the special problems that may arise. The case shown in Fig. 8-39 will serve as an example. Four links of a rather complicated mechanism have been shown, the other links having been omitted in order to simplify the drawing. The velocities of the three points *D*, *E*, and *F* are assumed to be known, and it is required to determine the velocity of any point on link 2.

In attempting to determine the velocity of the point *A* by the method of relative velocities, it is at once seen that the direction of V_A is unknown and the method breaks down. The same statement applies to the points

B and C. It would seem, therefore, that the problem could not be solved.

On link 2, however, let a line be drawn from the point A in the same direction as the line DA. In the same way let a line be drawn from B in the direction EB. The velocity of the intersection S of these two lines is given by the simultaneous solution of the two vector equations

$$V_S = V_A \nrightarrow V_{SA} = V_D \nrightarrow V_{AD} \nrightarrow V_{SA}$$
$$V_S = V_B \nrightarrow V_{SB} = V_E \nrightarrow V_{BE} \nrightarrow V_{SB}$$

Now let it be noted that the vectors V_{AD} and V_{SA} are both perpendicular to the line DAS. They must therefore have the same direction and

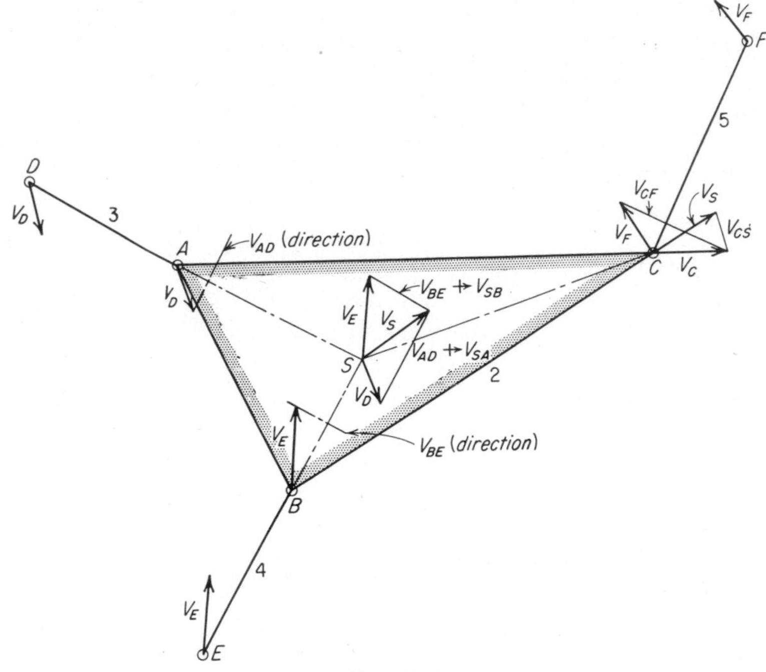

Fig. 8-39

hence their vector sum ($V_{AD} \nrightarrow V_{SA}$) will also have this direction. The latter, therefore, may be treated as a single vector perpendicular to the line DAS. A similar statement can be made for the vector sum ($V_{BE} \nrightarrow V_{SB}$). If, therefore, the vectors ($V_{AD} \nrightarrow V_{SA}$) and ($V_{BE} \nrightarrow V_{SB}$) are laid off (in the polygon at S) from the extremities of the vectors V_D and V_E, respectively, their intersection will give the vector representing the velocity V_S. The reader should carefully check the vector diagrams against the equations which the diagrams represent. After V_S has been determined, the velocity of any other point of link 2, such as C, can be determined.

It sometimes happens that a combination of the method of relative velocities and the method of instantaneous centers is required for the solution of a given problem. A case of this kind is the mechanism shown in Fig. 8-40. The velocity of the point A is given, and it is required to determine the velocity of the point E.

In attempting to solve the problem by the method of relative velocities, it is at once seen that, since the directions of the velocities of B and C are unknown, further progress is impossible. In order to apply the method of instantaneous centers, the centers 13, 16, and 36 should be

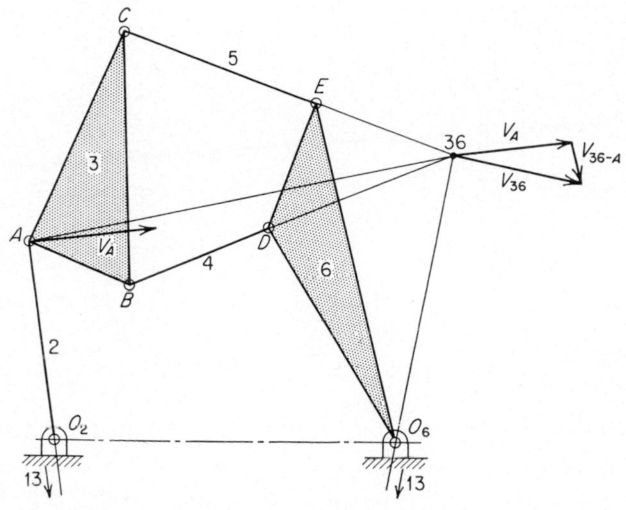

Fig. 8-40

known. The position of 16 is at O_6, and the position of 36 may readily be determined; the center 13, on the other hand, is situated at a point far outside the limits of the paper and hence its determination and use will be difficult, if not impossible.

In order to solve the problem, let it first be noted that 36 is the position of a pair of coincident points on 3 and 6 having the same velocity. As a point on 6 the direction of its velocity must be perpendicular to the line O_6-36. As a point on 3 the following vector equation must hold:

$$V_{36} = V_A \nrightarrow V_{36A}$$

The direction of V_{36A} is perpendicular to the line $36A$ on link 3 (extended). The velocity V_{36} may therefore be determined as indicated in the figure. V_{36} having been determined, the velocity of the point E may be determined by either the method of relative velocities or the method of instantaneous centers.

ANALYTICAL METHODS

The methods described in the preceding articles have been in the main graphical. It must not be inferred from this, however, that graphical methods are always to be preferred to analytical methods. Both methods have their particular applications, and one should not be used where the other is simpler or more convenient. As examples of the analytical approach, two mechanisms will be analyzed in the following sections.

8-27. The Slider-crank Mechanism. As a first example let the slider-crank mechanism shown in Fig. 8-41 be considered. The expression for

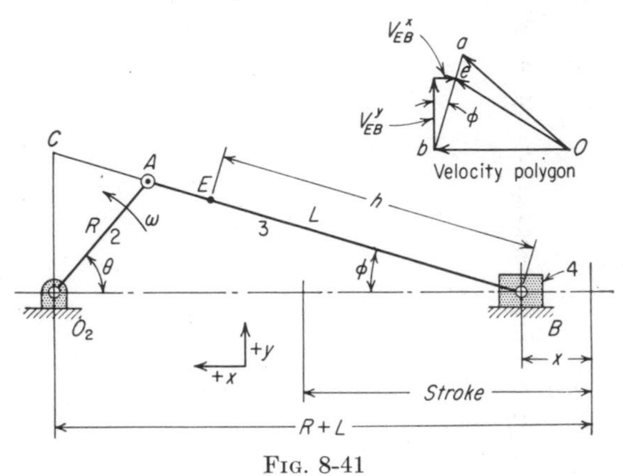

FIG. 8-41

displacement of the slider has already been derived in the chapter on linkwork, but the steps will be repeated here for completeness. For the slider, the displacement x from the extreme right position is

$$x = R + L - R\cos\theta - L\cos\phi$$
$$= R(1 - \cos\theta) + L(1 - \cos\phi)$$

From the geometry of the figure,

$$R\sin\theta = L\sin\phi$$
$$\sin\phi = \frac{R}{L}\sin\theta$$

Also,
$$\cos\phi = \sqrt{1 - \sin^2\phi} = \sqrt{1 - \left(\frac{R}{L}\sin\theta\right)^2}$$

Substituting this value of $\cos\phi$ in the equation for x leaves θ, the crank angle, as the only variable.

$$x = R(1 - \cos\theta) + L\left[1 - \sqrt{1 - \left(\frac{R}{L}\sin\theta\right)^2}\right] \qquad (8\text{-}15)$$

This expression can be simplified for further manipulation by using the binomial theorem to expand the radical in an infinite series. When this is done,

$$x = R(1 - \cos\theta) + R\left(\frac{1}{2}\frac{R}{L}\sin^2\theta + \frac{1}{2 \times 4}\frac{R^3}{L^3}\sin^4\theta + \frac{1 \times 3}{2 \times 4 \times 6}\frac{R^5}{L^5}\sin^6\theta + \frac{1 \times 3 \times 5}{2 \times 4 \times 6 \times 8}\frac{R^7}{L^7}\sin^8\theta \cdots\right)$$

If this equation is differentiated with respect to time, the velocity V_B of the slider will be obtained.

$$V_B = \frac{dx}{dt} = R\sin\theta\frac{d\theta}{dt} + R\left(\frac{2}{2}\frac{R}{L}\sin\theta\cos\theta\frac{d\theta}{dt} + \frac{1 \times 4}{2 \times 4}\frac{R^3}{L^3}\sin^3\theta\cos\theta\frac{d\theta}{dt} \cdots\right)$$

$$= R\omega_2\left[\sin\theta + \frac{1}{2}\frac{R}{L}\sin 2\theta\left(1 + \frac{1}{2}\frac{R^2}{L^2}\sin^2\theta + \frac{1 \times 3}{2 \times 4}\frac{R^4}{L^4}\sin^4\theta + \frac{1 \times 3 \times 5}{2 \times 4 \times 6}\frac{R^6}{L^6}\sin^6\theta \cdots\right)\right] \quad (8\text{-}16)$$

where ω_2, the angular velocity of the crank, is written for the derivative $d\theta/dt$.

The foregoing is what might be styled the *fundamental analytical method.* It consists of first determining the displacement of one member in terms of the known displacement of another member and then in differentiating to determine the velocity of the first member in terms of the known velocity of the second. As an alternative it is possible to make use of the fundamental principles that were brought out in the discussion of the graphical methods.

Consider again Fig. 8-41. From a relative velocity analysis, and the resulting similar triangles oab in the velocity polygon and O_2AC in the mechanism,

$$V_B = V_A\frac{O_2C}{O_2A}$$

This expression may be put in a form suitable for calculation as follows:

$$V_B = V_A\frac{O_2C}{O_2A} = V_A\frac{\overline{O_2B}\tan\phi}{R} = V_A\frac{(L\cos\phi + R\cos\theta)\tan\phi}{R}$$

$$= R\omega_2\left(\frac{L}{R}\sin\phi + \frac{\sin\phi\cos\theta}{\cos\phi}\right)$$

and since

$$\sin\phi = \frac{R}{L}\sin\theta$$

$$V_B = R\omega_2\left(\sin\theta + \frac{R}{L}\frac{\sin\theta\cos\theta}{\cos\phi}\right) = R\omega_2\left(\sin\theta + \frac{1}{2}\frac{R}{L}\frac{\sin 2\theta}{\cos\phi}\right) \quad (8\text{-}17)$$

Comparing this expression with Eq. (8-16) previously developed for V_B it is evident that $1/\cos\phi$ is equal to the infinite series in the parentheses of Eq. (8-16). An expression for $1/\cos\phi$ in terms of R, L, and θ may be obtained as follows:

$$\frac{1}{\cos\phi} = \frac{1}{\sqrt{1 - \sin^2\phi}} = \frac{1}{\sqrt{1 - R^2/L^2 \sin^2\theta}}$$

If this expression is expanded, the infinite series obtained will be identical with the one obtained formerly.

The angular velocity of the connecting rod may be determined as follows:

$$\sin\phi = \frac{R}{L}\sin\theta$$

$$\cos\phi \frac{d\phi}{dt} = \frac{R}{L}\cos\theta \frac{d\theta}{dt}$$

$$\omega_3 = \frac{d\phi}{dt} = \frac{R}{L}\frac{\cos\theta}{\cos\phi}\omega_2 \tag{8-18}$$

Now let it be required to determine the velocity of a point E on the connecting rod situated at a distance h from the wrist pin B.

The horizontal and vertical components of V_E are

$$V_E^x = V_B^x - V_{EB}^x = V_B - V_{EB}\sin\phi$$
$$V_E^y = V_B^y + V_{EB}^y = 0 + V_{EB}\cos\phi = V_{EB}\cos\phi$$

The relative velocity V_{EB} is given by

$$V_{EB} = h\omega_3 = h\frac{R}{L}\frac{\cos\theta}{\cos\phi}\omega_2$$

Therefore,

$$\begin{aligned} V_E^x &= R\omega_2\left(\sin\theta + \frac{1}{2}\frac{R}{L}\frac{\sin 2\theta}{\cos\phi}\right) - h\frac{R}{L}\frac{\cos\theta}{\cos\phi}\omega_2\sin\phi \\ &= R\omega_2\left(\sin\theta + \frac{1}{2}\frac{R}{L}\frac{\sin 2\theta}{\cos\phi}\right) - R\omega_2\frac{h}{L}\frac{R}{L}\frac{\sin\theta\cos\theta}{\cos\phi} \\ &= R\omega_2\left[\sin\theta + \left(1 - \frac{h}{L}\right)\frac{1}{2}\frac{R}{L}\frac{\sin 2\theta}{\cos\phi}\right] \end{aligned} \tag{8-19}$$

and

$$V_E^y = h\frac{R}{L}\frac{\cos\theta}{\cos\phi}\omega_2\cos\phi = \frac{h}{L}R\omega_2\cos\theta \tag{8-20}$$

If h is equal to zero, the horizontal component will be equal to V_B and the vertical component will be zero. If h is equal to L, the horizontal and vertical components will be the same as those of the crankpin.

8-28. The Crank-shaper Mechanism. As a further application of the analytical method let the crank shaper discussed in Art. 8-14 be again considered. Referring to Fig. 8-18, let it be assumed that the angular velocity of the crank is ω_2, and let it be required to determine the velocity V_{B_5} of the ram. This may be accomplished by means of the following analysis, which is self-explanatory:

$$V_{B_5} = \frac{V_{B_4}}{\cos \phi} = \frac{V_{A_4}(O_4B/O_4A)}{\cos \phi}$$

$$V_{A_4} = V_{A_2} \cos (\theta - \phi) = V_{A_2}(\cos \theta \cos \phi + \sin \theta \sin \phi)$$

$$\frac{O_4B}{O_4A} = \frac{h_2}{h_1 + R \cos \theta}$$

Therefore

$$V_{B_5} = \frac{V_{A_2} h_2}{h_1 + R \cos \theta} \frac{\cos \theta \cos \phi + \sin \theta \sin \phi}{\cos \phi}$$

$$V_{A_2} = R\omega_2 \qquad \text{and} \qquad \tan \phi = \frac{R \sin \theta}{h_1 + R \cos \theta}$$

Therefore

$$V_{B_5} = R\omega_2 \frac{h_2}{h_1 + R \cos \theta} \left(\cos \theta + \frac{R \sin^2 \theta}{h_1 + R \cos \theta} \right) \tag{8-21}$$

CHAPTER 9

ACCELERATIONS IN MACHINES

In the preceding chapter methods were outlined for the determination of velocities in machines. In this chapter methods will be developed for the determination of accelerations. It will be found that a velocity analysis is a necessary accompaniment of an acceleration analysis. Similarly, both the velocity and acceleration analyses are essential in an inertia-force analysis. The study of inertia forces will be taken up in Chap. 11.

9-1. Acceleration of a Point. The acceleration of a point is the time rate of its vector change of velocity. Referring to Fig. 9-1*a* let the point

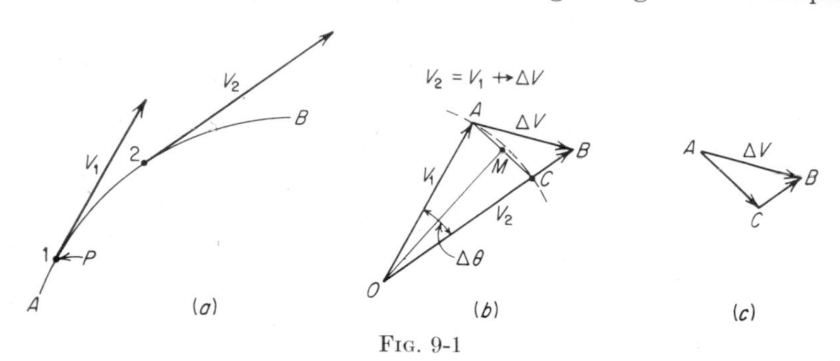

Fig. 9-1

P be moving along the plane curve AB, and let it be assumed that, as the point moves from position 1 to position 2, its velocity changes from V_1 to V_2. This change is completely represented by the vector difference ΔV, as shown at *b* in the figure. The ratio $\Delta V/\Delta t$ of the vector change of velocity to the length of time required for the point to pass from position 1 to position 2 is the magnitude of the *average acceleration* of the point during the interval Δt. The direction of the average acceleration is that of the vector ΔV. If now the segment 1-2 and the time required for P to traverse it are infinitesimals, then the ratio $\Delta V/\Delta t$ becomes dV/dt and is the magnitude of the instantaneous acceleration of the point. The term *acceleration* usually means *instantaneous acceleration*.

Referring again to the vector polygon of Fig. 9-1*b*, let OC be laid off equal to OA and let the chord AC be drawn. It is evident from *b* and *c* in the figure that the vector change of velocity ΔV is the vector sum of AC and CB, and may be represented by the following vector equation:

$$\Delta V = AC \nrightarrow CB$$

The vector length CB obviously represents the change of the *magnitude* (scalar change) of the velocity of the point P as it passes from position 1 to position 2. AC is the vector that would represent the total change of velocity if the scalar change were zero, i.e., if the velocity were changed in *direction* only.

If we divide both sides of the vector equation by the time interval Δt, we obtain

$$A_{\text{avg}} = \frac{\Delta V}{\Delta t} = \frac{AC}{\Delta t} \nrightarrow \frac{CB}{\Delta t} \tag{9-1}$$

giving the average acceleration of P in terms of the vector components $AC/\Delta t$ and $CB/\Delta t$.

Before proceeding further it is well to recognize that some confusion is possible when using a symbol for velocity or acceleration as to whether the complete vector or just its magnitude is intended. When there is a possibility of such confusion in the discussion which follows the use of lower-case letters will indicate that magnitude only is intended. Thus

$$v = \text{mag } V \qquad \text{and} \qquad a = \text{mag } A$$

Returning to our equation and the figure, we note

$$\text{mag} \frac{AC}{\Delta t} = v_1 \frac{2 \sin \Delta\theta/2}{\Delta t} \tag{9-2}$$

$$\text{mag} \frac{CB}{\Delta t} = \frac{v_2 - v_1}{\Delta t} = \frac{\Delta v}{\Delta t} \tag{9-3}$$

In the limit, as Δt approaches zero, it is noted that the directions of the vectors $AC/\Delta t$ and $CB/\Delta t$ become respectively normal (perpendicular) to and the same as the direction of the vector V_1, that is, respectively *normal and tangential to the path of motion at position* 1. These vectors will therefore be called the *normal* and *tangential* components of acceleration A, and will be represented by A^n and A^t, respectively. Finally, we have for the *magnitudes* of A^n and A^t,

$$a^n = \lim_{\Delta t \to 0} v_1 \frac{2 \sin \Delta\theta/2}{\Delta t} = v \frac{d\theta}{dt} \tag{9-4}$$

$$a^t = \lim_{\Delta t \to 0} \frac{\Delta v}{\Delta t} = \frac{dv}{dt} \tag{9-5}$$

and

$$A = A^n \nrightarrow A^t \tag{9-6}$$

An inspection of Fig. 9-1 will bring out the fact that the component A^n is always directed toward the center of curvature of the path, and that the direction of the component A^t is either the same as or opposite that of the velocity depending upon whether the magnitude of the latter is increasing or decreasing.

The quantity $d\theta$ is the angular displacement of the velocity vector V during the time interval dt. The quantity $d\theta/dt$, therefore, may be conveniently thought of as the angular velocity of the velocity vector V. If the radius of curvature of the path of the point is imagined to move *with* the point, then the angular velocity of the vector V is also that of the radius.

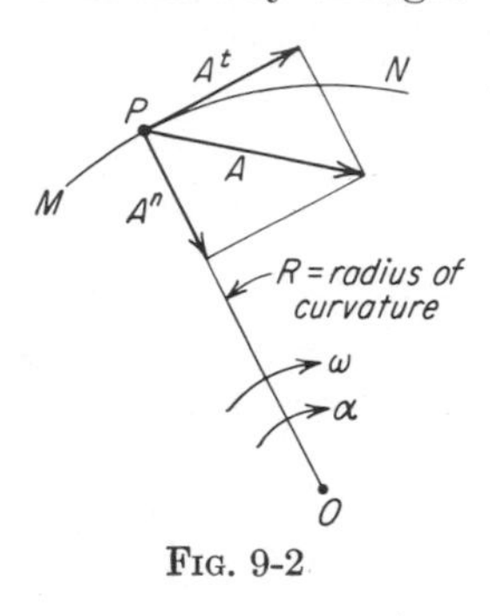

FIG. 9-2

There are a number of interesting relations among the quantities V, A^n, A^t and the radius of curvature R. Referring to Fig. 9-2, let the point P be moving along the plane curve MN with a velocity V and an acceleration A, and let the radius of curvature OP $(= R)$ be assumed to move with the point. It has been shown that the magnitude of the velocity V is equal to the product $R\omega$, where ω is the angular velocity of the radius of curvature. It follows that

$$a^n = v\frac{d\theta}{dt} = v\omega = R\omega \cdot \omega = R\omega^2 = \frac{v^2}{R} \tag{9-7}$$

and

$$a^t = \frac{dv}{dt} = \frac{d}{dt}R\omega = R\frac{d\omega}{dt} + \omega\frac{dR}{dt} = R\alpha + \omega\frac{dR}{dt} \tag{9-8}$$

where α is the angular acceleration of the velocity vector and hence of the radius of curvature. The quantity α is obviously equal to $d^2\theta/dt^2$. It is to be noted that, if the radius of curvature is a constant (i.e., if the point is moving along a circular arc), the quantity $\omega\, dR/dt$ becomes equal to zero.

Since A^n and A^t are perpendicular to each other, the magnitude of the total acceleration may be expressed as follows:

$$a = \sqrt{(a^n)^2 + (a^t)^2} \tag{9-9}$$

9-2. Relative Acceleration of Any Two Points. The relative displacement of two points has been defined as the vector change of their relative position, and their relative velocity as the time rate of the vector change of their relative position. By analogy the relative acceleration of two points will be defined as the time rate of the vector change of their relative velocity.

In Fig. 9-3a the points A and B are moving along the curves there shown. During the same interval of time Δt, the point A moves from A_1 to A_2, and the point B from B_1 to B_2. The resulting vector changes of velocity are shown by the polygons.

In Fig. 9-3*b* the relative velocity V_{AB} at the beginning of time interval Δt is shown; and at *c* the relative velocity at the end of the interval. Finally, at *d*, the vector change ΔV_{AB} of V_{AB}, the velocity of A relative to B, is shown.

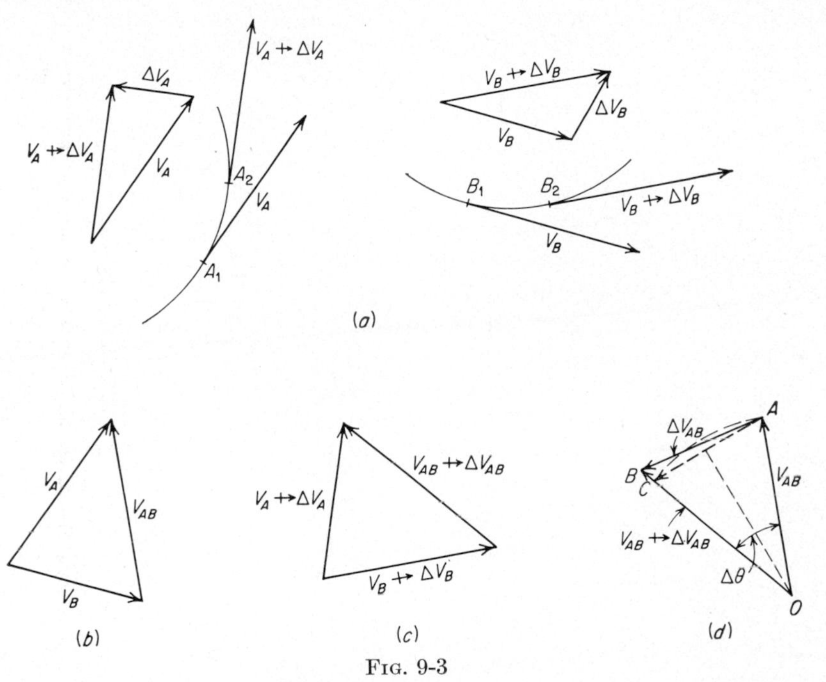

FIG. 9-3

If we express the foregoing in terms of vector equations, we obtain

$$
\begin{aligned}
\Delta V_{AB} &= (V_{AB} \nrightarrow \Delta V_{AB}) \rightarrow V_{AB} && \text{(9-10)}\\
&= [(V_A \nrightarrow \Delta V_A) \rightarrow (V_B \nrightarrow \Delta V_B)] \rightarrow V_{AB} &&\\
&= V_A \rightarrow V_B \nrightarrow \Delta V_A \rightarrow \Delta V_B \rightarrow V_{AB} &&\\
&= \Delta V_A \rightarrow \Delta V_B && \text{(9-11)}
\end{aligned}
$$

Dividing both sides by Δt,

$$\frac{\Delta V_{AB}}{\Delta t} = \frac{\Delta V_A}{\Delta t} \rightarrow \frac{\Delta V_B}{\Delta t} \tag{9-12}$$

For an infinitesimal interval of time this becomes

$$A_{AB} = A_A \rightarrow A_B \tag{9-13}$$

where A_{AB} is the desired relative acceleration. That is, *the acceleration of A relative to B is equal to the vector difference of the absolute accelerations of A and B.*

In Fig. 9-3*d*, ΔV_{AB} has been resolved into components representing the change in direction (AC) and in magnitude (CB) of V_{AB}, as was done in Fig. 9-1. In a manner analogous to that employed in Art. 9-1, we obtain

$$a_{AB}^n = v_{AB} \frac{d\theta}{dt} \tag{9-14}$$

$$a_{AB}^t = \frac{dv_{AB}}{dt} \tag{9-15}$$

and

$$A_{AB} = A_{AB}^n \nrightarrow A_{AB}^t \tag{9-16}$$

The normal component A_{AB}^n is perpendicular (normal) to the relative velocity V_{AB}, with its sense along that perpendicular dictated by the sense of $d\theta/dt$, the absolute angular velocity of the V_{AB} vector. The direction of the tangential component A_{AB}^t is the same as or opposite that of V_{AB} depending upon whether v_{AB} is increasing or decreasing.

9-3. Relative Acceleration of Two Points on the Same Rigid Body. The relative acceleration of two points on the same rigid body possesses certain special characteristics that correspond in a general way to those possessed by the relative velocity of two such points. In order to bring out these characteristics, let the case shown at *a* in Fig. 9-4 be considered.

It was shown in Art. 9-2 that the relative acceleration of any two points may be broken up into components perpendicular and parallel to the direction of the *relative velocity*. It was also shown that the magnitudes of these components are

$$a_{BA}^n = v_{BA} \frac{d\theta}{dt} \qquad \text{and} \qquad a_{BA}^t = \frac{dv_{BA}}{dt}$$

where $d\theta/dt$ is the angular velocity of the relative-velocity vector V_{BA}. Since in this case the vector V_{BA} is always perpendicular to the line AB, it is evident that the direction of the normal component is along this line, from B toward A; never from B away from A.[1] Its magnitude is

$$a_{BA}^n = v_{BA} \frac{d\theta}{dt} = v_{BA}\omega = (AB \cdot \omega)\omega = (AB)\omega^2 \tag{9-17}$$

where $d\theta/dt$ has been replaced by its obvious equivalent ω, the angular velocity of the body.

The direction of the tangential component is parallel to V_{BA} and hence perpendicular to the line AB. The sense of its direction is such as to make B turn about A in the same direction as the angular acceleration of the body. Its magnitude is

$$a_{BA}^t = \frac{d(v_{BA})}{dt} = \frac{d}{dt}(AB \cdot \omega) = AB\frac{d\omega}{dt} + \omega\frac{d(AB)}{dt} = (AB)\alpha \tag{9-18}$$

[1] The proof of this statement makes a good exercise. Let the rigid body be drawn in two positions, and then let the fundamental method discussed in Art. 9-2 be applied.

{The term $\omega[d(AB)/dt]$ is zero because the distance AB is constant.} It should be noted that the line AB enters into the expressions for A_{BA}^t and A_{BA}^n in the same way that the radius of curvature of the path of a point moving in a circle enters into the corresponding expressions for its absolute acceleration. This was to be expected, as the motion of the point B *relative to the point* A is a rotation of B about A, the relative path, so to speak, being a circle of radius AB. Obviously, as indicated by the dashed vectors at B,

$$A_B = A_A +\!\!\!\!\rightarrow A_{BA}^n +\!\!\!\!\rightarrow A_{BA}^t$$

It has been seen that the direction of the relative velocity of two points on the same rigid body is perpendicular to the line joining the two points. A similar condition is true for the relative acceleration of two points

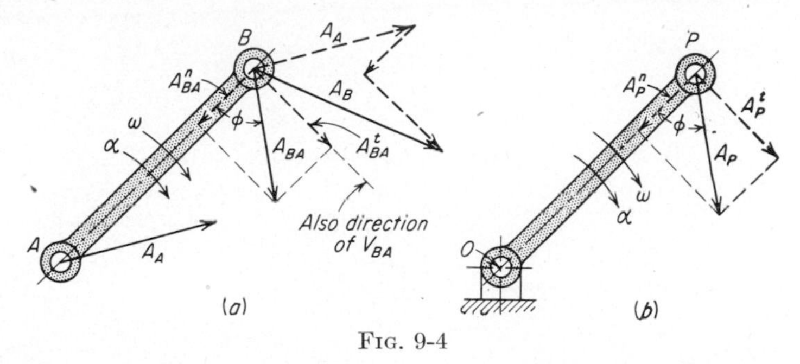

FIG. 9-4

on the same rigid body. The angle ϕ in Fig. 9-4a is given by the following relation:

$$\tan \phi = \frac{a_{BA}^t}{a_{BA}^n} = \frac{AB \cdot \alpha}{AB \cdot \omega^2} = \frac{\alpha}{\omega^2} \tag{9-19}$$

It follows that the angle ϕ is independent of the location of the two points A and B and is dependent only upon the angular velocity and angular acceleration of the body.

When a rigid body is turning about a fixed center O as shown in Fig. 9-4b, the absolute acceleration of any point P of the body is merely its relative acceleration with respect to the fixed center O. It follows, therefore, that the normal and tangential components of the absolute acceleration of the point P are

$$a_P^n = OP \cdot \omega^2$$
$$a_P^t = OP \cdot \alpha$$

This is the same situation as that discussed in connection with Fig. 9-2, for a constant radius of curvature. The total acceleration of the point is

$$A_P = A_P^n +\!\!\!\!\rightarrow A_P^t = OP \cdot \omega^2 +\!\!\!\!\rightarrow OP \cdot \alpha$$

Very often in acceleration problems the angular velocity ω of a link is known, but the angular acceleration is unknown. As an illustration of this type of problem, let the motion of the rigid body shown in Fig. 9-5 be considered. The acceleration of the point A is completely known, the acceleration of the point B is known in direction only, and the angular velocity ω of the rigid body is known. It is required to determine the

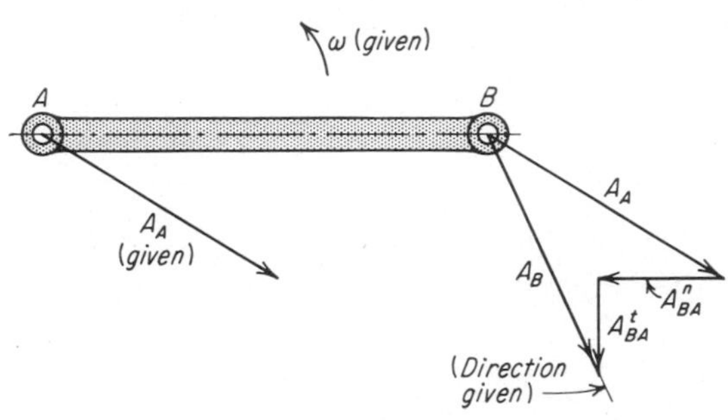

Fig. 9-5

magnitude of A_B and the angular acceleration α. The acceleration A_B is given by the following vector equation:

$$A_B = A_A \nrightarrow A^n_{BA} \nrightarrow A^t_{BA}$$

The magnitude of A^n_{BA} may be calculated as follows:

$$a^n_{BA} = AB \cdot \omega^2$$

and it may then be laid off from the extremity of A_A as shown. Although the magnitude of A^t_{BA} is unknown, its direction is known to be perpendicular to A^n_{BA}, and it may therefore be laid off from the extremity of A^n_{BA} as shown. The intersection of this line with the line representing the direction of A_B will determine the magnitudes of the vectors A_B and A^t_{BA}. The angular acceleration of the rigid body is $\alpha = a^t_{BA}/AB$. It is evident from the construction that the angular acceleration is clockwise for the particular case shown.

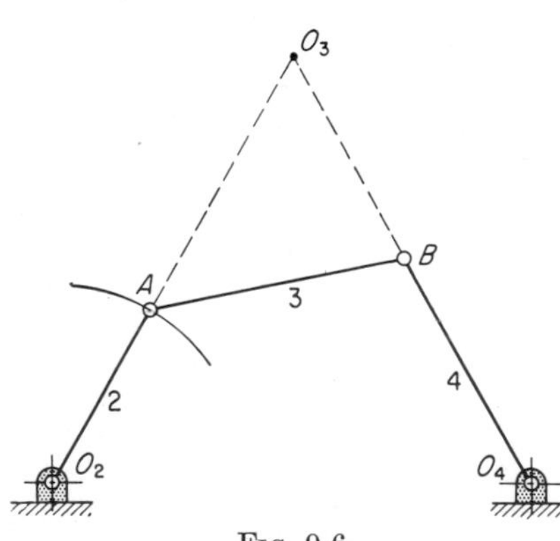

Fig. 9-6

Attention is called to the fact that in dealing with acceleration problems the instantaneous center for velocity analysis must not be confused with the center of curvature of the path of motion. For example, in Fig. 9-6 point A of the four-link mechanism moves in a circular path with center at O_2. Point A is, however, also a point on link 3, for which O_3 is the instantaneous center for velocity analysis. The normal component of the acceleration of A obviously is directed toward the center of curvature O_2 of its path and *not* toward the instantaneous center O_3.

It should be noted that the instantaneous center as defined and used for velocity analysis cannot be used for acceleration analysis. At the point at which two links have the same velocity, they do not necessarily have the same acceleration.

9-4. The Acceleration Polygon. In the preceding article it was shown how accelerations could be determined by means of vector additions on the figure. It is the purpose of this article to outline the procedure to be followed in the construction of an acceleration polygon. In Fig. 9-7a is shown a simple four-link mechanism that has been drawn

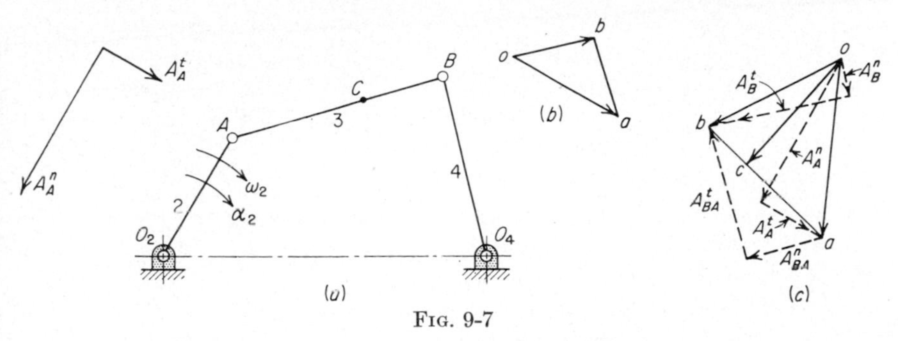

FIG. 9-7

to a convenient scale. The angular velocity ω_2 and the angular acceleration α_2 of the driving link are given and the complete acceleration polygon is required. The first step in the solution is to draw the velocity polygon. Since the velocity of A is known from the relation $v_A = O_2A \cdot \omega_2$, the vector oa representing this velocity is drawn to a convenient scale and the velocity polygon completed as shown at b in the figure.

The next step is to calculate the acceleration of A on link 2 and start the construction of the acceleration polygon at c. The acceleration of A is

$$A_A = A_A^n +\!\!\!\!\rightarrow A_A^t$$
$$a_A^n = O_2A \cdot \omega_2^2$$
$$a_A^t = O_2A \cdot \alpha_2$$

After selecting a pole o and a convenient scale the vectors A_A^n and A_A^t are drawn and the vector oa determined as shown in the figure.

Since A represents a pair of coincident points on links 2 and 3, which have the same acceleration, the acceleration of one point on link 3 has now been determined. The acceleration of point B on link 3 is, therefore,

$$A_B = A_A +\!\!\!\!\rightarrow A_{BA} = A_A +\!\!\!\!\rightarrow (A_{BA}^n +\!\!\!\!\rightarrow A_{BA}^t) \qquad (9\text{-}20)$$

The normal component A_{BA}^n is readily determined. Its magnitude $(v_{BA})^2/AB$ $(= AB \cdot \omega_3^2)$ can be calculated since either V_{BA} or ω_3 can be determined from the velocity polygon. Its direction and sense are BA. The tangential component A_{BA}^t is known to be perpendicular to AB. Its magnitude $AB \cdot \alpha_3$ cannot be calculated since α_3 is unknown. However, the vector A_{BA}^n can be drawn from the extremity of *oa* in the direction BA, and A_{BA}^t can be represented in direction as shown in the figure at *c*.

In order to complete the acceleration polygon, it is necessary to consider B as a point on link 4.

$$A_B = A_B^n \nrightarrow A_B^t$$

The normal component A_B^n is known in both magnitude and direction, since its magnitude v_B^2/O_4B $(= O_4B \cdot \omega_4^2)$ can be calculated (since V_B or ω_4 can be determined from the velocity polygon) and its direction is BO_4. The tangential component A_B^t is known only to be perpendicular to O_4B. However, the vector A_B^n can be drawn from the pole *o* (since this is an absolute acceleration) in the direction BO_4. The component A_B^t is represented in direction as shown in the figure. The point *b* and hence the vector *ob* are determined by the intersection of the components A_{BA}^t and A_B^t, and the polygon is completed by the final step of drawing in the relative-acceleration vector *ab*. At this stage it is wise to check the vector polygon just completed against the vector equation (9-20) which it represents.

As in the case of the velocity polygon, *oa* in the acceleration polygon is the image of link O_2A, *ob* the image of link O_4B, and *ab* the image of link AB. Hence the acceleration of any other point on AB, say C, is represented by the vector *oc* drawn from the pole *o* to a point *c* such that *ac* in the polygon is proportional to AC in the figure of the mechanism.

It is observed that the acceleration polygon proper, *oab*, has been drawn in solid lines. As in the case of the velocity polygon, arrowheads on relative-acceleration vectors are not essential, and often will be omitted for convenience in reading them in either sense as desired. The normal and tangential components have been drawn in broken lines and identified by the symbols and arrowheads in the manner indicated in the figure.

9-5. Accelerations in the Four-link Mechanism. Figure 9-8 shows a four-link mechanism very similar to the one which has already been analyzed. In this section, the numerical aspects of obtaining a solution will be emphasized. The angular velocity and angular acceleration of link 2 are given as

$$\omega_2 = 20 \text{ radians/sec} \qquad \alpha_2 = 150 \text{ radians/sec}^2$$

and it is required to make a complete acceleration determination for the mechanism. The scale of the original drawing was 3 in. = 1 ft.

The first step in the solution is to determine the normal and tangential components of the acceleration of the point A. Since link 2 is rotating about the fixed center O_2, these components are

$$a_A^n = O_2A \cdot \omega_2^2 = 6/12 \times 20^2 = 200 \text{ ft/sec}^2$$
$$a_A^t = O_2A \cdot \alpha_2 = 6/12 \times 150 = 75 \text{ ft/sec}^2$$

The magnitude and direction of total acceleration A_A may, after a suitable scale has been chosen (a scale of 1 in. = 150 ft/sec² was chosen

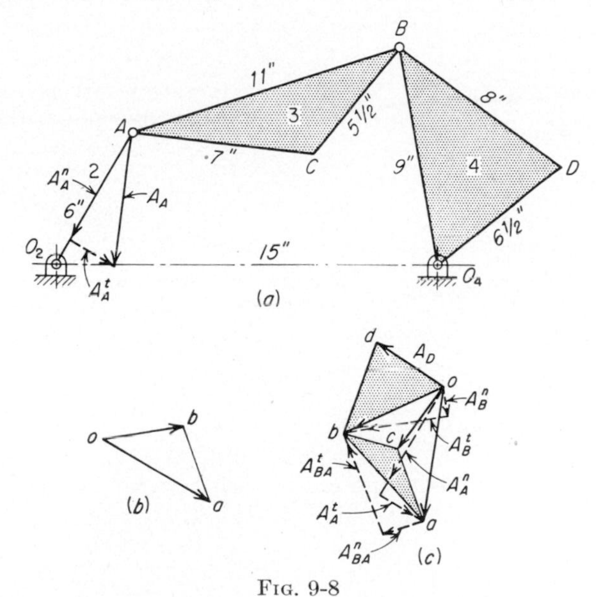

FIG. 9-8

for the original drawing), be determined by the vector addition of A_A^n and A_A^t.

Since A is a point on both links 2 and 3, the acceleration of one point on link 3 has been determined. The acceleration of the point B is represented by the vector equation

$$A_B = A_A \nrightarrow A_{BA}$$
$$A_B^n \nrightarrow A_B^t = A_A \nrightarrow A_{BA}^n \nrightarrow A_{BA}^t \qquad (9\text{-}21)$$

The determination of A_{BA}^n requires that the angular velocity of link 3 be known. To determine ω_3, let the magnitude of V_A be calculated and let the partial velocity polygon shown at b in the figure be drawn.

$$v_A = O_2A \cdot \omega_2 = 6/12 \times 20 = 10 \text{ fps}$$

The scale of the polygon in the original drawing was 1 in. = 8 fps. The length of the vector ab in the polygon was 0.96 in. The relative velocity v_{BA} is therefore 7.68 fps. The angular velocity of link 3 is

$$\omega_3 = \frac{v_{BA}}{AB} = \frac{7.68}{11/12} = 8.38 \text{ radians/sec} \qquad \text{(counterclockwise)}$$

The magnitude of the normal component A^n_{BA} can now be calculated.

$$a^n_{BA} = AB \cdot \omega_3^2 = 11/12 \times 8.38^2 = 64.3 \text{ ft/sec}^2$$

This vector is directed from B toward A.

The remaining acceleration vectors in Eq. (9-21) are known to be either along or normal to the corresponding velocity vectors. One more item must be determined before the vector polygon can be drawn. The magnitudes of the remaining tangential accelerations are as yet unknown. However, the normal acceleration of B can be calculated.

$$a^n_B = v_B \frac{d\theta}{dt} = v_B\omega_4 = \frac{v_B^2}{O_4B}$$

Scaling the vector ob in the polygon, it is found that v_B is 6.55 fps. Therefore,

$$a^n_B = \frac{(6.55)^2}{9/12} = 57.0 \text{ ft/sec}^2$$

This vector is directed from B toward O_4.

The only remaining unknowns are the magnitudes of A^t_B and A^t_{BA}, and so the vector polygon representing Eq. (9-21) now can be drawn. Both sides of the equation start at the pole o, with vectors adding outward. Known vectors are placed first, and finally the intersection of the directional lines of A^t_B and A^t_{BA} establishes point b. All vectors in the equation are now determined. The acceleration of point B is represented by the vector ob which, when scaled, gives

$$a_B = 165 \text{ ft/sec}^2$$

The accelerations of the points C and D may be determined most conveniently by drawing figures (images) in the acceleration polygon similar to links 3 and 4. The accelerations thus determined are

$$a_C = 117 \text{ ft/sec}^2 \qquad a_D = 121 \text{ ft/sec}^2$$

If desired, the angular accelerations of links 3 and 4 may be determined as follows:

$$\alpha_3 = \frac{a^t_{BA}}{AB} = \frac{167}{11/12} = 182 \text{ radians/sec}^2 \qquad \text{(counterclockwise)}$$

$$\alpha_4 = \frac{a^t_B}{O_4B} = \frac{159}{9/12} = 212 \text{ radians/sec}^2 \qquad \text{(counterclockwise)}$$

9-6. Application to Slider-crank Mechanism. As a further application of the principles developed in the foregoing articles, consider the slider-crank mechanism shown in Fig. 9-9. The crank is assumed to have a constant angular velocity of 3,000 rpm, and it is required to determine the acceleration of the wrist pin B and the acceleration of the center of gravity G of the connecting rod. The scale of the original drawing was 6 in. = 1 ft.

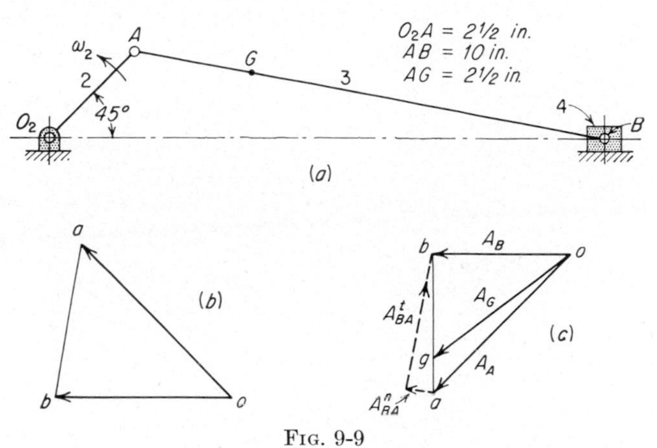

FIG. 9-9

Since the angular acceleration of the crank O_2A is zero, the total acceleration of the crankpin is the same as the normal acceleration. That is,

$$a_A = O_2A \cdot \omega_2^2 = \frac{2.5}{12}\left(2\pi\,\frac{3{,}000}{60}\right)^2 = 20{,}570 \text{ ft/sec}^2$$

The acceleration of point B is given by the following vector addition:

$$A_B = A_A \nrightarrow A_{BA} = A_A \nrightarrow (A^n_{BA} \nrightarrow A^t_{BA})$$

Before A^n_{BA} can be determined, the angular velocity ω_3 of the connecting rod (link 3) must be known. But in order to determine ω_3 it will be necessary to draw the velocity polygon for the mechanism. From the given data

$$v_A = O_2A \cdot \omega_2 = \frac{2.5}{12}\left(2\pi\,\frac{3{,}000}{60}\right) = 65.5 \text{ fps}$$

A velocity scale of 1 in. = 30 fps was chosen for the original drawing and the velocity polygon drawn as shown at b in the figure. From this polygon the relative velocity V_{BA} is found to be 46.80 fps. The angular velocity of link 3 is

$$\omega_3 = \frac{v_{BA}}{AB} = \frac{46.80}{10/12} = 56.16 \text{ radians/sec}$$

The magnitude of the normal component A^n_{BA} is

$$a^n_{BA} = AB \cdot \omega_3^2 = {}^{10}\!/_{12}(56.16)^2 = 2{,}624 \text{ ft/sec}^2$$

The acceleration polygon can now be drawn. Let the vector *oa*, representing A_A, be laid off from a pole *o* as shown at *c* in the figure. The scale chosen for the original drawing was 1 in. = 10,000 ft/sec². Next the vector A^n_{BA} is added in the direction *B* toward *A*. A line through the extremity of the vector A^n_{BA} and perpendicular to it will represent the direction of the tangential component A^t_{BA}. The direction of the acceleration of the point *B* is determined by the fact that this point moves in a straight line. It follows that a horizontal line through the pole *o* will represent the direction of A_B. The intersection of this line with the line representing the direction of A^t_{BA} determines the magnitudes of A_B and A^t_{BA}.

The line *ab* in the acceleration polygon is the image of the connecting rod *AB*. Hence the following proportion must hold:

$$\frac{ag}{ab} = \frac{AG}{AB}$$

Hence $$ag = ab\,\frac{AG}{AB} = 1.44 \times \frac{2.5}{10} = 0.36 \text{ in.}$$

If, therefore, the distance *ag* is laid off in the acceleration polygon, the vector *og* represents the acceleration of point *G*. Using the acceleration scale chosen above, the accelerations of points *B* and *G* were found to be

$$a_B = 14{,}400 \text{ ft/sec}^2 \qquad \text{and} \qquad a_G = 18{,}000 \text{ ft/sec}^2$$

Attention is called to the fact that these accelerations are many times greater than the acceleration of gravity (32.2 ft/sec²). It is evident, therefore, that in an engine of this size running at 3,000 rpm the inertia forces of its moving members must be many times greater than their weights.

9-7. Graphical Constructions. Scale Relations. Although the methods employed in the preceding articles were in the main graphical, it was nevertheless necessary to make a number of intermediate calculations before a full solution of the problem could be obtained. The present article deals with a method that will materially reduce such calculations.

In any acceleration problem it is possible to choose the space, velocity, and acceleration scales so that the vector representing the *normal component* of the acceleration of any point may be determined by means of a graphical construction. This construction, which will be explained, involves the normal component vector, the vector representing the velocity of the point, and the length on the drawing of the radius of curvature of the path of the point.

1. *The Scales Defined.* Before proceeding further it will be necessary to make sure that the precise meaning of the three scales just mentioned is clearly understood. Velocities and accelerations are usually expressed in feet per second and feet per second per second, respectively, and in order, therefore, to make use of the various motion formulas, lengths must also be expressed in feet. The space, velocity, and acceleration scales (k_s, k_v, and k_a, respectively) used in this text will be defined as follows (although any consistent set of units could be used):

1 in. on drawing = a distance of k_s ft on mechanism.
1 in. of velocity vector = k_v fps.
1 in. of acceleration vector = k_a ft/sec^2.

For example:

For a space scale of 1 in. = ⅓ ft (3 in. = 1 ft), $k_s = \frac{1}{3}$.
For a velocity scale of 1 in. = 100 fps, $k_v = 100$.
For an acceleration scale of 1 in. = 150 ft/sec^2, $k_a = 150$.

2. *Graphical Method.* The graphical method mentioned above for determining A^n will now be developed. Referring to Fig. 9-10, let the point P be moving along a plane curve with the velocity V. Let this velocity be represented in the figure by the vector length PQ in., and let R, the actual radius of curvature of the path of the point in feet, be represented on the drawing by the length OP in. The normal component of the acceleration is

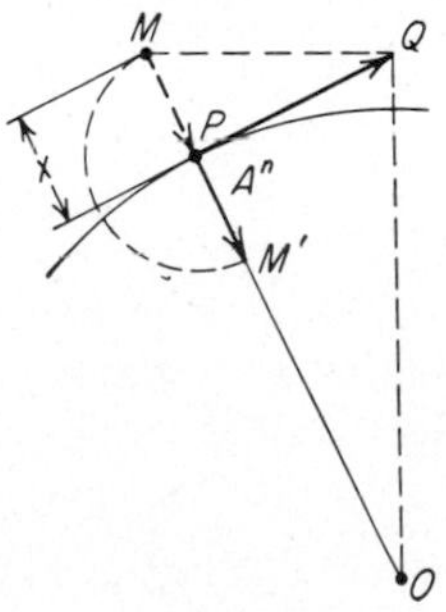

$$a^n = \frac{v^2}{R} = \frac{(PQ)^2 k_v^2}{OP \cdot k_s} = \frac{(PQ)^2}{OP}\frac{k_v^2}{k_s}$$

FIG. 9-10

where k_v and k_s are the velocity and space scales, respectively. Now let it be assumed that the acceleration scale is k_a and that the vector length representing A^n is x in. Then, from the preceding expression

$$a^n = xk_a = \frac{(PQ)^2}{OP}\frac{k_v^2}{k_s}$$

If the three scales k_s, k_v, and k_a are chosen so that

$$k_a = \frac{k_v^2}{k_s} \tag{9-22}$$

then

$$x = \frac{(PQ)^2}{OP} \quad \text{or} \quad \frac{x}{PQ} = \frac{PQ}{OP} \tag{9-23}$$

That is, PQ is a mean proportional between x and OP, and hence any of the various geometrical constructions pertaining to a mean proportional may be employed to determine x from PQ, or vice versa.

In the construction employed in Fig. 9-10 to determine x, the line OQ is drawn, and then the line QM is drawn at right angles to OQ, intersecting OP (extended) at M. The distance MP is the desired vector length x. That is, from similar triangles,

$$\frac{MP}{PQ} = \frac{PQ}{OP}$$

$$MP = x = \frac{(PQ)^2}{OP} = \frac{a^n}{k_a}$$

The proper *direction* for the vector representing A^n is toward the center of curvature O as shown.

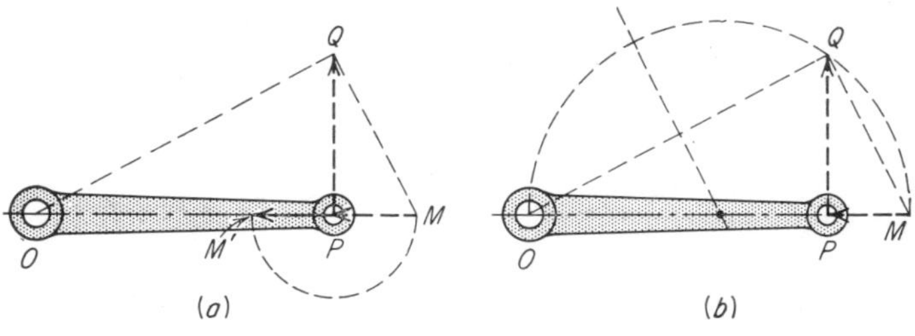

Fig. 9-11

When P and O are two points on a rigid link and the PQ vector represents the velocity of P *relative* to O (V_{PO}), then the foregoing construction gives the corresponding *relative* normal acceleration A^n_{PO}.

Of the various geometrical constructions that could be used, the two shown in Fig. 9-11 will be found convenient. The construction shown at a has already been explained in connection with Fig. 9-10.

The construction at b makes use of the fact that any angle inscribed in a semicircle is a right angle. The perpendicular bisector of chord OQ cuts the line OP at the center of the circle which will pass through points O and Q. This circle is drawn, establishing point M and the vector length MP. If it happens that the acceleration scale is chosen first, then OP and the vector MP are known and the proper length of the vector PQ representing the velocity must be found. This is easily done by locating the center of the line MO, which is the proper center for the circle already discussed. This circle intersects the line drawn perpendicular to OP through P at point Q and thereby establishes the proper vector length for the velocity of P.

It is important to keep in mind that the use of the above constructions requires that the space, velocity, and acceleration scales satisfy the relation $k_a = k_v^2/k_s$. Any two of the scales may be selected at will, but the third scale must be calculated from the above relation.

3. *Uniformly Rotating Crank.* When the motion of one of the members of a mechanism is uniform rotation about a fixed center, it is convenient to choose the velocity scale so that the vector representing the velocity of any point of the member is equal in length to the radius of the point. When this is done, and when the acceleration scale is chosen in accordance with the relation $k_a = k_v^2/k_s$, the radius of the point also represents its acceleration. Referring to Fig. 9-12 let the velocity scale be so chosen that the vector PQ representing the velocity of the point P is equal in length to the *distance OP on the drawing.* Then let the line OQ be drawn and the right angle OQM constructed. PM is the length of the vector that represents the normal component of the acceleration of the point P; and since the tangential component is zero, owing to the uniform rotation, it is evident that in this case the length PM is also that of the total acceleration vector. Because $PQ = OP$, the angle PQO is 45 deg and hence PQM is also 45 deg from which it follows that $PQ = PM$. That is, PM is equal to the length OP on the drawing.

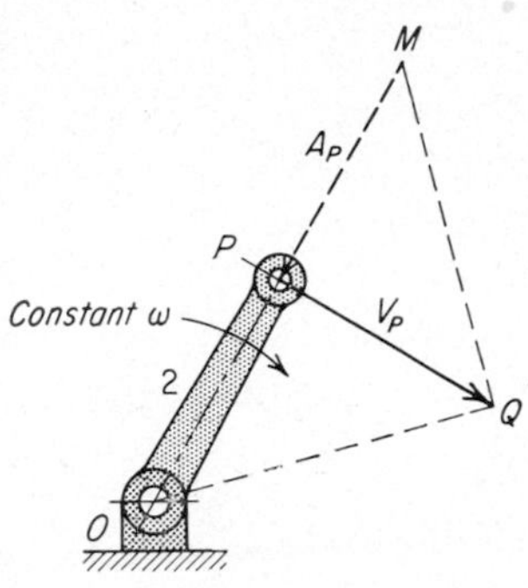

FIG. 9-12

9-8. Four-link Mechanism. Complete Graphical Method. As an application of the methods described in the preceding article, let the four-link mechanism again be considered. Referring to Fig. 9-13, let the angular velocity and angular acceleration be given as follows:

$$\omega_2 = 22 \text{ radians/sec} \qquad \alpha_2 = 180 \text{ radians/sec}^2$$

and let it be required to determine the accelerations of the points C and D.

The first step is to calculate the magnitudes of the normal and tangential components of the acceleration of the point A.

$$a_A^n = O_2A \cdot \omega_2^2 = 5/12 \times 22^2 = 201.7 \text{ ft/sec}^2$$
$$a_A^t = O_2A \cdot \alpha_2 = 5/12 \times 180 = 75.0 \text{ ft/sec}^2$$

Before proceeding with the graphical work, attention must be given to the selection of scales. The space scale, as is usually the case, has already been selected, that of the original drawing having been 3 in. = 1 ft; or 1 in. = 1/3 ft; i.e., $k_s = 1/3$. If the methods of Art. 9-7 are to be followed, only one of the other two scales may be selected arbitrarily, and for the purposes of the present problem it will be more convenient to select the acceleration scale. In the case of the original drawing the scale selected was 1 in. = 100 ft/sec²; i.e., $k_a = 100$. For the purposes of the present problem it is not necessary to make use of the velocity scale. It could, however, be calculated in the following manner:

$$k_a = \frac{k_v^2}{k_s}$$

Therefore $k_v = \sqrt{k_a k_s} = \sqrt{100/3} = 5.79$; i.e., 1 in. = 5.79 fps. The acceleration scale having been selected, the vector A_A may be determined from its normal and tangential components and laid off in the polygon as shown at *c*.

The next step is to determine the length of the vector representing the velocity of the point *A*. This has been accomplished on the drawing

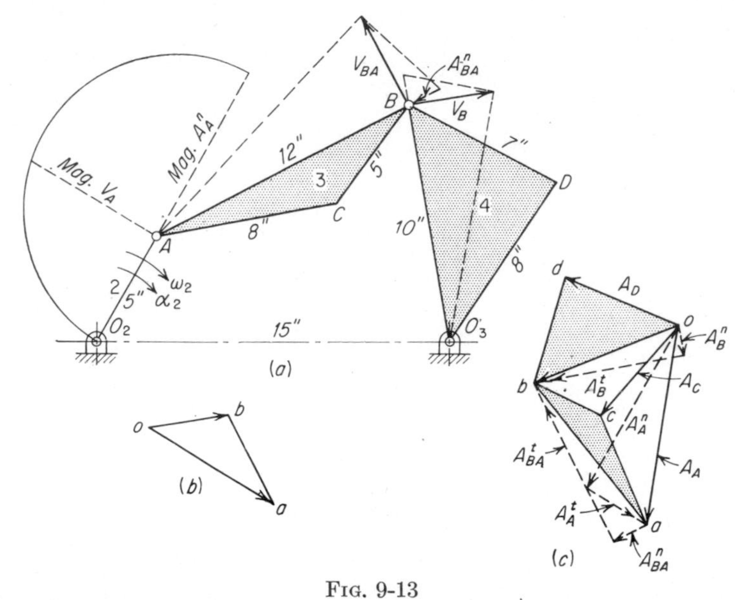

Fig. 9-13

of the mechanism itself by the methods of Art. 9-7. The partial velocity polygon may then be drawn as shown at *b*.

Using the vector length V_{BA}, the normal component A^n_{BA} may be determined by means of the construction shown on the figure of the mechanism. The vector length A^n_B may be determined in a like manner. Both these vectors are then laid off in the polygon. Lines representing the directions of the tangential components A^t_{BA} and A^t_B may also be laid off in the polygon, the intersection of these lines determining the position of the extremity of the vector *ob* representing A_B. The acceleration images of links 3 and 4 are then drawn and the accelerations of the points *C* and *D* obtained.

$$A_C = 118 \text{ ft/sec}^2 \qquad A_D = 134 \text{ ft/sec}^2$$

For the solution of the foregoing problem only one calculation had to be made, namely, the calculation by means of which the acceleration of the point A was determined.

9-9. Slider-crank Mechanism. The Ritterhaus Construction. For mechanisms that are of sufficient importance that they require frequent analysis, special rapid methods are sometimes developed. As an example, the slider-crank mechanism will be discussed. In Fig. 9-14, it is assumed that:

1. The angular velocity of the crank OA is constant.
2. The scales used satisfy the relation $k_a = k_v^2/k_s$.
3. The vector length representing a_A^n is chosen to be equal to the length of the line OA which represents the crank on the paper.

A consequence of the last two items, considered together, is that the length of vector representing the velocity of A is also equal to the length

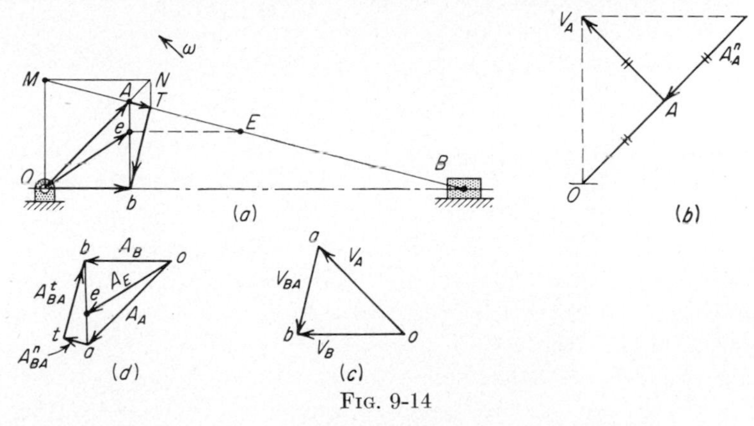

FIG. 9-14

of line OA on the paper. Figure 9-12, and the graphical construction previously discussed, show this to be true.

In Fig. 9-14 the velocity and acceleration polygons shown at c and d, respectively, have been drawn in accordance with the methods of the preceding articles but the construction has been omitted from the figure. The triangle OAM on the figure of the mechanism is equal to the triangle oab in the velocity polygon. Let the polygon oab be turned through 90 deg and shifted so that the pole o coincides with the center O of the crank and so that the velocity vector oa falls along the line OA. It follows that OA and OM represent the magnitudes of the velocities of A and B, respectively, and that AM represents the magnitude of the relative velocity of these points.

Now let the acceleration polygon at d be turned through 180 deg and shifted so that the pole o of the polygon coincides with the center O of the crank, and so that the acceleration vector oa falls along the line OA. It is evident that the vector ob will fall along the line OB, that the normal component A^n_{BA} will fall along the connecting rod AB, and that the tangential component A^t_{BA} will be perpendicular to the connecting rod. It is to be observed that the directions of the acceleration vectors are now (Fig. 9-14a) directly *opposite* their true directions.

The revolved vector polygon shown at a in the figure can be obtained by a construction which utilizes only the drawing of the mechanism. This construction is known as the *Ritterhaus construction.* The method will be proved for the engine of Fig. 9-14, but it is also adaptable to the situation in which the direction of motion of B does not pass through the crank center O. The procedure is as follows: Draw OM perpendicular to the direction OB of motion of B, intersecting AB (or its extension) at M; draw MN parallel to OB, intersecting OA (or its extension) at N; draw NT parallel to OM; draw Tb perpendicular to AB; draw Ob parallel to the direction of motion of B.

It must now be proved that the figure $OATb$ just constructed is, in fact, the acceleration polygon. The pertinent vector equation is

$$A_B = A_A \nrightarrow A^n_{BA} \nrightarrow A^t_{BA}$$

The direction lines of these acceleration vectors (see d, Fig. 9-14) correspond to the direction of lines in the Ritterhaus construction polygon $OATb$. Further, by choice of scales, the length OA in the Ritterhaus polygon is the same as the length of the vector oa representing A_A in d of the figure. If it can now be proved that the length AT in the Ritterhaus polygon is the vector length representing A^n_{BA}, then the Ritterhaus polygon will have been proved to be equal to the acceleration polygon. (From geometrical considerations two four-sided figures are equal if four angles and two sides in one figure are equal to the corresponding four angles and two sides of the other.) The proof that AT is equal to the vector length of A^n_{BA} follows.

The triangle ANT is similar to the triangle AOM; and the triangle AMN is similar to the triangle ABO. Therefore

$$\frac{AT}{AM} = \frac{AN}{AO} \qquad \text{and} \qquad \frac{AM}{AB} = \frac{AN}{AO}$$

from which

$$\frac{AT}{AM} = \frac{AM}{AB} \qquad \text{or} \qquad AT = \frac{(AM)^2}{AB}$$

It has already been established that $AM \cdot k_v = v_{BA}$ and $AB \cdot k_s =$ actual length L of the connecting rod. Therefore,

$$AT = \frac{v^2_{BA}}{k^2_v}\frac{k_s}{L} = \frac{v^2_{BA}}{L}\frac{k_s}{k^2_v}$$

$$AT = \frac{a^n_{BA}}{k_a}$$

$$AT \cdot k_a = a^n_{BA}$$

Therefore, the length AT represents, to the scale k_a, the value of a^n_{BA}, and the Ritterhaus polygon is now proved to be the acceleration polygon. The image of AB in the Ritterhaus polygon is the line drawn from A to b. For the particular position of the mechanism shown this line appears to be nearly perpendicular to OB, but this in general is not true. The acceleration image of the point E may be determined by drawing a line through E parallel to the horizontal line BO. The proof of this construction follows from the fact that the triangles AEe and ABb are similar. The vector Oe in the acceleration image (Fig. 9-14a) represents A_E (reversed).

It has been established that the vectors shown in the Ritterhaus polygon at a of Fig. 9-14 are all opposite their true directions. It follows

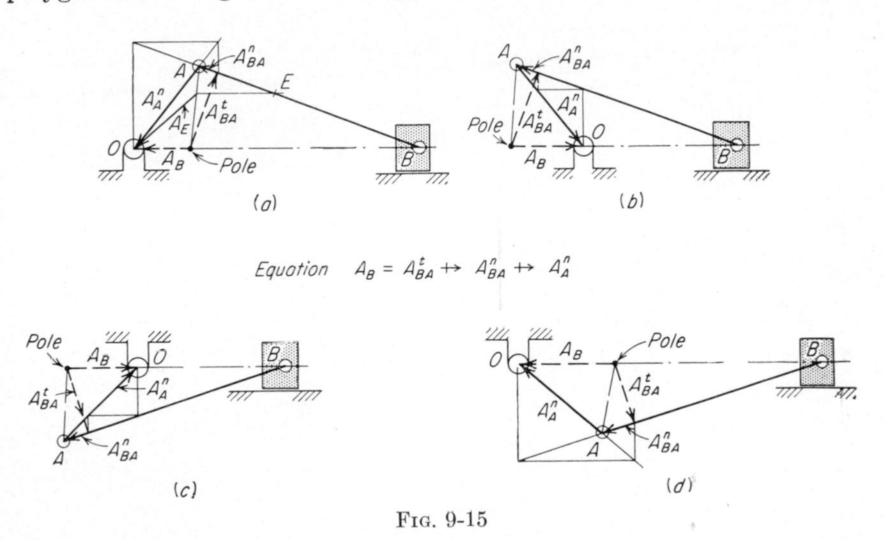

FIG. 9-15

that vectors AO, eO, and bO should all be pointed *toward* the main bearing O instead of away from it when using the Ritterhaus construction. Then the *proper* acceleration directions will be obtained.

Figure 9-15 shows examples of the Ritterhaus construction with the crank OA in each of the four quadrants. The vectors obtained are all in the proper directions, and they are labeled to correspond to the proper basic vector equation

$$A_B = A^t_{BA} \nrightarrow A^n_{BA} \nrightarrow A_A$$

It might be noted that the Ritterhaus construction does not work for the top and bottom dead center positions, making necessary the use of the more usual methods for these positions.

In order to bring out the method of dealing with a problem where a number of acceleration determinations are required for different positions of the crank, consider Fig. 9-16. The outline of the mechanism is the same as that of Fig. 9-14. In Fig. 9-16 half the crank circle has been drawn to an enlarged scale, and the acceleration determination has been made for one position of the crank OA.

First, a circle of radius OQ was drawn such that

$$\frac{OQ}{OA} = \frac{\text{length of crank}}{\text{length of connecting rod}}$$

The line AO was then extended to meet this circle at Q, and a horizontal line was drawn through Q intersecting the crank circle at P. The line OP gives the direction of the connecting rod[1] for this position of the crank.

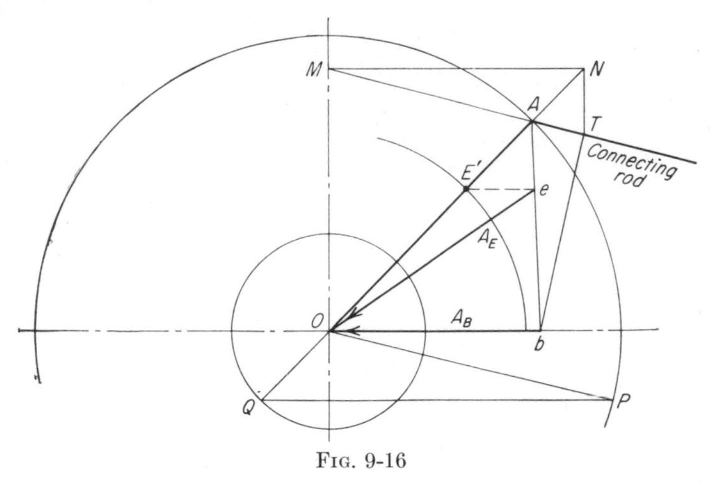

FIG. 9-16

Using this direction, the vector length TA, which represents A^n_{BA}, was determined by means of the Ritterhaus construction as shown. A line through T perpendicular to the connecting rod gave the vector length bo, which represents the acceleration of the wrist pin. The acceleration of a point E on the connecting rod (outside the limits of the drawing) was determined as follows: First, a circle of radius OE' was drawn such that

$$\frac{OE'}{OA} = \frac{\text{distance from wrist pin to the point } E}{\text{length of connecting rod}}$$

Through the intersection E' of the line OA with this circle, a horizontal line was drawn intersecting Ab (the acceleration image of AB) at e.

[1] This means of eliminating the full length of the connecting rod from the drawing permits of the use of a much larger scale in laying out the crank circle. This results in larger scales being used for the velocities and accelerations and thereby increases the accuracy of the determination. Note that the Ritterhaus construction does not require the full length of the connecting rod as does the usual construction given in Art. 9-7.

The point e is the acceleration image of the point E on the connecting rod and eO is the vector that represents the acceleration of point E, with direction e toward O.

The determinations for any other crank positions may be made in the same manner. This illustration brings out clearly the extent to which special methods can be developed for the routine solution of certain types of problems, provided the problems are of sufficient importance to warrant such a development.

9-10. Relative Acceleration of a Pair of Coincident Points. Coriolis' Law. The preceding articles have dealt with the relative acceleration of two points situated on the same rigid body. The present article deals with the relative acceleration of a pair of coincident points of two rigid bodies. The general characteristics of the *relative velocity* of two

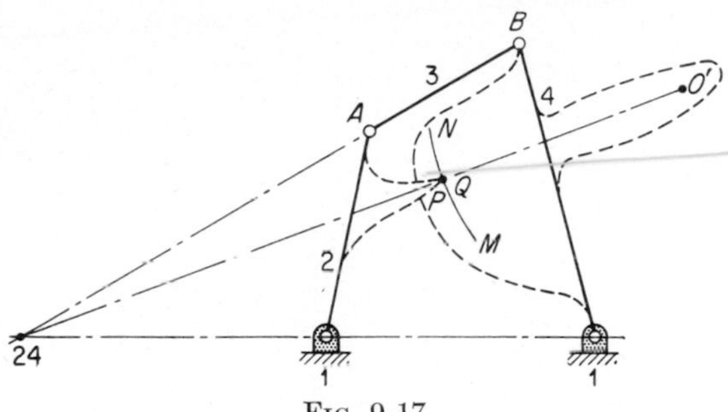

FIG. 9-17

such points have already been discussed in Chap. 8 by means of the theory of instantaneous centers; in discussing their relative acceleration a slightly different method of attack will be employed.

Referring to Fig. 9-17, let links 2 and 4 of the four-link mechanism there shown be assumed extended in the manner indicated. It is evident that, as the mechanism is set in motion, the point P on the extension of link 2 will trace out a curve such as MN on the extension of link 4. In the position shown, P is coincident with the point Q of link 4.

It is evident from the discussions of Chap. 8 that the magnitude of the relative velocity V_{PQ} is given by the product $(24\text{-}P) \times \omega_{24}$, where ω_{24} is the angular velocity of link 2 relative to link 4, and that the direction of V_{PQ} is perpendicular to the instantaneous radius $24\text{-}P$. This method of determining V_{PQ} does not involve any reference to the curve MN mentioned above.

The relative velocity V_{PQ} can, however, be expressed also in terms of the curve MN traced out by P on link 4. For, let ds be the length of an infinitesimal segment of the curve and dt the time required for the point P to trace out that segment. The length of the infinitesimal segment is given by

$$ds = R\,d\theta$$

where $R(= O'Q)$ is the average radius of curvature of the infinitesimal segment, and where $d\theta$ is the angle between the two radii drawn to the ends of the segment. If both sides of the equation are divided by dt,

$$\frac{ds}{dt} = R\frac{d\theta}{dt} = R\omega_{R4}$$

where ds/dt is obviously the magnitude of V_{PQ}, and where $d\theta/dt$ $(= \omega_{R4})$ is the angular velocity of the radius of curvature (assumed to move with P)

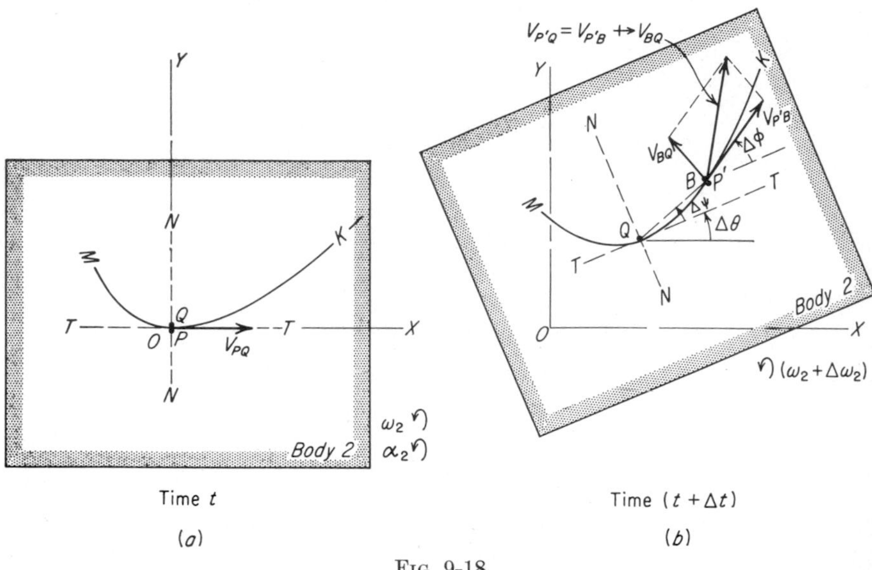

Fig. 9-18

relative to the rigid body 4. The direction of V_{PQ} is perpendicular both to the radius of curvature R and, as seen above, to the instantaneous radius 24-P. It follows that the two radii are collinear. It should be observed, however, that the center of curvature O' and the instantaneous center 24 are separate and distinct points, and that in this case they are even situated on opposite sides of the curve MN.

The foregoing discussion has shown how the relative velocity of a pair of coincident points may be discussed in terms of the path that one of the points traces out on the rigid body of the other. The relative acceleration of a pair of coincident points will be discussed in a similar manner.

In Fig. 9-18a, point P is moving so that it traces out curve MK on rigid body 2 which is also moving. At time t point P is coincident with point Q of rigid body 2. Rectangular axes QTN are fastened to the body and move with it; at time t they are coincident with *fixed* axes OXY. The velocity V_{PQ} of P relative to Q is tangent to the curve as shown. It is desired to determine A_{PQ}, the acceleration of P relative to Q.

At time $t + \Delta t$ the body will have moved to a new position, as shown in Fig. 9-18*b*, and point P will have moved along the curve to P', where it is coincident with a new point B of rigid body 2. Note that P and P' are two different positions of the *same* point, whereas Q and B are two *different* points of the same rigid body. The vector change of the velocity of P relative to Q during the interval Δt is

$$\begin{aligned} \Delta V_{PQ} &= V_{P'Q} \rightarrow V_{PQ} \\ &= V_{P'B} \nrightarrow V_{BQ} \rightarrow V_{PQ} \end{aligned} \tag{9-24}$$

Consider the magnitude of $V_{P'B}$. If $v_{P'B} = v_{PQ}$, then obviously there has been no change in the magnitude of velocity *along the path*. More generally, when such a velocity change *does* occur,

$$v_{P'B} = v_{PQ} + \Delta v_{PQ}$$

The magnitude of V_{BQ} is

$$v_{BQ} = (QB)(\omega_2 + \Delta\omega_2)$$

For infinitesimal changes, these magnitudes can be written

$$\begin{aligned} v_{P'B} &= v_{PQ} + dv_{PQ} \\ v_{BQ} &= (QB)(\omega_2 + d\omega_2) \end{aligned} \tag{9-25}$$

Now let the vectors in Eq. (9-24) be resolved into components parallel to the X and Y directions, and let time interval Δt become an infinitesimal $(= dt)$. The angles shown as $\Delta\theta$, $\Delta\psi$, and $\Delta\phi$ also become infinitesimals, and will be denoted by $d\theta$, $d\psi$, and $d\phi$, respectively. We obtain first:

$$dv^x_{PQ} = v^x_{P'B} + v^x_{BQ} - v^x_{PQ} \tag{9-26}$$

$$dv^y_{PQ} = v^y_{P'B} + v^y_{BQ} - v^y_{PQ} \tag{9-27}$$

To evaluate the quantities on the right-hand sides of these equations we shall consider each vector in turn, dropping higher-order infinitesimals (terms involving products of infinitesimal quantities) as they occur.

1. For the vector $V_{P'B}$:

$$\begin{aligned} v^x_{P'B} &= (v_{PQ} + dv_{PQ}) \cos (d\phi + d\theta) = v_{PQ} + dv_{PQ} \\ v^y_{P'B} &= (v_{PQ} + dv_{PQ}) \sin (d\phi + d\theta) = v_{PQ}(d\phi + d\theta) \end{aligned}$$

2. For the vector V_{BQ}, first note that

$$QB = ds = \frac{ds}{dt}\, dt = v_{PQ}\, dt$$

Then:

$$\begin{aligned} v_{BQ}^x &= -(QB)(\omega_2 + d\omega_2) \sin (d\psi + d\theta) \\ &= -v_{PQ}\, dt\, (\omega_2 + d\omega_2)(d\psi + d\theta) = 0 \\ v_{BQ}^y &= (QB)(\omega_2 + d\omega_2) \cos (d\psi + d\theta) = v_{PQ}\, dt\, \omega_2 \end{aligned}$$

3. For the vector V_{PQ}:

$$v_{PQ}^x = v_{PQ} \qquad v_{PQ}^y = 0$$

If we substitute these values in Eqs. (9-26) and (9-27) for dv_{PQ}^x and dv_{PQ}^y we obtain:

$$\begin{aligned} dv_{PQ}^x &= (v_{PQ} + dv_{PQ}) + 0 - v_{PQ} = dv_{PQ} \\ dv_{PQ}^y &= v_{PQ}(d\phi + d\theta) + v_{PQ}\, dt\, \omega_2 - 0 \end{aligned}$$

If now we note that dv_{PQ}^y is the *normal* component of the vector change of V_{PQ}, and that dv_{PQ}^x is the *tangential* component, we obtain, after dividing through by dt:

$$\begin{aligned} a_{PQ}^n &= \frac{dv_{PQ}^y}{dt} = v_{PQ}\left(\frac{d\phi}{dt} + \frac{d\theta}{dt}\right) + v_{PQ}\omega_2 \\ &= v_{PQ}\frac{d\phi}{dt} + 2v_{PQ}\omega_2 \\ a_{PQ}^t &= \frac{dv_{PQ}^x}{dt} = \frac{dv_{PQ}}{dt} \\ A_{PQ} &= A_{PQ}^n \nrightarrow A_{PQ}^t \end{aligned}$$

The normal component A_{PQ}^n, which is perpendicular to V_{PQ}, is composed of two parts which may have opposite directions. The part corresponding to $v_{PQ}(d\phi/dt)$ is directed away from the concave side of curve MK; that is, toward the instantaneous center of curvature of the path MK. The part corresponding to $2v_{PQ}\omega_2$ is associated with the movement of P *along* curve MK and with the angular motion *of* the curve; a little study will show that it must be directed so as to tend to turn V_{PQ} about its origin in the same angular direction as that of ω_2.

Now note that $d\phi/dt$ is the angular velocity of the radius of curvature R *relative to body* 2. If we set $d\phi/dt = \omega_{R2}$, we may write

$$\begin{aligned} v_{PQ}\frac{d\phi}{dt} &= v_{PQ}\omega_{R2} = R\omega_{R2}^2 = \frac{v_{PQ}^2}{R} \\ a_{PQ}^n &= v_{PQ}(\omega_{R2} + 2\omega_2)^1 \\ a_{PQ}^t &= \frac{d}{dt}(R\omega_{R2}) = R\alpha_{R2} \qquad \text{(if } R \text{ is constant)} \end{aligned}$$

[1] Note that the angular velocity of the vector V_{PQ} is $(\omega_{R2} + 2\omega_2)$ and not $(\omega_{R2} + \omega_2)$, as it might appear to be at first glance. It is not always easy to visualize the angular velocity of a velocity vector.

It is to be observed that, if rigid body 2 were standing still, and if P were moving along curve MK with an absolute velocity $V_P = V_{PQ}$, the magnitudes of the normal and tangential components of the absolute acceleration of P would be $v_P(d\phi/dt)$ and dv_P/dt. The quantities $v_{PQ}(d\phi/dt)$ and dv_{PQ}/dt may therefore be considered as the magnitudes

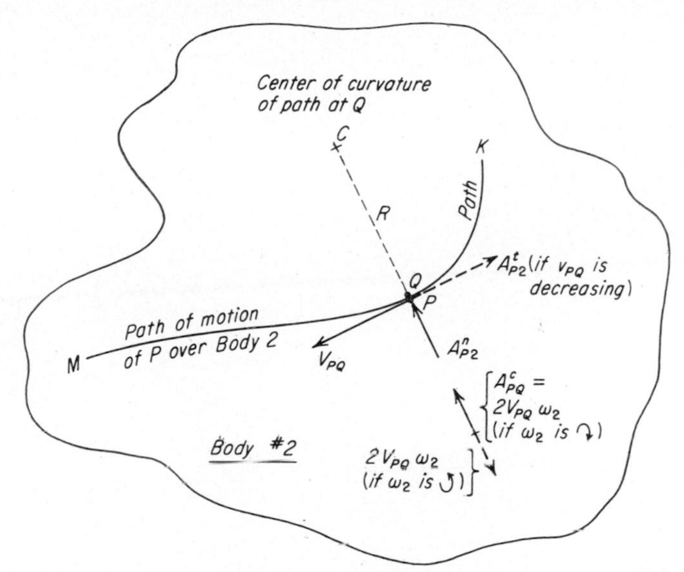

Rule for direction of coriolis component:
The direction of the vector $2V_{PQ}\omega_2$ is such that if placed at the extremity (tip) of vector V_{PQ} it would turn this vector in the sense ω_2.

FIG. 9-19

of the acceleration components of P *relative to rigid body* 2, and we may write:

$$a^n_{P2} = v_{PQ}\frac{d\phi}{dt} \qquad a^t_{P2} = \frac{dv_{PQ}}{dt} \qquad A_{P2} = A^n_{P2} \nrightarrow A^t_{P2}$$

The term $2v_{PQ}\omega_2$ is the magnitude of a vector known as the *Coriolis component.* If we designate this vector by A^c_{PQ},

$$A_{PQ} = A^n_{P2} \nrightarrow A^t_{P2} \nrightarrow A^c_{PQ} \tag{9-28}$$

For reference, the rules regarding the directions of the various acceleration vectors discussed in this article have been illustrated in Fig. 9-19.

Finally, the vector equation relating the complete accelerations of contacting points P and Q appears:

$$A_P = A_{PQ} \nrightarrow A_Q$$
$$A_P = A^n_{P2} \nrightarrow A^t_{P2} \nrightarrow A^c_{PQ} \nrightarrow A_Q \tag{9-29}$$

Example situations involving the use of the analysis of this article will be presented in the articles which follow.

9-11. Shaper Mechanism. In Fig. 9-20 the crank O_2P, which is 6 in. in length, makes an angle of 15 deg with the horizontal. The crank rotates counterclockwise at a constant rate of 10.5 rpm. It is required to determine the acceleration of point B.

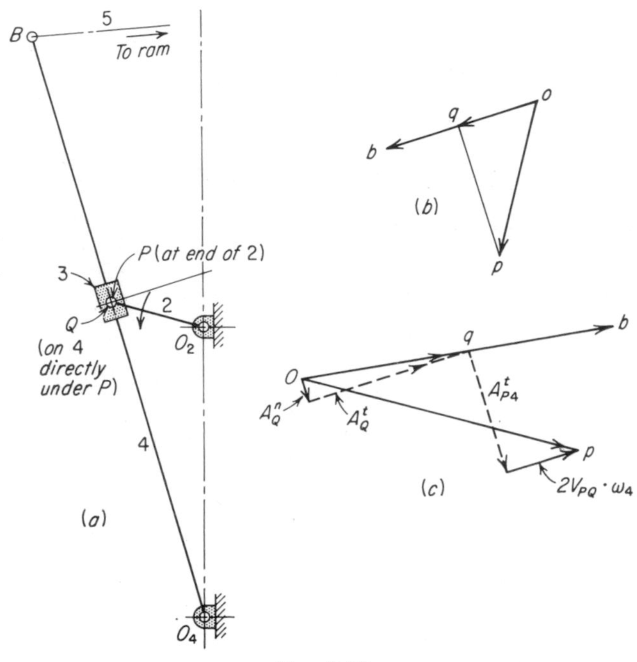

FIG. 9-20

The first step in the analysis is to draw the velocity polygon. The velocity of the crankpin is

$$v_P = (O_2P)\omega_2 = 6/12 \times 2\pi \times 10.5 = 33 \text{ fpm}$$

After a suitable scale is chosen, the velocity polygon may be drawn as shown at b, relating the velocities of points Q and P.

It is observed that point P is constrained to trace a path on link 4 which is a straight line coincident with O_4Q. The proper acceleration equation is

$$A_P = A_{PQ} \nrightarrow A_Q$$
$$A^n_P \nrightarrow A^t_P = A^n_{P4} \nrightarrow A^t_{P4} \nrightarrow A^c_{PQ} \nrightarrow A^n_Q \nrightarrow A^t_Q$$

The magnitudes of the several components of acceleration can be determined as follows, the velocities being scaled from the velocity polygon as required:

$$a_P^n = \frac{v_P^2}{O_2P} = \frac{(33/60)^2}{6/12} = 0.605 \text{ ft/sec}^2$$

$$a_P^t = 0$$

$$a_Q^n = \frac{v_Q^2}{O_4Q} = \frac{(17.2/60)^2}{20.4/12} = 0.0483 \text{ ft/sec}^2$$

$a_Q^t = (O_4Q)\alpha_4$ (unknown, since α_4 is unknown; but A_Q^t is known to be perpendicular to O_4Q)

$a_{P4}^n = \frac{v_{PQ}^2}{R} = 0$ (since R, the radius of curvature, is infinite for a straight path)

$a_{P4}^t = \frac{dv_{PQ}}{dt}$ (unknown in magnitude; but A_{P4}^t is known to be along link 4)

$$a_{PQ}^c = 2v_{PQ}\omega_4 = 2 \times \frac{28.2}{60} \times \frac{17.2/60}{20.4/12} = 0.1585 \text{ ft/sec}^2$$

The vector polygon representing the vector equation can now be drawn and a solution obtained for the two remaining unknown magnitudes.

Starting at the pole O in c of Fig. 9-20, the vectors on the left side of the vector equation above are laid off. There is only one, A_P^n, shown as Op. The right side of the equation must now also be started from the pole O. Lay off A_Q^n and the known direction of A_Q^t. From the tip of A_P^n, where the right side of the vector equation must finish, lay off the component $2V_{PQ}\omega_4$ in the position shown. Applying the rules given in Art. 9-10, this vector is perpendicular to V_{PQ} and is directed to the right. The component shown as A_{P4}^t in the figure is along the line of V_{PQ}. The intersection of the tangential components A_Q^t and A_{P4}^t at q is the image of Q. Hence Ob, determined by proportion, represents A_B. By scaling the vector Ob the acceleration of point B is obtained as 0.630 ft/sec.^{2} The reader should, as usual, check the vector equation against the diagram which represents it, to be certain that the two correspond.

9-12. Sliding-block Linkage. In Fig. 9-21, link 2 (of length 5 in.) is assumed to rotate at constant angular velocity at the instant shown, giving point P a velocity of 60 in./sec. The original drawing was one-third actual size. The problem is to find the angular velocity and angular acceleration of body 4.

It is first recognized that the velocity and acceleration of point P at the end of link 2 are both known, that point P traces a known (dotted) path over body 4, and that the point Q is a point on 4 coincident with P. The vector equations relating the motions of P and Q are

$$V_P = V_{PQ} \nrightarrow V_Q$$

$$A_P^n \nrightarrow A_P^t = A_{P4}^n \nrightarrow A_{P4}^t \nrightarrow 2V_{PQ}\omega_4 \nrightarrow A_Q^n \nrightarrow A_Q^t$$

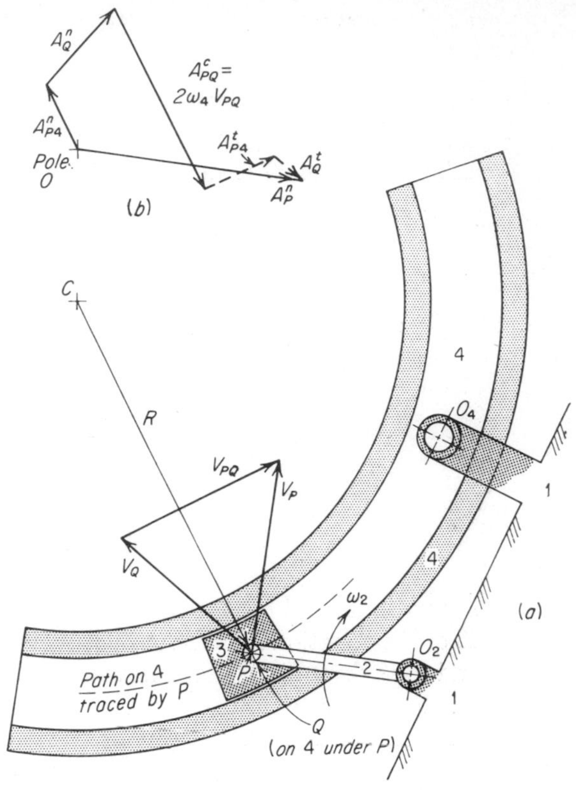

FIG. 9-21

The velocity polygon (scale of 1 in. = 30 in./sec in original drawing) is shown on the figure, with the pole at point P. The angular velocity of 4 can now be determined.

$$\omega_4 = \frac{v_Q}{QO_4} = \frac{53.5}{8.95} = 5.98 \text{ radians/sec} \qquad \textbf{(clockwise)}$$

The magnitudes of the several components of acceleration can next be computed, scaling velocities from the vector diagram as required.

$$a_P^n = v_P\omega_2 = \frac{v_P^2}{PO_2} = \frac{60 \times 60}{5} = 720 \text{ in./sec}^2$$

$$a_P^t = 0 \qquad \text{because } \omega_2 \text{ is constant}$$

$$a_{P4}^n = \frac{v_{PQ}^2}{CQ} = \frac{53 \times 53}{12.2} = 230 \text{ in./sec}^2$$

$$a_{P4}^t = \frac{dv_{PQ}}{dt} \qquad \text{(unknown in magnitude; but } A_{P4}^t \text{ is along line of } V_{PQ}\text{)}$$

$$2v_{PQ}\omega_4 = 2 \times 5.98 \times 53 = 634 \text{ in./sec}^2$$

$$a_Q^n = v_Q\omega_4 = \frac{v_Q^2}{QO_4} = \frac{53.5 \times 53.5}{8.95} = 320 \text{ in./sec}^2$$

$$a_Q^t = \frac{dv_Q}{dt} \qquad \text{(unknown in magnitude; but } A_Q^t \text{ is along line of } V_Q\text{)}$$

The vector polygon representing the acceleration equation is shown at *b* in the figure, and the two unknown magnitudes above have been determined. In the original drawing, the acceleration scale was

$$1 \text{ in.} = 300 \text{ in./sec}^2$$

The final required item, α_4, can now be calculated.

$$\alpha_4 = \frac{a_Q^t}{QO_4} = \frac{125}{8.95} = 13.97 \text{ radians/sec}^2 \qquad \text{(counterclockwise)}$$

9-13. Acceleration Analysis of the Scotch Yoke. In Fig. 9-22, the block 3 slides in a slot in link 4. The complete acceleration and velocity of point P at the end of link 2 are given. The problem is to find the acceleration of link 4.

Point P traces a known path on link 4—a straight vertical line. The vector equation relating the velocity of point P to the velocity of coincident point Q on 4 is

$$V_P = V_{PQ} \nrightarrow V_Q$$

The vector diagram is shown on the figure, with the pole at P. The acceleration equation is

$$A_P^n \nrightarrow A_P^t = A_{P4}^n \nrightarrow A_{P4}^t \nrightarrow 2V_{PQ}\omega_4 \nrightarrow A_Q^n \nrightarrow A_Q^t$$

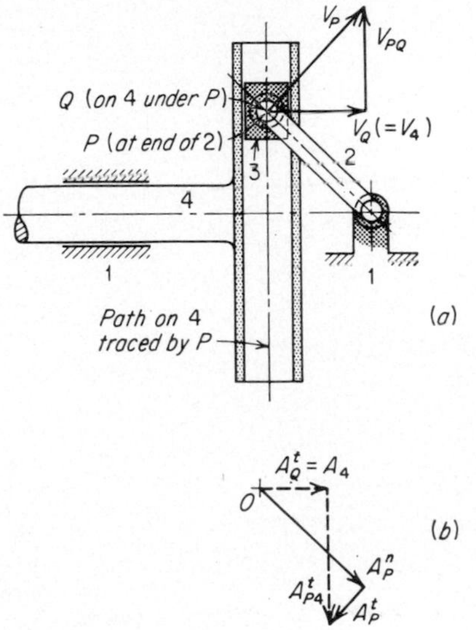

Fig. 9-22

No numerical values will be cited here, but each vector in the equation will be discussed. A_P^n and A_P^t are known (given).

$$a_{P4}^n = \frac{v_{PQ}^2}{R} = \frac{v_{PQ}^2}{\infty} = 0$$

$$a_{P4}^t = \frac{dv_{PQ}}{dt} \qquad \text{(unknown in magnitude; but } A_{P4}^t \text{ is along the line of } V_{PQ}\text{)}$$

$$2v_{PQ}\omega_4 = 2v_{PQ}(0) = 0$$

$$a_Q^n = v_Q\omega_4 = 0$$

$$a_Q^t = \frac{dv_Q}{dt} \qquad \text{(unknown in magnitude; but } A_Q^t \text{ is along the line of } V_Q\text{)}$$

The acceleration equation can now be simplified.

$$A_P^n \nrightarrow A_P^t = A_{P4}^t \nrightarrow A_Q^t$$

The vector diagram in which the two unknown magnitudes above are determined for a specific instance is shown at *b* in the figure. The required answer, the acceleration of link 4, is obtained by evaluating the vector A_Q^t to the same scale which was chosen when the vector diagram was drawn.

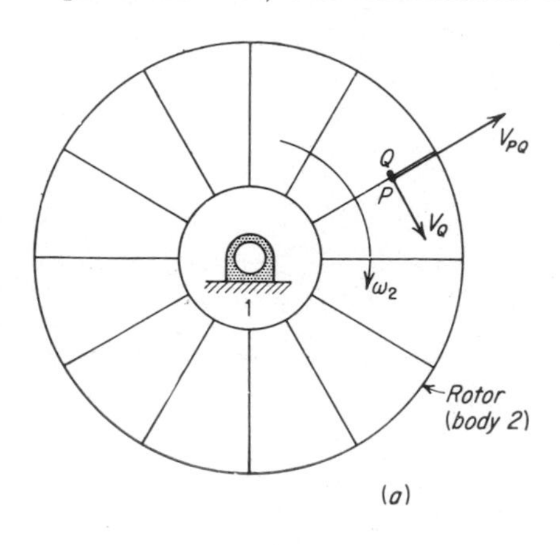

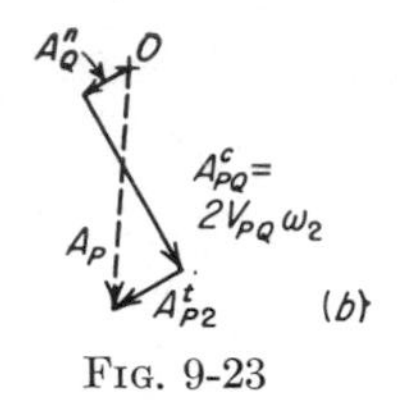

FIG. 9-23

In this example a sliding connection exists between the links 3 and 4, and yet no Coriolis component of acceleration appeared in the analysis. There is motion *along* a path (V_{PQ}) but no rotation *of* the path ($\omega_4 = 0$). If link 4 were to be constrained in a different manner resulting in angular velocity of 4, then a Coriolis component of acceleration would be present.

9-14. Flow of Fluid in a Centrifugal Pump. In Fig. 9-23 is shown the rotor of a centrifugal pump having radial vanes. A particle of fluid at P travels radially outward along the vane, having a velocity V_{PQ} relative to point Q on the vane. This velocity is assumed to be decreasing at a known rate. The rotor turns at a constant angular velocity ω_2. The velocity and acceleration equations are

$$V_P = V_{PQ} \nrightarrow V_Q$$
$$A_P = A_{PQ} \nrightarrow A_Q$$
$$A_P = A_{P2}^n \nrightarrow A_{P2}^t \nrightarrow 2V_{PQ}\omega_2 \nrightarrow A_Q^n \nrightarrow A_Q^t$$

The magnitudes of the terms in the acceleration equation can be evaluated as follows:

$a_P = \text{unknown}$ (and direction of A_P also is unknown)

$a_{P2}^n = \dfrac{v_{PQ}^2}{R} = \dfrac{v_{PQ}^2}{\infty} = 0$

$a_{P2}^t = \dfrac{dv_{PQ}}{dt}$ (assumed to be known in magnitude, and A_{P2}^t assumed to be opposite V_{PQ})

$2v_{PQ}\omega_2 = \text{known}$ (because the individual terms are known)

$$a_Q^n = v_Q\omega_2$$

$$a_Q^t = \frac{dv_Q}{dt} = 0 \qquad \text{(because } \omega_2 = \text{const)}$$

A possible acceleration polygon representing the equation is shown at *b* in Fig. 9-23. The vector $2V_{PQ}\omega_2$ (the Coriolis component) requires that the vane exert a force on the particle of fluid P in the direction of $2V_{PQ}\omega_2$. The reaction force of the particle on the vane requires that a clockwise torque be supplied to the shaft of the rotor.

Forces and torques will not be discussed further at this point, but it can be noted that a consideration of all the particles in the pump leads to determination of the total torque supplied to the rotor shaft. The particular item of interest in this discussion has been the analysis for acceleration, resulting in a brief indication as to how the requirement for shaft torque arises in this radial-vaned type of machine. It is noted that only the Coriolis component of acceleration ($2V_{PQ}\omega_2$) requires a torque in the shaft (when it is assumed that all particles of fluid move *radially* in the rotor).

Many fans, pumps, compressors, and various fluid-drive transmissions make use of this basic centrifugal type of machine.

9-15. Acceleration Analysis of a Circular-arc Cam with Roller Follower. In Fig. 9-24*a*, ω_2 is constant at a given value, and the problem is to find the acceleration of point P on link 4, so that α_4 can be determined. First, it is visualized that point P traces the dotted path shown on link 2 (extended). Thus, it is recognized that a point (P) moves along a path on another link (2) which has angular velocity. The path is a circular arc (at this instant) with center of curvature at C, because the portion of the cam surface from A to B is a circular arc with center at C. The velocity and acceleration equations for the coincident points P and Q are as follows:

$$V_P = V_{PQ} \nrightarrow V_Q$$

$$A_P = A_{PQ} \nrightarrow A_Q$$

$$A_P^n \nrightarrow A_P^t = A_{P2}^n \nrightarrow A_{P2}^t \nrightarrow 2V_{PQ}\omega_2 \nrightarrow A_Q^n \nrightarrow A_Q^t$$

The details of the calculation of each of the acceleration terms will be omitted, as the procedure has already been outlined several times. However, particular attention should be given to the fact that Q is visualized as a point at a fixed location on the cam 2, and so is constrained to move around O_2 as a permanent center. For example, the normal acceleration of Q ($a_Q^n = v_Q\omega_2$) is directed from Q toward O_2.

In the acceleration equation only the magnitudes of A_P^t and A_{P2}^t remain unknown, and the vector polygon containing the solution for these two magnitudes is shown at *b* in Fig. 9-24.

The diagram shown at *c* will be discussed in the following article.

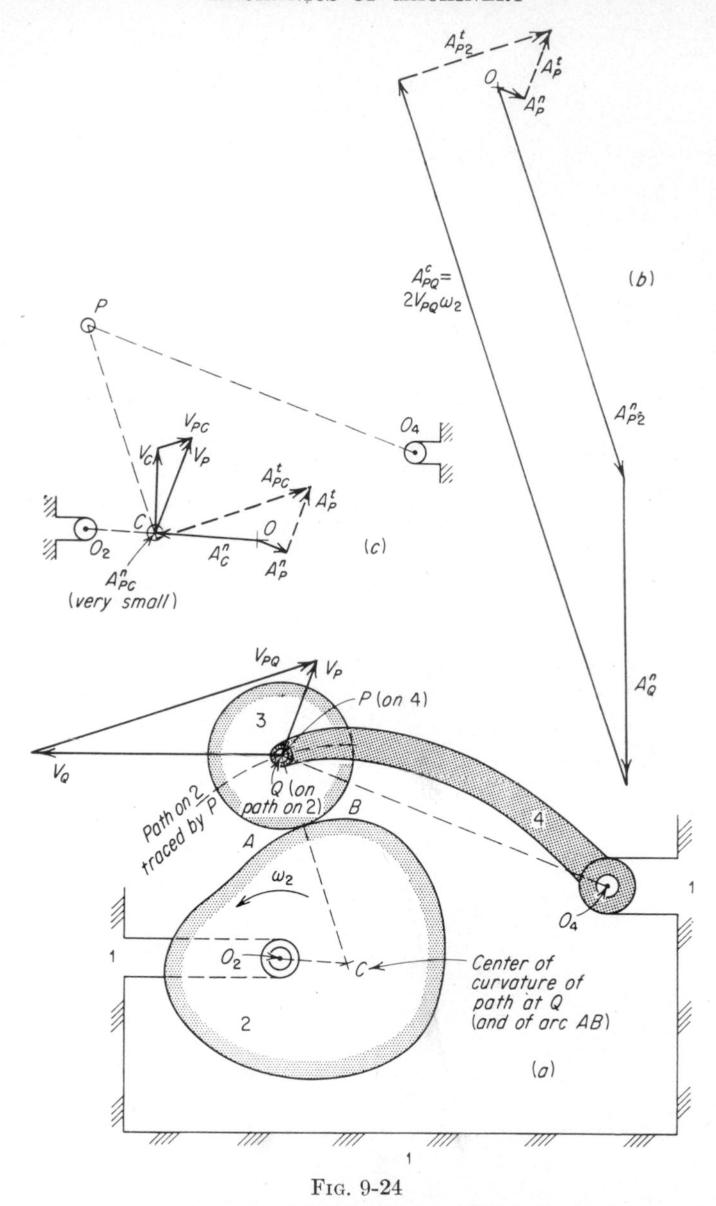

FIG. 9-24

9-16. Acceleration Analysis by the Use of Equivalent Linkages. The analysis of certain types of devices (usually those involving the Coriolis component of acceleration) can be simplified somewhat by recognizing that a pin-connected (hinge-jointed) mechanism can be drawn which is kinematically equivalent to a more complicated-appearing device. As an example, consider the cam and follower of the previous article (Fig.

9-24). It is obvious that the distances O_4P and O_2C remain constant as the mechanism operates. Also, the distance QC (or PC) remains constant while the roller contact falls between A and B. Therefore, at the instant shown, the simple four-bar linkage shown dotted in a of Fig. 9-24

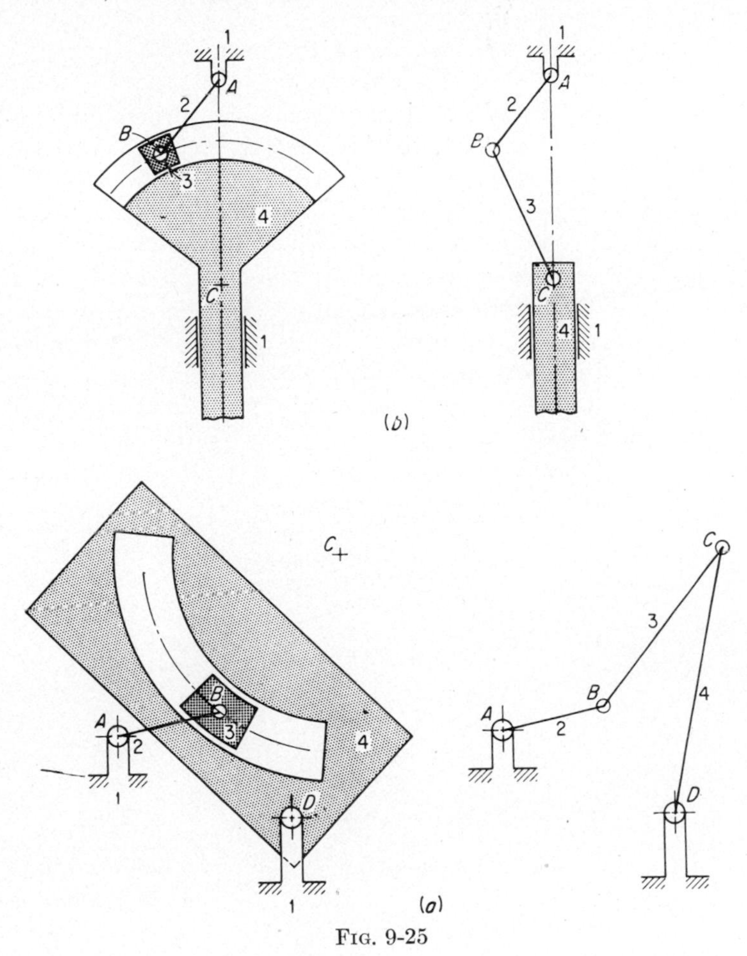

FIG. 9-25

and reproduced at c is an equivalent linkage, kinematically, to the original mechanism. For this linkage,

$$V_P = V_{PC} \nrightarrow V_C$$
$$A_P = A_{PC} \nrightarrow A_C$$
$$A_P^n \nrightarrow A_P^t = A_{PC}^n \nrightarrow A_{PC}^t \nrightarrow A_C^n \nrightarrow A_C^t$$

No sliding members exist, and no Coriolis component of acceleration is present. The velocity and acceleration polygons are shown at c in Fig. 9-24. The given information and scales used are the same as those in the original mechanism. It is seen that the required acceleration of point P is the same for both methods of analysis.

It is not necessary in general that a path of slide such as the dotted path in *a* of Fig. 9-24 be a *circular* arc, in order for the equivalent-link idea to be valid. It is only necessary (the proof will not be given here) that C be at the *center of curvature* of the path at the instant under consideration.

In Fig. 9-25, examples of additional kinematically equivalent linkages are shown. Also, in Fig. 9-21, the kinematically equivalent linkage is recognized to be O_2PCO_4.

It is readily visualized (consider again Fig. 9-25) that when the path of slide is a straight line one of the equivalent links (BC) becomes infinitely long. For such instances, and for those in which the radius of curvature (BC) of the path is inconveniently large, the equivalent linkage approach is not feasible. The original unmodified mechanism must then be analyzed, and the analysis often involves the conditions in which the Coriolis component of acceleration appears.

9-17. Analytical Methods. Most of the acceleration analyses of this chapter have been concerned with graphical methods. However, it

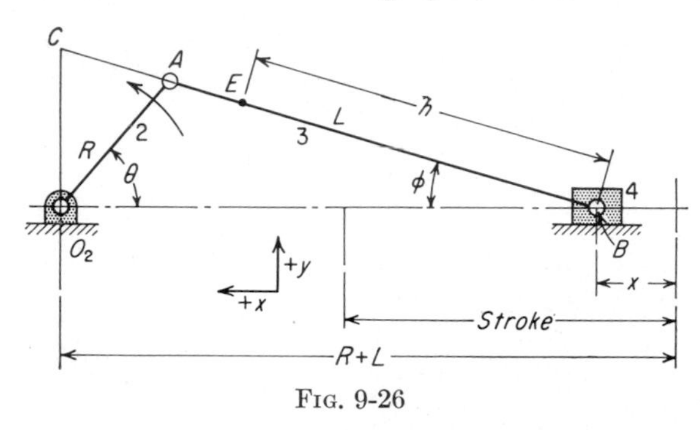

FIG. 9-26

should be recognized that a consideration of the trigonometry and geometry of the acceleration polygons could lead to *computed* solutions as contrasted with solutions obtained graphically by intersection of lines.

The present article presents an example of the analytical method, in which algebraic equations are set up for the x and y components of the pertinent vector quantities.

In Fig. 9-26, the crank is turning with a constant angular velocity ω_2, and the acceleration of the point E on the connecting rod is required.

It was shown near the end of the chapter on Velocities that the x component of the velocity of the point E is given by

$$v_E^x = R\omega_2 \left[\sin\theta + \left(1 - \frac{h}{L}\right) \frac{1}{2} \frac{R}{L} \frac{\sin 2\theta}{\cos\phi} \right]$$

Differentiating with respect to time,

$$a_E^x = \frac{d}{dt}(v_E^x) = R\omega_2 \left[\cos\theta \frac{d\theta}{dt} + \left(1 - \frac{h}{L}\right) \frac{1}{2} \frac{R}{L} \left(\frac{\cos 2\theta}{\cos\phi} \frac{d(2\theta)}{dt} + \frac{\sin 2\theta \tan\phi}{\cos\phi} \frac{d\phi}{dt} \right) \right]$$

In the section of the text mentioned above it was also shown that

$$\frac{d\phi}{dt} = \frac{R}{L} \frac{\cos\theta}{\cos\phi} \omega_2 \qquad \text{and} \qquad \sin\phi = \frac{R}{L} \sin\theta$$

Therefore

$$\begin{aligned} a_E^x &= R\omega_2^2 \left[\cos\theta + \left(1 - \frac{h}{L}\right) \frac{1}{2} \frac{R}{L} \left(2 \frac{\cos 2\theta}{\cos\phi} + \frac{2 \sin\theta \cos\theta \tan\phi}{\cos\phi} \frac{R}{L} \frac{\cos\theta}{\cos\phi} \right] \right) \\ &= R\omega_2^2 \left[\cos\theta + \left(1 - \frac{h}{L}\right) \frac{R}{L} \left(\frac{\cos 2\theta}{\cos\phi} + \frac{R^2}{L^2} \frac{\sin^2\theta \cos^2\theta}{\cos^3\phi} \right) \right] \end{aligned} \qquad (9\text{-}30)$$

The vertical component of the velocity of the point E is

$$v_E^y = R\omega_2 \frac{h}{L} \cos\theta$$

Differentiating,

$$a_E^y = R\omega_2 \frac{h}{L} (-\sin\theta) \frac{d\theta}{dt} = -R\omega_2^2 \frac{h}{L} \sin\theta \qquad (9\text{-}31)$$

If h is equal to zero, the expression a_E^x becomes the acceleration of the crosshead

$$a_B = R\omega_2^2 \left[\cos\theta + \frac{R}{L} \left(\frac{\cos 2\theta}{\cos\phi} + \frac{R^2}{L^2} \frac{\sin^2\theta \cos^2\theta}{\cos^3\phi} \right) \right] \qquad (9\text{-}32)$$

This is the exact expression. The following approximate expression is, however, sufficiently accurate for most purposes.

$$a_B = R\omega_2^2 \left[\cos\theta + \frac{R}{L} \cos 2\theta \right] \qquad \text{approx} \qquad (9\text{-}33)$$

Similarly

$$a_E^x = R\omega_2^2 \left[\cos\theta + \left(1 - \frac{h}{L}\right) \frac{R}{L} \cos 2\theta \right] \qquad \text{approx} \qquad (9\text{-}34)$$

CHAPTER 10

STATIC FORCES IN MACHINES

The analysis of the forces acting in any machine is based upon the fundamental principle which states that the system composed of all the external forces and all the inertia forces acting upon any single member of the machine is a system in equilibrium. The forces acting in machines having plane motion are for the most part situated in parallel planes. In such cases it is customary to analyze the system as though all its forces were situated in the same plane, the moments due to the offset of the forces often being disregarded or taken into account by a second analysis in a plane perpendicular to the first plane. The assumption of plane forces will be employed in the following articles unless otherwise stated. In the *static force analysis* of a machine the inertia forces (caused by acceleration) of the machine members are disregarded. If the inertia forces are taken into account, the analysis is called a *dynamic force analysis.* In either case the effect of friction may be taken into account or disregarded depending upon the requirements of the problem. The present chapter deals only with the static force analysis of machines.

10-1. Reactions between Members with Friction Disregarded. In the present article and in the next few articles the effect of friction will be disregarded. An analysis in which friction is disregarded is useful in obtaining preliminary results, and in the case of well-lubricated machinery these results are often all that is required. In such an analysis the results obtained represent ideal conditions, i.e., 100 per cent efficiency.

The forces acting upon a machine member are in general applied at its points of contact with other members of the machine, the two most important exceptions being the dead-weight forces and the inertia forces. This contact is usually made by means of turning connections or sliding pairs, and sometimes by means of higher pairs. Whatever the type of connection, the forces exerted by one member upon another are, if friction is disregarded, normal to the surfaces of the members at their points of contact.

When the contact between two members of a machine is made by means of a turning or sliding joint, the force exerted by one of the members upon the other is not applied at one point alone, but is the resultant of a number of elementary forces distributed over a considerable area.

These elementary forces have directions normal to the surfaces at their corresponding points of contact. In the case of the turning joint such as a shaft in a bearing the line of action of each elementary force passes through the center of the joint, and hence the resultant of all the elementary forces will also pass through the center of the joint. In the case of the sliding joint such as a crosshead and guide the elementary forces are all normal to the same plane, and hence their resultant will also be normal to the plane. The line of action of this resultant, however, cannot be determined from the characteristics of the joint alone. In order to simplify the analysis, the position of the line of action of the resultant is sometimes assumed to be situated at the center of gravity of the contact area, but it is usually determined from the conditions of equilibrium.

10-2. Analysis of a Bell Crank. A simple example of a static force analysis is shown in Fig. 10-1. The bell crank 2 is part of a system of levers, and it is connected to the rest of the system by means of the rods 3 and 4. The force P is assumed to be known, and it is required to determine the other forces acting on the bell crank. The forces exerted by the rods are assumed to act along their center lines. With regard to the notation, F_{42} means the force exerted by link 4 upon link 2; F_{24} would be the equal and opposite force exerted by link 2 upon link 4. This notation will be employed very extensively in the force analyses to follow.

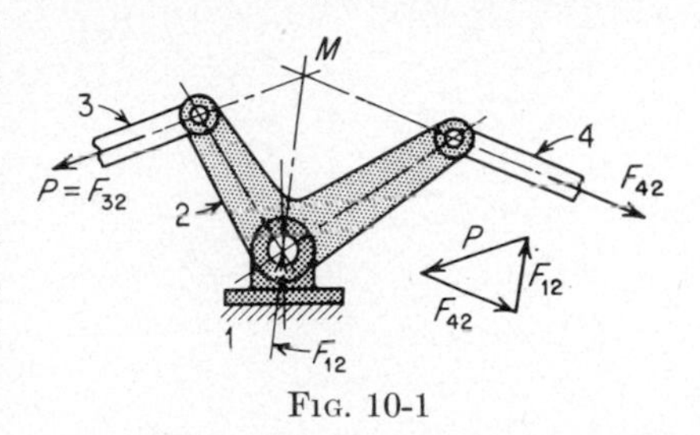

Fig. 10-1

Three forces act upon the bell crank 2: the known force P, the force exerted by rod 4, and the force exerted by the bearing 1. No external couple or torque is applied to 2. The line of action of F_{42} is completely known. The resultant of P and F_{42} must pass through the intersection M of their lines of action. The remaining force F_{12}, which passes through the center of the bearing, must be equal and opposed to the resultant of P and F_{42}, and also must pass through the point M. If it did not, a couple would remain. But this cannot be, because no external torque is exerted on 2 to oppose such a couple. The directions of all three forces being known, the force polygon may be drawn as shown in the figure and the magnitudes of F_{12} and F_{42} thus determined. The equation of force vectors which the diagram portrays is

$$F_{32} \nrightarrow F_{12} \nrightarrow F_{42} = 0$$

10-3. Engine Mechanism. In the force analysis of a machine, it is necessary to assume either that the effort is known or that the useful resistance is known. In the engine mechanism shown in Fig. 10-2*a*, the effort P due to the pressure of the steam or gas against the piston is the known force. The useful resistance is the unknown couple Qq, which acts upon the crankshaft and which is equal and opposed to the torque exerted *by* the latter.

The first step in the analysis is to consider the forces acting upon the crosshead (link 4). These are as follows:

P = the known effort.
F_{14} = the reaction of the guides.
F_{34} = the force exerted by the connecting rod.

Of these three forces only the effort is known. Owing to the fact, however, that only two forces act upon the connecting rod, namely, a force at A and a force at B, it is evident that the line of action of F_{34} must be coincident with the line AB. The resultant of P and F_{34} must pass

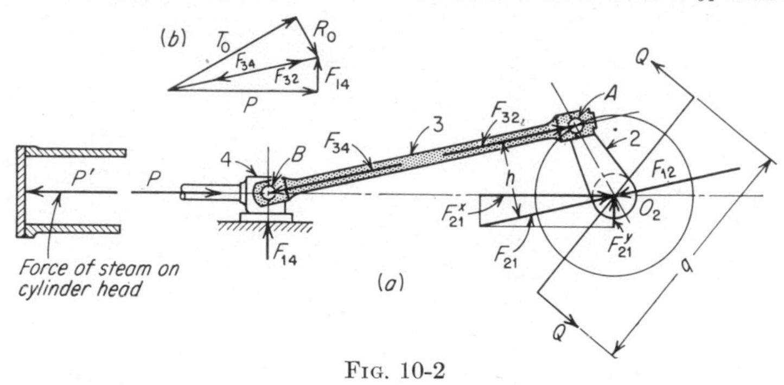

FIG. 10-2

through the intersection B of their lines of action; and since three forces only act upon the crosshead, F_{14} must also pass through the point B. The direction of F_{14} must be normal to the guides as shown. All three forces, therefore, being known as to direction, the force polygon can be drawn as shown at b in the figure and the magnitudes of F_{34} and F_{14} thus determined.

The force F_{32} exerted by the connecting rod upon the crank is obviously equal and opposed to F_{34}. The other forces acting upon the crank are F_{12}, the main bearing pressure, and the forces of the unknown couple Qq. In order to hold the couple Qq in equilibrium, the force F_{12} must form a couple with F_{32}; that is, F_{12} must be equal and parallel to F_{32} and must have the opposite direction. The moment of the couple Qq is equal to $F_{32}h$; the latter is usually called the turning moment. If F_{32} is resolved at the crankpin into the components T_0 and R_0, perpendicular and parallel respectively to the crank, then the perpendicular component T_0 is called the *turning effort*. It is evident from the figure that $\dfrac{F_{32}}{T_0} = \dfrac{O_2A}{h}$; therefore $F_{32}h = T_0 \cdot O_2A$.

During any one revolution of the crank the effort P is a varying quantity, and the turning effort T_0 is subject to corresponding variations. On the other hand the actual useful resistance overcome by the engine is ordinarily fairly constant, and in order to prevent violent fluctuations in the speed of the crankshaft, it is in general necessary to make use of a flywheel. The function of the latter is to store up energy when P is too great and to deliver this energy to the crankshaft when P is too small. It is evident that, where a flywheel is used, the couple Qq is not the actual useful resistance but is rather the combined effect of this resistance and the inertia of the flywheel.

For completeness an analysis of the forces acting upon the frame should be made. These forces are as follows:

1. The force P', which is equal and opposed to P.
2. The force F_{21}, which is equal and opposed to F_{12}.
3. The force F_{41}, which is equal and opposed to F_{14}.
4. The forces due to the weight of the frame and foundation.

Since the vector sum of P, F_{14}, and F_{12} is zero, it is evident that the vector sum of P', F_{41}, and F_{21} will also be zero. These three forces, however, are not in equilibrium, as may be seen from the following analysis. Let F_{21} be resolved at the center of the main bearing into its horizontal and vertical components F_{21}^x and F_{21}^y. The horizontal component is equal and opposed to P' and is thus neutralized, but the vertical component F_{21}^y forms a couple with the force F_{41}. This couple, which is equal to the couple Qq, tends to rotate the engine as a whole about the crankshaft and must be held in equilibrium by the forces due to the weight of the frame and the foundation.

10-4. Drag-link Mechanism. The static force analysis of a somewhat more complicated mechanism is shown in Fig. 10-3a. It is required to determine the effort P that must be applied in the form of force upon the teeth of gear 3 in order to overcome the known resistance Q. The first part of the analysis is the same as that of the engine mechanism discussed in the preceding article. The forces acting upon the ram (link 7) are Q, F_{17}, and F_{67}. The magnitudes of F_{17} and F_{67} may be determined from the force polygon shown at b. The corresponding vector equation for force equilibrium is

$$Q \nrightarrow F_{17} \nrightarrow F_{67} = 0$$

The forces acting upon link 5 are F_{65}, F_{45}, and F_{15}. The lines of action of F_{65} and F_{45} are completely known, and hence the intersection M of these lines can be determined. F_{15}, applied to 5 at point O_5, must pass through the point M because there is no couple or torque applied to link 5. The directions of all three forces and one magnitude (F_{65}) being known, the remaining magnitudes may be determined from the force polygon.

The forces acting upon link 3 may be determined in a similar manner, and the tooth force P thus obtained.

10-5. Friction. Part of the energy supplied to every machine must be used in overcoming the frictional forces occurring within the machine. These motion-resisting forces can be classified according to the manner in which they are set up, the most important being resistance to sliding

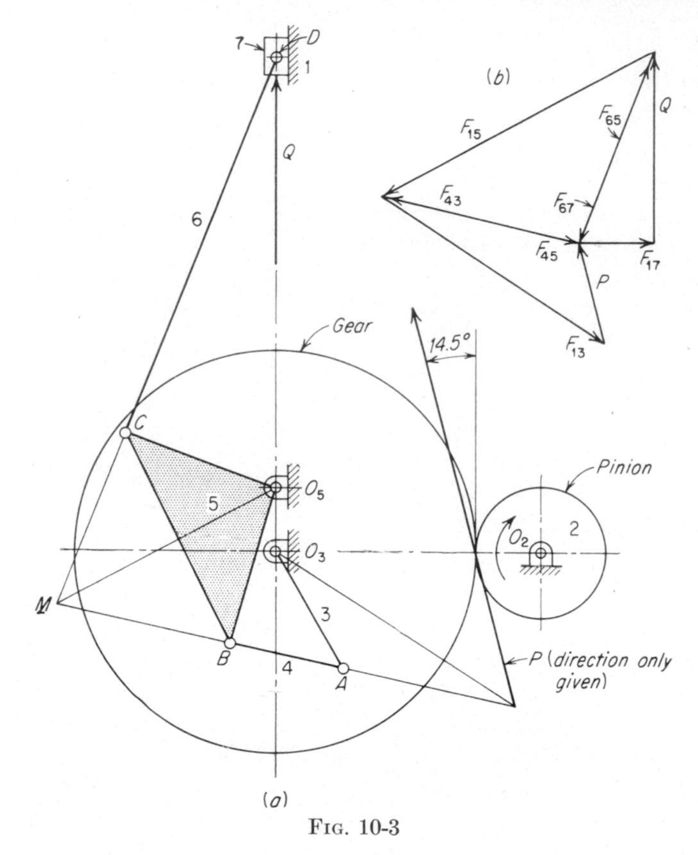

Fig. 10-3

and resistance to rolling. There is also the frictional resistance that a wrapping connector offers to being bent around a sheave or pulley. In pumps and in those cases where a fluid is used for the transmission of power, part of the energy is used in overcoming fluid friction.

Sliding friction is the resistance offered to motion, or the tendency to motion, when two surfaces are rubbed on each other. This type of friction may be reduced materially by the use of lubricants, the action of the latter being partially to separate the rubbing surfaces, thus substituting fluid friction for a part of the dry sliding friction.

Rolling resistance (also called rolling friction) is the resistance offered to motion when two surfaces roll on each other. The fact that this kind of friction is much less than sliding friction is the reason for substituting ball, roller, or needle bearings for the ordinary turning joint in cases where it is desired to reduce the frictional resistance to a minimum.

The frictional resistance of a wrapping connector is an internal friction caused by the fibers of the connector sliding over one another.

10-6. Sliding Friction. In dealing with sliding friction, it is customary to assume that the so-called *law of proportionality* is strictly true. This law, which is only an approximation, states that the friction force is proportional to the normal force pressing the rubbing surfaces together. It may be represented by the equation $F = \mu N$, where F is the force of friction, N the normal force, and μ a quantity called the coefficient of friction. This coefficient is usually assumed to be the same for all the points of contact of any one joint, but it is not necessarily assumed to be the same for all joints. The force of friction at any point of contact has a direction and line of action tangent to the rubbing surface at that point.

The coefficient of sliding friction depends, in general, upon the materials and condition of the rubbing surfaces and upon the lubricants used. With no lubricant, *dry friction* exists. For *thick-film* (other names used are *complete*, *fluid*, *viscous*, or *perfect*) *lubrication*, the surfaces are completely separated by a thick fluid film under pressure produced by hydrostatic or hydrodynamic action. For example, a rotating journal in a bearing generously supplied with oil can develop a pressure in the oil surrounding the journal sufficient to float the load and prevent actual surface contact.

Some situations are intermediate between that with dry friction and that with thick-film lubrication. Lubricant is present, but hydrodynamic action is not sufficient to support the load. The fluid film is very thin, and the adhesion of the lubricant to the surfaces plus other physical and chemical aspects of the lubricant's "oiliness" are depended upon to prevent or minimize surface-to-surface contact while allowing slipping to occur within the lubricant. Such lubrication is called *thin-film* or *boundary* lubrication. Greased surfaces and journal bearings at the start of rotation are examples of boundary lubrication. It appears likely that at times there may be contact of the small projections which are present even upon "smooth" surfaces. The amount of friction is usually intermediate between that for dry surfaces and that for surfaces having thick-film lubrication.

Table 10-1 presents a few representative values of the coefficient of sliding friction for dry surface contact and for greasy or boundary lubrication.

For thick-film lubrication, values of equivalent sliding coefficient of friction lie roughly in the range of 0.001 to 0.03. It is seen that these values are appreciably lower than those given in Table 10-1.

TABLE 10-1. REPRESENTATIVE VALUES OF COEFFICIENT OF SLIDING FRICTION*

Materials	μ	
	Dry	Boundary lubrication
Hard steel on hard steel	0.42	0.029–0.108
Mild steel on mild steel	0.57	0.09 –0.19
Hard steel on babbitt	0.33–0.35	0.06 –0.16
Mild steel on phosphor bronze	0.34	0.173
Mild steel on cast iron	0.23	0.133
Mild steel on lead	0.95	0.3
Glass on glass	0.40	0.09 –0.116
Cast iron on cast iron	0.15	0.064–0.070
Bronze on cast iron	0.22	0.077
Cast iron on oak	0.49	0.075

* Selected values from L. S. Marks (ed.), "Mechanical Engineers' Handbook," 5th ed., p. 218, McGraw-Hill Book Company, Inc., New York, 1951.

It should be mentioned here that coefficients of *static* friction are in general much higher than coefficients of *sliding* friction, particularly for dry or nearly dry surfaces. Thus to *start* a block moving from rest on a plane surface takes much more force than is necessary to *maintain* the motion.

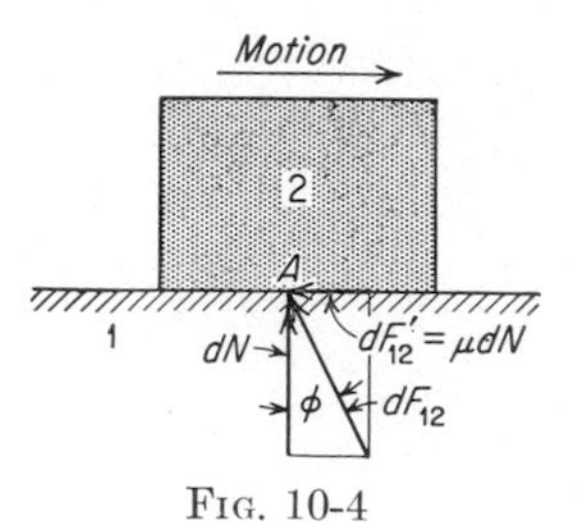

FIG. 10-4

1. *Sliding Joint.* A simple sliding joint consisting of a block moving along a plane surface is shown in Fig. 10-4. The block 2 is assumed to be under the action of a system of forces (not shown) that cause it to move in the direction indicated and that press it against the flat surface 1. It is convenient to assume for the purposes of analysis that the law of proportionality holds, or that applicable equivalent values of the coefficient of friction are available. The discussion will proceed on this basis.

Referring to Fig. 10-4, the normal unit pressure exerted by the flat surface upon the block will not in general be uniform; i.e., it will vary in magnitude at different points of the contact area. Let dN be the elementary normal force acting at A and μ the coefficient of friction. The elementary force of friction at A is

$$dF'_{12} = \mu \, dN$$

The resultant of the elementary force of friction and the elementary normal force dN is the total elementary force dF_{12} exerted upon the block 2 by the flat surface 1 at the point A. The angle ϕ that the direction of dF_{12} makes with the normal is called the angle of friction. This angle may be determined in the following manner:

$$\tan \phi = \frac{dF'_{12}}{dN} = \frac{\mu \, dN}{dN} = \mu$$

$$\phi = \arctan \mu \qquad (10\text{-}1)$$

It is evident from the equation that, if μ is the same for all points of the contact area, the angle ϕ is also the same. It follows that the resultant of all the elementary total forces will make an angle ϕ with the resultant normal force, and that the lines of action of these two forces will pass through the same point of the contact area.

Now let the case shown in Fig. 10-5 be considered. The block 2 is pressed against the flat surface 1 by the known force Q and is moved along the flat surface under the influence of the force P, the latter being known as to line of action but unknown as to magnitude. For equilibrium the force exerted upon the sliding block by the flat surface must pass through the point A. If friction is taken into account, the force F_{12} must be inclined to the normal by an amount equal to the angle of friction. If ϕ is this angle, the direction of F_{12} is as shown. The magnitudes of P and F_{12} may be determined from the force polygon shown. If F_{12} is resolved into horizontal and vertical components at its point of application, it is seen at once that the horizontal component is equal to P and has the opposite direction, and that the vertical component is equal and opposite Q. If there were no friction, the force P would be zero, and the resultant normal force F^y_{12} would be directly opposed to Q; i.e., it would have the same line of action as Q. It is evident, therefore, that the effect of friction has been to shift the line of action of F^y_{12} in the same direction as that of the motion of the block.

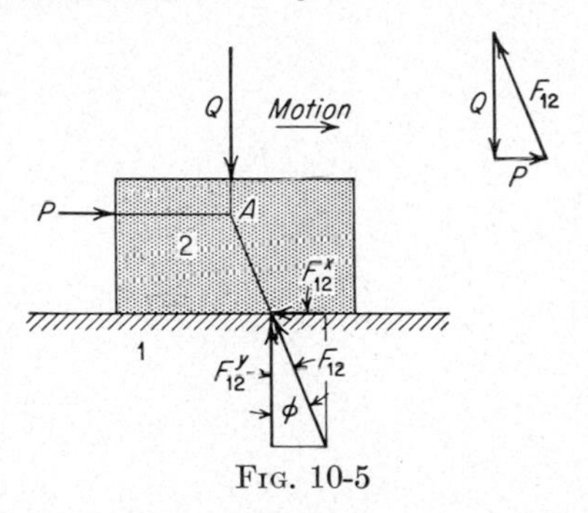

FIG. 10-5

2. *Turning Joint.* The action here is *sliding* action, as already discussed. Many bearings or turning joints are not perfectly lubricated, and this is particularly true of many of the turning joints found in linkages. The statements made previously concerning sliding also apply to the turning joint, and it will be assumed as before for purposes of analysis that frictional force *is* dependent upon normal force or that there are applicable equivalent coefficients of friction.

A journal rotating in its bearing is shown in Fig. 10-6. Because of the clearance the journal will roll or climb into the contact position at A.

The resultant force F_{12} makes an angle ϕ with the radius as shown in the figure. This force may be considered the resultant of the resultant normal force N and the resultant tangential force μN. The force F_{12} is tangent to the small circle of radius h. The radius of this circle is $h = R \sin \phi$. But since angle ϕ is very small, $\sin \phi = \tan \phi$ (approximately). Hence $h = R \tan \phi = \mu R$. This approximation (assuming the sine and tangent to be equal) is permissible in the case of small angles and is sufficiently accurate in this case because the angle ϕ is very small and also because the coefficient μ is a quantity that may vary between wide limits. It is evident from the foregoing equation that, if μ remains constant in accordance with the law of proportionality, the distance h will remain constant also and will be independent of the load. It follows that the line of action of the resultant force F_{12} on the journal due to a load force Q of any size will be tangent to a circle of radius h drawn about the center of the journal. This circle is called the *friction circle* and is dependent only upon the radius of the journal and the coefficient of journal friction. Obviously, the force F_{21} exerted by the journal upon the bearing also must be tangent to the friction circle.

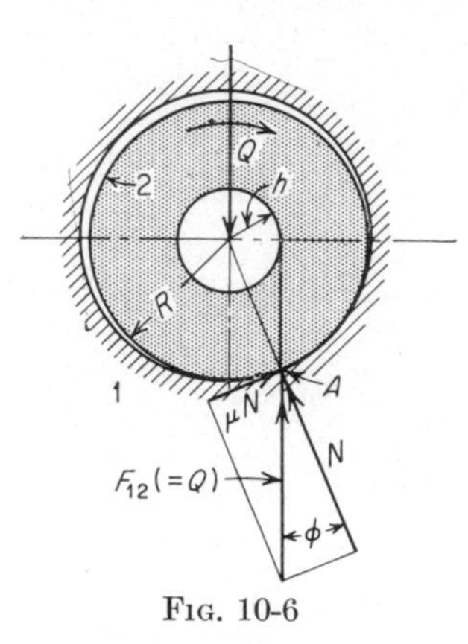

Fig. 10-6

The friction circle is of considerable service in a graphical analysis. Its radius may be calculated as soon as the coefficient of journal friction is known or assumed, after which it may be drawn in and used to locate the line of action of the force exerted by the journal upon the bearing, or vice versa. Whatever the direction of the force, its line of action must be tangent to the friction circle.

3. *Higher Pair.* The friction analysis of a higher pair is almost identical with that of a sliding pair. An example is the locomotive driving wheel shown in Fig. 10-7. The normal reaction of the rail upon the wheel is directed toward the center of the wheel. The total resultant reaction of the rail upon the wheel has a direction differing from that of the normal by an amount equal to the angle ϕ. It is evident that the total reaction of the rail exerts a turning moment on the wheel that opposes the turning moment due to the force exerted by the driving rod.

4. *Magnitude of Friction.* The magnitude of the coefficient of friction in any of the cases discussed is very much dependent upon the conditions of lubrication. Representative values have been cited, but a brief summary is in order. In a well-lubricated bearing, the coefficient may be as low as 0.001, or even less, and in the case of a dry flat slide, it may be as high as 0.30 or 0.4, or even higher (see Table 10-1). In actual practice

the coefficient rarely approaches either extreme. In a well-lubricated joint with straight-line sliding, the value of μ usually varies from about 0.02 to 0.04, while for a well-lubricated journal bearing it varies from about 0.01 to 0.02. For gears and cams running in an oil bath, the value of μ is about the same as it is for a joint with straight-line sliding. With the intermittent lubrication often found in the turning and sliding joints of linkages, these values are considerably higher, say about 0.09 or more for sliding joints, cams, and gears, and about 0.06 for bearings. Between a locomotive driving wheel and the rail μ varies from 0.15 to 0.25. For rubber tires on dry pavement, values of the sliding coefficient are of the order of 0.65 to 0.85.

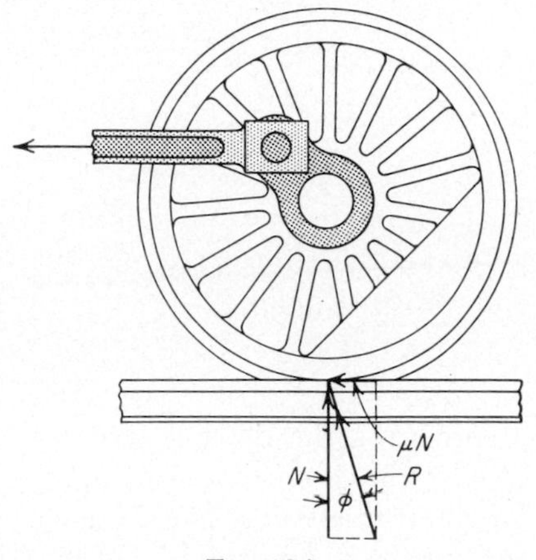

FIG. 10-7

10-7. Friction Analysis of Bell Crank. An application of the methods described in the preceding article is given in Fig. 10-8. The effort P acts through the bell crank there shown, overcoming the useful resistance Q. The only other force acting upon the bell crank is F_{12}, the force exerted by the bearing. The line of action of F_{12} must pass through the intersection of the lines of action of P and Q. It must also be tangent to the friction circle drawn about the center of the journal. The radius of the friction circle is given by $r = \mu R$, where μ is the coefficient of journal friction and R is the radius of the journal. In order to decide on which side of the friction circle the force F_{12} should be drawn, the direction of the motion, or the tendency to motion, should be known. Then since F_{12} is the force exerted by link 1 on link 2, it should be drawn so as to exert a moment about the pin center to oppose the motion of link 2 relative to link 1. The directions of the three forces and one magnitude (P) being known, the force polygon can be drawn and the remaining magnitudes determined.

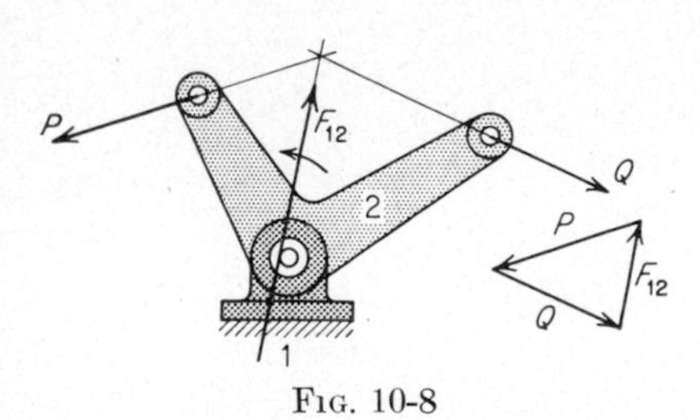

FIG. 10-8

10-8. Efficiency. The efficiency of a machine is defined in general as the ratio of the energy that a machine delivers in a given interval of time to the energy that it receives during the same interval of time. In many machines the motions of the members, as well as the forces acting on them, vary for different positions of the machine. In such cases the efficiency at any instant, called the instantaneous efficiency, is determined by the ratio of the energy supplied and delivered during an infinitesimal interval of time. The average efficiency is that corresponding to the time required for the machine to make one complete cycle. The word *efficiency* as ordinarily used means *average efficiency*.

The efficiency of a machine may be expressed algebraically as follows:

$$Eff. = \frac{W_1 - W_f}{W_1} = \frac{W_2}{W_2 + W_f} \tag{10-2}$$

where W_1 is the input, W_2 the output, and W_f the lost work converted to heat by friction, all for the same interval of time. If, for a given interval, the effort and the useful resistance remain constant, then the quantities W_1 and W_2 may be expressed as follows:

$$W_1 = Pp \qquad W_2 = Qq$$

where P is the driving force, p is the distance through which it acts during the given interval of time, Q is the resisting force at the point where the net energy of the machine is being delivered, and q is the distance through which the force Q acts. Since the input is equal to the output plus the lost work,

$$Pp = Qq + W_f$$

The foregoing equation leads to a means of expressing efficiency in terms of forces alone. Let Q be the useful resistance overcome by the driving force P in the actual machine, and let Q_0 be the useful resistance which could be overcome by P if the machine were able to operate without losing any work. Then

$$Pp = Qq + W_f = Q_0q$$

$$Eff. = \frac{W_1 - W_f}{W_1} = \frac{Pp - W_f}{Pp} = \frac{Qq}{Q_0q} = \frac{Q}{Q_0} \tag{10-3}$$

Again let Q be the useful resistance overcome by the driving force P in the actual machine, and let P_0 be the force that would be required to overcome this resistance were the machine able to operate without loss of work. Then

$$P_0p = Pp - W_f = Qq$$

$$Eff. = \frac{W_2}{W_2 + W_f} = \frac{Qq}{Qq + W_f} = \frac{P_0p}{Pp} = \frac{P_0}{P} \tag{10-4}$$

If the forces P and Q are varying, the instantaneous ratios Q/Q_0 and P_0/P represent the instantaneous efficiency of the machine.

10-9. Effect of Friction in the Engine Mechanism. The friction analysis of the engine mechanism is shown for one position of the mechanism in Fig. 10-9. Here the force P acts on the piston (link 4), and it is required to determine the other forces acting in the machine. The first step in the analysis is to assume a coefficient of journal friction, not necessarily the same for each joint, and to calculate the radii of the friction circles for the turning joints O, A, and B. Next let the forces acting on the connecting rod be considered. These are F_{43} and F_{23}. The angular motion of the connecting rod relative to the piston is counterclockwise, and to oppose this motion the force F_{43} must be tangent to the

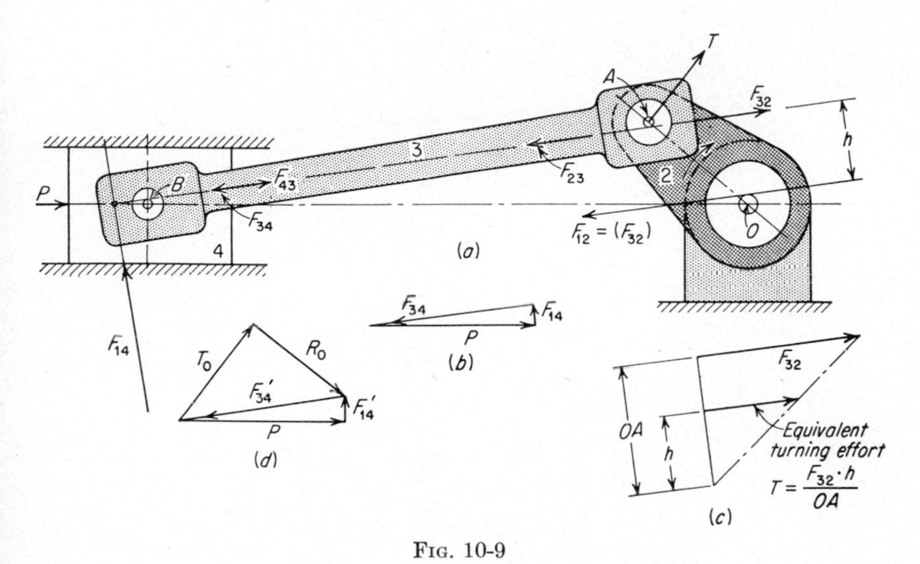

FIG. 10-9

upper part of the friction circle at B. Similarly, the motion of the connecting rod relative to the crank is counterclockwise, and hence F_{23} must be tangent to the lower part of the friction circle at A. The common line of action of F_{43} and F_{23} is thus determined.

The forces acting on the piston are P, F_{14}, and F_{34}. The force P and the direction of the force F_{34} are known and may be laid out in the force polygon at b. The polygon shows that the guide reaction F_{14} must be directed upward, and the direction of the motion of the piston shows that it must be inclined to the left, the angle of this inclination being determined by the value assumed for the coefficient of sliding friction. The direction of F_{14} being determined, the polygon may be completed as shown. Attention is called to the fact that the sense of the inclination of F_{14} is such as to decrease the force transmitted through the connecting rod to the crank. The line of action of F_{14} is determined by the fact that it must pass through the intersection of the lines of action of P and F_{34}.

The forces acting upon the crank are F_{32}, F_{12}, and the forces of the couple exerted on the main shaft by the useful resistance. In order to balance the couple, the force F_{12} must be equal and parallel to F_{32} and must have the opposite direction. In order to oppose the motion of the crank, F_{12} must be tangent to the upper part of the friction circle at O. The moment arm h of the couple formed by the forces F_{32} and F_{12} is therefore determined. The *equivalent turning effort* is the force that, if applied at the crankpin in a direction perpendicular to the crank, would have a moment with respect to the point O equal to the moment of the couple $F_{32}h$. If T is this force and OA is the crank radius, then

$$T = \frac{F_{32}h}{OA}$$

A graphical scheme for obtaining the vector length corresponding to T is shown at c.

The *instantaneous efficiency* of the mechanism may be determined by making a static analysis as shown at d, disregarding friction and determining the turning effort T_0 that would be given by the force P if there were no friction. The efficiency for the position shown is then given by the ratio T/T_0. In the original drawing the length of the vector representing T was 1.18 in.; the length of the vector representing T_0 was 1.52 in. The corresponding efficiency is 77.6 per cent. The reason for the low efficiency is that the effect of friction was purposely exaggerated in order to facilitate the demonstration. In an actual engine, the value would be of the order of 95 per cent.

10-10. Stone Crusher. In Fig. 10-10 the useful resistance Q is known, and it is required to determine the equivalent turning effort that must be applied at the crank in order to overcome resistance Q.

The radii of the various friction circles having been determined, the lines of action of the forces acting in the mechanism may be determined as shown. The force Q may then be laid off as shown at b and the remainder of the polygon completed. The equivalent turning effort is determined by means of the construction at c.

In order to determine the efficiency, let the static force analysis, disregarding friction, be made on the skeleton outline of the mechanism as shown in Fig. 10-11, and the turning effort T_0 corresponding to 100 per cent efficiency thus determined. The ratio T_0/T is equal to the instantaneous efficiency. In the determination made on the original drawing this was found to be 68 per cent. Here again the low efficiency is explained by the fact that the effect of friction has been purposely exaggerated.

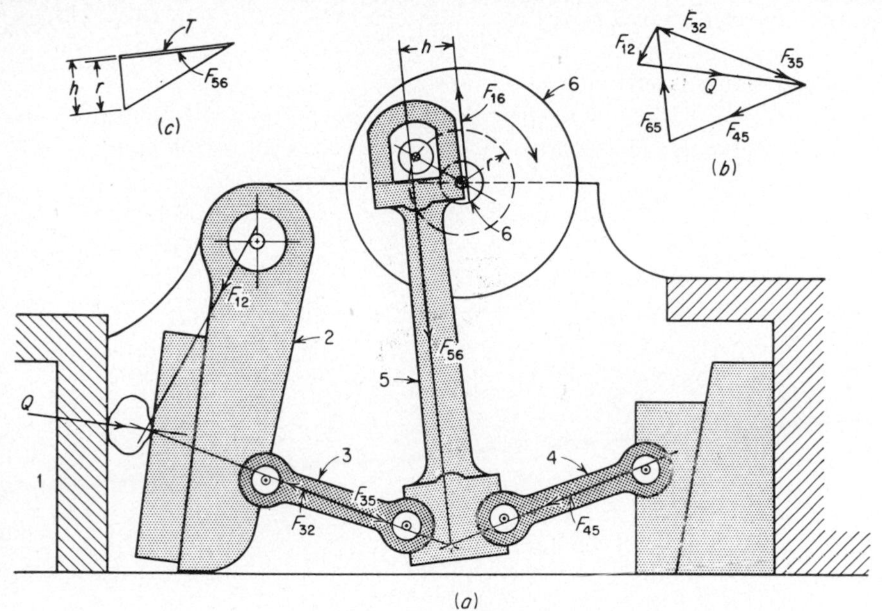

FIG. 10-10

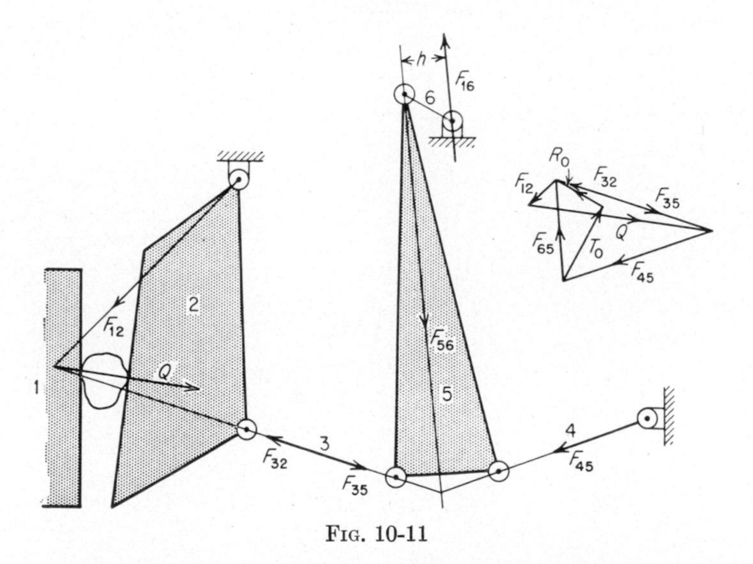

FIG. 10-11

10-11. Cam Friction. As stated in Art. 10-6 the friction analysis of a higher pair is very similar to that of a flat sliding joint. An example where the higher pair consists of the working profiles of a pair of cams is shown in Fig. 10-12. The left-hand cam (link 2) drives the right-hand cam (link 3) by means of force exerted at the point of contact A. The total force F_{23} exerted by cam 2 upon cam 3 is inclined to the common normal by an amount equal to ϕ, the angle of friction, and the sense of this inclination is such as to oppose the relative motion of the two cams. The direction of the relative motion, which is determined from kinematic considerations, is indicated by the two arrows shown at the point A.

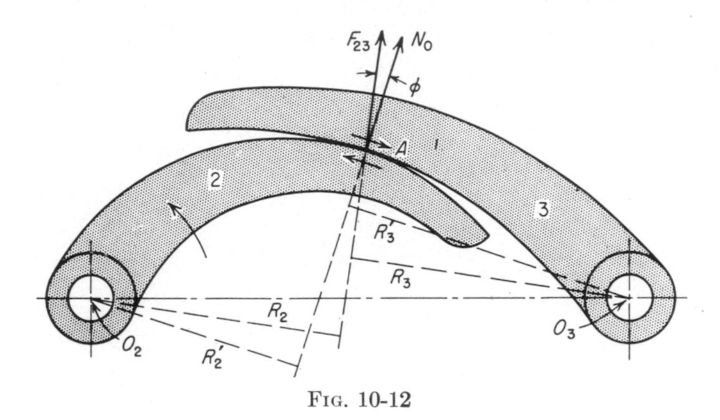

FIG. 10-12

If Pl is the turning moment exerted on the driving cam by the shaft to which it is keyed, it is evident that

$$Pl = F_{23}R_2$$

Likewise, if Qh is the resisting moment exerted on cam 3,

$$Qh = F_{23}R_3$$

These two equations may be combined, thus expressing the resisting moment in terms of the effort:

$$Qh = Pl\frac{R_3}{R_2}$$

If friction were eliminated from the mechanism, the force exerted by cam 2 upon cam 3 would act along the common normal. If N_0 were this force and Q_0h the corresponding resisting moment then

$$Pl = N_0R_2' \qquad Q_0h = N_0R_3'$$

and

$$Q_0h = Pl\frac{R_3'}{R_2'}$$

The instantaneous efficiency of the mechanism is

$$Eff. = \frac{Q}{Q_0} = \frac{R_3}{R_2}\frac{R_2'}{R_3'}$$

An application of this expression to the original figure gave

$$Eff. = \frac{3.18}{2.82}\frac{2.45}{3.31} = 0.805 = 80.5 \text{ per cent}$$

Attention is called to the fact that in the above analysis the friction of the bearings has not been taken into account.

10-12. Spur Gears. The friction analysis of a pair of spur gears is fundamentally that of a pair of cams, which has been discussed in the preceding article. Let it suffice to note that the efficiency of pairs of spur or bevel gears produced by modern manufacturing methods is of the order of 99 per cent (excluding bearing friction).

10-13. Screw Threads. The force analysis of a square-thread screw is merely that of a block sliding on an inclined plane, the latter being

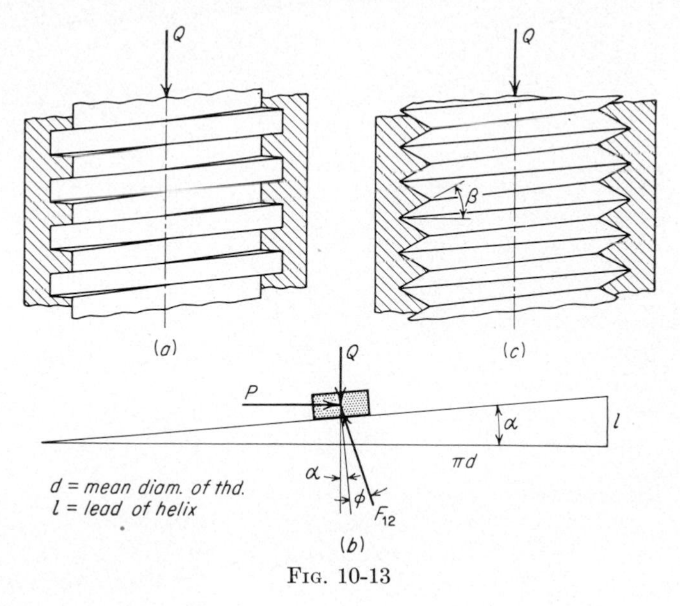

Fig. 10-13

wrapped around a cylinder. The screw itself is illustrated in Fig. 10-13*a*. The analysis is based upon the assumption that the resultant effect of the forces exerted upon the screw by the nut in which it turns is equivalent to that of a single force concentrated at the mean thread radius. A development of one thread of the nut is shown at *b*, and the block may be taken to represent the screw. The block is subjected to the vertical load Q and is moved up the inclined plane (thread of nut) under the action of the horizontal force P, which is the turning effort exerted upon the screw reduced to the mean thread radius. The force of the plane upon the block is inclined to the normal by an amount ϕ, the latter being the angle of friction. If α is the mean lead angle, then

$$P = F_{12} \sin (\alpha + \phi) \qquad Q = F_{12} \cos (\alpha + \phi)$$

$$P = Q \frac{F_{12} \sin (\alpha + \phi)}{F_{12} \cos (\alpha + \phi)} = Q \tan (\alpha + \phi) \tag{10-5}$$

The efficiency may be determined by assuming ϕ equal to zero for the ideal case of no friction. Then

$$P_0 = Q \tan \alpha$$

$$Eff. = \frac{P_0}{P} = \frac{\tan \alpha}{\tan (\alpha + \phi)} = \frac{1 - \mu \tan \alpha}{1 + \mu \cot \alpha} \quad \text{where } \mu = \tan \phi \tag{10-6}$$

The corresponding expressions[1] for the case where the threads are not *square* as, for example, as shown at *c* in the figure, are as follows:

$$P = Q \frac{\tan \alpha + \mu \sqrt{1 + \dfrac{\tan^2 \beta}{1 + \tan^2 \alpha}}}{1 - \mu \tan \alpha \sqrt{1 + \dfrac{\tan^2 \beta}{1 + \tan^2 \alpha}}} \tag{10-7}$$

where β is half the thread angle as indicated in the figure. A very close approximation is to disregard $\tan^2 \alpha$ under the radical. This gives

$$P = Q \frac{\cos \beta \tan \alpha + \mu}{\cos \beta - \mu \tan \alpha} \tag{10-8}$$

The corresponding expression for the efficiency is

$$Eff. = \frac{\cos \beta - \mu \tan \alpha}{\cos \beta + \mu \cot \alpha} \tag{10-9}$$

10-14. Resistance to Rolling. Frictional resistance to motion ordinarily can be reduced greatly by substituting rolling for sliding. The invention of the wheel probably was the first recognition of this fact. Prehistoric men learned that a load could be moved much more easily if it were moved by means of rollers or wheels than if it were dragged along the ground. Today the same principle is applied in using roller and ball bearings in machines in which minimum resistance to motion is desired. In general, the resistance to rolling in a ball or roller bearing is too small to be given consideration in a graphical force analysis, and this is usually true for any rolling of steel on steel. An example of a situation in which rolling resistance is much larger is the rolling of a loaded rubber tire on dry pavement.

[1] See C. W. Ham and D. G. Ryan, An Experimental Investigation of the Friction of Screw Threads, *Univ. Ill. Eng. Expt. Sta. Bull.* 247, June, 1932.

Pictures such as those of Figs. 10-14 to 10-16 may be helpful in visualizing the resistance to rolling. As the roller moves toward the right, the deformation of the contacting surfaces can be visualized as causing the resultant force of the plane on the roller to be shifted to the right as shown. The two components are the normal reaction N, which must be equal to but opposite the load Q, and the rolling resistance component

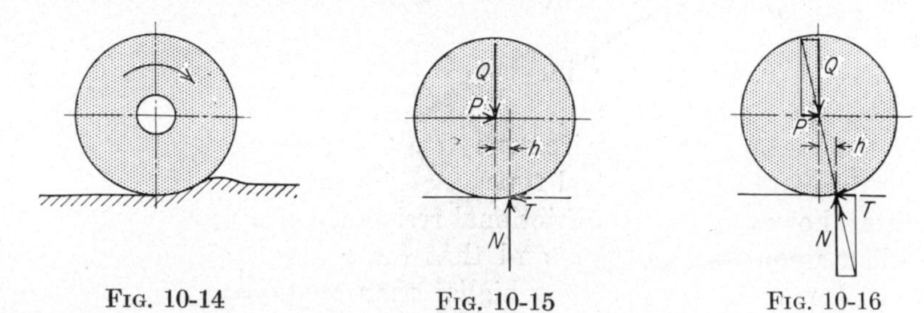

FIG. 10-14 FIG. 10-15 FIG. 10-16

T, which is overcome by the force P at the center of the roller in this case. Obviously the couple formed by Q and N with offset h must be opposed by the couple formed by T and P, which are a distance apart very nearly equal to the radius r.

$$hQ = Pr$$

or

$$P = \frac{h}{r} Q \qquad (10\text{-}10)$$

Therefore, for the same load Q, as rolling resistance increases h increases. Values of the coefficient of rolling friction h are usually given in inches, and are of the order of 0.005 or less to 0.03 for metals.

10-15. Friction of Wrapping Connectors. The action of a rope in passing over a sheave, as in hoisting tackle, is illustrated in Fig. 10-17. Owing to its internal friction, the rope first resists being bent around the sheave and then resists being straightened again. The net result is that the rope tends to force itself away from the running-on side of the sheave and tends to hug the running-off side. The combined effect of these two actions is to make the moment arm of the force Q greater than that of the force P. For equilibrium, assuming that the magnitude of the offset is the same on both sides of the sheave,

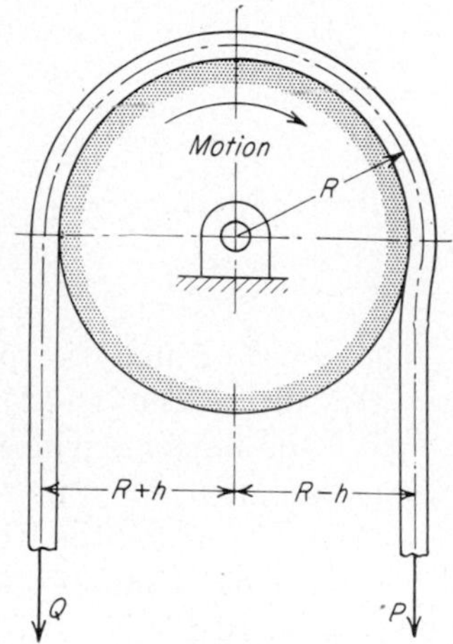

FIG. 10-17

$$P(R - h) = Q(R + h) \qquad \text{therefore } P = Q\frac{R + h}{R - h} \qquad (10\text{-}11)$$

The foregoing relation does not take account of the bearing friction of the sheave.

Very few data are available as to the value of the offset h. Professor Heck gives the following formula:

$$h = \left(0.10 + \frac{kd^2}{P}\right) d^2 \qquad (10\text{-}12)$$

where P = load, lb
d = diameter of rope, in.
k = 10 for fiber rope
= 60 for soft-iron wire rope
= 120 for strong steel rope

He states, however, that the frictional resistance of a rope is very much dependent upon its condition and that for a new stiff hemp rope the value of h may be 50 per cent higher than that given by the above equation, whereas for an old and soft rope it may be one-half or even one-third the value given by the equation.

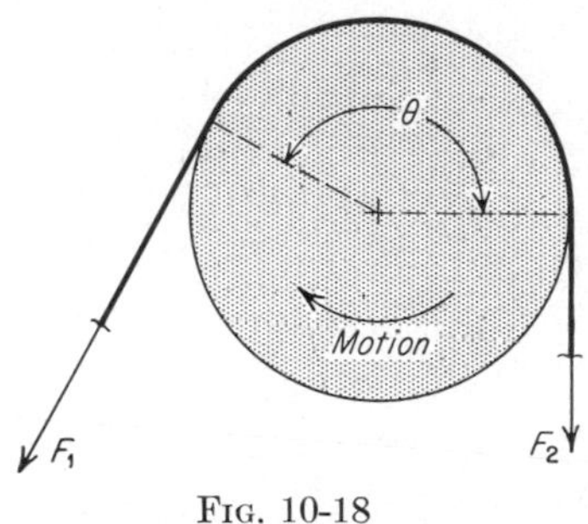

Fig. 10-18

Professor Leutwiler, in his text on "Machine Design," has summarized the results of some experiments on the efficiency of hoisting tackle. He has, however, included the effect of the friction of the sheave bearings, and this should be taken into account in making comparisons with the above formula. If proper allowance is made for bearing friction, it will be found that his results check the formula fairly closely.

The foregoing analysis applies in situations in which the loads Q and P are very nearly equal so that the tangential surface frictional force between the flexible connector and the pulley is negligible. These conditions are essentially met in hoisting tackle, for example, but *not* in belt drives, in which the friction of belt on pulley must carry the load.

The formula[1] relating the forces (tension) in a belt on two sides of a pulley (Fig. 10-18) is

$$\frac{F_1}{F_2} = e^{\mu\theta} \qquad (10\text{-}13)$$

where F_1 = force in the belt on the tight side
F_2 = force in the belt on the slack side
μ = coefficient of friction between belt and pulley surfaces
θ = angle of contact of belt and pulley, radians
e = base for natural logarithms

[1] For derivation, see any textbook on mechanics or machine design.

Equation (10-13) assumes uniform contact over the active arc of contact, and neglects centrifugal forces which tend to separate the belt from the pulley and reduce the load which can be carried at high speeds. The difference in forces F_1 and F_2 is, of course, responsible for the power transmitted by the belt.

Values for the friction coefficient μ lie roughly in the range 0.1 to 0.5 for the usual belting materials, depending upon the natures of the contacting surfaces.

10-16. Forces in Parallel Planes. In all the preceding articles the couples due to the offset of the forces have been disregarded. The reason for this is that the forces acting in machines are usually situated in parallel planes, and such a system may, by disregarding the couples, be analyzed as though all the forces were situated in the same plane. As a result the force analysis is very much simplified. Usually, these couples are of little importance. If the offset is of sufficient magnitude to make necessary the consideration of the couples, they may be determined by means of a very simple secondary analysis, once the coplanar analysis has been made.

CHAPTER 11

INERTIA FORCES

11-1. Inertia Force of a Particle. The resultant of all the forces that act on a particle (a "point mass") is a force having a magnitude equal to the product of the mass and acceleration of the particle. The direction of the force is the same as that of the acceleration. Expressed in equation form,

$$\Sigma F = F_{\text{resultant}} = F$$

and

$$F = mA \tag{11-1}$$

in which F and A must have the same direction (including the same sense).

The following discussion reviews the essential dimensions and units and their relationship necessary for proper understanding and use of the above expression.

Weighing a quantity of matter by the usual method of using a beam balance is really a process of comparing the *mass* of the matter with that of a standard mass. The number obtained is actually the *mass* of the body, although it is commonly called the *weight*. In the English system there are two common *units* of mass—the pound mass (this is the number of pounds read from the scale on the beam balance) and the slug mass. The slug unit is 32.17 times larger than the pound unit, by definition.

It is necessary to be specific concerning units and dimensions in quantitative work. In Newton's law, $F = mA$, the dimensions are

$$\text{Force} = (\text{mass})(\text{length})/(\text{time})^2$$

This equation really serves to define force when mass, length, and time are already defined (as by the Bureau of Standards). Taking the unit of mass as the slug, the unit of length as the foot, and the unit of time as the second, we obtain the definition of the unit of force called the pound. Thus, 1 lb of force applied to 1 slug of mass produces an acceleration of 1 ft/sec². Therefore the *units* of pounds force are equivalent to units of slug-ft/sec².

$$F = mA$$
$$\text{lb}_F = (\text{slug})(\text{ft/sec}^2)$$

In the force analyses in this book, the following symbols and units ordinarily will apply: F for pounds force, m for *slugs* mass, M for *pounds* mass, A for acceleration in feet per second per second, W for weight in pounds force. Masses measured in pounds must be converted to slugs by dividing the pounds of mass by 32.17 before using in the equation $F = mA$.

If weight W of a body (in pounds) is regarded as a *force* of attraction between the mass of the body and the mass of the earth, then at any location on the earth,

$$W = \frac{\mathrm{lb}_m}{32.17} g$$

where g is the local acceleration of gravity. Thus, the force W lb acting on the mass produces an acceleration of g ft/sec². Rearranging the above equation gives

$$\frac{W}{g} = \frac{\mathrm{lb}_m}{32.17} = m$$

Thus, if the *weight* of a body is determined, it must be divided by the *local acceleration of gravity* to determine m, the mass in slugs required in the equation representing Newton's law. It is repeated here, however, that the "pounds" usually determined for a body is *mass*, and must be divided by the *constant* 32.17 to determine m, rather than by the local acceleration of gravity. Numerically, g is usually very nearly equal to 32.17 for the usual small differences in elevation on the surface of the earth, and so it makes little difference numerically whether or not one distinguishes critically in assigning a name mass or force to a measurement of what is commonly called the "weight" of a body. However, in some applications the distinction is absolutely necessary to obtain a correct numerical answer. For example, in the analysis of rocket machines which fly at extremely high altitudes (possibly in outer space), it must be recognized that the *mass* of a part is invariant, but the *weight* (regarded as a force of mass attraction for the earth) can change appreciably and even approach zero.

In Fig. 11-1*a* are illustrated three forces acting on a particle. At *b* the resultant is found by graphical addition of the vectors. At *c* is shown the resultant force acting on the particle, and the direction of the acceleration which must result. At *d* a very important principle is illustrated which will be used throughout this book in the study of dynamic forces. Belief in Newton's laws allows us to state, as before,

$$F = mA \tag{11-2}$$

In the figure at *d* is shown, in addition to force F, another force $-mA$, which has the same magnitude as the force F $(= mA)$ but which is opposite in direction; this force is called the *inertia force* of the particle. The negative algebraic sign of the inertia force vector indicates merely that its direction is opposite that of the acceleration A produced by the force F.[1] In analyzing the forces acting on a particle, the use of the inertia force $(-mA)$, as shown at *e*, reduces the problem to an equivalent static force analysis, for the sum of all forces applied to the particle (including the inertia force) must be zero.

$$F_2 \leftrightarrow F_1 \leftrightarrow F_3 \leftrightarrow (-mA) = 0 \tag{11-3}$$

This equation really defines the inertia force vector as $-mA$, as before.

[1] This follows the usual convention of vector analysis which, if V is a vector, defines $-V$ as another vector having a magnitude the *same* as that of V, and a direction *opposite* that of V.

A rigid body may be considered to be composed of a large number of particles (or point masses) held in fixed positions relative to each other. These particles are called the *elements* of the rigid body. The inertia force of any element is dependent upon its mass and its acceleration, and the inertia force of the body as a whole is the resultant of all its elementary inertia forces.

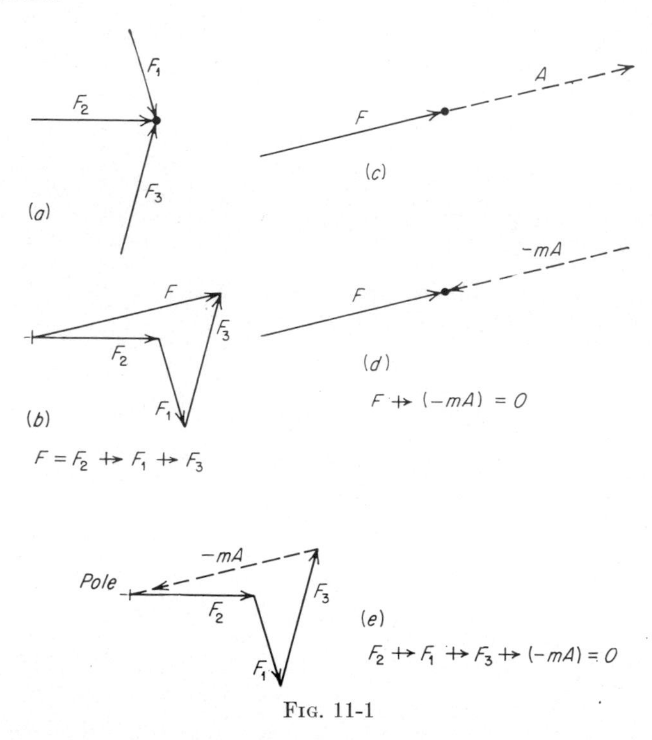

FIG. 11-1

As stated in the preceding chapter, the force analysis of a machine taking account of inertia forces is called a dynamic force analysis. In order to simplify the discussion, the effect of friction will be disregarded in the dynamic force analyses that follow. This is further justified by the fact that in most machines having inertia forces large enough to be taken into account the journals are well lubricated and the frictional forces are small.

11-2. Inertia Forces of a Rigid Body Having Plane Motion. The elements of a rigid body having plane motion move in parallel planes. Hence the inertia forces of these elements have lines of action situated in parallel planes, and, as a result, the inertia forces may conveniently be treated as though they formed a coplanar system. An analysis based on such an assumption neglects certain couples that are usually unimportant but may, if desired, be determined by means of a simple secondary analysis. The assumption of a coplanar system is, in dealing with inertia forces, also equivalent to assuming that the mass of the body has been concentrated in the plane of motion.

In Fig. 11-2, let G be the center of gravity of the rigid body, the mass m of which is assumed to be concentrated in one plane, and let P be any element of the body, the mass of P being dm. The acceleration of the element P is

$$A_P = A_G \nrightarrow A^n_{PG} \nrightarrow A^t_{PG}$$

$$a^n_{PG} = r\omega^2 \qquad a^t_{PG} = r\alpha$$

(recall that $a = \text{mag } A$)

where ω and α are, respectively, the angular velocity and angular acceleration of the body. Multiplying the above equation by the mass dm of the element,

$$A_P\, dm = A_G\, dm \nrightarrow A^n_{PG}\, dm \nrightarrow A^t_{PG}\, dm \tag{11-4}$$

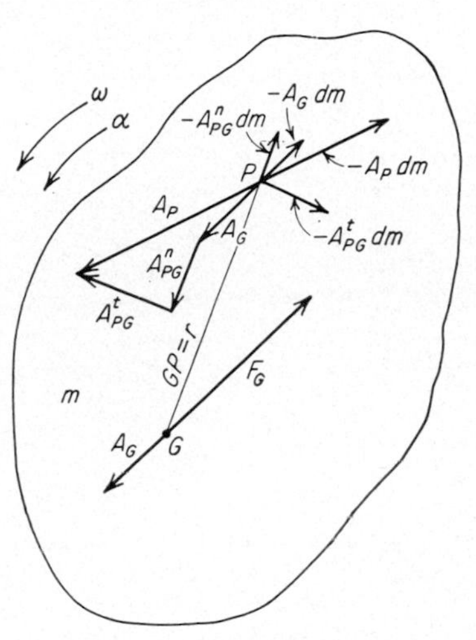

Fig. 11-2

The inertia force of the element P is $-A_P\, dm$. This force may be resolved, as shown by Eq. (11-4) and in Fig. 11-2, into the components $-A_G\, dm$, $-A^n_{PG}\, dm$, and $-A^t_{PG}\, dm$. The resultant inertia force of the body as a whole is made up of:

1. The resultant of all forces like $-A_G\, dm$.
2. The resultant of all forces like $-A^n_{PG}\, dm$.
3. The resultant of all forces like $-A^t_{PG}\, dm$.

The forces like $-A_G\, dm$ are all parallel and opposite to the acceleration of the center of gravity G, and the magnitude of each is proportional to the mass of its corresponding element. They form, therefore, a system of parallel forces, the resultant of which, F_G, passes through the center of gravity of the body as indicated in the figure, this last characteristic being due to the definition of the center of gravity.[1] The force F_G is

$$F_G = \Sigma - A_G\, dm = -mA_G \tag{11-5}$$

m being the mass of the body.

The forces like $-A^n_{PG}\, dm$ ($a^n_{PG} = r\omega^2$) all pass through the center of gravity G. The vector summation of all such forces gives a resultant force on the body as follows:

$$F = \Sigma - r\omega^2\, dm = -\omega^2 \Sigma r\, dm$$

[1] The properties of the center of gravity which are of interest here are discussed in elementary textbooks on mechanics.

But r is the radius from the center of gravity to any of the elements such as P, and the first moment of mass about an axis through the center of gravity of a rigid body is always zero ($\Sigma r\, dm = 0$). Therefore the vector sum of the concurrent system of forces is zero and they cannot produce a net torque or moment about any axis because all the forces pass through the same point. Hence the resultant of all forces like $-A^n_{PG}\, dm$ is zero.

The resultant of the forces like $-A^t_{PG}\, dm$ is $\Sigma - r\alpha\, dm = -\alpha\Sigma r\, dm = 0$ because $\Sigma r\, dm$ is zero, as before. In this case, however, as the forces obviously do not pass through the same point, yet their sum is zero, their

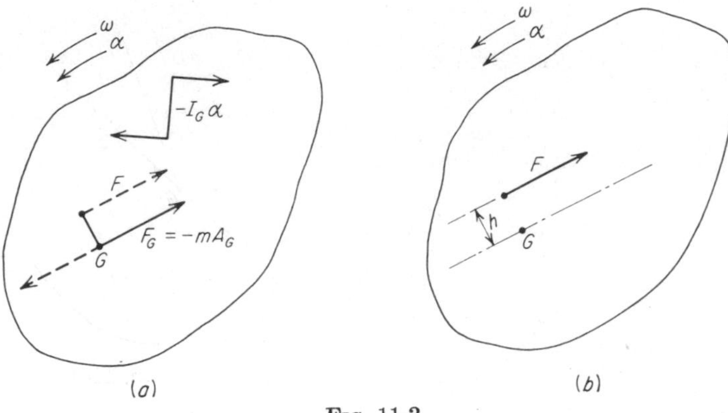

FIG. 11-3

resultant must be equivalent to a couple. To find this couple, it is convenient to take moments about the point G. Then, if T is the sum of the moments of all forces like $-A^t_{PG}\, dm$,

$$T = \Sigma - r\alpha\, dm(r) = -\alpha\Sigma r^2\, dm = -I_G\alpha \qquad (11\text{-}6)$$

where I_G is the moment of inertia of the body with respect to the center of gravity G.

In summary, the resultant of all the inertia forces of a rigid body is therefore *an inertia force through the center of gravity of magnitude* ma_G, *and an inertia couple of magnitude* $I_G\alpha$. The corresponding directions are opposite A_G and opposite α, respectively. Similarly to the inertia force of a particle, but now expressed for an entire rigid body, *the inertia force and the inertia couple are defined, therefore, as* $-mA_G$ *and* $-I_G\alpha$, *respectively*. The force and couple may be combined as indicated at a and b in Fig. 11-3, giving thus a single force as the resultant inertia force of the body. The magnitude and direction of the resultant inertia force F are the same as those of F_G and the line of action is situated at a distance $h = I_G\alpha/F_G$ from the center of gravity G.

Attention is again called to the fact that the above analysis is based upon the assumption that the mass of the rigid body is concentrated in one plane, namely, a plane parallel to the plane of motion and containing the center of gravity G of the body. Actually, however, the mass is distributed in three dimensions and in order completely to determine the net effect of the elementary inertia forces, it is necessary to take this fact into account. It can be shown that the total inertia force thus obtained is made up of the force and couple determined by the coplanar analysis described above, and a secondary couple tending to rotate the rigid body about an axis in the plane of motion. This secondary couple usually may be disregarded, in which case the analysis can be made on the coplanar basis described above. If the reference plane is a plane of symmetry and if the density of the body is uniform throughout, then the resultant secondary couple disappears. In the following articles it is always assumed that the mass is concentrated in one plane unless otherwise stated.

11-3. Inertia Forces of a Floating Link. The engine connecting rod, having a motion of combined rotation and translation, is a link that affords an ideal illustration of the general case of inertia forces in a link having plane motion, as discussed in the preceding article. Moreover, the connecting rod is a typical example of what may be termed a "floating link" (i.e., a link not fixed at any point), and the method outlined for the determination of its resultant inertia force will cover "floating members" in general.

In Fig. 11-4 is shown a connecting rod of mass m with the acceleration A_G of its center of gravity and its angular acceleration α given as indicated in the figure. From the discussion in the preceding article it is known that the resultant of the inertia forces of the link is made up of the inertia force $F_G = -mA_G$ passing through the center of gravity G in a direction opposite that of the acceleration A_G, and the inertia couple $-I_G\alpha$, where I_G is the moment of inertia with respect to G. This is indicated at b in the figure. Since the angular acceleration is clockwise, the inertia couple must be counterclockwise. The couple may be assumed, of course, to act on the link anywhere in its plane of motion.

If it is desired to represent the resultant of the inertia forces as a single force, rather than a force and a couple, this can be accomplished by letting the forces of the couple each equal the magnitude of F_G. Then the arm of the couple is

$$h = \frac{I_G\alpha}{F} = \frac{I_G\alpha}{F_G} \tag{11-7}$$

Placing the couple with one force F equal and opposite the force F_G through G as shown at c in the figure, the resultant is the single force F,

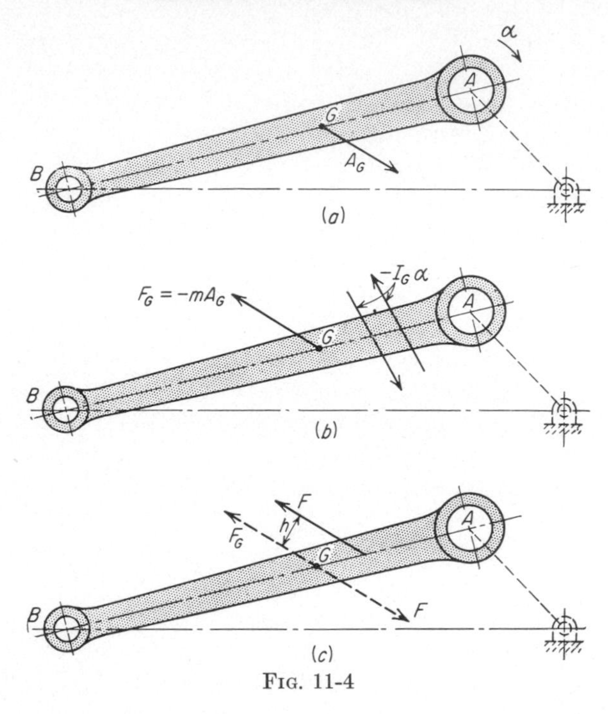

FIG. 11-4

which is parallel to A_G, opposite in direction to A_G, and displaced a distance h from A_G. It is obvious that the moment Fh must oppose the angular acceleration of the link.

11-4. Inertia Force of a Link Rotating about a Fixed Center. Figure 11-5 represents a link rotating about center O with an angular velocity ω and an angular acceleration α in the sense indicated. The mass of the link is m and its moment of inertia about its center of gravity G is I_G.

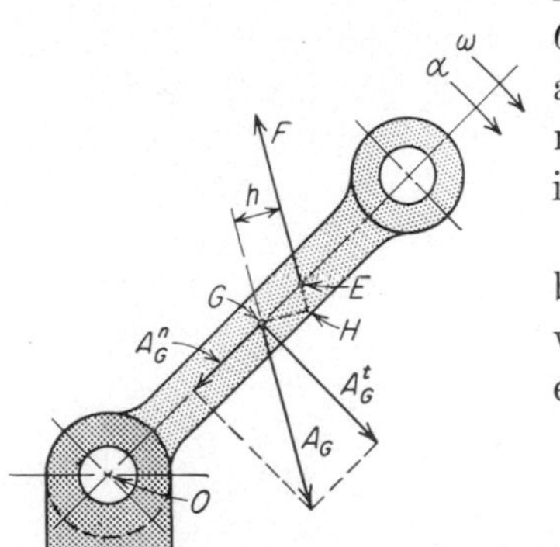

FIG. 11-5

The resultant of the inertia forces must be a force $F_G = -mA_G$, and a couple whose moment is $T = -I_G\alpha$. The acceleration of the center of gravity is

$$A_G = A_G^n \leftrightarrow A_G^t$$

$$a_G^n = (OG)\omega^2 \qquad a_G^t = (OG)\alpha$$

From the vector A_G the magnitude and direction of the resultant inertia force F_G can be determined. The resultant F obtained by combining the inertia force and the inertia couple has a line of action situated at a distance h

such that

$$h = \frac{I_G \alpha}{F_G}$$

the distance h being measured from G in a direction such that the moment of F with respect to G tends to oppose α.

The inertia force F always passes through the same point E on the line OG extended regardless of the values of ω and α. This is shown by the following considerations:

$$\frac{GE}{GH} = \frac{A_G}{A_G^t}$$

$$GE = GH \frac{A_G}{A_G^t} = h \frac{A_G}{A_G^t} = \frac{I_G \alpha}{F} \frac{A_G}{(OG)\alpha} = \frac{I_G}{m A_G} \frac{A_G}{OG} = \frac{I_G}{m(OG)}$$

From the above expression it may be observed that the distance GE is independent of ω and α and depends only on I_G, m, and OG. Therefore, with a *fixed* axis of rotation, the inertia force of the link always passes through a fixed point E on the link. This point E is called the *center of percussion of the link with respect to the fixed axis of rotation O.*

If k is the principal radius of gyration (radius about the center of gravity),

$$GE = \frac{I_G}{m(OG)} = \frac{mk^2}{m(OG)} = \frac{k^2}{OG} \qquad \text{therefore}$$

$$k^2 = OG(GE) \qquad (11\text{-}8)$$

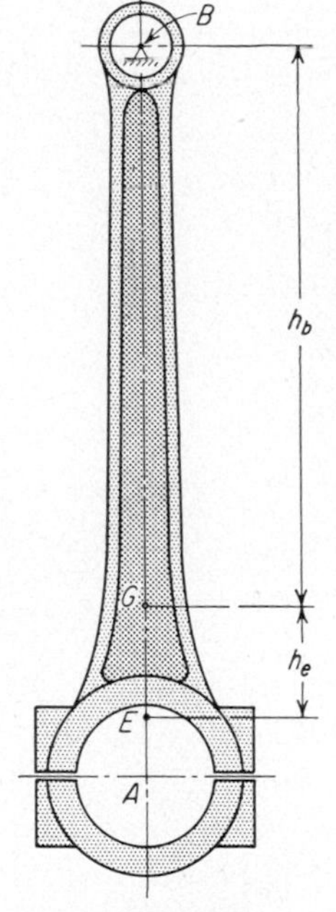

FIG. 11-6

If a rigid body is permitted to swing freely as a pendulum, its period of vibration is the same as that of a simple pendulum having a length equal to the distance between the axis of oscillation and the center of percussion with respect to that axis, i.e., a length equal to the distance OE for the case just discussed. This suggests an experimental method of determining the moment of inertia of a machine member. Referring to Fig. 11-6, let it be required to determine the moment of inertia of the connecting rod there shown. It is assumed that the position of the center of gravity G has been previously determined, that I_G is the moment of inertia with respect to G, and that E is the center of percussion with respect to the axis of oscillation. The connecting rod is suspended on the knife-edge B, about which it is permitted to oscillate as a pendulum. The time required for a given number of complete oscillations (over and back) is noted, thus determining the time T for one complete oscillation. The center of percussion and moment of inertia of the rod are then determined from the pendulum formula (see any textbook on analytical mechanics) as follows:

$$T = 2\pi \sqrt{\frac{BE}{g}} = 2\pi \sqrt{\frac{h_b + h_e}{g}} \tag{11-9}$$

where g = local acceleration of gravity = approximately 32.2 ft/sec^2. The distance of the center of percussion E from the center of gravity G is then

$$h_e = \frac{T^2 g}{4\pi^2} - h_b \tag{11-10}$$

The moment of inertia I_G with respect to the center of gravity is

$$I_G = mk^2 = mh_b h_e = mh_b \left(\frac{T^2 g}{4\pi^2} - h_b \right) \tag{11-11}$$

11-5. Transverse and Radial Components. In force problems it is often necessary to deal with a single force that acts upon a two-joint link and has a line of action crossing the link at some point between the centers

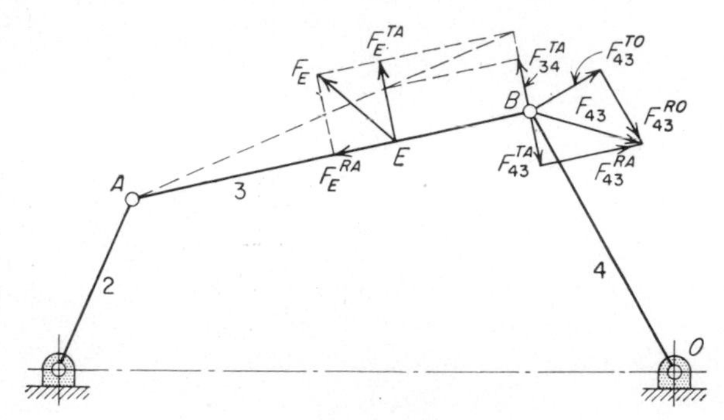

FIG. 11-7

of the two joints. An example is the force F_E in Fig. 11-7. In order to determine the other forces acting on link 3, let F_E be resolved at E into components perpendicular and parallel to the line AB. This is done in order to facilitate the taking of moments with respect to the point A. The perpendicular component F_E^{TA} is called the transverse component of F_E with respect to the point A, and F_E^{RA} is called the radial component of F_E with respect to A. It is evident that in taking moments about A the moment of the radial component F_E^{RA} is zero. The moment of the transverse component is F_E^{TA} (AE), and it follows from the principle of moments that this product is also the moment of the original force F_E with respect to the point A.

Besides F_E, only two other forces act upon link 3, namely, the force F_{23} of 2 acting on 3 at A (not shown), and a force F_{43} of 4 acting on 3 at B. The moment about A of the force at A is zero. That of the force at B is equal, in accordance with the scheme outlined above, to the product of its transverse component F_{43}^{TA} and the distance AB. Since for equilibrium the sum of all moments about A must be zero,

$$F_E^{TA}(AE) + F_{43}^{TA}(AB) = 0$$

and

$$F_{43}^{TA} = -F_E^{TA}\left(\frac{AE}{AB}\right) \tag{11-12}$$

The significance of the negative sign is that the direction of F_{43}^{TA} is opposite that of F_E^{TA}. A graphical method of applying the moment relation to determine F_{43}^{TA} is shown in the figure.

The radial component F_{43}^{RA} is unknown, and unless something further is known regarding the total force F_{43}, no further progress is possible. Let it be supposed, however, that the force F_{34}^{TO} has been determined by taking moments about O of forces on link 4 (construction not shown). Reversing F_{34}^{TO} gives F_{43}^{TO}. Then, since F_{43}^{RO} and F_{43}^{RA} are perpendicular to F_{43}^{TO} and F_{43}^{TA}, respectively, it is evident that the extremity of the total-force vector F_{43} is situated at the intersection of the perpendiculars drawn through the extremities of F_{43}^{TO} and F_{43}^{TA}. In vector equation form (the corresponding vector diagram appears in Fig. 11-7),

$$F_{43}^{TA} \nrightarrow F_{43}^{RA} = F_{43}^{TO} \nrightarrow F_{43}^{RO} = F_{43}$$

After the total force F_{43} has been determined, the force F_{23} acting at A may be determined in the usual manner by setting the vector summation of all forces on link 3 equal to zero.

11-6. Four-link Mechanism. In the present article the dynamic force analysis of the four-link mechanism shown in Fig. 11-8 is discussed. The angular velocity and angular acceleration of link 2 are given as follows:

$$\omega_2 = 20 \text{ radians/sec} \qquad \alpha_2 = 160 \text{ radians/sec}^2$$

The necessary linear dimensions are given in the figure. The masses and moments of inertia of the various members are as follows:[1]

$$M_2 = 4.65 \text{ lb} \qquad M_3 = 2.17 \text{ lb}$$
$$I_{G2} = 0.01379 \text{ slug-ft}^2 \qquad I_{G3} = 0.00814 \text{ slug-ft}^2$$
$$M_4 = 5.27 \text{ lb}$$
$$I_{G4} = 0.02040 \text{ slug-ft}^2$$

[1] See discussion in small type at beginning of this chapter.

It is required to determine (1) the inertia forces of the moving members, (2) the torque which must be applied to link 2, and (3) the effect of the inertia forces upon the frame.

1. *Determination of Inertia Forces.* The first step is to make the complete acceleration determination. This, of course, requires the use of the

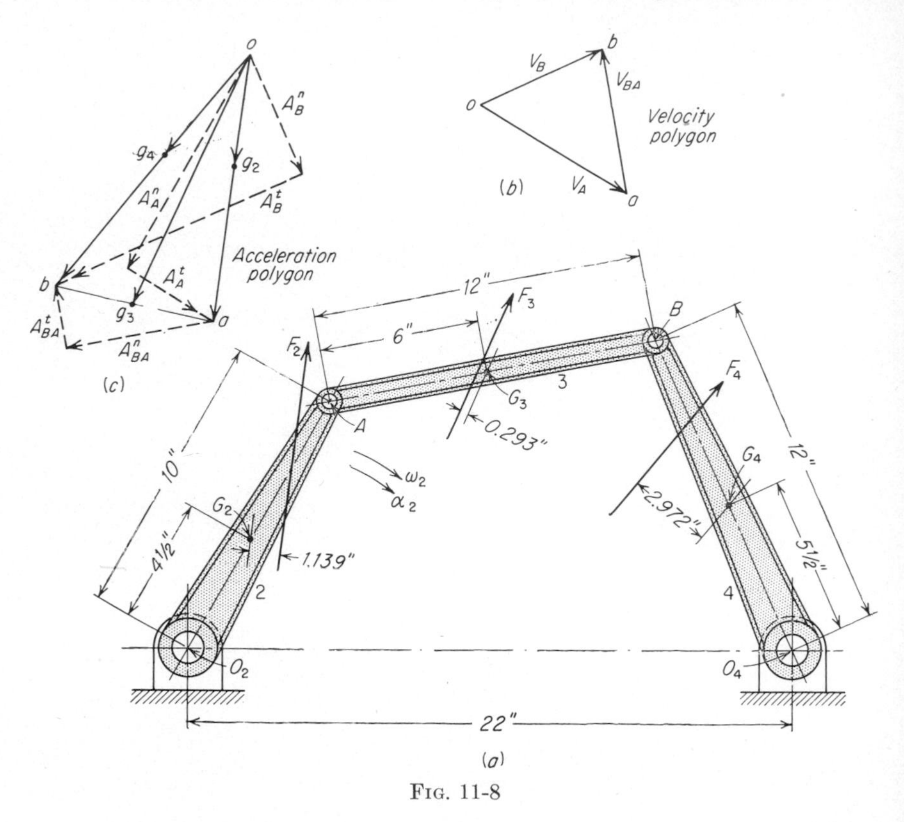

FIG. 11-8

velocity polygon at b. The normal and tangential components of the acceleration of point A can be calculated as follows:

$$a_A^n = (O_2A)\omega_2^2 = 10/12 \times 20^2 = 333 \text{ ft/sec}^2$$
$$a_A^t = (O_2A)\alpha_2 = 10/12 \times 160 = 133 \text{ ft/sec}^2$$

Then, after a suitable scale has been chosen, the acceleration polygon may be drawn as shown at c by means of the methods of Chap. 9. The scale chosen for the original drawing was 1 in. = 150 ft/sec². The magnitudes of the accelerations of the several centers of gravity as determined from the polygon are

$$a_{G_2} = 162 \text{ ft/sec}^2 \qquad a_{G_3} = 371 \text{ ft/sec}^2 \qquad a_{G_4} = 189 \text{ ft/sec}^2$$

The magnitudes of the inertia forces are

$$F_2 = \frac{M_2}{32.17} a_{G_2} = \frac{4.65}{32.17} \times 162 = 23.25 \text{ lb}$$

$$F_3 = \frac{M_3}{32.17} a_{G_3} = \frac{2.17}{32.17} \times 371 = 25.01 \text{ lb}$$

$$F_4 = \frac{M_4}{32.17} a_{G_4} = \frac{5.27}{32.17} \times 189 = 30.89 \text{ lb}$$

In order to determine the lines of action of the various inertia forces, it is first necessary to determine the angular accelerations of the corresponding links. The angular acceleration of link 2 is given, and those of links 3 and 4 may be determined as follows:

$$\alpha_3 = \frac{a^t_{BA}}{AB} = \frac{75}{12/12} = 75 \text{ radians/sec}^2$$

$$\alpha_4 = \frac{a^t_B}{O_4B} = \frac{375}{12/12} = 375 \text{ radians/sec}^2$$

the numerical values of a^t_{BA} and a^t_B being determined from the acceleration polygon. The distances of the lines of action of the various inertia forces from the centers of gravity of their corresponding links may now be calculated as follows:

$$h_2 = \frac{I_{G_2}\alpha_2}{F_2} = \frac{0.01379 \times 160}{23.25} = 0.0949 \text{ ft} = 1.139 \text{ in.}$$

$$h_3 = \frac{I_{G_3}\alpha_3}{F_3} = \frac{0.00814 \times 75}{25.01} = 0.0244 \text{ ft} = 0.293 \text{ in.}$$

$$h_4 = \frac{I_{G_4}\alpha_4}{F_4} = \frac{0.02040 \times 375}{30.89} = 0.2467 \text{ ft} = 2.972 \text{ in.}$$

The inertia forces F_2, F_3, and F_4 may then be laid off on the drawing of the mechanism as shown in Fig. 11-8, the direction of each force being opposite that of the corresponding acceleration of G and the moment of each force about G opposing the angular acceleration. Attention is called to the fact that the lines of action of the inertia forces of links 2 and 4 might have been located by determining the centers of percussion of these links with respect to their fixed centers of rotation.

2. *Accelerating Effort.* The effort required to accelerate the mechanism may be determined by applying the conditions of equilibrium. The underlying fundamental principle is the fact that the system composed of all the inertia forces and all the external forces acting on any one member is a system in equilibrium. Let it be assumed that the required accelerating torque applied to the shaft to which link 2 is keyed is represented by a couple PL.

In Fig. 11-9, where the mechanism has been redrawn in skeleton form, let the forces acting on link 4 be first considered. These are

F_4 = the inertia force of link 4.
F_{14} = the force of 1 on 4 at O_4.
F_{34} = the force of 3 on 4 at B.

A suitable force scale having been chosen (a scale of 1 in. = 30 lb was chosen for the original drawing), the inertia force F_4 is laid off at E_4

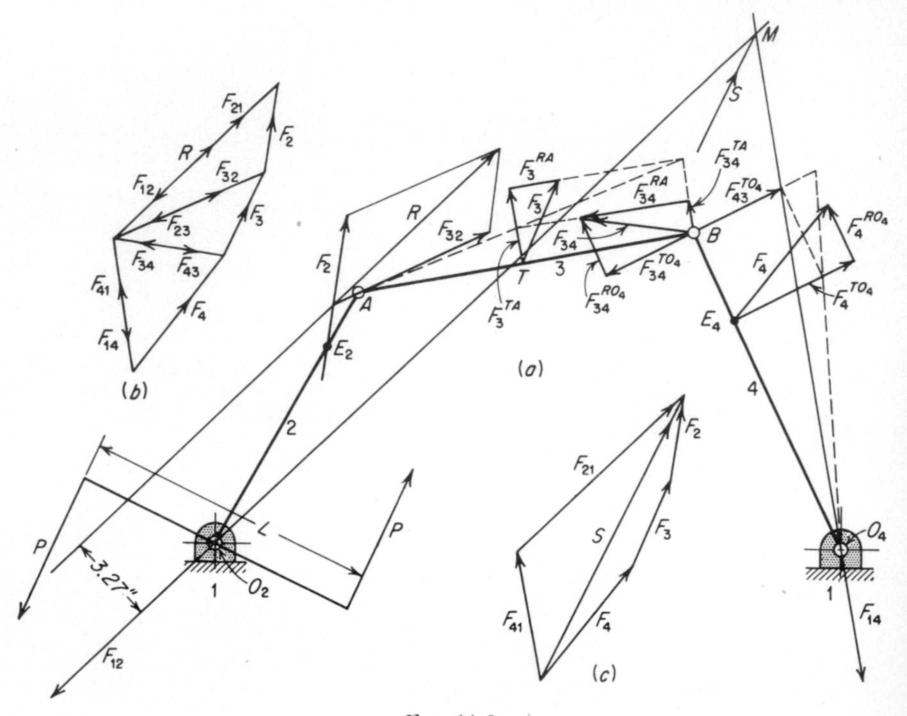

FIG. 11-9

and resolved into transverse and radial components with respect to the point O_4. Moments are then taken about O_4, and the transverse component $F_{34}^{TO_4}$ is determined in accordance with the methods of the preceding article. No further progress is possible by considering the forces acting on link 4 alone.

The forces acting on link 3 are

F_3 = the inertia force of link 3.
F_{43} = the force of 4 on 3 at B.
F_{23} = the force of 2 on 3 at A.

Let the inertia force F_3 be resolved at T into transverse and radial components with respect to the point A. The transverse component F_{43}^{TA}, or rather its opposite F_{34}^{TA}, may then be determined by taking moments about A. The extremity of the vector representing the total force F_{34} is situated at the intersection of the perpendiculars drawn through the extremities of the vectors $F_{34}^{TO_4}$ and F_{34}^{TA}. The vector equation representing this step is

$$F_{34} = F_{34}^{TO_4} \mathrel{+\!\!\!\!\rightarrow} F_{34}^{RO_4} = F_{34}^{TA} \mathrel{+\!\!\!\!\rightarrow} F_{34}^{RA}$$

The reaction F_{34} (and the reverse F_{43}) having been determined, the only unknown force remaining to act on link 3 is F_{23}.

$$F_{43} \mathrel{+\!\!\!\!\rightarrow} F_3 \mathrel{+\!\!\!\!\rightarrow} F_{23} = 0$$

The graphical solution of this equation for F_{23} is shown at b. Also shown is the solution for F_{14} of

$$F_{34} \mathrel{+\!\!\!\!\rightarrow} F_4 \mathrel{+\!\!\!\!\rightarrow} F_{14} = 0$$

The forces acting on link 2 are

F_2 = the known inertia force of link 2.
F_{32} = the force of 3 on 2 at A that has just been determined.
F_{12} = the unknown force of 1 on 2 at O_2.
PL = the unknown accelerating couple.

The solution for F_{12} is shown in the vector polygon at b.

$$F_{32} \mathrel{+\!\!\!\!\rightarrow} F_2 \mathrel{+\!\!\!\!\rightarrow} F_{12} = 0$$

The couple PL obviously has no effect on the force polygon.

To determine the couple PL, the sum of the moments of all forces and torques on 2 about O_2 are equated to zero. F_{12} acts through O_2 and produces no moment. The sum of F_{32} and F_2 is shown as the resultant R in the figure. The magnitude of R is determined from the polygon as 60.8 lb, and the moment arm about O_2 can be scaled from the drawing as 3.27 in. The moment of the couple PL is, therefore,

$$PL = 60.8 \times 3.27 = 198.8 \text{ lb-in.} = 16.57 \text{ lb-ft}$$

It should be noted that the couple PL is exactly equal (but opposite) to the couple formed by F_{12} and the resultant R of F_{32} and F_2. It is interesting to note that the accelerating effort PL, in this particular case, is resisting the motion of the mechanism and that the actual driving force is largely supplied by the kinetic energy of link 4.

3. *Effect of Inertia Forces on the Frame.* The only forces acting on the frame of the mechanism are the reactions at O_2 and O_4, i.e., the forces F_{21} and F_{41}. The resultant S of these two forces passes through the intersection M of their lines of action. The force S is called the resultant

shaking force ($S = F_{41} \nrightarrow F_{21}$). It is the force that tends to shake or lift the frame of the machine from the foundation to which it is fastened. The magnitude and direction of S are given by the vector sum of the forces F_{21} and F_{41}. An inspection of the partially redrawn polygon at c will bring out the fact that the magnitude and direction of S are also given by the vector sum of the three inertia forces F_2, F_3, and F_4. This could have been anticipated by considering links 2, 3, and 4 to compose a single system in equilibrium.

$$\Sigma F \text{ (applied to a system)} = \Sigma mA \text{ (for the mass particles within the system)}$$

$$F_{14} \nrightarrow F_{12} = -(F_2 \nrightarrow F_3 \nrightarrow F_4) \qquad \text{or} \qquad F_{41} \nrightarrow F_{21} = F_2 \nrightarrow F_3 \nrightarrow F_4$$

This type of relation is true in the case of all machines, and because of this fact the magnitude and direction of the resultant shaking force of any machine may be determined directly from the inertia forces of its members without any reference to the manner in which these inertia forces are transmitted through the machine. The position of the line of action of S, however, can be determined only by means of a complete analysis.

11-7. Combined Static and Inertia Force Analysis. In the preceding article, the inertia forces alone have been considered. In most problems, however, it is necessary to consider also the static forces and reactions by means of which the effort is transmitted through the machine. In such cases the reaction between any two members of the machine may be determined by combining the results of two separate analyses, one in which the inertia forces are taken into account and the static forces are disregarded, and another in which the inertia forces are disregarded and only the static forces are considered.

This will be discussed briefly, using as an illustration the mechanism shown in Fig. 11-10. For simplicity, links 3 and 4 will be assumed essentially weightless so that their inertia forces are negligible. The known inertia force f_2 and the known resistance Q are assumed to be held in equilibrium by the unknown effort P. It is required to determine the effort P and the other forces acting in the machine. The following notation will be used:

f will denote inertia forces and the reactions that are due to inertia forces.

$\mathfrak{F}$ will denote static forces and the reactions that are due to static forces.

F will denote the forces and reactions due to the combined effect of the inertia forces and the static forces.

The static force analysis is based upon the assumption that the inertia

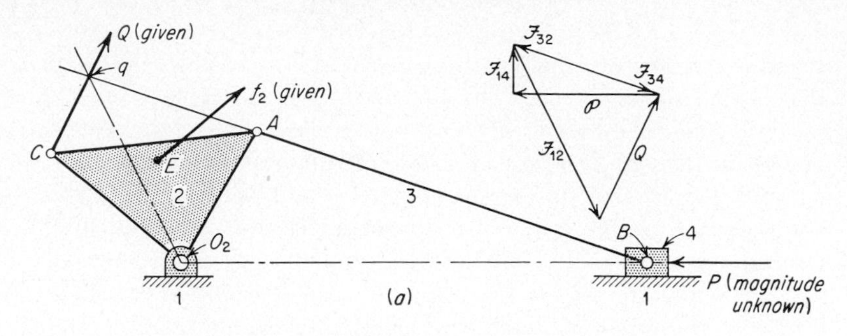

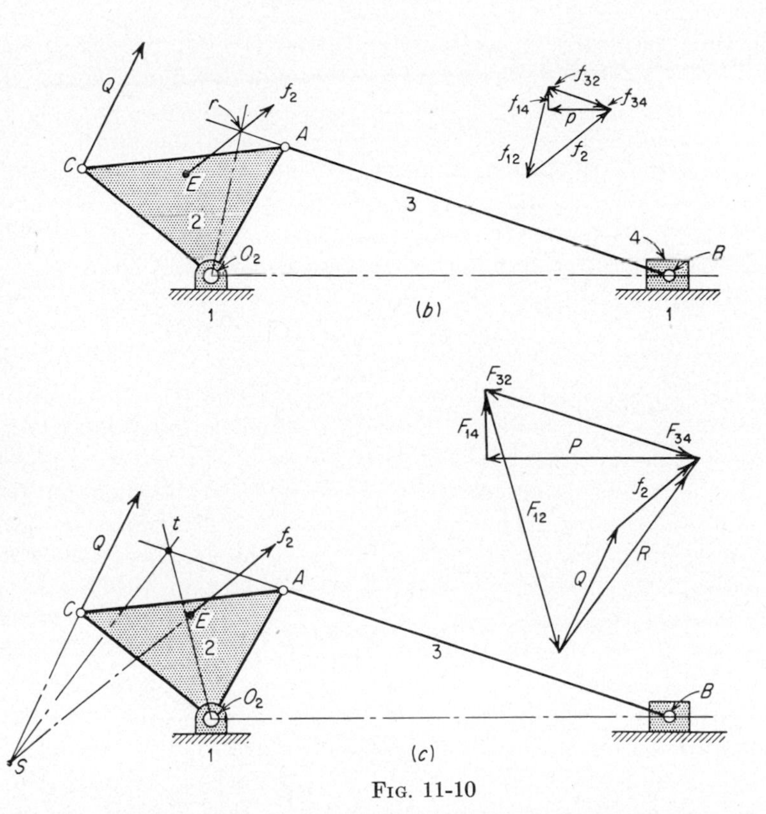

FIG. 11-10

forces are nonexistent, and that the resistance Q is held in equilibrium by means of the static effort $\mathcal{P}$, which is applied on link 4 in a horizontal direction. On this assumption, the only forces acting on link 2 are

Q = the known resistance.
$\mathfrak{F}_{12}$ = the reaction at O_2.
$\mathfrak{F}_{32}$ = the reaction at A.

It is emphasized that the forces $\mathfrak{F}_{12}$ and $\mathfrak{F}_{32}$ are not the actual reactions at O_2 and A but are merely those parts of the actual reactions that are due to the resistance Q. The line of action of $\mathfrak{F}_{32}$ must be coincident with the line AB, since only two forces act on link 3. The force $\mathfrak{F}_{12}$ must pass through the intersection q of the lines of action of Q and $\mathfrak{F}_{32}$: if it did not, a pure torque (or couple) would be present on 2, with no counteracting torque. The magnitudes of $\mathfrak{F}_{12}$ and $\mathfrak{F}_{32}$ may, therefore, be determined by means of the polygon shown at a ($Q \nrightarrow \mathfrak{F}_{32} \nrightarrow \mathfrak{F}_{12} = 0$), and the remainder of the polygon may then be drawn without special difficulty, the force $\mathfrak{F}_{14}$ being that part of the guide reaction that is due to the resistance Q and the force $\mathcal{P}$ being the static effort required to hold Q in equilibrium.

The method employed in making the inertia force analysis is similar in every respect to that employed above, the only difference being that the inertia force f_2 is taken into account and the resistance Q is disregarded. The inertia force analysis is shown at b in the figure. The force p is the accelerating effort required to overcome the inertia force f_2.

After the preceding two analyses have been made, the total effort P and the total reaction between any two members may be determined by combining the corresponding vectors of the polygons at a and b. The total effort, for example, is

$$P = \mathcal{P} \nrightarrow p = \mathcal{P} + p$$

since $\mathcal{P}$ and p are parallel. Likewise, the total guide reaction is given by the vector sum of $\mathfrak{F}_{14}$ and f_{14}; the total reaction of link 1 on link 2, by the vector sum of $\mathfrak{F}_{12}$ and f_{12}; etc.

The total force determination might, of course, have been made in the first place as shown at c. The method would be to determine first the resultant R in the polygon, and then to draw st, the line of action of R, on the figure of the mechanism. The succeeding steps are the same as those of the above analyses. In some instances this method may be advantageous, but in general it tends to obscure the separate effects of the two classes of forces.

This is probably the proper place to call attention to the fact that, in most practical problems involving the determination of inertia forces, it is usually assumed that the motion of one member of the machine is completely known, and that either the effort or the useful resistance, but not both, is also known. This usually involves either the assumption of a very large flywheel, or the assumption that power is received from a constant-speed source.

There is one class of problems, however, in which both the effort and the useful resistance are assumed to be known, namely, problems dealing with the speed fluctuations in machinery. In such cases it is usual to assume that the velocities of the machine members are known but that the accelerations are not known. Problems of this nature are considerably more complicated than those in which the motion of one member is assumed to be completely known.

11-8. Kinetically Equivalent System. In determining the line of action of the inertia force F from the expression $Fh = I_G\alpha$, it is necessary to determine first the moment of inertia I_G and the angular acceleration α of the link under consideration. A determination wholly in terms of forces, thus avoiding the preceding moment equation, will now be developed. The procedure is to reduce the distributed mass of a link such as shown in Fig. 11-11 to a simple concentrated system, namely, two masses. As finally applied this procedure may be found more convenient in certain applications, such as will be discussed later, than that outlined in the preceding articles.

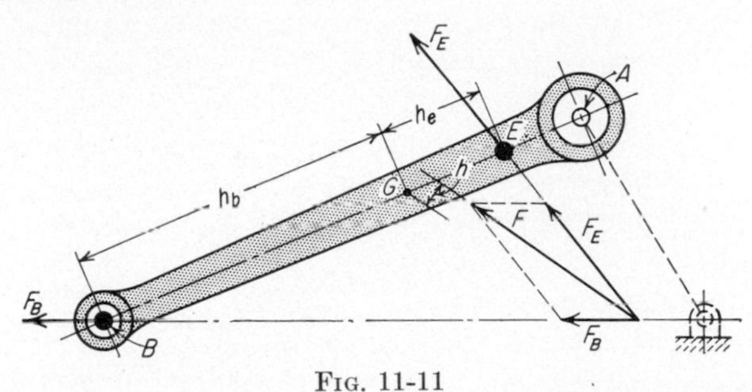

FIG. 11-11

The complete inertia effects of a rigid body are given by a force $-mA_G$ and a couple $-I_G\alpha$. If the body is to be replaced by a kinetically equivalent system of masses, it is evident that m, A_G, and I_G must be the same for the new system as for the original one.

In Fig. 11-11 let G be the center of gravity of the rigid link there shown, and let the mass m be replaced by the two masses at B and E. One of the masses may be placed at any convenient point such as B, but the other must be so located as to satisfy three conditions, the first of which is that the total mass must not be changed. That is,

$$m_B + m_E = m$$

where m_B and m_E are the masses at B and E, respectively. As a second condition, the position of the center of gravity must be unchanged, and therefore the masses at B and E must be on a line through that center at distances such as h_b and h_e from G, satisfying the moment equation

$$m_B h_b = m_E h_e$$

The third condition is that the moment of inertia of the system must remain unchanged. Therefore

$$m_B h_b^2 + m_E h_e^2 = mk^2 = I_G$$

and

$$m_B h_b h_b + m_E h_e h_e = m_E h_e h_b + m_B h_b h_e = mk^2$$

Therefore

$$(m_E + m_B) h_b h_e = mk^2, \text{ or } h_b h_e = k^2$$

The masses concentrated at B and E may be thought of as being connected by a weightless rod, and, since this system has the same mass, same center of gravity, and same moment of inertia as that of the original body, it is known as a *kinetically equivalent system.*

In order to determine the characteristics of the kinetically equivalent system, the mass, center of gravity, and moment of inertia of the original body must be known. Then the following relations, obtained by simultaneous solution of the foregoing equations, will determine the two-mass system to be substituted for the actual body.

$$h_b h_e = k^2 = \frac{I_G}{m} \qquad (11\text{-}13)$$

$$m_E = \frac{h_b}{h_b + h_e} m \qquad (11\text{-}14)$$

$$m_B = \frac{h_e}{h_b + h_e} m \qquad (11\text{-}15)$$

Obviously the two systems must have a resultant inertia force F having the same magnitude, direction, and line of action. In any given case the acceleration polygon for the mechanism would be drawn. The force determination for the original system has been covered in the preceding articles. That for the kinetically equivalent system is shown in the figure. Forces F_B and F_E are determined in the usual way by means of data from the acceleration polygon. These forces are then the two components of the resultant force F, as indicated in the figure.

In the case of the connecting rod shown in Fig. 11-11, it is convenient to locate one of the masses at B, the axis of the wrist pin. Thus, h_b is *selected*, and the equations then determine uniquely h_e, m_E, and m_B. Position E falls at the center of percussion of the rod with respect to position B.

It may be well at this point to consider the advantage in using a kinetically equivalent system. There is obviously no advantage if the inertia-force determination is to be made for one crank position only. The advantage becomes evident only when determinations must be made for a number of crank positions, as, for example, in the gasoline engine, where it may be desirable to make a complete force analysis involving 48 crank positions in the four-stroke cycle (see Chap. 13). In such a case the line of action of the inertia force of the piston is definitely located in the path of reciprocation. That of the unbalanced crank, assuming the general case of angular acceleration, passes always through a fixed point on the crank, the center of percussion (see Art. 11-4). For the connecting rod, however, the distance $h = I_G\alpha/F$, which locates the line of action of the inertia force of the rod, is different for each of the 48 crank positions, since this is a floating link. If, therefore, the kinetically equivalent system is substituted for the connecting rod, all the inertia forces in the mechanism will have their lines of action through fixed points on the links, regardless of the positions of the links. This will be found a decided convenience when making the analysis for a large number of crank positions.

11-9. Application to Engine Mechanism. In order to illustrate the application of the kinetically equivalent system, consider the engine mechanism shown in Fig. 11-12. The gas force $\mathfrak{F}_4$ applied to the piston is assumed to be known, and it is required to determine the reactions at O, A, and B, taking into account not only the force $\mathfrak{F}_4$ but also the inertia forces of the moving members of the mechanism. The crank is assumed to be turning counterclockwise with a uniform angular velocity of ω radians/sec. The masses of the three moving members are m_2, m_3, and m_4, respectively. The center of gravity of the crank is at G_2; that of the connecting rod is at G_3. It is also assumed that E, the center of percussion of the connecting rod with respect to the point B, has been determined experimentally by the method described in Art. 11-4.

In determining the *turning effort* (the component of the force of 3 on 2 which is perpendicular to 2), the guide pressure, and the reactions at O, A, and B, two separate analyses will be made: a static force analysis and an inertia force analysis. The results of the two analyses will then be combined to determine the required forces.

1. *Static Force Analysis.* In the present problem, the effort $\mathfrak{F}_4$ is the known quantity and is therefore the starting point in drawing the static force polygon shown at b in the figure. The static guide pressure is $\mathfrak{F}_{14}$, and the static reactions at A and B are $\mathfrak{F}_{32}$ $(= -\mathfrak{F}_{23})$ and $\mathfrak{F}_{34}$ $(= -\mathfrak{F}_{43})$, respectively. The static force $\mathfrak{F}_{32}$ is the force that would be transmitted along the connecting rod to the crank (and thence to a useful application) if part of the effort $\mathfrak{F}_4$ were not required to overcome inertia forces. Furthermore, $\mathfrak{F}_{32}$ is the static force at A on the crankpin (assumed integral with the crank), and $\mathfrak{F}_{34}$ is the static force at B on the wrist pin (assumed fastened to the piston or crosshead). The equal and opposite forces, $\mathfrak{F}_{23}$ and $\mathfrak{F}_{43}$, are obviously the static forces exerted on the bearings of the connecting rod.

If it is assumed that power is taken off the shaft in the form of a pure torque or couple, the static reaction at O, i.e., $\mathfrak{F}_{12}$ (which is not shown in the polygon at b) is equal, parallel, and opposite to the reaction $\mathfrak{F}_{32}$ at A ($\mathfrak{F}_{32} \mathrel{+\!\!\!\!\rightarrow} \mathfrak{F}_{12} = 0$). This force, $\mathfrak{F}_{12}$, is the static force exerted by the main bearing at O upon the crankshaft; the force $\mathfrak{F}_{21}$ is the equal and opposite force exerted upon the bearing. The *static turning effort* $\mathfrak{F}_{32}^{TO}$ is determined

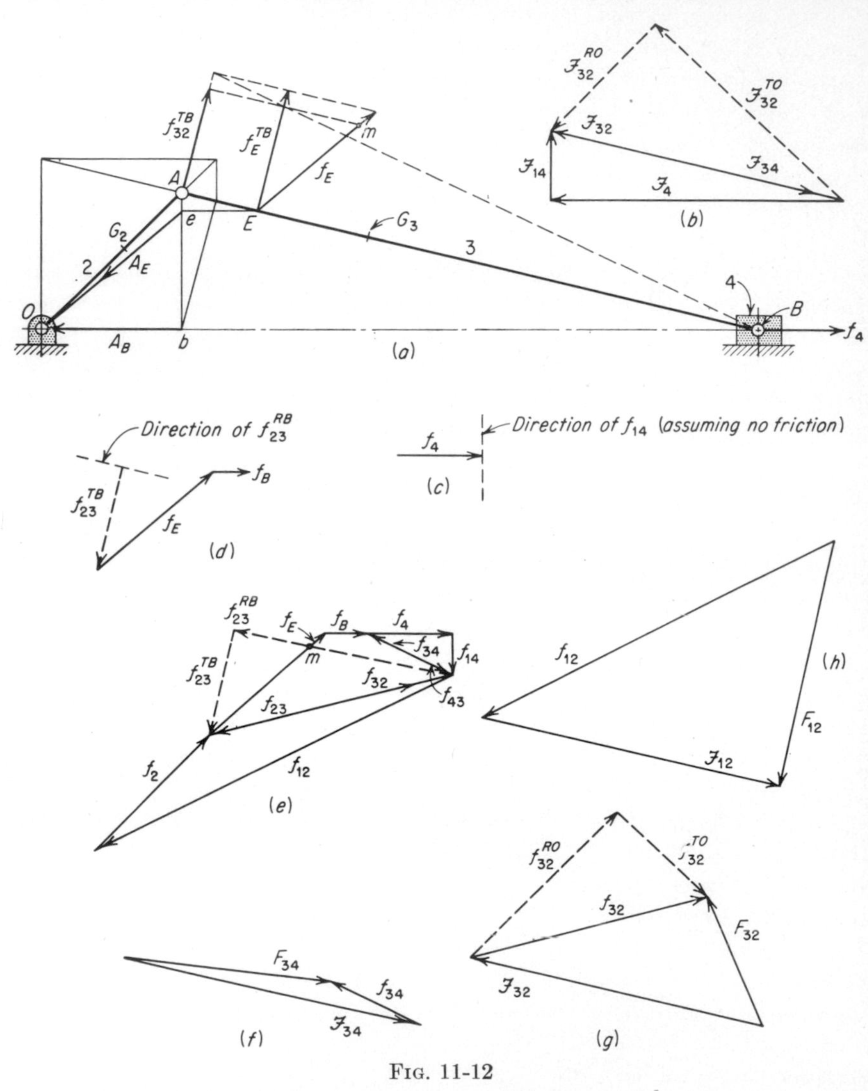

FIG. 11-12

by resolving $\mathfrak{F}_{32}$ into components perpendicular and parallel to the crank as shown.

2. *Inertia Force Analysis.* Preliminary to the inertia force analysis, it is necessary to make the acceleration determination. Since the crank has uniform rotation, this can be most easily accomplished by means

of the Ritterhaus construction, as shown at a in Fig. 11-12. It should be recognized that the line drawn from A to b is not, in general, perpendicular to OB, as it appears to be for the particular position of the engine mechanism shown.

The next step is to determine a kinetically equivalent system for the connecting rod (link 3), since this member has a fairly complex motion. The most convenient system for our purposes consists of two masses, one concentrated at B and the other at E, the magnitudes of the masses being given as follows:

$$m_B = \frac{G_3E}{BE} m_3 \qquad m_E = \frac{BG_3}{BE} m_3$$

As link 2 is rotating about the fixed center O with a uniform angular velocity, there is no advantage to be gained in replacing this link by its kinetically equivalent system. The same statement applies to link 4, in this case because of the fact that the link has a motion of translation.

The various inertia forces are:

Inertia force of crank.......................... $f_2 = -m_2A_{G2}$
Inertia force of connecting-rod mass at E.......... $f_E = -m_EA_E$
Inertia force of connecting-rod mass at B.......... $f_B = -m_BA_B$
Inertia force of link 4.......................... $f_4 = -m_4A_B$

It will be assumed in what follows that the inertia forces have been calculated in accordance with the above equations and that they are represented in the polygons by the vectors f_2, f_E, f_B, and f_4. The determinations of the reactions due to the inertia forces will now be discussed in detail, the analysis being based upon the fact that the inertia force of any single link together with the inertia reactions (not the static reactions) of the other links upon it constitute a system of forces in equilibrium.

The forces acting upon link 4 will be considered first. Since only three forces (f_4, f_{14}, and f_{34}) and no torque act upon the link, the forces intersect at B. As indicated at c in the figure, the first step in drawing the polygon is to lay off the known force f_4, after which the known direction of f_{14} may be laid off.

$$f_4 \nrightarrow f_{14} \nrightarrow f_{34} = 0 \tag{11-16}$$

As nothing is known as yet concerning the third force f_{34}, no further progress can be made in this direction.

The forces acting on link 3 are f_E, f_B, f_{23}, and f_{43}.

$$f_E \nrightarrow f_B \nrightarrow f_{43} \nrightarrow f_{23} = 0 \tag{11-17}$$

The known forces f_E and f_B may be laid off as shown at d in Fig. 11-12. By taking moments about B, the transverse component of f_{23} with respect to B may be determined, the construction being made on the drawing of the mechanism, as shown at *a*.* The force f_{23}^{TB} may then be laid off in the polygon and a line drawn perpendicular to it through its origin to represent the general direction of f_{23}^{RB}.

$$f_E +\!\!\!\!\rightarrow f_B +\!\!\!\!\rightarrow f_{43} +\!\!\!\!\rightarrow f_{23}^{RB} +\!\!\!\!\rightarrow f_{23}^{TB} = 0 \tag{11-18}$$

No further progress is possible by considering the forces acting on link 3 alone.

The two solutions attempted above have both failed because of a seeming lack of known quantities. That sufficient data are given for the solution of the problem is at once evident, however, if the two polygons are combined as shown at *e* in the figure. The vector equation depicted there is obtained by substituting f_{43} from Eq. (11-16) into Eq. (11-18), thus solving these two vector equations simultaneously.

$$f_E +\!\!\!\!\rightarrow f_B +\!\!\!\!\rightarrow f_4 +\!\!\!\!\rightarrow f_{14} +\!\!\!\!\rightarrow f_{23}^{RB} +\!\!\!\!\rightarrow f_{23}^{TB} = 0$$

The vectors f_{34} and f_{23} $(= -f_{32})$ may then be drawn as shown, and the two force polygons thus completed. It may be observed that the force f_{34} could also have been found by solving the following vector equation:

$$f_4 +\!\!\!\!\rightarrow f_{14} +\!\!\!\!\rightarrow f_{34}^{TA} +\!\!\!\!\rightarrow f_{34}^{RA} = 0$$

The force f_{34}^{TA} could have been determined by taking moments of forces on 3 about point A. Only two unknown magnitudes then would have remained in the equation, and a solution would have been possible.

Since only three forces, f_{32} $(= -f_{23})$, f_2, and f_{12}, act on link 2, and since the first two are known, the third force f_{12} forms the closing side of the triangle.

$$f_{32} +\!\!\!\!\rightarrow f_2 +\!\!\!\!\rightarrow f_{12} = 0$$

The polygon of forces at *e* can now be completed for the entire mechanism. The force at B on the wrist pin due to inertia is f_{34}, the force at A on the crankpin is f_{32}, and the force at O on the crankshaft is f_{12}. The corresponding equal and opposite forces, namely, f_{43}, f_{23}, and f_{21}, are obviously the forces on the bearings at B, A, and O, respectively. It should be observed that, although the *force* polygon at *e* closes, the inertia forces and reactions acting on the crank are *not* in equilibrium, since the lines of action of f_2, f_{12}, and f_{32} do not intersect in a common point. For equilibrium, there must be an external counterclockwise torque applied to the crankshaft, the magnitude of which can be determined by taking moments about some point on the crank, such as point O.

If the force f_{32} is resolved into components perpendicular and parallel

* The letter *m* in the construction at *a*, and in the polygon at *e*, has no significance here, being shown in the figure only to facilitate the discussion in the next article.

to the crank, as shown at g where f_{32} from e has been redrawn, the perpendicular component f_{32}^{TO} will represent the *inertia or dynamic turning effort.* The inertia force f_2 of the crank (a centrifugal force) has a line of action passing through the center of the main bearing, and has, therefore, no effect upon the turning effort. It is to be noted that the dynamic turning effort f_{32}^{TO} opposes $\mathfrak{F}_{32}^{TO}$ for the position of the mechanism shown. Later in the stroke this condition reverses itself, and the kinetic energy previously given to the connecting rod and piston is regained.

3. *Total Forces.* Corresponding static and inertia force vectors can now be combined to give any desired total force. For example, the total force at B on the wrist pin is $F_{34} = \mathfrak{F}_{34} \nrightarrow f_{34}$, as shown at f; the total force at A on the crankpin is $F_{32} = \mathfrak{F}_{32} \nrightarrow f_{32}$, as shown at g; and the total force at O on the crankshaft is $F_{12} = \mathfrak{F}_{12} \nrightarrow f_{12}$, as shown at h. The total turning effort F_{32}^{TO} is obviously the algebraic sum of components $\mathfrak{F}_{32}^{TO}$ and f_{32}^{TO}, or it may be obtained directly from the total force F_{32} shown at g.

It could be shown in a manner entirely analogous to that employed in Art. 11-6 that the total shaking force exerted on the frame of the machine is equal to the vector sum of the three inertia forces. The line of action of this force would pass through the intersection of the lines of action of f_{21} and f_{41}. There is also a static couple exerted on the frame, the moment of this couple being equal to the moment of the static turning effort.

11-10. Engine Mechanism. Special Methods. In the chapter on Accelerations special methods were developed for dealing with the engine mechanism in cases where acceleration determinations are required for a number of different positions of the crank. In the present article similar methods will be developed for expediting the routine work of analyzing the forces in the engine mechanism for a number of crank positions. Data will be given for the solution of a numerical example, and, in order to illustrate the method, the analysis will be worked out in detail for one position of the crank.

It is required to determine the total turning effort and the various reactions of the engine mechanism (Fig. 11-13) covered by the following description:

Length of crank OA	4 in.
Length of connecting rod AB	12 in.
Distance of center of percussion E of connecting rod from center of wrist pin B	10¾ in.
Speed of crank, link 2 (constant)	1,800 rpm
Total mass of connecting rod, link 3	3.59 lb
Mass assumed concentrated at center of percussion E	2.84 lb
Mass assumed concentrated at wrist pin	0.75 lb
Mass of piston, link 4	2.00 lb
Force on piston due to gas pressure (45-deg position)	1,800 lb

The crank is assumed to be fully countervalanced; i.e., weights have been added to the crankshaft opposite the crankpin such that the center of gravity of the resulting system is situated at the center line of the main bearing. As a result, the inertia force of the crank is zero. It is observed that a kinetically equivalent system has been substituted for the connecting rod.

In Fig. 11-13*a*, the first step in the solution of the problem is to make the acceleration determination, and this has been accomplished by the methods of Art. 9-19. In the original drawing, the crank circle was laid out full size, and the distance OA on the drawing was therefore 4 in. The acceleration scale k_a was then calculated as follows:

$$a_A = OA\omega_2^2 = \frac{4}{12}\left(\frac{2\pi \times 1{,}800}{60}\right)^2 = 11{,}850 \text{ ft/sec}^2$$

$$k_a = \frac{11{,}850}{4} = 2{,}963$$

The scale having been calculated, the accelerations of the wrist pin B and the center of percussion E are determined from the polygon to be

$$a_B = 8{,}500 \text{ ft/sec}^2$$
$$a_E = 11{,}260 \text{ ft/sec}^2$$

The magnitudes of the inertia forces f_E, f_B, and f_4 are then calculated as follows:

$$f_E = \frac{M_E a_E}{32.17} = \frac{2.84 \times 11{,}260}{32.17} = 993 \text{ lb}$$

$$f_B = \frac{M_B a_B}{32.17} = \frac{0.75 \times 8{,}500}{32.17} = 198 \text{ lb}$$

$$f_4 = \frac{M_4 a_B}{32.17} = \frac{2.00 \times 8{,}500}{32.17} = 528 \text{ lb}$$

Next, the equations of force equilibrium for the connecting rod and piston are written.

$$f_E \nrightarrow f_B \nrightarrow f_{43} \nrightarrow f_{23} = 0 \qquad \text{(for connecting rod 3)}$$
$$f_4 \nrightarrow f_{14} \nrightarrow f_{34} = 0 \qquad \text{(for piston 4)}$$

Combining,

$$f_E \nrightarrow f_B \nrightarrow f_4 \nrightarrow f_{14} \nrightarrow f_{23} = 0$$

The forces f_E, f_B, and f_4 may now be laid off in the polygon as shown. It is convenient to use O, the center of the main bearing, as the origin for the vector f_E when drawing the force polygon, for the reason that it will bring the vector f_E always along the line Oe and in the sense Oe. The direction of f_{14} may be laid off from the extremity of f_4 as shown. In order to determine the magnitude of f_{14}, let a point m be located on the vector f_E such that

$$Om = f_E \frac{BE}{BA} = 993 \times \frac{10.75}{12} = 896 \text{ lb}$$

The line mn drawn parallel to the connecting rod will then determine the length of the vector f_{14}. The proof of this statement is as follows. In Fig. 11-12*e* the intersection of the vector f_{23}^{RB} (which is parallel to the connecting rod) and the vector f_E is denoted by the letter m. The same

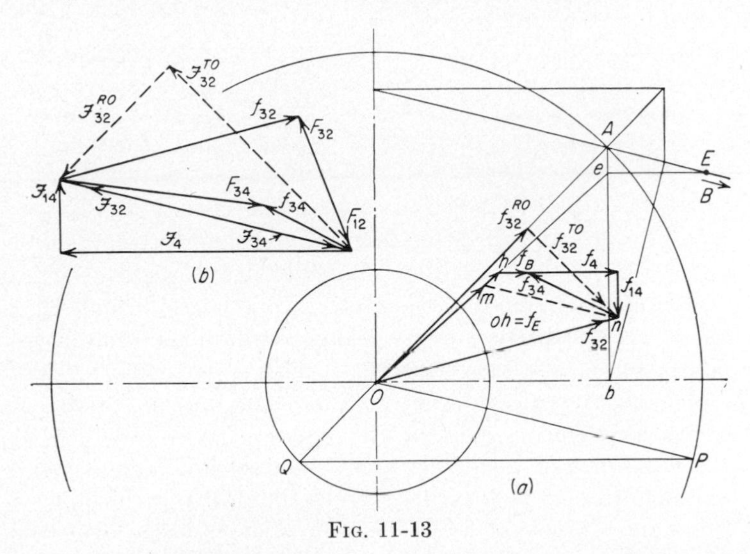

FIG. 11-13

letter is used to designate the corresponding point on the drawing of the mechanism (on the vector f_E drawn from the point E in Fig. 11-12*a*). It is evident from the geometry of the figure that

$$\frac{Em}{f_E} = \frac{f_{32}^{TB}}{f_E^{TB}} = \frac{BE}{BA} \qquad \text{therefore } Em = f_E \frac{BE}{BA}$$

Referring again to Fig. 11-13 and the equilibrium equations, it is to be noted that the point n (intersection of f_{14} and the line mn parallel to the connecting rod) also locates the extremity of the vector f_{32} (origin of f_{23}). The force polygon can now be completed by drawing in the vectors f_{34} and f_{32}. With the exception that $f_2 = 0$, and the omission of f_{23}^{TB} and f_{23}^{RB}, it can be seen that the arrangement of vectors in this superimposed force polygon is the same as that shown at *e* in Fig. 11-12. If the vector f_{32} is resolved into components perpendicular and parallel to the crank, the component f_{32}^{TO} will give the turning effort due to inertia.

The procedure outlined above is of great convenience when determinations must be made for a large number of crank positions.

The next step in the solution is to make the static force analysis; i.e., to determine the forces due to the gas pressure. These forces are represented in the triangle shown in Fig. 11-13*b*, in which piston pressure $\mathfrak{F}_4$ forms the base. If the static turning effort $\mathfrak{F}_{32}^{TO}$ is required, it may be determined as shown in the figure. Any total force may be determined by combining the inertia force vector with the corresponding static force vector, as shown in Fig. 11-13*b*, where, for convenience, the vector addition has been made on the figure. The solution may be completed and the required forces obtained as follows:

Total force on crankpin.......	$F_{32} = \mathfrak{F}_{32} \leftrightarrow\!\!\!\!+ f_{32} =$	890 lb
Total turning effort...........	$F_{32}^{TO} = \mathfrak{F}_{32}^{TO} \leftrightarrow\!\!\!\!+ f_{32}^{TO} =$	810 lb
Total force on wrist pin.......	$F_{34} = \mathfrak{F}_{34} \leftrightarrow\!\!\!\!+ f_{34} =$	1,300 lb
Total force on crankshaft......	$F_{12} = \mathfrak{F}_{12} \leftrightarrow\!\!\!\!+ f_{12} =$	890 lb
Total force on cylinder wall....	$F_{41} = \mathfrak{F}_{41} \leftrightarrow\!\!\!\!+ f_{41} =$	160 lb
Total effective piston force....	$F_4 = \mathfrak{F}_4 \leftrightarrow\!\!\!\!+ f_4 =$	1,270 lb

The directions of the above forces are determined from the vector diagrams.

Force Diagrams. In the force analysis of a machine, where the speed is such as to set up inertia forces of appreciable magnitude, it may be desirable to plot certain force diagrams. This is particularly true in the case of high-speed engines, where diagrams of piston forces, turning effort, bearing pressures, etc., for the complete cycle serve a valuable purpose. Examples of such diagrams, for a gasoline engine where the forces were determined in accordance with the principles outlined in this chapter, are given in Chap. 13. The student should read over Chap. 13 where it will be found that all the forces represented in the static and inertia force polygons (Fig. 11-12) have been plotted in the form of diagrams for the complete cycle. Force determinations were made for 48 crank positions (every 15 deg) in the four-stroke cycle. Also shown are the total turning-effort diagram for the six cylinders combined and the mean turning-effort diagram, which is the basis for the design of the engine flywheel.

11-11. Inertia Forces in a Cam Mechanism. The inertia force analyses thus far have dealt with linkwork. The methods outlined can readily be adapted to the great majority of cases that may arise. An example of the application to direct-contact mechanisms, such as cams or gears, is the object of the present article.

Assume a cam mechanism in which the center of the roller follower moves along a straight radial line drawn outward from the cam center. The cam rotates at a constant rate of 180 rpm, the follower moves with simple harmonic motion, and the follower receives its maximum displacement of 2 in. while the cam rotates through an angle of 120 deg. In Fig. 11-14*a*, the follower total rise, or total displacement, is represented by h. The rise s at any point is determined by the projection of the extremity of the rotating radius r $(= h/2)$.

$$s = r(1 - \cos\theta)$$

The magnitude of the velocity of the follower is

$$v = \frac{ds}{dt} = r\omega_r \sin\theta$$

The tangential acceleration of the follower is

$$a = \frac{dv}{dt} = r\omega_r^2 \cos\theta \tag{11-19}$$

where $r = h/2$ and ω_r = the angular velocity of the rotating radius r. There is no normal acceleration because the follower moves along a straight line. The time for the total displacement h of the follower is

$$t = {}^{60}\!/\!_{180} \times {}^{120}\!/\!_{360} = {}^{1}\!/\!_{9} \text{ sec}$$

The angular velocity of the rotating radius r is

$$\omega_r = \frac{\pi}{1/9} \text{ radians/sec} = 9\pi \text{ radians/sec}$$

From Eq. (11-19) the accelerations of the follower at positions 0 to 6 are, respectively, 66.62, 57.69, 33.31, 0, -33.31, -57.69, -66.62 ft/sec^2.

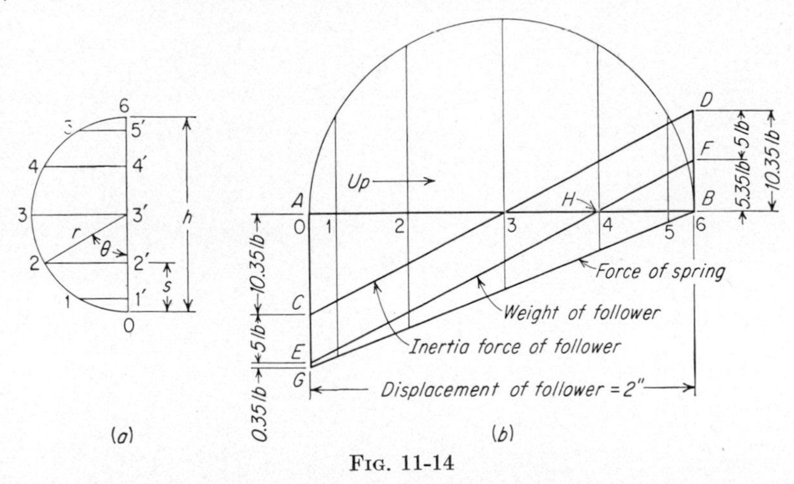

FIG. 11-14

The mass of the follower is 5 lb. From the expression $-mA$ the inertia forces of the follower at positions 0 to 6 are, respectively, -10.35, -8.96, -5.18, 0, $+5.18$, $+8.96$, $+10.35$ lb.

In Fig. 11-14b the inertia forces for the six follower positions have been plotted to a scale of 1 in. = 10 lb on base line AB representing

2-in. stroke of the follower. These inertia forces are represented by line CD. The weight of the follower is plotted to the same scale and is represented by line EF in the figure; ordinates below base line AB represent vertical forces of the follower against the cam, and ordinates above AB represent vertical forces away from the cam. That is, to the right of point H the acceleration of the follower is such that its weight is not sufficient to keep it in contact with the cam, and it becomes necessary to use a spring. At the end of the follower stroke the difference between the inertia force of the follower and its weight is $10.35 - 5 = 5.35$ lb. Hence, a spring with a scale of $2\frac{1}{2}$ lb/in. set up under an initial load of 0.35 lb will be just sufficient to keep the follower in contact with the cam when it reaches the end of its stroke, as indicated by the line GB in the figure. It should be stated, however, that the effect of friction has been neglected in the foregoing analysis.

11-12. Scotch Yoke Device. This article discusses another example involving dynamic force analysis. In Fig. 11-15, the crank (link 2) of

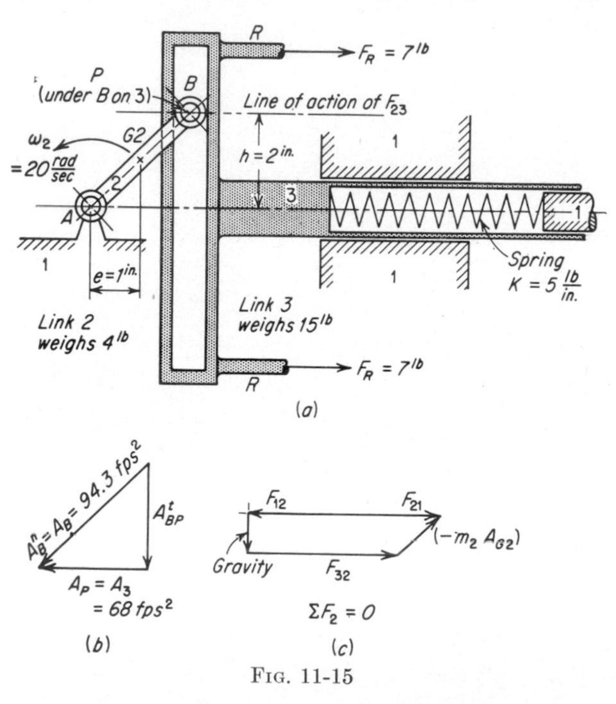

FIG. 11-15

length 2.83 in. rotates at constant angular velocity of 20 radians/sec as shown. The spring is compressed 6 in. and assists the driving torque applied to the shaft of link 2 to overcome a force of 7 lb applied to each rod R by connected apparatus (not shown). For the instant shown, the required items are:

1. The force F_{23} transmitted from 2 to 3 along the common normal.
2. The force F_{21} exerted on the crank bearing.
3. The external torque supplied to the crank.

Friction forces are to be neglected in the analysis, and all forces will be combined into one analysis, without separation into static and inertia effects.

For equilibrium of body 3, the vector sum of forces on 3 must be zero.

$$F_{23} +\!\!\!\!\rightarrow F_{\text{spring}} +\!\!\!\!\rightarrow 2F_R +\!\!\!\!\rightarrow (-m_3A_3) = 0$$

The vector diagram containing the solution of the equation

$$A_B = A_{BP} +\!\!\!\!\rightarrow A_P$$

for A_P (= A_3) is shown at *b*. The forces on 3 are as follows:

F_{23} = unknown in magnitude, but direction is horizontal.
$F_{\text{spring}} = 5 \times 6 = 30$ lb to the left.
$2F_R = 14$ lb to the right.
$(-m_3A_3) = 15 \times 68/32.17 = 31.73$ lb to the right.

As the downward pull of gravity is balanced by the upward force of 1 on 3 these forces cancel and are not shown. Therefore,

$$F_{23} = 14 + 31.73 - 30 = 15.73 \text{ lb}$$

to the left. For force equilibrium of link 2,

$F_{32} +\!\!\!\!\rightarrow F_{12} +\!\!\!\!\rightarrow \text{weight} +\!\!\!\!\rightarrow (-m_2A_{G2}) = 0.$
$F_{32} = 15.73$ lb to right (opposite F_{23}).
F_{12} = unknown completely.
Weight = 4 lb down.
$(-m_2A_{G2}) = 4 \times 94.3/(2 \times 32.17) = 5.86$ lb opposite A_{G2}.

The force polygon for the equation above is shown at *c*. The required force F_{21} (opposite F_{12}) is scaled to be 20.0 lb. The last required item, the external torque supplied to the crank, is determined by taking moments of forces on 2 about point A.

$F_{32} \times 2 +\!\!\!\!\rightarrow \text{weight} \times 1 +\!\!\!\!\rightarrow T_{EXT} = 0.$
$F_{32} \times 2 = 15.73 \times 2 = 31.46$ lb-in. clockwise.
Weight $\times 1 = 4 \times 1 = 4$ lb-in. clockwise.
Therefore, $T_{EXT} = 35.46$ lb-in. counterclockwise.

The *weight* (gravitational attraction) has been included in this analysis, but of course it can be omitted with negligible error in analyses of many high-speed machines. Disregard of the weight should not become *automatic*, however, and should come only after determination of its relative importance in any particular problem.

11-13. Rocket Sled. A very interesting example involving inertia forces is a rocket sled which has been used by the U.S. Air Force in studying the effects of rapid accelerations and decelerations on human beings. Figure 11-16*a* is a sketch showing the two main units—the passenger sled carrying the subject and instruments, and the propulsion sled. The sled moves in a straight line, sliding on rails several thousand feet long. At *b*, *c*, and *d* of the figure are shown the force equations of equilibrium along the line of motion, and the forces acting to produce dynamic equilibrium of the several elements while the sled is accelerating. Forces in the vertical direction are not shown, but it must be recognized

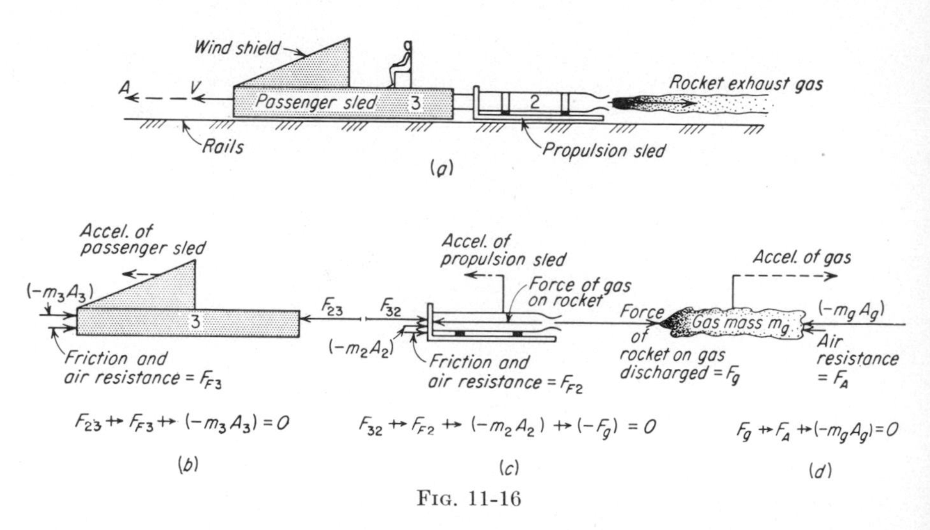

FIG. 11-16

that gravitational attraction (weight), force of the rails on the sleds, and aerodynamic lift (or depression) must be in equilibrium *without* net vertical acceleration. If the line of action of the external resultant force on each sled passes through the center of gravity of the sled, then there is no rotational tendency, of course. Perfection here is improbable, and so constraints both above and below the rails are necessary.

In actual tests, the 1½-ton sled combination and its human passenger were accelerated to a speed above 600 mph by a 40,000-lb thrust from the rockets. Then the rockets were cut off, and the sled was decelerated rapidly by dipping scoops attached to the sled into water troughs along the rails. The figure has been quoted that 2 lb (of force) on the sled was obtained from each pound of water accelerated. Thus the sled was *decelerated* by receiving the reaction from water which was *accelerated*. A deceleration in excess of $25g$ (25×32.2 ft/sec^2) acted for a period of about 1.1 sec in some of the tests, with instantaneous values as high as $40\,g$ (with a human passenger). Thus, assuming the passenger in the seat to weigh 180 lb, he was subjected to a force of (180/32.17)(25 × 32.2) lb, or approximately 4,500 lb, applied to him by the harness used to keep him from being catapulted forward out of his seat.

The black eyes suffered by the test subject during this rapid deceleration were also the result of inertia forces. The subject's head was held rigidly against the sled, but his eyes continued to move forward, restrained mainly by connecting tissue. This tissue suffered damage because of the large forces applied to it, resulting in the black eyes already mentioned. For equilibrium of an eyeball E of mass m_E,

$$F \nrightarrow (-m_E A_E) = 0$$

Force F is the force on the eyeball because of contacting tissue (and bones, perhaps).

In this article an interesting situation has been discussed in which the inertia forces are all-important. The proper positions of the centers of gravity, the strength required of the supports for the rockets, the load carried by the connection between the propulsion sled and the passenger sled, the design of the seat and harness, proper control of the accelerations during tests, etc., all required dynamic force analysis.

11-14. Other Examples of Inertia Forces. 1. *High-speed Rotor.* Consider a small gas-turbine rotor which revolves at 24,000 rpm. At this speed a blade breaks, and a particle at a 4 in. radius weighing 0.2 oz leaves the tip. Figure 11-17 illustrates the direction of motion of the blade particle, and also the force applied suddenly to the rotor bearings because of the equivalent unbalanced mass directly opposite the particle which has left the rotor. The magnitude of this force is

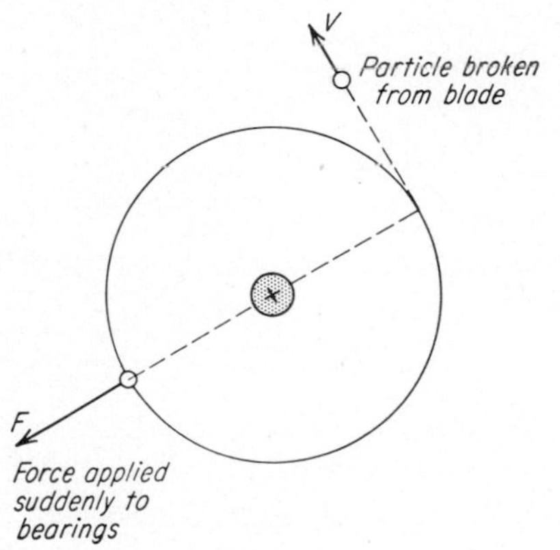

FIG. 11-17

$$F = mr\omega^2 = \frac{0.2}{16 \times 32.17} \times \frac{4}{12} \times \left(\frac{24{,}000 \times 2\pi}{60}\right)^2$$
$$= 818 \text{ lb}$$

2. *Rivers and Wind.* Figure 11-18 shows a simplified diagram of the earth revolving about its axis. Point P represents a particle of water in a river which flows south in the Northern Hemisphere, V_{PQ} representing the velocity of P relative to a point Q on the riverbed. The top view shows a component of V_{PQ} (this component depends on the latitude of Q)

acting along a path which is rotating. Thus a Coriolis component of acceleration is present, and the accompanying force on the river bank tends to erode the right bank. The force is small, perhaps (ω of the earth is only $\pi/43{,}200$ radians/sec), but persistent.

If the particle P is assumed to be air blowing south, it is easily visualized that the Coriolis forces can play an appreciable part in the weather, which involves large air masses often traveling at high speeds.

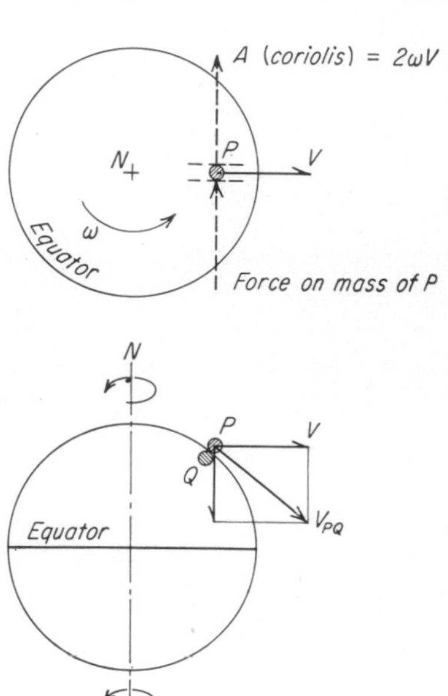

FIG. 11-18

It is interesting to note that, if P were a 3,600-lb car at 45° north latitude traveling south at 60 mph, the sideways force due to Coriolis acceleration would be only 1.01 lb. This force would be produced mostly by friction between the tires and the road. Usually, of course, other forces acting would make such a small force unnoticeable.

11-15. Analytical Methods. The analytical determination of the forces acting in even a simple mechanism is likely to become somewhat involved. In certain cases, however, it will be found more convenient to use analytical methods than to use the graphical methods discussed in preceding articles. For this reason it is worthwhile to give some consideration to the general analytical method of attack. In the present article, a brief outline will be given of the analytical determination of the inertia forces and dynamic reactions that act in the engine mechanism.

Referring to Fig. 11-19, the various accelerations are given as follows (see chapter on Accelerations):

$$a_B = R\omega_2^2\left[\cos\theta + \frac{R}{L}\left(\frac{\cos 2\theta}{\cos\phi} + \frac{R^2}{L^2}\frac{\sin^2\theta\cos^2\theta}{\cos^3\phi}\right)\right]$$

$$a_E^x = R\omega_2^2\left[\cos\theta + \left(1 - \frac{h}{L}\right)\frac{R}{L}\left(\frac{\cos 2\theta}{\cos\phi} + \frac{R^2}{L^2}\frac{\sin^2\theta\cos^2\theta}{\cos^3\phi}\right)\right]$$

$$a_E^y = -R\omega_2^2\frac{h}{L}\sin\theta$$

$$a_A^x = R\omega_2^2\cos\theta \qquad \text{and} \qquad a_A^y = -R\omega_2^2\sin\theta$$

$$a_{G2} = r\omega_2^2 = \frac{r}{R}R\omega_2^2 = \frac{r}{R}a_A$$

Therefore

$$a_{G2}^x = \frac{r}{R} a_A^x \qquad \text{and} \qquad a_{G2}^y = \frac{r}{R} a_A^y$$

In order to reduce the algebraic work, let

$$a_B' = R\omega_2^2 \cos \theta$$

and

$$a_B'' = R\omega_2^2 \frac{R}{L}\left(\frac{\cos 2\theta}{\cos \phi} + \frac{R^2}{L^2}\frac{\sin^2\theta \cos^2\theta}{\cos^3\phi}\right) = \frac{R^2}{L}\,\omega_2^2 \cos 2\theta \qquad \text{approx}$$

The quantities a_B' and a_B'' are called respectively the primary and

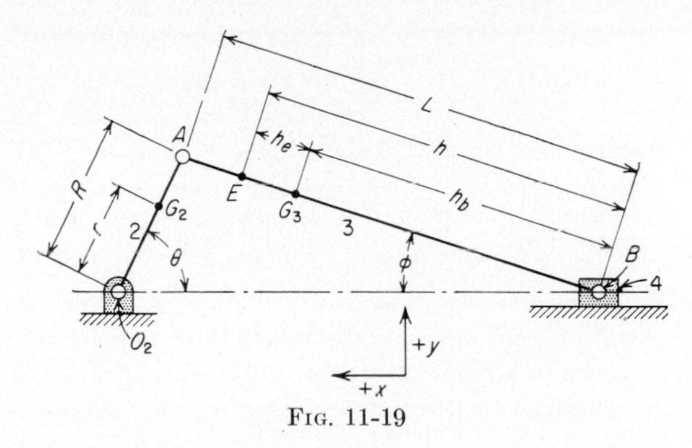

FIG. 11-19

secondary acceleration of the wrist pin. Substituting in the above equations,

$$a_B = a_B' + a_B'' \tag{11-20}$$

$$a_E^x = a_B' + \left(1 - \frac{h}{L}\right) a_B'' \tag{11-21}$$

Also

$$a_A^x = a_B' \qquad \text{and} \qquad a_E^y = \frac{h}{L} a_A^y \tag{11-22}$$

The expressions for the various inertia forces may then be written as follows, the notation being the same as in Arts. 11-9 and 11-10 except that the letters x and y are used to denote horizontal and vertical components respectively.

$$f_4 = -m_4(a_B' + a_B'') \qquad f_B = -m_B(a_B' + a_B'')$$

$$f_E^x = -m_E\left[a_B' + \left(1 - \frac{h}{L}\right) a_B''\right] \qquad f_E^y = -m_E \frac{h}{L} a_A^y$$

$$f_2^x = -m_2 \frac{r}{R} a_A^x \qquad f_2^y = -m_2 \frac{r}{R} a_A^y$$

The inertia force of the piston is usually given special prominence because of its importance in connection with the balancing of engines. As ordinarily written, this force is

$$f_4 = -m_4 R\omega_2^2 \left(\cos\theta + \frac{R}{L}\cos 2\theta\right) \qquad (11\text{-}23)$$

In order to determine any of the dynamic reactions, it is necessary to set up the equations for the equilibrium of one or more members of the mechanism. The method will be illustrated by determining the expression for the reaction at A. The first step is to set the sums of the horizontal and vertical components of the forces acting on links 3 and 4 equal to zero. The equations for link 3 are

$$f_{23}^x + f_E^x + f_B + f_{43}^x = 0$$
$$f_{23}^y + f_E^y + f_{43}^y = 0 \qquad \text{(not used)}$$

The corresponding equations for link 4 are

$$f_4 + f_{34}^x = 0$$
$$f_{14} + f_{34}^y = 0 \qquad \text{(not used)}$$

In order to obtain a solution, one more relation is necessary, since two of the above equations are of no direct use in determining the reaction at A. The most convenient one to apply is obtained by setting the sum of the moments about B of the forces acting on link 3 equal to zero.

$$f_{23}^x L\sin\phi + f_E^x h\sin\phi - f_{23}^y L\cos\phi - f_E^y h\cos\phi = 0$$

The simultaneous solution of the preceding equations will give the following expression for the dynamic reaction at A:

$$f_{32}^x = f_E^x + f_B + f_4 \qquad (11\text{-}24)$$

$$f_{32}^y = f_{32}^x \tan\phi - \frac{h}{L}(f_E^x \tan\phi - f_E^y) \qquad (11\text{-}25)$$

$$= (f_E^x + f_B + f_4)\tan\phi - \frac{h}{L}(f_E^x \tan\phi - f_E^y)$$

Stated in terms of acceleration, this becomes

$$f_{32}^x = -(m_3 + m_4)(a_B' + a_B'') + m_E\frac{h}{L}a_B'' \qquad (11\text{-}26)$$

$$f_{32}^y = f_{32}^x \tan\phi + m_E\frac{h}{L}\left\{\left[a_B' + \left(1 - \frac{h}{L}\right)a_B''\right]\tan\phi - \frac{h}{L}a_A^y\right\} \qquad (11\text{-}27)$$

The reactions f_{34} and f_{14} could be determined in the same way and would be represented by similar expressions.

The dynamic turning effort T is given by

$$T = f_{32}^x \sin \theta + f_{32}^y \cos \theta \tag{11-28}$$

The total shaking force exerted on the frame is the resultant of the inertia forces of the members. If the shaking force is represented by f_s, then

$$f_s^x = -(m_3 + m_4)(a_B' + a_B'') + m_E \frac{h}{L} a_B'' - m_2 \frac{r}{R} a_A^x \tag{11-29}$$

$$f_s^y = -m_E \frac{h}{L} a_A^y - m_2 \frac{r}{R} a_A^y \tag{11-30}$$

The horizontal component passes through the center of the main bearing, and the vertical component has a line of action situated at a distance $\bar{x}$ to the right of the main bearing, where $\bar{x}$ is given by the following equation:

$$\bar{x} = \frac{f_{14}(R \cos \theta + L \cos \phi)}{f_s^y} \tag{11-31}$$

CHAPTER 12

BALANCING OF MACHINERY

12-1. Balancing and Vibration. One of the important applications of the study of inertia forces is in the balancing of machinery. Many moving parts of machines have reciprocating motion similar to that of the piston of an engine, or rotating motion such as that of the crankshaft of an engine or the rotor of a turbine, generator, or electric motor. If the moving parts are not in perfect balance, or if the parts have variable motion or are subject to acceleration, inertia forces (also called shaking forces) are set up that tend to produce vibrations in the frame of the machine, and hence in the foundations to which the frame is attached. Such vibrations, particularly if they occur at high speeds, may produce excessive noise, cause undue wear and tear on the machinery and its supports or on adjacent equipment or personnel, or result in the faulty performance of the work for which the machine was designed. Furthermore, if the natural period of vibration of any part of the supporting framework or foundation should happen to coincide with the period of vibration of the moving part, the disturbances set up could become dangerous. The purpose of balancing, therefore, is to neutralize or minimize these unpleasant and injurious vibratory effects as far as may be practicable.

The moving parts may be in (1) static or standing balance or (2) dynamic or running balance.

Static balance exists if the parts are in equilibrium among themselves when not running, regardless of the position in which the parts may be placed, i.e., if the center of gravity of all the moving parts remains in a fixed position relative to the frame of the machine regardless of the positions of the parts. A system is in dynamic balance when the inertia forces and couples exerted by the moving masses are in equilibrium among themselves.

Much of the discussion in this chapter concerns the balancing of engines, but the methods and principles involved are fundamental and may be applied to machinery of all kinds. The reactions between the component parts of an engine in motion may be analyzed in two main groups, namely, the forces due to the working fluid and the forces due to the accelerations of the parts. The former are classified as static forces, and their effect upon the engine frame is dependent upon the manner in which power is taken off the crankshaft. When the useful resistance overcome by the crankshaft is in the form of a pure torque or couple, there is a corresponding static torque or couple that tends to rotate the engine frame about the crankshaft. The forces due to acceleration are of course, the inertia forces, and these may be combined to form a force or a couple, or both, which ultimately acts upon the foundation and tends to cause vibration.

The inertia forces in high-speed engines are of great magnitude, and the balancing of these forces has received a great deal of attention in the development of marine, locomotive, aircraft, and automotive types of engines. In the following discussions, all parts are assumed to be rigid, so that, for example, centers of gravity are not shifted by deflections.

12-2. Balancing of Rotating Masses. The effect of rotating masses is to produce centrifugal forces or kinetic loads on the shaft to which the masses are connected. The balancing of such a system consists in rearranging the masses forming the system, or in introducing into the system additional masses, so that the forces acting on the shaft in consequence of rotation form a system in equilibrium.

The forces set up by a system of rotating masses are always situated in planes perpendicular to the axis of the shaft to which the masses are connected. If the rotation is at constant speed, as will be assumed here and in the succeeding articles, the lines of action of the forces all pass through the axis of the shaft; that is, they are pure centrifugal forces.

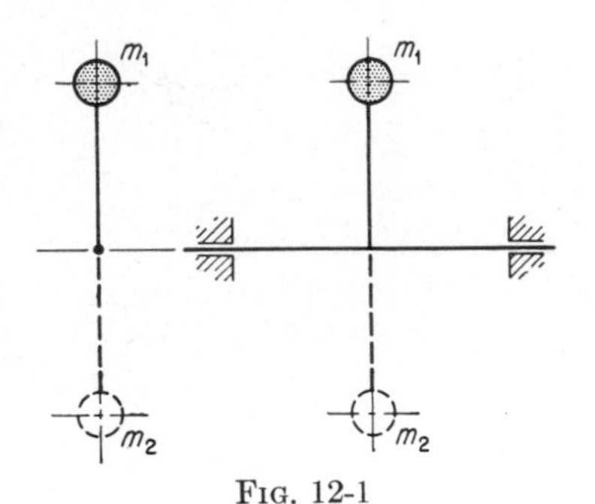

FIG. 12-1

12-3. Effect of a Single Rotating Mass. If a shaft (Fig. 12-1) rotating at an angular velocity of ω radians/sec carries a single mass of m_1 slugs[1] whose center of gravity is R_1 ft from the axis of rotation, the shaft will be subjected to a centrifugal force of magnitude $m_1R_1\omega^2$ lb. This force, which is continually changing in direction, causes the shaft to exert forces on the bearings, and these forces in turn are transmitted to the frame and foundation of the machine. The effect of such a kinetic load is to set up vibrations that are particularly noticeable at high speed.

The effect of the centrifugal force due to the single rotating mass may be eliminated by the addition of another mass m_2, whose center of gravity

[1] Refer to the beginning of Chap. 11 for review of dimensions and units. The number of slugs is equal to the weight of the mass (in pounds) divided by 32.2.

is in the plane of m_1 diametrically opposite m_1 and at a distance R_2 from the axis of rotation, such that

$$m_1R_1\omega^2 = m_2R_2\omega^2$$

It will be noted that, with these two masses in the same plane, the shaft is in both static and running balance. The center of gravity is in the axis of the shaft and remains therefore in a fixed position relative to the frame of the machine regardless of whether the shaft is turning or standing still.

12-4. Effect of Two Rotating Masses Not in the Same Plane of Rotation. If a shaft (Fig. 12-2) carries two rotating masses m_1 and m_2 in different planes of rotation but in the same axial plane, i.e., the same plane passing through the axis of the shaft, and if, further, the centrifugal forces $m_1R_1\omega^2$ and $m_2R_2\omega^2$ are equal. then the shaft is subjected to the action of an unbalanced *centrifugal couple.* This unbalanced couple tends to turn the shaft in an axial plane and is therefore resisted by an equal and opposite couple applied to the shaft at the bearings.

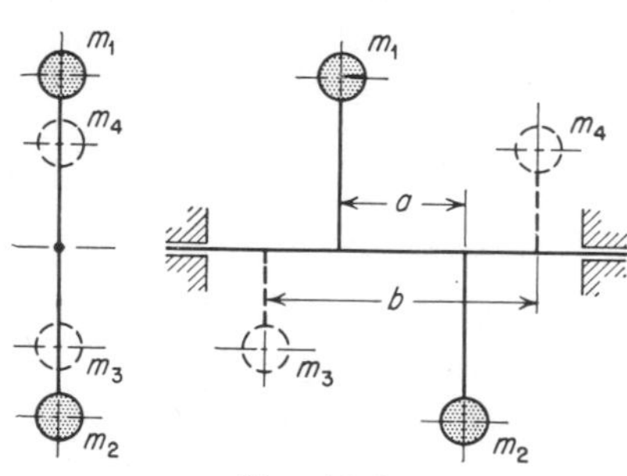

FIG. 12-2

Although this system is obviously not in running balance, it is, nevertheless, in standing balance. That is, the center of gravity of the system is in the axis of the shaft and is therefore stationary.

The two rotating masses may be balanced by the introduction of two additional masses m_3 and m_4 in the same axial plane such that the centrifugal couple that they set up equals the centrifugal couple of m_1 and m_2. That is,

$$m_1R_1\omega^2a = m_3R_3\omega^2b \qquad \text{and} \qquad m_3R_3\omega^2 = m_4R_4\omega^2$$

12-5. Several Rotating Masses in a Single Transverse Plane. If several masses m_1, m_2, m_3, etc., lie in the same transverse plane, as in Fig. 12-3*a*, the shaft is subject to a concurrent system of centrifugal forces, $m_1R_1\omega^2$, $m_2R_2\omega^2$, etc.

Let F_R be the *vector* sum (the resultant) of such a system of forces. Then (the small arrow above the summation sign $\vec{\Sigma}$ indicates *vector* summation),

$$\begin{aligned} F_R &= \vec{\Sigma} mR\omega^2 \\ &= m_1R_1\omega^2 \nrightarrow m_2R_2\omega^2 \nrightarrow \cdots \nrightarrow m_nR_n\omega^2 \\ &= \omega^2\vec{\Sigma} mR \end{aligned} \qquad (12\text{-}1)$$

The condition that such a system shall balance is that $F_R = 0$, or that $\Sigma m\vec{R}\omega^2 = 0$ and $\omega^2\Sigma m\vec{R} = 0$. Thus, the products m_1R_1, m_2R_2, etc., may be summed vectorially rather than summing the forces themselves. The equation $\Sigma m\vec{R} = 0$ requires that the center of gravity of the system lie in the axis of the shaft.

In Fig. 12-3*b*, starting from point A, the vectors m_1R_1, m_2R_2, etc., are added. The additional vector (E to A) closes the polygon and makes $F_R = 0$. This balancing vector is obtained from a mass m_0 at a radius R_0, mounted to the shaft with the radius R_0 in the proper direction, as

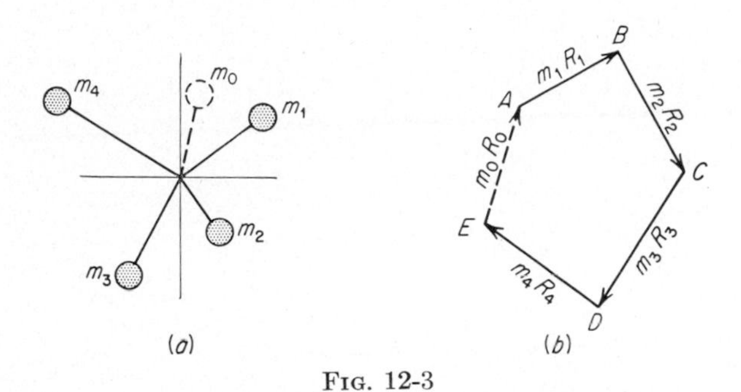

FIG. 12-3

shown. Individual values for m_0 or R_0 may be adjusted because of space or other requirements, but the product of the two (m_0R_0) must be constant at the value determined from the vector polygon. It should be noted that the gap EA may be closed by two or more vectors chosen at will, if desired, and the system may thus be balanced by two or more masses instead of one.

12-6. Several Rotating Masses in Different Transverse Planes. Referring to Fig. 12-4, let the masses m_1 and m_2 be connected to a shaft the axis of which is XX; and let the centers of gravity of the two masses be situated in the planes A_1 and A_2. The forces set up by the rotation of the system are F_1 and F_2. At some point O in the shaft axis, pass a transverse plane A_0 (called the reference plane, or briefly, the R.P.). At O in the reference plane introduce two equal and opposite forces F_1 parallel to the original force F_1 in plane A_1. The force F_1 at O and the opposite force F_1 at O_1 form a centrifugal couple whose moment is $m_1R_1\omega^2a_1$, where ω is the angular velocity of the shaft. Hence the single force F_1 acting at O_1 in plane A_1 may be replaced by an equal and parallel force acting at O in the reference plane and a couple whose moment is F_1a_1. Likewise the force F_2 acting at O_2 in plane A_2 may be replaced by an equal and parallel force F_2 acting in the reference plane and a couple whose moment is $F_2a_2 = m_2R_2\omega^2a_2$.

Therefore it is evident that a reference plane can be chosen arbitrarily, and the system of forces set up by the rotating system may be reduced to a system of concurrent forces acting in the reference plane and a system of couples acting in various axial planes. The forces, if not balanced, have a single resultant in the reference plane, and the couples, if not balanced, can be reduced to a single couple in some axial plane. Hence, in general, the system of forces, if not balanced, may be reduced to a single force and a single couple. The magnitude of the couple will depend on the position chosen for the reference plane.

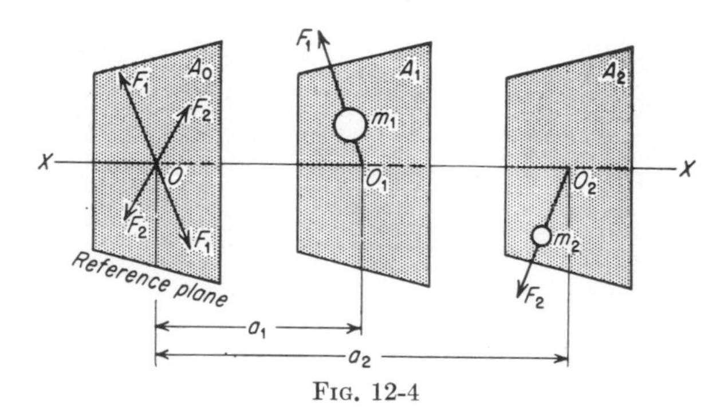

Fig. 12-4

The conditions of balance, then, for the system of rotating masses in different transverse planes are evidently the following:

1. The vector sum of all the centrifugal forces must be zero. That is,

$$\overrightarrow{\Sigma m R \omega^2} = 0 \tag{12-2}$$

2. The vector sum of the moments of all the forces with respect to any arbitrarily chosen reference plane must be zero. That is,

$$\overrightarrow{\Sigma m R \omega^2 a} = 0 \tag{12-3}$$

In order to balance such a system, at least two additional masses in different transverse planes are required. That is, the forces must be balanced by a mass in the reference plane, and the moments by a mass situated in some other transverse plane.

It should be mentioned that it is not always necessary, or desirable, to restrict the balancing masses to only two planes, although this is a minimum number for dynamic balance. In long rotors, it is sometimes desirable to add masses in *several* planes, to minimize bending moments. This topic will not be elaborated here, but it is evident that a resultant balancing force can be replaced by two (or more) forces in planes situated on either side of the plane of the balancing force, provided that the resultant of these forces is identical with the original force which is replaced. Thus the balancing forces can be distributed along the shaft.

12-7. General Graphical Method of Balancing Any Number of Rotating Masses. Referring to Fig. 12-5, let it be required to balance the system

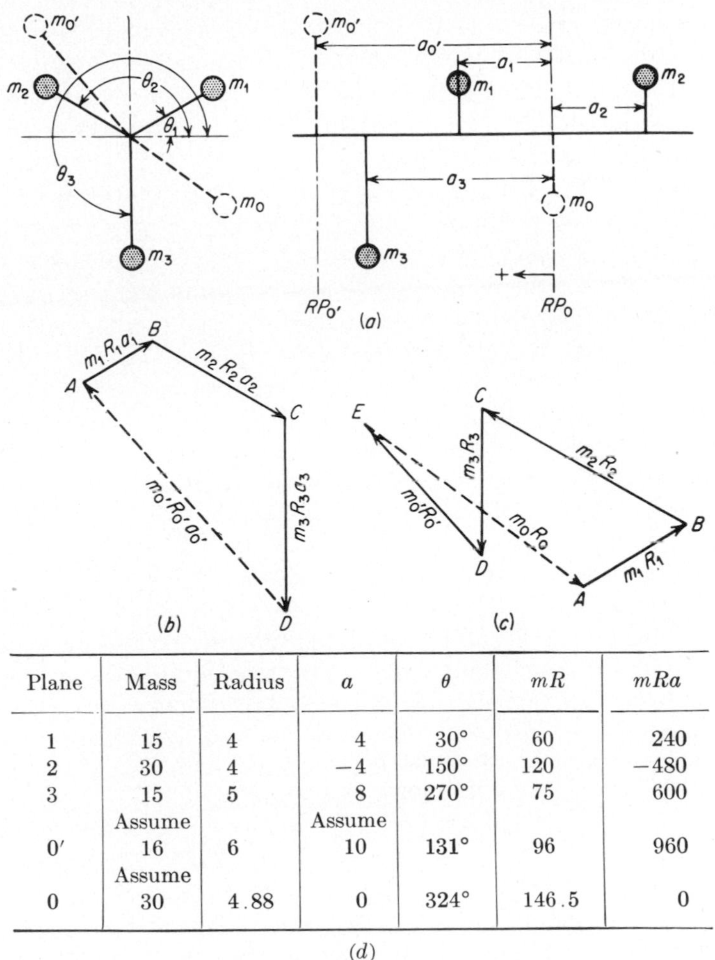

Plane	Mass	Radius	a	θ	mR	mRa
1	15	4	4	30°	60	240
2	30	4	−4	150°	120	−480
3	15	5	8	270°	75	600
	Assume		Assume			
0′	16	6	10	131°	96	960
	Assume					
0	30	4.88	0	324°	146.5	0

(*d*)

FIG. 12-5

composed of the three rotating masses m_1, m_2, and m_3. The solution consists merely in applying the conditions of balance developed in the preceding article. These conditions are represented by the equations

$$\overrightarrow{\Sigma m R \omega^2} = 0$$

$$\overrightarrow{\Sigma m R \omega^2 a} = 0$$

or, if the common term ω^2 be dropped,

$$\overrightarrow{\Sigma mR} = 0$$
$$\overrightarrow{\Sigma mRa} = 0 \qquad (12\text{-}4)$$

Two methods of solving the problem will be given, the first of which will be made to depend upon both the force and the moment relations, and the second of which will depend upon the moment relation alone. In both methods, the planes in which the balancing masses are to be added are shown as RP_0 and $RP_{0'}$.

1. *First Method.* Let the masses in slugs and distances in feet of the system, "m_1, m_2, m_3," be given in the table shown at d in the figure, and let the reference planes be chosen as shown. Let the balancing masses be m_0 and $m_{0'}$ and let the mass m_0 be situated in the reference plane RP_0. The left side of this reference plane is taken as the positive side.

The first step is to calculate the terms mR and mRa for the three masses m_1, m_2, and m_3. This has been done, and the numerical values of these quantities have been set down in the table. Let it be noted that since the mass m_2 is on the negative side of the reference plane, its moment is negative.

The next step is to draw the moment polygon which expresses the fact that the sum of moments with respect to the plane RP_0 is zero. The polygon is shown at b in the figure. On the positive side of the reference plane, the directions of the moment vectors are the same as those of the corresponding forces, namely, outward from the axis of the shaft and parallel to the corresponding radii. The moment $m_2R_2a_2$, being negative, is represented by a vector having a direction opposite that of the centrifugal force of the mass m_2. The unknown balancing mass m_0, being situated in the reference plane, has zero for a moment. The moment $m_{0'}R_{0'}a_{0'}$ of the unknown balancing mass $m_{0'}$ is given by the closing side of the polygon. Assuming that $m_{0'}$ is situated on the left-hand (positive) side of the reference plane, the direction of its radius will be the same as that of the moment vector $m_{0'}R_{0'}a_{0'}$. Let it also be assumed that its distance to the left of the reference plane is 10 ft. The quantity $m_{0'}R_{0'}$ may then be calculated as shown, and if it is decided to use a balancing mass of 16 slugs, the proper radius to use is found to be 6 ft.

It is seen, therefore, that by applying the moment relation the characteristics of the balancing mass $m_{0'}$ have been determined. It remains to determine the characteristics of the balancing mass m_0 in the reference plane RP_0. This is accomplished by applying the force relation, namely, that the vector sum of *all* the centrifugal forces must be zero.

The quantities mR are first calculated for all the known masses including the balancing mass $m_{0'}$. The force polygon is then drawn as shown at c in the figure, the closing side of the polygon being the vector that represents the quantity m_0R_0. A convenient value of m_0 is then assumed, and the value of R_0 calculated as shown in the table.

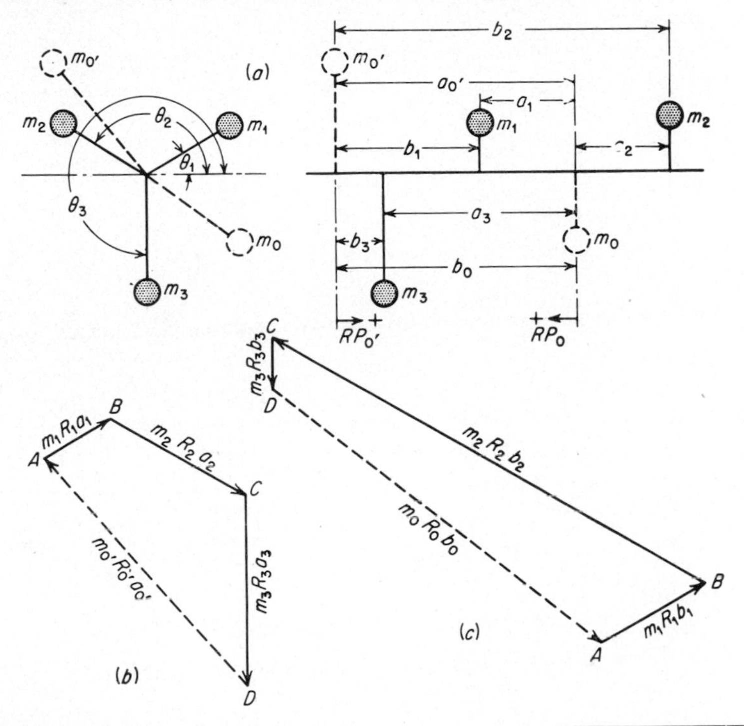

Plane	Mass	Radius	a	b	θ	mR	mRa	mRb
1	15	4	4	6	30°	60	240	360
2	30	4	−4	14	150°	120	−480	1,680
3	15	5	8	2	270°	75	600	150
0′	Assume 16	6	Assume 10	0	131°	96	960	0
0	Assume 30	4.88	0	10	324°	146.5	0	1,465

(d)

FIG. 12-6

2. *Second Method.* The second method for the solution of the problem (refer to Fig. 12-6) is exactly like the first up to the point where the moment polygon has been drawn and the characteristics of the balancing mass $m_{0'}$ determined. The remainder of the problem consists merely of taking moments with respect to the plane of the balancing mass

$m_{0'}$ and drawing a second moment polygon as shown at c in the figure. In this case the right side of the plane 0′ has been taken as the positive side. The closing side of the polygon is the vector $m_0R_0b_0$ from which the characteristics of the balancing mass m_0 can be determined as shown in the table at d.

This method is analogous to the determination of the reactions of a simple beam by taking moments first with respect to one support and then with respect to the other. It is evident by comparing the tables in Figs. 12-5 and 12-6 that the results of the two methods are identical. The second method, however, has the advantage of requiring one vector less in the layout, and in addition it eliminates the necessity of remembering whether to draw the moment polygon or the force polygon first. It should be kept in mind that these are not the true moment and force polygons because of the omission of ω^2.

Also, it should be recognized that *any* units can be used in the mR and mRa terms provided that the *same* units are used consistently in any one problem. In the equilibrium equations $\Sigma m\vec{R} = 0$ and $\Sigma m\vec{Ra} = 0$, any unit conversion factors which are common to all terms can be factored out and brought outside the summation sign. Thus, in $\Sigma m\vec{R} = 0$, units of ounces and inches are just as proper as slugs and feet, and may often give numerical values which are more convenient for computations when unbalances are small.

12-8. Reduction of Masses to a Common Radius. It has been assumed in the discussion of the preceding articles that the masses have unequal radii. The centrifugal force $mR\omega^2$ is proportional to the product mR, and the individual factors may have any value provided their product remains constant. It is therefore possible to choose some convenient radius, say that of an engine crank, or unity in a general problem, and all the masses that have different radii can be reduced to masses having this common radius. In this case the factor R can be dropped and the reduced masses m_1, m_2, etc., can be used for the sides of the force polygon, and the reduced mass moments m_1a_1, m_2a_2, etc., become the sides of the moment polygon.

12-9. Analytical Method of Balancing a System of Rotating Masses. It is often convenient to use analytical methods in the solution of balancing problems instead of the graphical methods discussed in the preceding articles. In applying the analytical method to a system of rotating masses, the forces are resolved into vertical and horizontal components as shown in Fig. 12-7. Each group of components, as well as the two groups formed by taking the moments of the components with respect to any arbitrarily chosen reference plane, must be balanced separately. The conditions of balance are expressed by the following equations, the masses being reduced to a common radius:

$$\begin{aligned}
&\text{Horizontal forces: } (m_1 \cos \theta_1 + m_2 \cos \theta_2 + \cdots \\
&\qquad\qquad + m_n \cos \theta_n) R\omega^2 = 0 \\
&\text{Vertical forces: } (m_1 \sin \theta_1 + m_2 \sin \theta_2 + \cdots \\
&\qquad\qquad + m_n \sin \theta_n) R\omega^2 = 0 \\
&\text{Horizontal moments: } (m_1 a_1 \cos \theta_1 + m_2 a_2 \cos \theta_2 + \cdots \\
&\qquad\qquad + m_n a_n \cos \theta_n) R\omega^2 = 0 \\
&\text{Vertical moments: } (m_1 a_1 \sin \theta_1 + m_2 a_2 \sin \theta_2 + \cdots \\
&\qquad\qquad + m_n a_n \sin \theta_n) R\omega^2 = 0
\end{aligned} \tag{12-5}$$

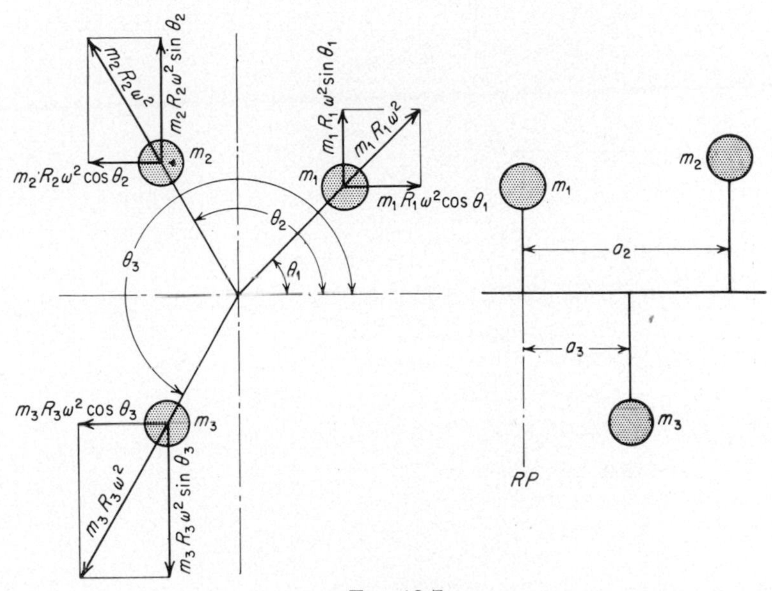

FIG. 12-7

Dropping the common factor $R\omega^2$, the four equations may be written as follows:

$$\Sigma m \cos \theta = 0 \tag{12-6a}$$

$$\Sigma m \sin \theta = 0 \tag{12-6b}$$

$$\Sigma ma \cos \theta = 0 \tag{12-6c}$$

$$\Sigma ma \sin \theta = 0 \tag{12-6d}$$

With Eqs. (12-6*a*) and (12-6*b*) satisfied, the system will be in static balance; but for running balance, all four equations must be satisfied. The significance of either of the moment equations not being satisfied is that a centrifugal couple exists that will set up vibrations when the system is rotating but will disappear and not affect the equilibrium when the system is standing still.

As brought out in the preceding articles, the characteristics of the masses required to balance any rotating system may be determined either by applying both the moment and force relations, or by applying the moment relation twice; i.e., either by using all four of the preceding equations or by making use of Eqs. (12-6*c*) and (12-6*d*) twice. It should be kept in mind that the true moments and forces are not involved because of the omission of $R\omega^2$.

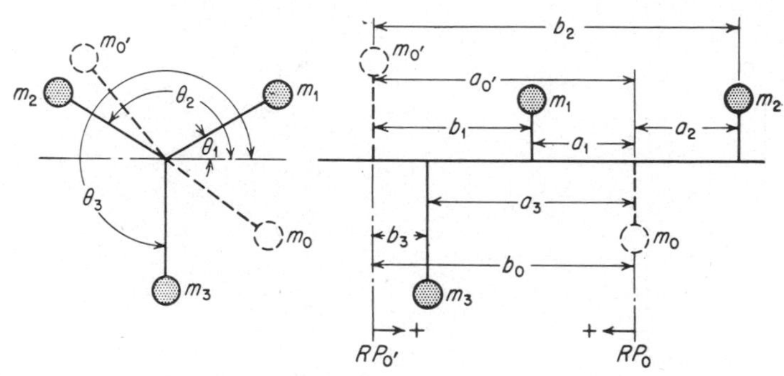

Plane	m	θ	a	ma	ma $\cos\theta$	ma $\sin\theta$	b	mb	mb $\cos\theta$	mb $\sin\theta$
1	60	30°	4	240	207.8	120	6	360	311.7	180
2	120	150°	−4	−480	415.7	−240	14	1,680	−1,455	840
3	75	270°	8	600	0	−600	2	150	0	−150
0′	[95.25]	[130.9°]	10	[952.5]	[−623.5]	[720]	0	0	0	0
0	[143.6]	[322.7°]	0	0	0	0	10	[1,436]	[1,143]	−870

$$\tan\theta_0{}' = \frac{720}{623.5} \qquad \tan\theta_0 = \frac{870}{1{,}143}$$

FIG. 12-8

As an example of the applications of the analytical method, let it be required to balance the system composed of the three rotating masses shown in Fig. 12-8. This system is the same as that shown in Figs. 12-5 and 12-6, except that the masses have been reduced to a unit radius. The second of the two schemes outlined above is used in the solution of the problem. The method of calculation is outlined in the table. The unknown quantities that were determined by taking moments are enclosed in brackets. The horizontal and vertical components of $m_{0'}a_{0'}$ were obtained from

$$\begin{aligned} m_{0'}a_{0'}\cos\theta_{0'} &= -\Sigma ma\cos\theta \\ m_{0'}a_{0'}\sin\theta_{0'} &= -\Sigma ma\sin\theta \end{aligned} \qquad (12\text{-}7)$$

where $\Sigma ma \cos\theta$ is written for the algebraic sum of the moments of the horizontal components of the three original masses with respect to the plane 0; and $\Sigma ma \sin\theta$ for the moments of the vertical components. The quadrant of the angle $\theta_{0'}$ was determined from the signs of $m_{0'}a_{0'} \cos \theta_{0'}$ and $m_{0'}a_{0'} \sin \theta_{0'}$, after which its numerical value was calculated from the tangent as indicated. The quantity $m_{0'}a_{0'}$ was obtained from

$$m_{0'}a_{0'} = \sqrt{(m_{0'}a_{0'} \cos \theta_{0'})^2 + (m_{0'}a_{0'} \sin \theta_{0'})^2} \tag{12-8}$$

The characteristics of the mass m_0 were determined in a similar manner.

12-10. The Function of Balancing Machines. The discussion of the preceding pages has shown that a rigid rotor having any number of unbalanced rotating masses can be completely balanced by the addition of one mass in each of two arbitrarily chosen transverse planes. It follows that the total unbalance present can be considered to be only two equivalent unbalanced masses—one in each of the chosen planes. For an actual rotor, then, the problem of balancing involves:

1. The selection of the two planes in which it is deemed convenient to add balancing weights (or *remove* unbalance).
2. The determination of the amount and location of the equivalent unbalanced mass in each of the selected planes, these two masses to represent the total of all the unbalance actually present.

It is the function of balancing machines to accomplish the determination of item 2, and so to determine the amount and location of the two additional masses required to balance the rotor in question. Balancing machines are discussed in more detail later in this chapter.

12-11. The Balancing of Reciprocating Masses. The discussion in the preceding articles shows that it is always possible to balance a system of rotating masses by means of two properly placed masses in any two desired planes of rotation. The following articles deal with the more complex problems of balancing masses that do not revolve in a circular path, but have either a motion of translation at variable speed, such as the piston of an engine, or else a swinging motion such as the connecting rod of an engine.

In an engine mechanism, when the motion of the crankpin is given, the corresponding acceleration of the piston can be found for any phase of the mechanism. A certain amount of force is required to produce this acceleration, i.e., to overcome the inertia of the piston. This force must be taken from or added to the total fluid pressure acting on the piston, to find the force transmitted along the connecting rod to the crank, accordingly as the acceleration is positive or negative. Hence there arises an unbalanced force (inertia force) equal and opposite to the force required to accelerate the piston, and, since this force must be absorbed by the frame and foundation, the effect is to set up vibrations.

The effect of the accelerating force on the piston may be compared with that of the motion of a shot from a gun. The acceleration of the shot is always accompanied by the recoil of the gun. Similarly, if the engine frame were free to move in the direction of the stroke, a recoil would take place just as if the piston were a shot and the cylinder a gun. It should be distinctly borne in mind that the recoil of the frame depends only upon the acceleration of the reciprocating mass and has nothing to do with the agent that causes the acceleration. That is, for a given acceleration of the reciprocating mass, the recoil of the framework is the same, whether the motion is due to the action of steam pressure on the piston, or to the explosion of a gas in the cylinder, or whether the turning effort on the crankshaft is accomplished by an outside agency as when the mechanism is that of an air compressor or pump.

12-12. Inertia Effects of the Reciprocating Masses in the Engine Mechanism. It was shown in the chapter on Inertia Forces that the

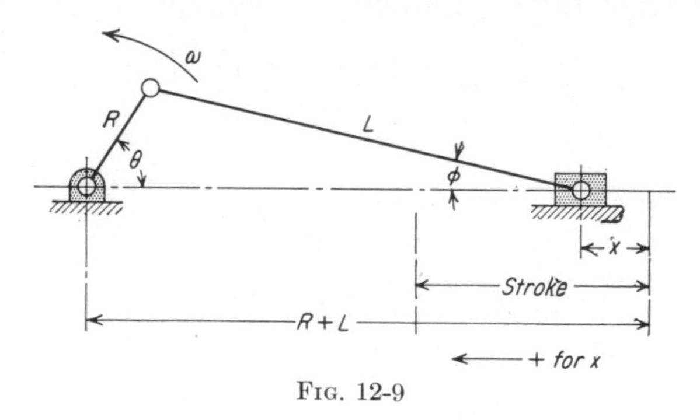

FIG. 12-9

inertia force of the reciprocating masses of the engine mechanism is given with very fair accuracy by the relation

$$F = mA = mR\omega^2 \left(\cos\theta + \frac{R}{L}\cos 2\theta\right)$$

where m and A are the mass and acceleration, respectively, of the reciprocating parts, and ω is the angular velocity of the crank. The various steps by means of which this relation was deduced will be repeated here in order to bring all the steps together.

Referring to Fig. 12-9, the displacement of the piston is given by

$$\begin{aligned} x &= R + L - R\cos\theta - L\cos\phi \\ &= R(1 - \cos\theta) + L(1 - \cos\phi) \end{aligned} \tag{12-9}$$

The cosine of the angle ϕ is given by

$$\cos\phi = \sqrt{1 - \sin^2\phi} = \sqrt{1 - \frac{R^2}{L^2}\sin^2\theta}, \text{ since } \sin\phi = \frac{R}{L}\sin\theta$$

The quantity under the radical may be expanded as follows:[1]

$$\sqrt{1 - \frac{R^2}{L^2}\sin^2\theta} = 1 - \frac{1}{2}\left(\frac{R}{L}\sin\theta\right)^2 - \frac{1}{2\times 4}\left(\frac{R}{L}\sin\theta\right)^4 - \frac{1\times 3}{2\times 4\times 6}\left(\frac{R}{L}\sin\theta\right)^6 \cdots \tag{12-10}$$

Substituting in Eq. (12-9), the displacement x of the piston becomes

$$x = R(1 - \cos\theta) + R\left(\frac{1}{2}\frac{R}{L}\sin^2\theta + \frac{1}{2\times 4}\frac{R^3}{L^3}\sin^4\theta + \frac{1\times 3}{2\times 4\times 6}\frac{R^5}{L^5}\sin^6\theta \cdots\right)$$

Since R/L is usually less than about 1/3.5 the terms beyond the square may be dropped without appreciable error. Then

$$x = R(1 - \cos\theta) + \frac{1}{2}\frac{R^2}{L}\sin^2\theta \qquad \text{approx} \tag{12-11}$$

Differentiating with respect to time,

$$\begin{aligned}\frac{dx}{dt} &= R\sin\theta\frac{d\theta}{dt} + \frac{1}{2}\frac{R^2}{L}2\sin\theta\cos\theta\frac{d\theta}{dt}\\ &= R\omega\left(\sin\theta + \frac{1}{2}\frac{R}{L}\sin 2\theta\right)\end{aligned} \tag{12-12}$$

where ω has been written for $d\theta/dt$ and the quantity $2\sin\theta\cos\theta$ has been replaced by its equal $\sin 2\theta$.

Differentiating a second time, assuming ω to be constant,

$$\begin{aligned}\frac{d^2x}{dt^2} &= R\omega\left(\cos\theta\frac{d\theta}{dt} + \frac{1}{2}\frac{R}{L}2\cos 2\theta\frac{d\theta}{dt}\right)\\ &= R\omega^2\left(\cos\theta + \frac{R}{L}\cos 2\theta\right)\end{aligned} \tag{12-13}$$

The magnitude of the inertia force due to a mass m having this acceleration is

$$F = m\frac{d^2x}{dt^2} = mR\omega^2\left(\cos\theta + \frac{R}{L}\cos 2\theta\right) \tag{12-14}$$

When this expression is positive, the inertia force will be directed away from the main bearing, and when it is negative, toward the main bearing (note positive direction for *motions* in Fig. 12-9).

[1] By use of the binomial theorem.

The expression for F may be written as follows:

$$F = mR\omega^2 \cos\theta + m\left(\frac{R^2}{4L}\right)(2\omega)^2 \cos 2\theta \qquad (12\text{-}15)$$

The term $mR\omega^2 \cos\theta$ is called the *primary* inertia force and would be the total inertia force if the connecting rod were infinite in length. The term $m(R^2/4L)(2\omega)^2 \cos 2\theta$ (or $mR\omega^2 R/L \cos 2\theta$) is called the *secondary* inertia force and is introduced (along with higher-order terms which are neglected here) as a result of the fact that the connecting rod is actually of finite length. It corresponds to the case of a mass m, an infinite connecting rod, and a crank of length $R^2/4L$ turning at a speed of 2ω.

Another common type of analysis makes use of the Fourier series, and leads to the following expression, which gives the inertia force of the reciprocating mass m as an infinite series of *harmonics*.

$$F = mR\omega^2(A \cos\theta + B \cos 2\theta + C \cos 4\theta + D \cos 6\theta + \cdots) \qquad (12\text{-}16)$$

It is noticed that only even harmonics occur (after the first). In this equation, the coefficient A is unity, and B, C, D, etc., each consists of an infinite series involving only R/L and constants. For example,

$$B = \frac{R}{L} + \frac{1}{4}\left(\frac{R}{L}\right)^3 + \frac{15}{128}\left(\frac{R}{L}\right)^5 + \cdots$$

$$C = -\left[\frac{1}{4}\left(\frac{R}{L}\right)^3 + \frac{3}{16}\left(\frac{R}{L}\right)^5 + \frac{35}{256}\left(\frac{R}{L}\right)^7 + \cdots\right]$$

Table 12-1 lists a few values in the range of practical interest.

TABLE 12-1. FOURIER COEFFICIENTS

Harmonic	Coefficient	$\frac{R}{L} = \frac{1}{3}$	$\frac{R}{L} = \frac{1}{4}$	$\frac{R}{L} = \frac{1}{5}$
First	A	1.0	1.0	1.0
Second	B	0.3431	0.2540	0.2020
Fourth	C	−0.0101	−0.0041	−0.0021
Sixth	D	0.0003	0.00007	0.00002

Correspondingly, for the *primary* and *secondary* forces of the approximate equations (12-14) and (12-15),

$$F = mR\omega^2\left(\cos\theta + \frac{R}{L}\cos 2\theta\right)$$

Inspection of Tables 12-1 and 12-2 shows that the primary force is truly the first harmonic given by the Fourier series but that the secondary force is not precisely the second harmonic. Thus the secondary force should *not* be called the second *harmonic*, although the numerical difference is small for the small values of R/L which are commonly used. The tables also show that higher harmonics (above the second) can be omitted with only small error.

TABLE 12-2. PRIMARY AND SECONDARY FORCE COEFFICIENTS

Force	Coefficient	$\frac{R}{L} = \frac{1}{3}$	$\frac{R}{L} = \frac{1}{4}$	$\frac{R}{L} = \frac{1}{5}$
Primary	Of $\cos \theta$	1.0	1.0	1.0
Secondary	Of $\cos 2\theta$	0.3333	0.2500	0.2000

It is interesting to note that lengthening the connecting rod reduces the importance of the harmonics above the first. In the limit, when L becomes infinite only the first harmonic remains, and the piston moves with simple harmonic motion, as already mentioned.

It should be noted that both the primary and secondary inertia forces may be considered as due to a mass m having harmonic motion. For the primary inertia force, the amplitude of the harmonic motion is equal to the length of the crank, and the period is the time required for the crank to turn through one complete revolution. For the secondary force, the amplitude is equal to the length $R^2/4L$ of an imaginary crank rotating at twice the speed of the actual crank; the period, therefore, is half that corresponding to the primary force.

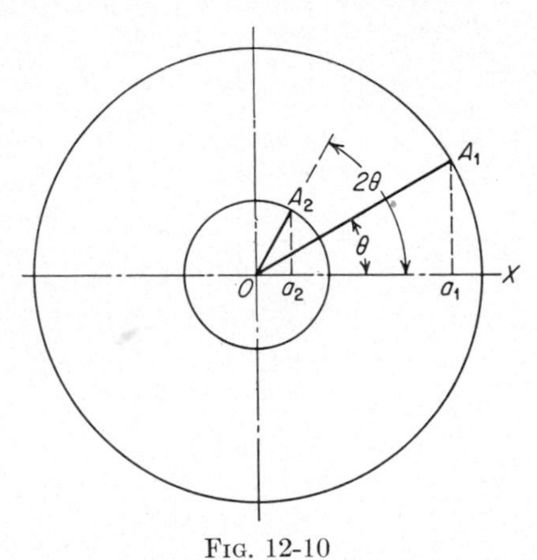

FIG. 12-10

The above inertia forces may be determined graphically for any position of the crank as follows: First, let a suitable force scale be chosen and then, referring to Fig. 12-10, let two concentric circles be drawn, one with a radius $OA_1 = mR\omega^2$ and the other with a radius $OA_2 = m(R^2/4L)(2\omega)^2$. Now let the crank angle θ be laid off as shown and let the line OA_2 be laid off making an angle 2θ with the horizontal line OX. Then if the points A_1 and A_2 are projected on the horizontal line OX the lines Oa_1 and Oa_2 will represent the primary and secondary inertia forces, respectively.

The relation between primary and secondary inertia forces will be made clear by the following example. For a six-cylinder gasoline engine the length of crank = $2\frac{1}{8}$ in.; length of connecting rod = $8\frac{1}{8}$ in.; weight of piston = 1.94 lb; and rpm of crankshaft = 2,500.

The primary inertia forces for a single cylinder have been calculated for each 30-deg crank position and plotted as the harmonic curve A in

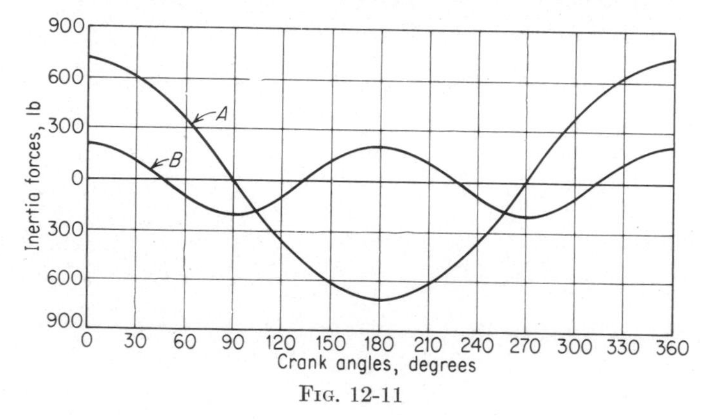

FIG. 12-11

Fig. 12-11. Similarly, the secondary inertia forces have been calculated and plotted as the harmonic curve B. It will be observed that the two curves are in phase at the head-end dead center, but that curve B has twice the frequency of curve A.

The inertia forces of the reciprocating masses are transmitted to the frame in the following manner, illustrated in Fig. 12-12. The inertia

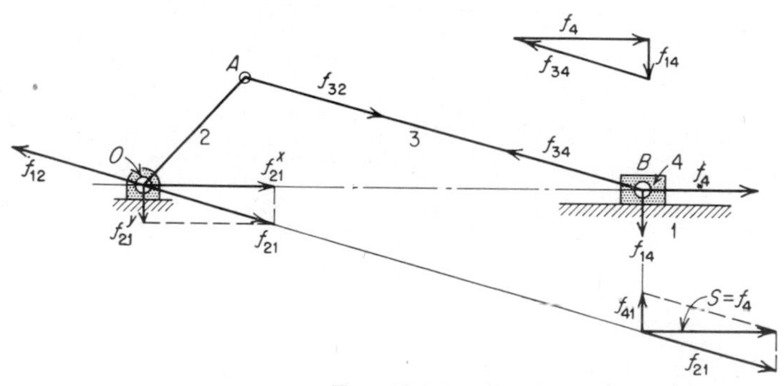

FIG. 12-12

force f_4 sets up the reactions f_{14} and f_{34}, the magnitudes of the latter being given by the polygon shown. The force f_{14} is the side pressure exerted upon the piston by the cylinder walls. An equal and opposite force f_{41} is exerted by the piston upon the cylinder walls, and this force tends to shake the frame. The reaction f_{34} is transmitted along the connecting rod and finally appears at the main bearing as f_{21}. The resultant shaking force S is obtained by combining f_{41} and f_{21}. The magnitude and direction of the force S are the same as those of the inertia force f_4. The line of action of S, however, is offset as indicated (except for the dead-center positions of the crank).

Instead of combining f_{21} and f_{41} to determine the resultant shaking force S, let the force f_{21} be resolved into horizontal and vertical components at the center of the main bearing O. The horizontal component f_{21}^x is equal to f_4 and has the same line of action. The vertical component f_{21}^y forms, with the force f_{41}, a couple that tends to rotate the frame of the engine about the crankshaft. The effect of this couple upon the frame is usually considered along with the other torque effects, such as that of the static turning effort, and it will therefore be disregarded in discussing the balancing of engines. This being the case, the inertia effect of the reciprocating masses upon the frame may be considered as equivalent to a force f_{21}^x, which is applied at the main bearing and has the same magnitude and direction as the inertia force f_4.

12-13. Inertia Effects of Crank and Connecting Rod. The inertia effect of the crank is a pure centrifugal force (assuming a uniform speed of rotation) and it may be calculated by multiplying the mass of the crank by the acceleration of its center of gravity. In analyzing the shaking forces of the engine, it is customary to replace the mass of the crank by an equivalent mass at the crankpin. Since the inertia force of the crank is a pure centrifugal force, it may be completely balanced by placing a counterbalancing mass of the proper characteristics opposite the crank.

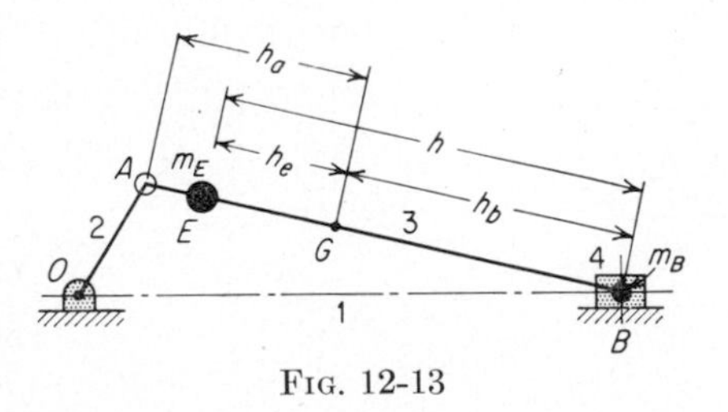

FIG. 12-13

The inertia effects of the connecting rod may be studied by replacing it by a kinetically equivalent system, i.e., by assuming that the mass of the connecting rod is concentrated at two points, at the wrist pin and, referring to Fig. 12-13, at the center of percussion E of the connecting rod with respect to the wrist-pin center. If the position of the center of gravity G of the rod is known, and if the moment of inertia of the rod with respect to G is also known, then the position of the center of percussion E may be calculated as follows (see Chap. 11 on Inertia Forces):

$$h_e = \frac{I_G}{m_3 h_b}$$

where I_G and m_3 are, respectively, the moment of inertia and mass of the rod. The masses assumed concentrated at B and E are given by

$$m_E = \frac{h_b}{h} m_3 \qquad \text{and} \qquad m_B = \frac{h_e}{h} m_3$$

The horizontal and vertical components of the inertia force of the mass concentrated at E are as follows (refer to Art. 11-15):

$$
\begin{aligned}
f_E^x &= m_E R\omega^2 \left[\cos\theta + \left(1 - \frac{h}{L}\right)\frac{R}{L}\cos 2\theta\right] \\
&= m_3 \frac{h_b}{h} R\omega^2 \left[\cos\theta + \left(1 - \frac{h}{L}\right)\frac{R}{L}\cos 2\theta\right] \qquad (12\text{-}17) \\
f_E^y &= m_E R\omega^2 \frac{h}{L}\sin\theta = m_3 \frac{h_b}{L} R\omega^2 \sin\theta
\end{aligned}
$$

The vertical component of the inertia force of the mass concentrated at B is zero. The horizontal component is

$$
\begin{aligned}
f_B^x &= m_B R\omega^2 \left(\cos\theta + \frac{R}{L}\cos 2\theta\right) \\
&= m_3 \frac{h_e}{h} R\omega^2 \left(\cos\theta + \frac{R}{L}\cos 2\theta\right)
\end{aligned} \qquad (12\text{-}18)
$$

The inertia forces determined by means of the above expression are exact in so far as concerns mass distribution, the only error being that introduced by the use of the approximate expression for the horizontal components of the accelerations at E and B.

Although a rigid analysis would require the mass of the connecting rod to be distributed as shown in Fig. 12-13, a convenient and much used approximate method of taking into account the inertia effect of the connecting rod is to assume that its mass is concentrated at the crankpin and wrist pin so that

$$m_A + m_B = m_3 \qquad \text{and} \qquad m_A h_a = m_B h_b$$

that is,

$$m_A = \frac{h_b}{L} m_3 \qquad \text{and} \qquad m_B = \frac{h_a}{L} m_3$$

This mass distribution might be termed a *pseudo kinetically equivalent system.* Then, in determining the inertia forces of the mechanism, the mass assumed concentrated at the wrist pin is simply added to the other reciprocating masses; and likewise the mass assumed concentrated at the crankpin is added to the crank mass.

The magnitude and direction of the resultant inertia force of the rod as obtained by assuming the mass concentrated at the crank and wrist pins are identical with the magnitude and direction as obtained using the precise kinetically equivalent system. The line of action, however, is different because of the change in I_G.

12-14. The Single-cylinder Engine. Counterbalancing. The inertia force of the crank and that part of the connecting rod mass which, by the approximate method, may be considered as concentrated at the crankpin can always be balanced by placing a mass of the proper characteristics opposite the crankpin. *Consideration of these forces will therefore be omitted in this and in the following articles unless otherwise stated.*

The situation in the case of the inertia forces of the *reciprocating* masses is altogether different; in the single-cylinder engine they cannot be balanced at all by a simple counterbalance opposite the crankpin. They can be *modified*, however, by this process of counterbalancing. The primary force is $mR\omega^2 \cos \theta$. If, as in Fig. 12-14, a counterbalancing mass m_0 is placed opposite the crank at a distance R_0 such that

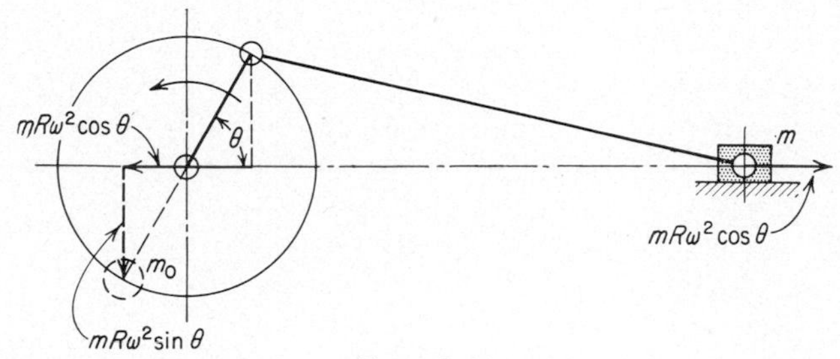

FIG. 12-14

$m_0R_0 = mR$, this mass will set up a centrifugal force equal to $mR\omega^2$ in a direction opposite the crank radius. If this force is resolved, at the center of the main bearing, into horizontal and vertical components, the result will be as follows:

$$\text{Horizontal force} = -mR\omega^2 \cos \theta$$
$$\text{Vertical force} = -mR\omega^2 \sin \theta$$

The primary shaking forces are thus balanced and vertical shaking forces substituted. Considering a complete revolution of the crank, the result is to shift the original primary shaking forces through 90 deg.

For engines of this type it is common practice to balance, by means of a revolving mass opposite the crankpin, all the revolving crank mass, including that part of the mass of the connecting rod assumed concentrated at the crankpin, and from one-half to two-thirds of the reciprocating mass. The balancing conditions will be made clear from the analysis of a typical case, as follows:

A single-crank gas engine of the stationary type (Fig. 12-15) runs at a speed of 230 rpm.

Length of crank = 10½ in.
Length of connecting rod = 52½ in.
Distance of center of gravity of connecting rod from crank end = 20.3 in.
Equivalent unbalanced weight of crank at 10½ in. radius = 162.5 lb
Weight of piston = 244 lb
Weight of connecting rod = 254.5 lb
Equivalent weight of counterbalance at 10½ in. radius = 533 lb

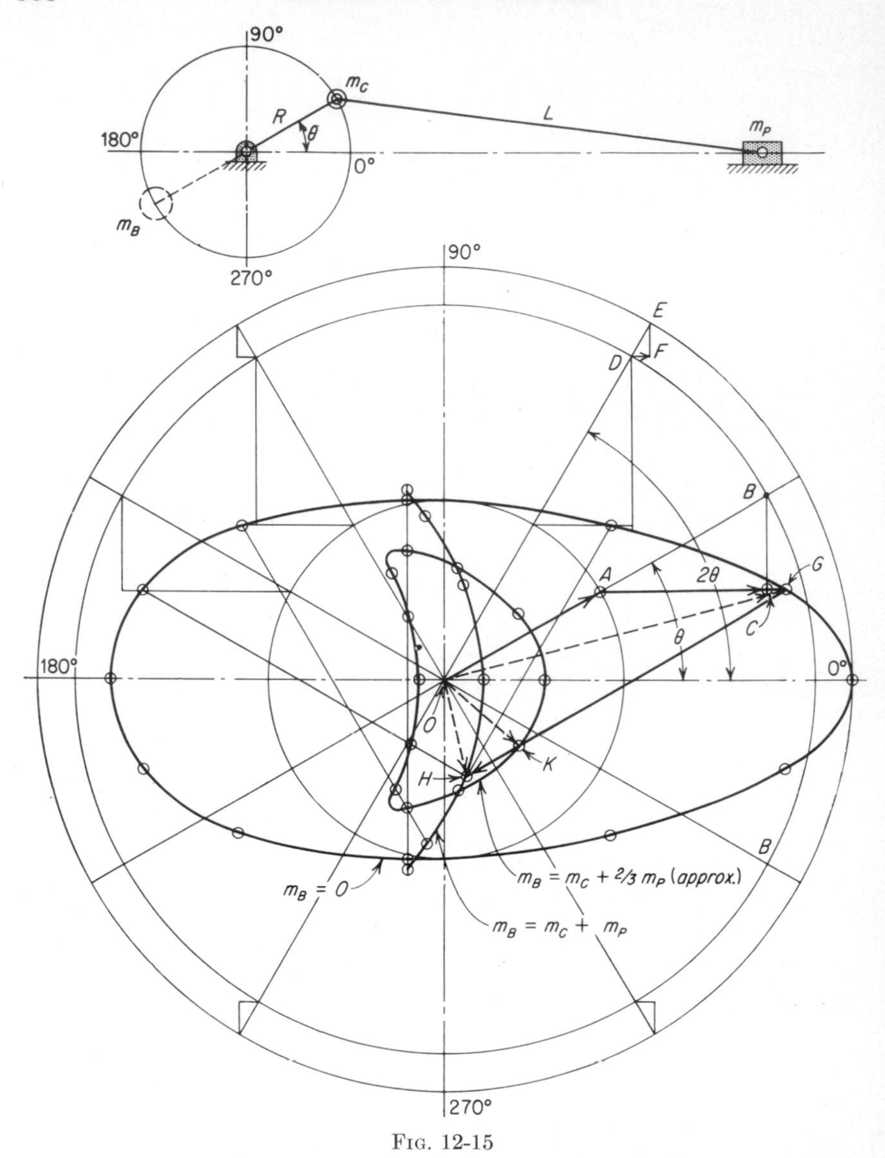

FIG. 12-15

From Art. 12-13 the proportion of the weight of the connecting rod that may be assumed to be concentrated at the crankpin is

$$\frac{32.2 \times 254.5}{52.5} = 156 \text{ lb}$$

The remainder, $254.5 - 156 = 98.5$ lb, is assumed to be concentrated at the crosshead pin. The total revolving crankpin weight at $10\frac{1}{2}$ in.

radius is then 162.5 + 156 = 318.5 lb, and the total reciprocating weight is 244 + 98.5 = 342.5 lb.

The total inertia forces are as follows:

For the revolving crank mass,

$$F_C = m_C R\omega^2 = 5{,}030 \text{ lb}$$

For the reciprocating mass (primary),

$$F'_P = m_P R\omega^2 \cos\theta = 5{,}400 \cos\theta \text{ lb}$$

For the reciprocating mass (secondary),

$$F''_P = m_P \frac{R^2}{4L} (2\omega)^2 \cos 2\theta = 1{,}080 \cos 2\theta \text{ lb}$$

For the counterbalancing mass,

$$F_B = m_B R\omega^2 = 8{,}400 \text{ lb}$$

It will be convenient, in studying the effect of the inertia forces, to plot curves (Fig. 12-15) showing the resultant inertia forces for each 30-deg position of the crank.

From a pole O (Fig. 12-15), draw the vector OA, representing to scale the inertia force (= 5,030 lb) of the crank mass. Next draw AB (= 5,400 lb). Then AC (= 5,400 cos θ) is the primary inertia force. If DE is now drawn to represent 1,080 lb, DF will represent 1,080 cos 2θ, the secondary force. Lay off $CG = DF$. OG is the resultant inertia force for this position of the crank (θ = 30 deg), and G is one point on the curve of inertia forces for the condition where the engine is not counterbalanced. The vector equation for this diagram is

$$F_C \nrightarrow F'_P \nrightarrow F''_P = F_S$$

where F_S is the resultant shaking force exerted on the frame of the engine.

The vector equation with counterbalance force F_B included is

$$F_C \nrightarrow F'_P \nrightarrow F''_P \nrightarrow F_B = F_S$$

The effect of using a counterbalancing weight equal to the sum of the crank weight and the reciprocating weight may be shown by bringing into the force polygon $OACG$ the vector $GH = OB$, giving a new resultant OH. The point H is then one point on the curve of inertia forces for this balancing condition. The curve clearly shows that, although the horizontal forces have been largely neutralized (only the secondary horizontal forces remain), rather large vertical forces have been substituted.

The effect of the actual counterbalancing weight used on the engine is shown by the curve correspondingly indicated in the figure. The point K on this curve for the 30-deg position of the crank is obtained by laying off $GK = 8{,}400$ lb. The counterbalancing weight is approximately equal to the rotating weight plus two-thirds of the reciprocating weight.

The circles of radii OA, OB, and OE have been drawn for convenience in making the graphical constructions for the remaining points on the the curves that have just been discussed.

It may be observed from the figure that these curves are symmetrical with respect to the horizontal axis. If the secondary forces were neglected, the resulting curves would be symmetrical with respect to the vertical axis also.

For *completely eliminating* the shaking forces, the scheme shown in Fig. 12-16 has been used in single-cylinder laboratory engines. The dummy balancing pistons exert a resultant inertia force always equal and opposite that of the engine piston and its associated reciprocating portion of the connecting rod.

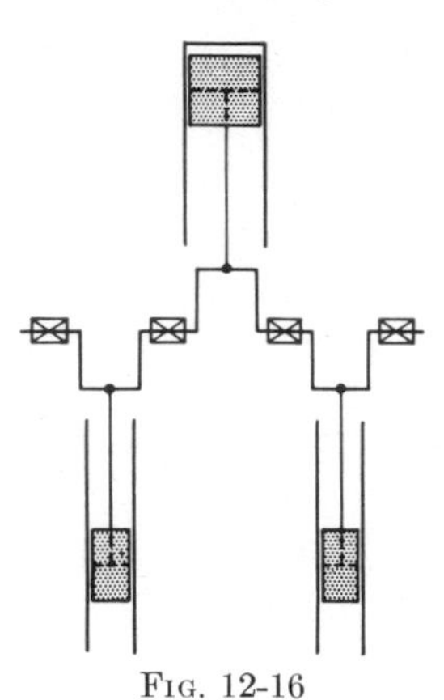

Fig. 12-16

It is possible also to make use of *pairs* of contrarotating weights, one pair rotating at crankshaft speed, a second pair at twice crankshaft speed, etc., to correspond to the first harmonic, second harmonic, etc., of the reciprocating mass at the piston. Each pair can be arranged so that the components of centrifugal force which lie *along* the direction of piston motion add to counteract the particular harmonic of the inertia force of the reciprocating mass, while the components *at right angles* oppose and thus cancel each other.

12-15. Conditions of Balance in a Multicylinder Engine. In an engine having more than one cylinder the rotating crank masses can easily be balanced, but the inertia forces of the reciprocating masses may or may not tend to neutralize each other, depending upon the arrangement of cranks and the proportions of the various links of the mechanism. In Fig. 12-17 the schematic outline of the three-cylinder engine has been shown, the inertia forces of the reciprocating masses being indicated by the arrows. These forces are transmitted through the connecting rods to the cranks and thence to the crankshaft. If the forces acting on the crankshaft are in equilibrium, no pressure due to the inertia of the reciprocating masses will be exerted on the main bearings; otherwise a force, a couple, or both will be exerted that will tend to shake the frame of the engine.

Assuming that the resultant shaking force due to the inertia of the reciprocating masses will pass through the axis of the crankshaft (i.e., disregarding the couple mentioned near the end of Art. 12-12), the problem of balancing the engine is reduced to the analysis of a system of vertical forces (for the case shown in Fig. 12-17), all of which are situated in the same axial plane.

In order to bring out the general conditions of balance for an engine

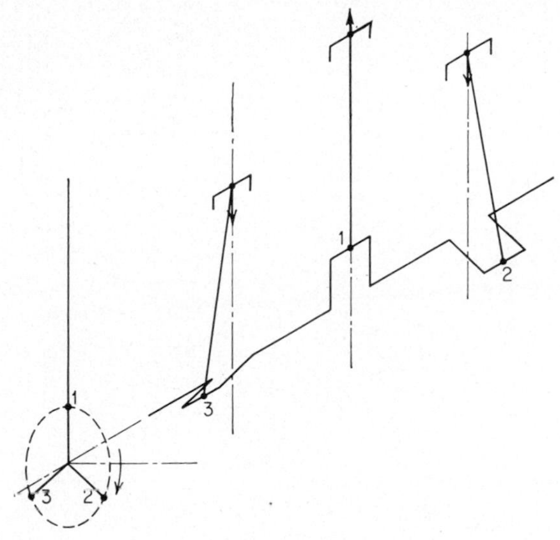

Fig. 12-17

of this type, let the case shown in Fig. 12-18 be considered. This is assumed to be an n-cylinder engine, although only four of the units are shown. The stroke line of each cylinder is assumed to be situated in a horizontal plane passing through the axis of the crankshaft, and the inertia forces of the reciprocating masses are of course also situated in

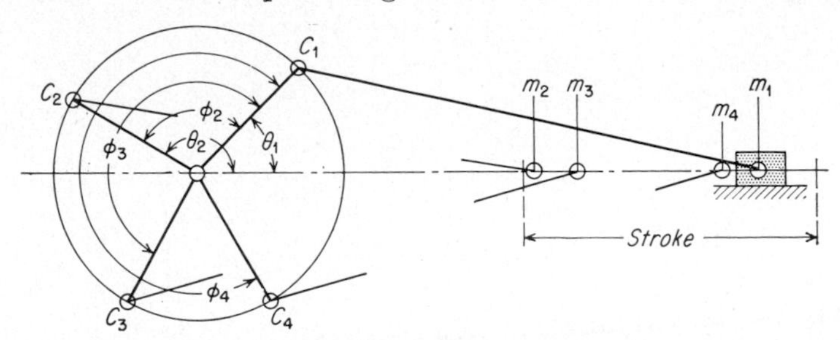

Fig. 12-18

this plane. It will be convenient to consider the equilibrium of the primary and secondary forces separately. Let R represent the crank, L the connecting rod, and m the reciprocating mass. The subscripts 1, 2, 3, etc., will refer to cylinders 1, 2, 3, etc.

1. The primary forces will be in complete equilibrium if their algebraic

sum is zero, and if the algebraic sum of their moments with respect to any transverse plane is also equal to zero. If F_p represents the *resultant* primary shaking force,

$$F_p = \Sigma mR\omega^2 \cos \theta$$
$$= m_1R_1\omega^2 \cos \theta_1 + m_2R_2\omega^2 \cos \theta_2 + \cdots \quad (12\text{-}19)$$

But (see Fig. 12-18) $\theta_2 = \theta_1 + \phi_2$, $\theta_3 = \theta_1 + \phi_3$, etc., where θ and ϕ are all measured in the same direction (counterclockwise in this instance). It is to be noted that the angle θ_1 is a variable whereas the values of the angles ϕ_2, ϕ_3, ϕ_4, etc., are constant, being merely the fixed angular positions of the various cranks relative to crank 1. The angle ϕ_1 is, of course, equal to zero. Rewriting Eq. (12-19),

$$F_p = \Sigma mR\omega^2 \cos (\theta_1 + \phi) \quad (12\text{-}20)$$

But $\cos (\theta_1 + \phi) = \cos \theta_1 \cos \phi - \sin \theta_1 \sin \phi$, and so

$$F_p = \Sigma mR\omega^2 \cos \theta_1 \cos \phi - \Sigma mR\omega^2 \sin \theta_1 \sin \phi \quad (12\text{-}21)$$
$$F_p = \omega^2 \cos \theta_1(\Sigma mR \cos \phi) - \omega^2 \sin \theta_1(\Sigma mR \sin \phi) \quad (12\text{-}22)$$

In order that the resultant primary force be zero for all values of θ_1, the quantities in the parentheses must be zero. Therefore, the criterion for balance of the primary forces in a multicylinder in-line engine is

$$\left.\begin{array}{l}\Sigma mR \cos \phi = 0\\ \Sigma mR \sin \phi = 0\end{array}\right. \quad (12\text{-}23)$$

Similarly,

$$\left.\begin{array}{l}\Sigma mRa \cos \phi = 0\\ \Sigma mRa \sin \phi = 0\end{array}\right. \quad (12\text{-}24)$$

where a is the distance from the center line of any cylinder to an arbitrarily chosen reference plane, the latter being perpendicular to the axis of the crankshaft.

Any engine for which Eqs. (12-23) and (12-24) are satisfied is said to be in primary balance.

It is to be noted that the preceding conditions of balance are precisely those of a rotating system, the masses of which are equal to the reciprocating masses and are situated at distances from the axis of rotation equal to the lengths of the cranks.

2. The secondary forces will be in complete equilibrium if their algebraic sum is equal to zero, and if the algebraic sum of their moments with respect to any transverse plane is also zero.

$$F_s = \sum mR\omega^2 \frac{R}{L} \cos 2\theta$$
$$= \sum mR\omega^2 \frac{R}{L} \cos 2(\theta_1 + \phi) \quad (12\text{-}25)$$

Using the identity $\cos (2\theta_1 + 2\phi) = \cos 2\theta_1 \cos 2\phi - \sin 2\theta_1 \sin 2\phi$,

$$F_s = \sum mR\omega^2 \frac{R}{L} \cos 2\theta_1 \cos 2\phi - \sum mR\omega^2 \frac{R}{L} \sin 2\theta_1 \sin 2\phi$$

$$= \omega^2 \cos 2\theta_1 \left(\sum \frac{mR^2}{L} \cos 2\phi \right) - \omega^2 \sin 2\theta_1 \left(\sum \frac{mR^2}{L} \sin 2\phi \right) \quad (12\text{-}26)$$

The criterion for balance of secondary forces is therefore

$$\sum \frac{mR^2}{L} \cos 2\phi = 0$$
$$\sum \frac{mR^2}{L} \sin 2\phi = 0 \quad (12\text{-}27)$$

Similarly,

$$\sum \frac{mR^2}{L} a \cos 2\phi = 0$$
$$\sum \frac{mR^2}{L} a \sin 2\phi = 0 \quad (12\text{-}28)$$

where a is the distance from the center lines of the cylinders to any transverse reference plane.

Any engine for which Eqs. (12-27) and (12-28) are satisfied is said to be in secondary balance.

12-16. The Eight Fundamental Equations of Balance. It is evident from the foregoing discussion that, for the complete balance of the primary and secondary forces and couples, eight equations must be satisfied. If we note that $\cos 2\phi = \cos^2 \phi - \sin^2 \phi$ and $\sin 2\phi = 2 \sin \phi \cos \phi$ and then let $x = \cos \phi$ and $y = \sin \phi$ (merely to reduce the amount of writing), these equations of balance may be written as follows:

Primary forces balanced:

$$\sum mRx = 0 \quad (\text{I})$$

$$\sum mRy = 0 \quad (\text{II})$$

Primary couples balanced:

$$\sum mRax = 0 \quad (\text{III})$$

$$\sum mRay = 0 \quad (\text{IV})$$

Secondary forces balanced:

$$\sum m \frac{R^2}{L} (x^2 - y^2) = 0 \quad (\text{V})$$

$$\sum m \frac{R^2}{L} xy = 0 \quad (\text{VI})$$

Secondary couples balanced:

$$\sum m \frac{R^2}{L} a(x^2 - y^2) = 0 \tag{VII}$$

$$\sum m \frac{R^2}{L} axy = 0 \tag{VIII}$$

These eight fundamental equations express completely[1] the analytical conditions of primary and secondary balance among the reciprocating parts for an engine with any number of cranks, the cylinders being arranged in a line on one side of the crankshaft with cylinder center lines all in the plane that contains the axis of the crankshaft.

Partial balance may be secured by satisfying part of the eight fundamental equations. It should be noted, however, that these equations must always be taken in pairs. That is, in order that the primary forces may be in balance *both* Eqs. (I) and (II) must be satisfied, and similarly for the other forces and couples.

Applications of the method will be illustrated in the articles which follow. In these applications the lengths of the connecting rods and cranks are assumed to be the same for all cylinders; and the reciprocating masses are all assumed to be equal. In this case the equations of balance can be written in the following form:

$$\begin{aligned} \Sigma x &= 0 & \Sigma(x^2 - y^2) &= 0 \\ \Sigma y &= 0 & \Sigma xy &= 0 \\ \Sigma ax &= 0 & \Sigma a(x^2 - y^2) &= 0 \\ \Sigma ay &= 0 & \Sigma axy &= 0 \end{aligned} \tag{12-29}$$

If, after the equations of balance have been applied, it is desired to determine the magnitudes of the unbalanced forces and couples for any position θ_1 of the mechanism, the following complete equations from Art. 12-15 may be applied:

$$\begin{aligned} F_p &= mR\omega^2(\cos\theta_1 \Sigma x - \sin\theta_1 \Sigma y) \\ C_p &= mR\omega^2(\cos\theta_1 \Sigma ax - \sin\theta_1 \Sigma ay) \\ F_s &= m\frac{R^2}{L}\omega^2[\cos 2\theta_1 \Sigma(x^2 - y^2) - 2\sin 2\theta_1 \Sigma xy] \\ C_s &= m\frac{R^2}{L}\omega^2[\cos 2\theta_1 \Sigma a(x^2 - y^2) - 2\sin 2\theta_1 \Sigma axy] \end{aligned} \tag{12-30}$$

where F_p, C_p, F_s, and C_s are, respectively, the unbalanced primary forces,

[1] Based upon the assumption that the inertia forces of the reciprocating masses are exerted at the main bearing and that the expression $F = mR\omega^2\left(\cos\theta + \frac{R}{L}\cos 2\theta\right)$ is sufficiently exact.

unbalanced primary couples, unbalanced secondary forces, and unbalanced secondary couples.

If an unbalanced force exists, its line of action is situated in the reference plane. If in addition, as is usually the case, an unbalanced couple also exists, it may be combined with the force to form a single force situated in some other transverse plane. It should be noted that, for an actual couple tending to tip the engine frame end over end to exist, the forces must be in balance. If the forces are not in balance, the couple equation merely shows the position of the line of action of the resultant shaking force (the force obtained by combining the unbalanced force and the unbalanced couple). In this case the unbalanced couple is simply the moment of the resultant force with respect to the reference plane. If for any position of the crank, however, the force becomes zero without the couple also becoming zero, then a true couple is exerted on the engine frame tending to tip it end over end; this is equivalent to the resultant shaking force (the resultant of the zero force and the couple) being situated in a plane at an infinite distance from the reference plane.

12-17. The Two-cylinder Engine. Consider Fig. 12-19. The cranks are at 180 deg. The *quantity* $mR\omega^2$ *is assumed equal to one*, and the

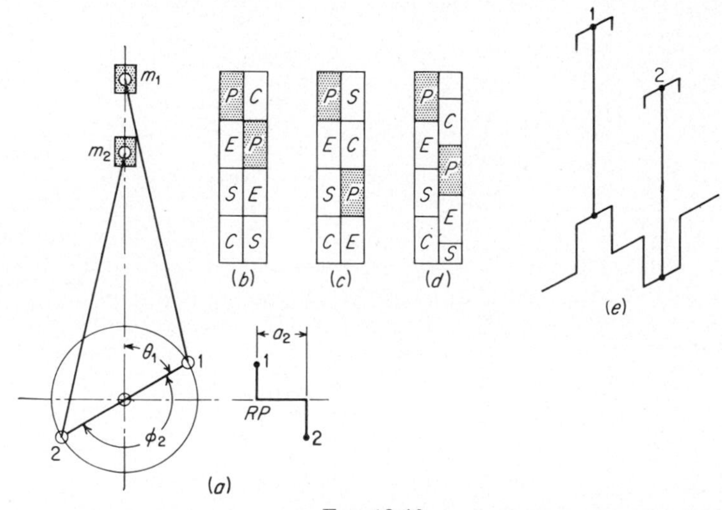

FIG. 12-19

quantity R/L is assumed equal to one-fourth. The distance between the center lines of any two adjacent cylinders will be assumed equal to unity, and the reference plane will be taken through the center line of the first cylinder. These same values will be maintained throughout the discussion of engine balance in order to compare the shaking forces of the various types of engines.

The first step is to apply the eight fundamental equations. A convenient tabular method of applying these equations is shown in Table 12-3. The results thus obtained indicate that the primary forces alone are balanced.

The unbalanced forces and couples for any crank position may be calculated as follows. Note that the summations Σ below are obtained from the Total column of Table 12-3.

$$F_p = mR\omega^2(\cos\theta_1\Sigma x - \sin\theta_1\Sigma y) = \cos\theta_1(0) - \sin\theta_1(0) = 0$$
$$C_p = mR\omega^2(\cos\theta_1\Sigma ax - \sin\theta_1\Sigma ay) = \cos\theta_1(-1) - \sin\theta_1(0) = -\cos\theta_1$$
$$F_s = mR\omega^2\frac{R}{L}[\cos 2\theta_1\Sigma(x^2 - y^2) - 2\sin 2\theta_1\Sigma xy]$$
$$= \tfrac{1}{4}[\cos 2\theta_1(2) - 2\sin 2\theta_1(0)] = \tfrac{1}{2}\cos 2\theta_1$$
$$C_s = mR\omega^2\frac{R}{L}[\cos 2\theta_1\Sigma a(x^2 - y^2) - 2\sin 2\theta_1\Sigma axy]$$
$$= \tfrac{1}{4}[\cos 2\theta_1(1) - 2\sin 2\theta_1(0)] = \tfrac{1}{4}\cos 2\theta_1$$

The total shaking force and the total shaking couple are, respectively,

$$F = F_p + F_s = \tfrac{1}{2}\cos 2\theta_1$$
$$C = C_p + C_s = -\cos\theta_1 + \tfrac{1}{4}\cos 2\theta_1$$

The resultant shaking force obtained by combining the force F in the reference plane and the couple has a magnitude equal to F and a line of action as given by

$$z = \frac{C}{F} = \frac{-\cos\theta_1 + \frac{1}{4}\cos 2\theta_1}{\frac{1}{2}\cos 2\theta_1} = \frac{1}{2} - \frac{2\cos\theta_1}{\cos 2\theta_1}$$

where z is the distance of the line of action of the resultant shaking force from the reference plane (i.e., from the center line of the first cylinder).

In order to study the manner in which the total unbalanced shaking forces and shaking couples vary, they may be calculated for each 15-deg position of the crank as shown in Table 12-4. These results may then be plotted as shown in Fig. 12-20. In this figure twice the total inertia

TABLE 12-3. TWO-CRANK ENGINE (CRANKS AT 180 DEG)

Crank	ϕ	x	y	a	ax	ay	x^2	y^2	$(x^2 - y^2)$	xy	$a(x^2 - y^2)$	axy
1	0	1	0	0	0	0	1	0	1	0	0	0
2	180	−1	0	1	−1	0	1	0	1	0	1	0
Total.....	...	0	0	..	−1	0	..	..	2	0	1	0

Let $mR\omega^2 = 1$; $R/L = \frac{1}{4}$; and distance between cranks = 1

$C = C_p + C_s = -\cos\theta_1 + \frac{1}{4}\cos 2\theta_1$ $\quad F = F_p + F_s = 0 + \frac{1}{2}\cos 2\theta_1 = \dfrac{\cos 2\theta_1}{2}$

TABLE 12-4. TWO-CRANK ENGINE (CRANKS AT 180 DEG)

θ	cos θ	cos 2θ	F_p	F_s	F	C_p	C_s	C
0	+1.000	+1.000	0.000	+0.500	+0.500	−1.000	+0.250	−0.750
15	+0.966	+0.866	.....	+0.433	+0.433	−0.966	+0.217	−0.749
30	+0.866	+0.500	.....	+0.250	+0.250	−0.866	+0.125	−0.741
45	+0.707	+0.000	.....	0.000	0.000	−0.707	0.000	−0.707
60	+0.500	−0.500	.....	−0.250	−0.250	−0.500	−0.125	−0.625
75	+0.259	−0.866	.....	−0.433	−0.433	−0.259	−0.217	−0.476
90	0.000	−1.000	.....	−0.500	−0.500	0.000	−0.250	−0.250
105	−0.259	−0.866	.....	−0.433	−0.433	+0.259	−0.217	+0.042
120	−0.500	−0.500	.....	−0.250	−0.250	+0.500	−0.125	+0.375
135	−0.707	0.000	.....	0.000	0.000	+0.707	0.000	+0.707
150	−0.866	+0.500	.....	+0.250	+0.250	+0.866	+0.125	+0.991
165	−0.966	+0.866	.....	+0.433	+0.433	+0.966	+0.217	+1.183
180	−1.000	+1.000	.....	+0.500	+0.500	+1.000	+0.250	+1.250
195	−0.966	+0.086	.....	+0.433	+0.433	+0.966	+0.217	+1.183
210	−0.866	+0.500	.....	+0.250	+0.250	+0.866	+0.125	+0.991
225	−0.707	0.000	.....	0.000	0.000	+0.707	0.000	+0.707
240	−0.500	−0.500	.....	−0.250	−0.250	+0.500	−0.125	+0.375
255	−0.259	−0.866	.....	−0.433	−0.433	+0.259	−0.217	+0.042
270	0.000	−1.000	.....	−0.500	−0.500	0.000	−0.250	−0.250
285	+0.259	−0.866	.....	−0.433	−0.433	−0.259	−0.217	−0.476
300	+0.500	−0.500	.....	−0.250	−0.250	−0.500	−0.125	−0.625
315	+0.707	0.000	.....	0.000	0.000	−0.707	0.000	−0.707
330	+0.866	+0.500	.....	+0.250	+0.250	−0.866	+0.125	−0.741
345	+0.966	+0.866	.....	+0.433	+0.433	−0.966	+0.217	−0.749
360	+1.000	+1.000	.....	+0.500	+0.500	−1.000	+0.250	−0.750

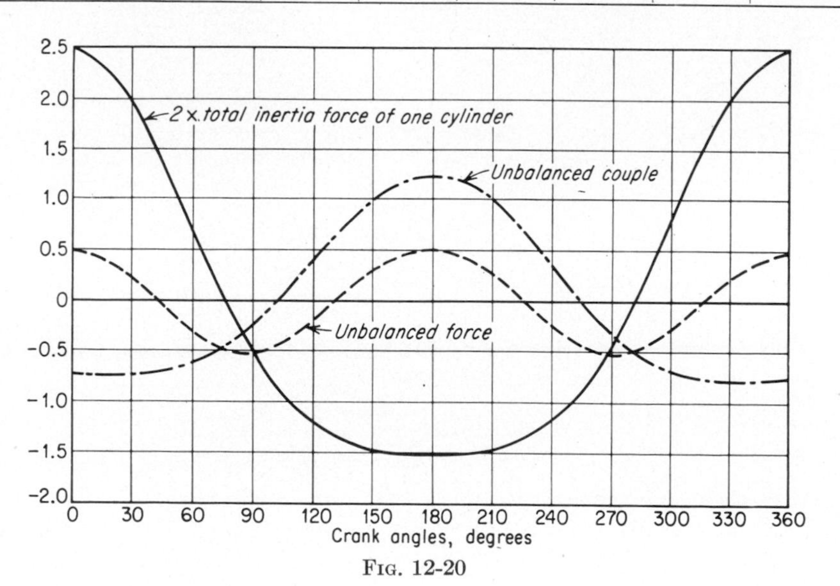

FIG. 12-20

force of a single cylinder has also been plotted, thus giving some idea of the extent to which the engine is balanced. This curve indicates the forces $(F_p + F_s)$ which would exist if the cranks were together instead of 180 deg apart.

TABLE 12-5. TWO-CRANK ENGINE (QUARTER CRANK)

Crank	ϕ	x	y	a	ax	ay	x^2	y^2	$(x^2 - y^2)$	xy	$a(x^2 - y^2)$	axy
1	0	1	0	0	0	0	1	0	1	0	0	0
2	90	0	1	1	0	1	0	1	−1	0	−1	0
Total.....	..	1	1	..	0	1	..	..	0	0	−1	0

Let $mR\omega^2 = 1$; $R/L = \frac{1}{4}$; and distance between cranks $= 1$

$F_p = \cos\theta_1 - \sin\theta_1$ $\qquad C_p = 0 - \sin\theta_1$

$F_s = \frac{1}{4}[0 - 0]$ $\qquad C_s = \frac{1}{4}[-\cos 2\theta_1 - 0]$

The arrangement with cranks at 180 deg is not ideal in a four-stroke-cycle gasoline engine because of the uneven turning effort, the two explosions of the complete cycle taking place in the same revolution of

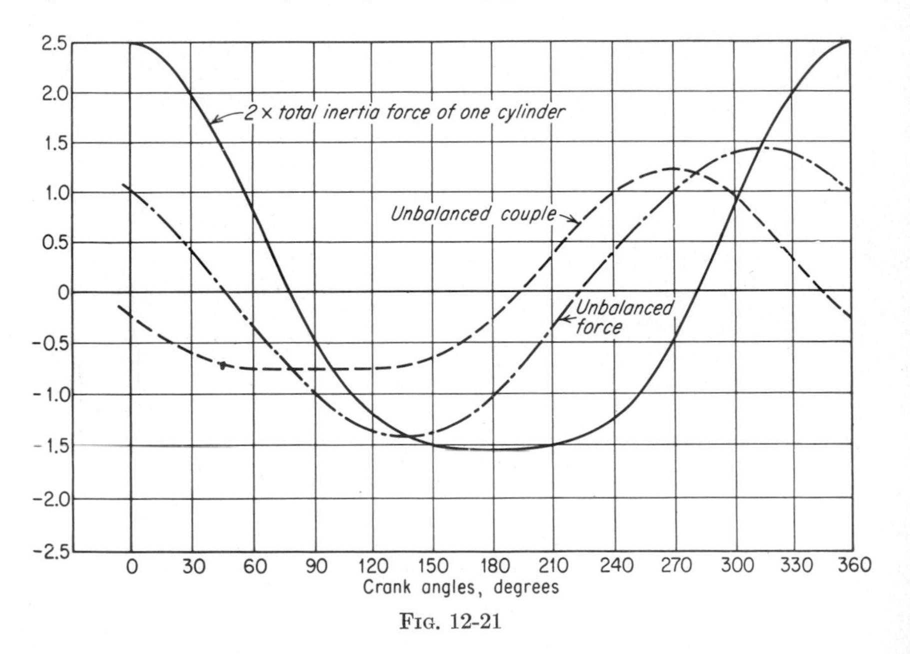

FIG. 12-21

the crank, as shown by the chart (Fig. 12-19b). If the two cranks are placed on the same side of the crankshaft, the engine becomes equivalent to two single-crank engines. In this case the balance is poor but the turning effort is good. If the cranks are placed at right angles, the balance is fair and the turning effort is good. The firing-order charts (Fig. 12-19c and 12-19d) show the character of the turning effort in the last two cases. Table 12-5 shows the balance with cranks at right angles, and the curves in Fig. 12-21 show how the unbalanced forces vary.

It must be remembered, in considering the arrangements of cranks in a multicylinder engine as regards balance, that the arrangement is

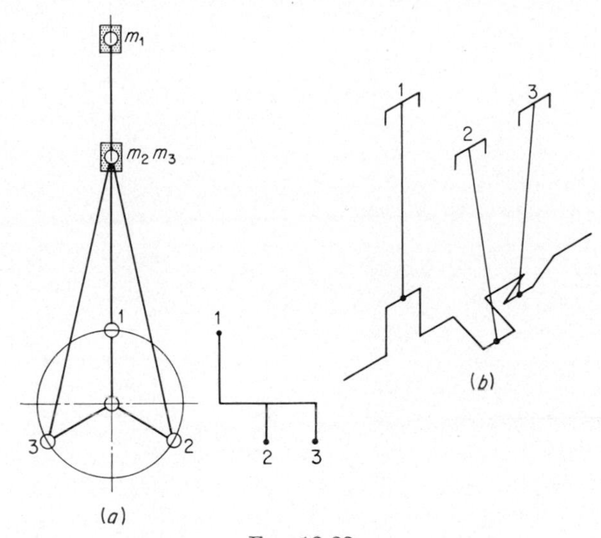

Fig. 12-22

governed primarily by the considerations of even turning effort. That is, the power strokes are so distributed as to produce as steady a torque as possible.

12-18. Three-cylinder Engine. The cranks of the three-cylinder engine are usually placed 120 deg apart, as shown in Fig. 12-22. The balance of this engine is shown by Table 12-6. The primary and second-

Table 12-6. Three-crank Engine

Crank	ϕ	x	y	a	ax	ay	x^2	y^2	$(x^2 - y^2)$	xy	$a(x^2 - y^2)$	axy
1	0	1.000	0	0	0	0	1.000	0	1.000	0	0	0
2	120	−0.500	+0.866	1	−0.500	+0.866	0.250	0.750	−0.500	−0.433	−0.500	−0.433
3	240	−0.500	−0.866	2	−1.000	−1.732	0.250	0.750	−0.500	+0.433	−1.000	+0.866
Total	...	0	0	...	−1.500	−0.866	.....	.....	0	0	−1.500	+0.433

Let $mR\omega^2 = 1$; $R/L = \frac{1}{4}$; and distance between cranks = 1

$F_p = 0$

$F_s = 0$

$F = F_p + F_s = 0$

$C_p = -1.500 \cos \theta_1 + 0.866 \sin \theta_1$

$C_s = \frac{1}{4}(-1.500 \cos 2\theta_1 - 0.866 \sin 2\theta_1)$

ary forces are in balance but the couples are not. The unbalanced couples have been plotted in Fig. 12-23, and a curve representing three times the inertia force of a single cylinder has also been drawn. This type of engine has been much used in marine work.

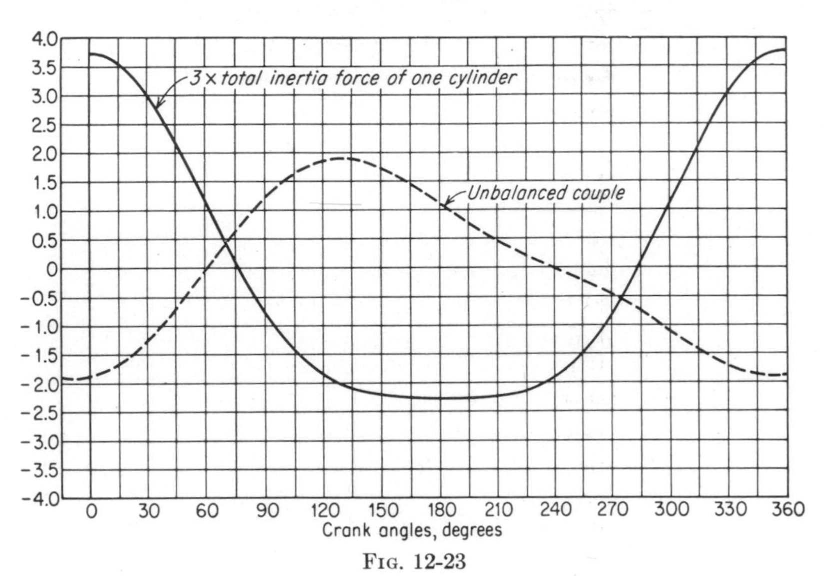

Fig. 12-23

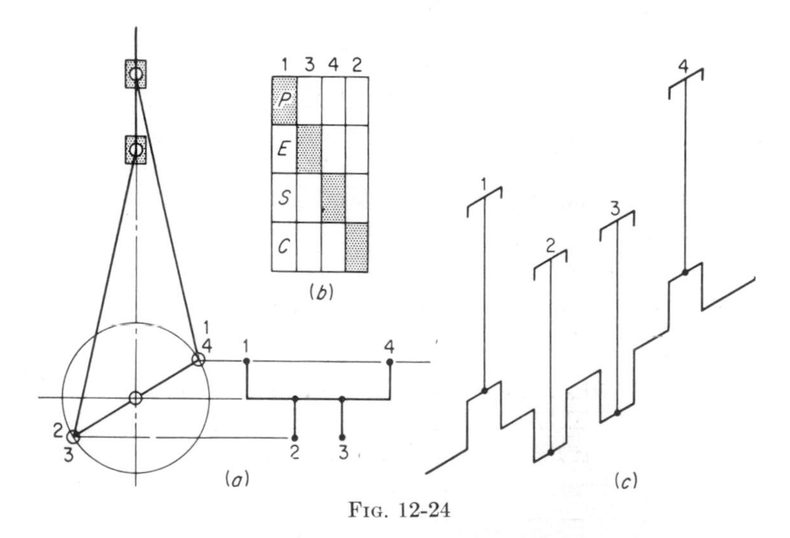

Fig. 12-24

12-19. The Four-cylinder Automotive-type Engine. The conventional crank arrangement of the four-cylinder automotive type engine is shown in Fig. 12-24, and the balance of this type of engine is shown in Table 12-7. The secondary forces are unbalanced, and when, as is the case

TABLE 12-7. FOUR-CRANK ENGINE

Crank	ϕ	x	y	a	ax	ay	x^2	y^2	$(x^2 - y^2)$	xy	$a(x^2 - y^2)$	axy
1	0	1	0	0	0	0	1	0	1	0	0	0
2	180	−1	0	1	−1	0	1	0	1	0	1	0
3	180	−1	0	2	−2	0	1	0	1	0	2	0
4	0	1	0	3	3	0	1	0	1	0	3	0
Total..........		0	0	...	0	0	...	...	4	0	6	0

Let $mR\omega^2 = 1$; $R/L = \frac{1}{4}$; and distance between cranks = 1

$$F_p = 0 \qquad\qquad C_p = 0$$
$$F_s = \tfrac{1}{4}[\cos 2\theta_1(4) - 0] \qquad\qquad C_s = \tfrac{1}{4}\cos 2\theta_1(6) = \tfrac{3}{2}\cos 2\theta_1$$
$$= \cos 2\theta_1$$
$$F = F_p + F_s = \cos 2\theta_1 \qquad\qquad C = C_p + C_s = \tfrac{3}{2}\cos 2\theta_1$$

here, the reference plane has been chosen through the center line of the first cylinder, it appears that the secondary couples are also unbalanced.

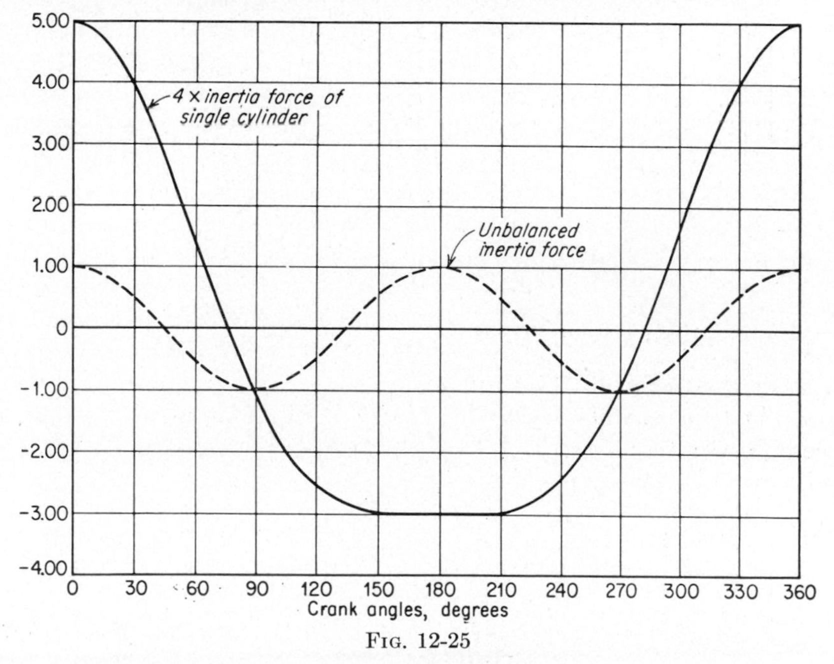

FIG. 12-25

The unbalanced forces and couples are given by

$$F = F_p + F_s = 0 + \tfrac{1}{4}[\cos 2\theta_1(4) - 2\sin 2\theta_1(0)] = \cos 2\theta_1$$
$$C = C_p + C_s = 0 + \tfrac{1}{4}[\cos 2\theta_1(6) - 2\sin 2\theta_1(0)] = \tfrac{3}{2}\cos 2\theta_1$$

The line of action of the resultant shaking force is given by

$$z = \frac{C}{F} = \frac{\tfrac{3}{2}\cos 2\theta_1}{\cos 2\theta_1} = \frac{3}{2}$$

That is, for every position of the crank the line of action of the resultant inertia force is situated midway between the two middle cylinders. This means that, if the plane midway between the two middle cylinders is chosen to be the reference plane, then secondary couples are in balance, so that the *total* unbalance of the engine is only the secondary force situated always in this plane.

The firing chart shown at *b* in Fig. 12-24 indicates that the turning effort with this arrangement of cranks is good; and the shaking-force curves in Fig. 12-25 show that the amount of unbalance is much less than what would be present if all four cranks were together so that the primary forces did not cancel.

A simple and effective device that has been used for balancing the secondary forces in the standard four-cylinder engine is represented kinematically in Fig. 12-26. Two gears driven at a speed twice that of the crankshaft are placed in an axial plane so that their pitch point is directly below the mid-point of the crankshaft, i.e., so that the line of action of the resultant shaking force passes through their pitch point. On these gears are placed the equal masses m_2 and m_3, set so that the horizontal components of their centrifugal forces always neutralize each other, leaving their combined vertical components to balance the otherwise unbalanced secondary forces. Since the masses rotate at a speed twice that of the crank, it is evident that the vertical components of their centrifugal forces will vary harmonically in the same period as that of secondary inertia forces.

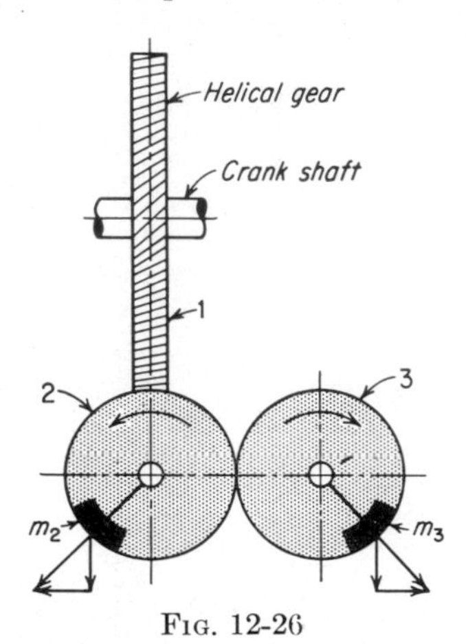

Fig. 12-26

Table 12-8. Six-crank Engine

Crank	ϕ	x	y	a	ax	ay	x^2	y^2	$x^2 - y^2$	xy	$a(x^2 - y^2)$	axy
1	0	+1.000	0	0	0	0	1.000	0.000	1.000	0.000	0.000	0.000
2	120	−0.500	+0.866	1	−0.500	+0.866	0.250	0.750	−0.500	−0.433	−0.500	−0.433
3	240	−0.500	−0.866	2	−1.000	−1.732	0.250	0.750	−0.500	+0.433	−1.000	+0.866
4	240	−0.500	−0.866	3	−1.500	−2.598	0.250	0.750	−0.500	+0.433	−1.500	+1.299
5	120	−0.500	+0.866	4	−2.000	+3.464	0.250	0.750	−0.500	−0.433	−2.000	−1.732
6	0	+1.000	0.000	5	+5.000	0.000	1.000	0.000	1.000	0.000	+5.000	0
Total.......		0	0	...	0	0	.....	.....	0	0	0	0

Let $mR\omega^2 = 1$; $R/L = 1/4$; and distance between cranks = 1

$F_p = 0$ $\quad$ $C_p = 0$

$F_s = 0$ $\quad$ $C_s = 0$

12-20. The Six-cylinder Automotive-type Engine. The crank arrangement in this type of engine is shown in Fig. 12-27. The balance of the engine is shown in Table 12-8. As indicated in the table, the six-cylinder engine is in complete balance with regard to both primary and secondary forces and couples. The six-cylinder engine arranged in this way is one of the most perfectly balanced engines of the usual multicylinder type that it is possible to construct.

12-21. The Eight-cylinder In-line Automotive Engine. In this type of engine the following constructions have been employed:

1. Two units, each composed of four cylinders arranged as described in Art. 12-19, with crankshafts 90 deg apart and constructed in a tandem arrangement.

2. The middle four cranks arranged in an axial plane making an angle of 90 deg with the plane of the outer four throws. This arrangement is shown in Fig. 12-28.

In case 1 the arrangement is such that the unbalanced forces cannot completely neutralize each other; and, therefore, as in the case of the four-cylinder engine, the engine is not balanced as far as the secondary forces are concerned.

In case 2 the engine is in complete balance as regards both primary and secondary forces and couples.

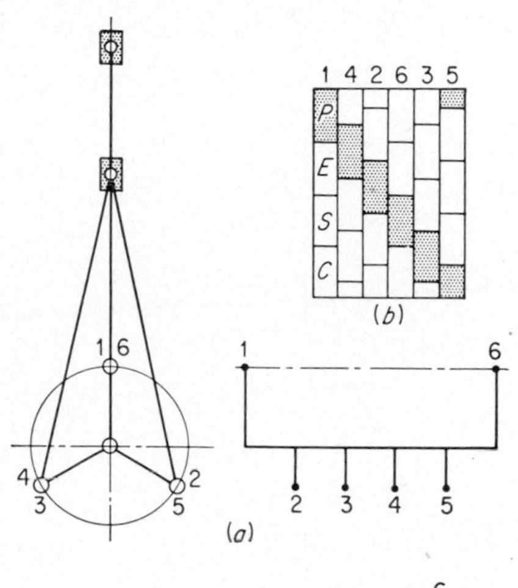

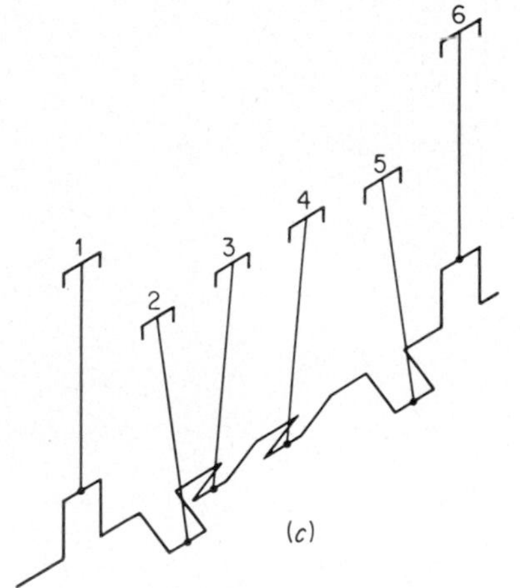

FIG. 12-27

12-22. The V-type Engine. Figure 12-29 shows a two-cylinder V engine with angle β between cylinder center lines. Both connecting rods connect to a single crank throw. An expression will be developed for the resultant of the primary and secondary forces. It will be assumed that reciprocating masses m_P are equal and connecting rods are of equal length. For the left bank,

$$F_L = m_P R\omega^2 \cos \theta_L + m_P \frac{R^2}{L} \omega^2 \cos 2\theta_L$$

For the right bank,

$$F_R = m_P R\omega^2 \cos \theta_R + m_P \frac{R^2}{L} \omega^2 \cos 2\theta_R$$

Adding vectorially,

$$\begin{aligned} F_L +\!\!\!\!\to F_R &= m_P R\omega^2[\cos \theta_L +\!\!\!\!\to \cos (\theta_L - \beta)] \\ &\qquad +\!\!\!\!\to m_P \frac{R^2}{L} \omega^2[\cos 2\theta_L +\!\!\!\!\to \cos (2\theta_L - 2\beta)] \\ &= F_p +\!\!\!\!\to F_s \end{aligned} \tag{12-31}$$

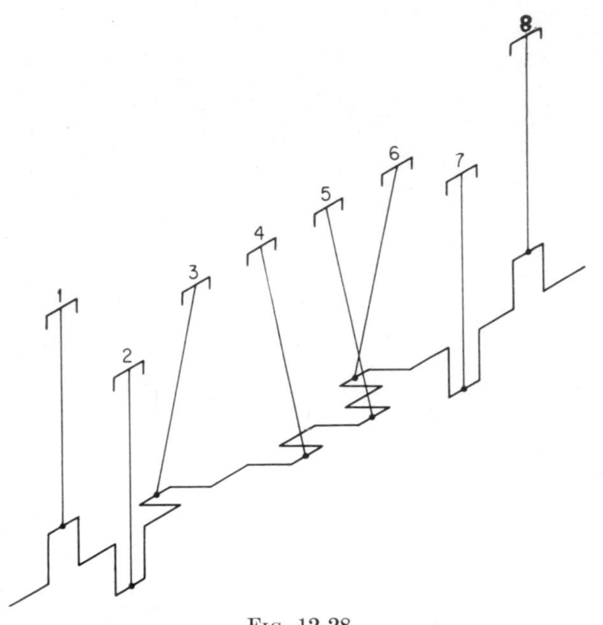

Fig. 12-28

If the bank angle β is 90 deg, as in Fig. 12-30,

$$\begin{aligned} F_p +\!\!\!\!\to F_s &= m_P R\omega^2(\cos \theta_L +\!\!\!\!\to \sin \theta_L) \\ &\qquad +\!\!\!\!\to m_P \frac{R^2}{L} \omega^2 [\cos 2\theta_L +\!\!\!\!\to (- \cos 2\theta_L)] \end{aligned} \tag{12-32}$$

Because the vector $m_P R\omega^2 \cos \theta_L$ lies along the left bank, and the vector $m_P R\omega^2 \sin \theta_L$ $(= m_P R\omega^2 \cos \theta_R)$ lies along the right bank, these two primary-force vectors are at right angles, and therefore their vector sum is always $m_P R\omega^2$, lying along the crank. Figure 12-30 illustrates this. Thus the *resultant primary force* is a *rotating* force lying always along the rotating crank, and it can therefore be canceled exactly by a rotating counterbalance opposite the crank. No such simple expedient exists

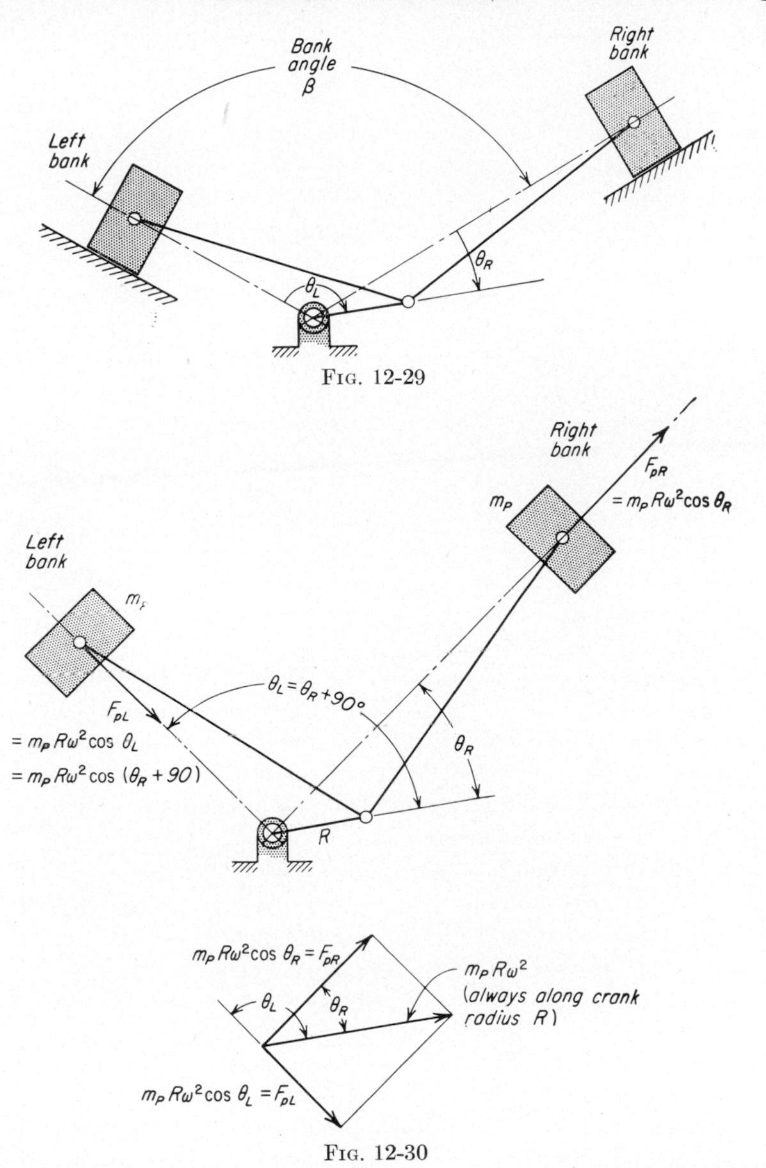

Fig. 12-29

Fig. 12-30

for the secondary forces, obviously. Of course, additional counterbalancing mass is required to balance the equivalent *rotating* mass situated at the crankpin.

In multicylinder V-type engines, the bank angle is not arbitrary but is chosen in conjunction with the number of cylinders so that the firing impulses are evenly spaced. To determine the condition of balance, each bank can be checked for balance as an in-line engine.

The V-8 with 90-deg bank angle is a widely used engine in automotive practice and will be discussed briefly as an important example. The usual crank arrangement is shown in Fig. 12-31, with the plane of the middle two crank throws perpendicular to the plane of the two end throws. An analysis of one bank as an in-line engine shows that only primary couples remain unbalanced. But a proper counterbalance opposite each crank throw can cancel the primary forces of the two cylinders connected to the throw, as discussed earlier in this article. Thus the use of four counterbalances can completely eliminate all primary forces and therefore also eliminate any moments or couples produced by those forces. Therefore, the addition of the four rotating counterweights results in a completely balanced engine, in so far as primary and secondary forces and couples are concerned. Of course, if desired, the four rotating counterweights can be replaced by any combination of two or more which produce the same resultant couple as the original four. This has already been discussed in the sections on the balancing of rotating masses. It is worth noting that counterbalancing each pair of cylinders separately eliminates shaft bending moments and bearing loads which otherwise would be present.

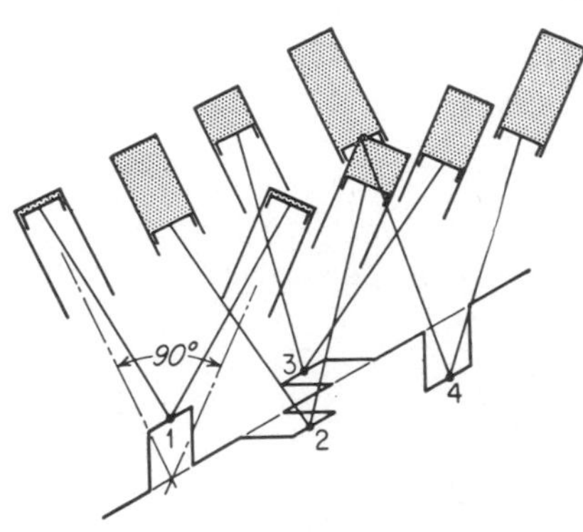

FIG. 12-31

12-23. Aircraft Engines. The smallest aircraft engines are generally of the two-cylinder opposed type. Various other types are built or have been built, such as the 4-cylinder opposed and vertical in-line; 6-cylinder opposed and vertical in-line; 12-cylinder opposed and V-type; 3, 5, 7, and 9-cylinder radial; 14- and 18-cylinder two-row radial; and 24-cylinder double V-type.

Balance of the in-line types has been discussed in the preceding articles. In the 2-cylinder opposed engine the primary and secondary forces are balanced, but the primary and secondary couples are unbalanced. The 4- and the 12-cylinder opposed engines are in complete balance as to primary and secondary forces and couples.

Radial engines of the four-stroke type have an odd number of cylinders (five, seven, and nine being the more common), permitting the firing of the odd-numbered cylinders in one revolution and the even-numbered cylinders in the next revolution and resulting in evenly spaced power impulses for equally spaced cylinders. The usual construction consists of a master connecting rod for one cylinder, the articulated rods for the other cylinders being connected to the crank end of the master rod by means of wrist pins, as indicated in Fig. 12-32. In making a force analysis of the engine or in analyzing it for balance, the usual simplifying procedure is to assume that all the articulated rod ends are on the same axis as the crankpin and that the articulated rods are of the same length as the master rod, as shown in Fig. 12-33. Although not strictly correct, this assumption gives results sufficiently accurate for ordinary purposes. In radial engines couples such as those discussed in connection with

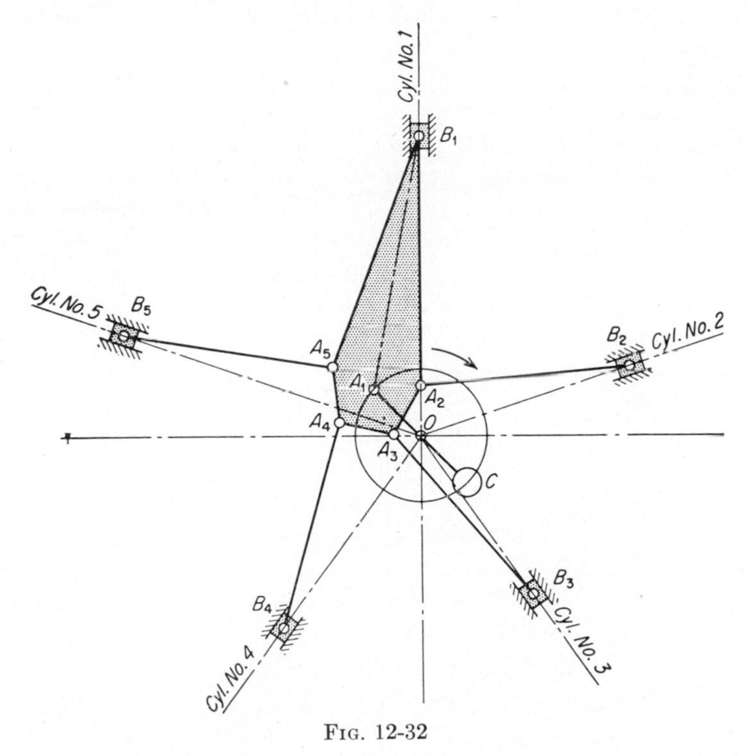

Fig. 12-32

in-line engines do not exist. Each connecting-rod weight may be considered as distributed at the wrist pin and crankpin in the manner suggested in Art. 12-13. Thus the problem of balancing is reduced to the consideration of the rotating weight at the crankpin and the reciprocating weights at the piston pins.

The three-cylinder single-crank radial engine is unbalanced as to both primary and secondary forces, but the unbalanced primary forces can be balanced by a suitable counterweight placed opposite the crank arm. It is not considered practical to balance the unbalanced secondary forces.

In single-crank radial engines of more than three cylinders, the secondary forces are balanced, but the primary forces are not. The unbalanced primary forces can be balanced by a suitable counterweight placed opposite the crank arm, and the engine thus placed in complete balance.

The following example will illustrate a procedure that may be followed in determining the extent to which a radial engine is in balance.

A five-cylinder single-crank radial aircraft engine of the type represented in Fig. 12-32 operates at a speed of 2,000 rpm. Length of crank = $2\frac{11}{16}$ in. Length of master rod = $10\frac{3}{4}$ in. It can be shown that the accelerations of the articulated-rod pistons do not differ much from the acceleration of the master-rod piston. This fact justifies the usual simplifying procedure of assuming the articulated rods to be equal in length to the master rod and that their crank ends are grouped together with the master-rod crank end on the master-rod crankpin axis, as shown in Fig. 12-33. The connecting-rod weights are assumed to be distributed at the crankpin and at the piston pins in the manner suggested in Art. 12-13 so that the estimated total rotating weight at the crankpin is 10 lb, and the total reciprocating weight at each piston pin is 4 lb. The above assumptions greatly simplify the analysis and give results sufficiently accurate for ordinary purposes.

The simplified mechanism of the engine (Fig. 12-33) is drawn to a scale of $\frac{1}{4}$ in. = 1 in., and the inertia-force vectors are drawn to a scale of 1 in. = 2,000 lb.

The inertia force (centrifugal force) F_A of the rotating weight at the crankpin is determined from the equation

$$F_A = m_A R \omega^2 \tag{12-33}$$

where m_A is the mass at the crankpin, R is the crank radius, and ω is the angular velocity of the crank in radians per second. The magnitude of this force is then

$$F_A = \frac{10}{32.2} \times \frac{2.6875}{12} \times \left(\frac{2\pi \times 2{,}000}{60}\right)^2 = 3{,}050 \text{ lb}$$

Since the centrifugal force is a constant, a circle with center at crankpin axis O and radius F_A will be the locus of the ends of centrifugal-force vector F_A for all positions of the crank.

The total inertia force F_T of each reciprocating mass at the piston pins is found by use of the equation

$$F_T = m_B R \omega^2 \left(\cos\theta + \frac{R}{L}\cos 2\theta\right) \tag{12-34}$$

where, for the reciprocating mass under consideration, m_B is the reciprocating mass, R is the radius of the crank, L is the length of the connecting rod, and θ is the angle that the crank makes with the center line of the cylinder (see Art. 12-12). Or, considering the primary and secondary reciprocating forces separately, the equation for the primary force F_p is

$$F_p = m_B R \omega^2 \cos \theta \tag{12-35}$$

and the equation for the secondary force F_s is

$$F_s = m_B \frac{R^2}{L} \omega^2 \cos 2\theta \tag{12-36}$$

The resultant inertia forces will be determined for crank positions

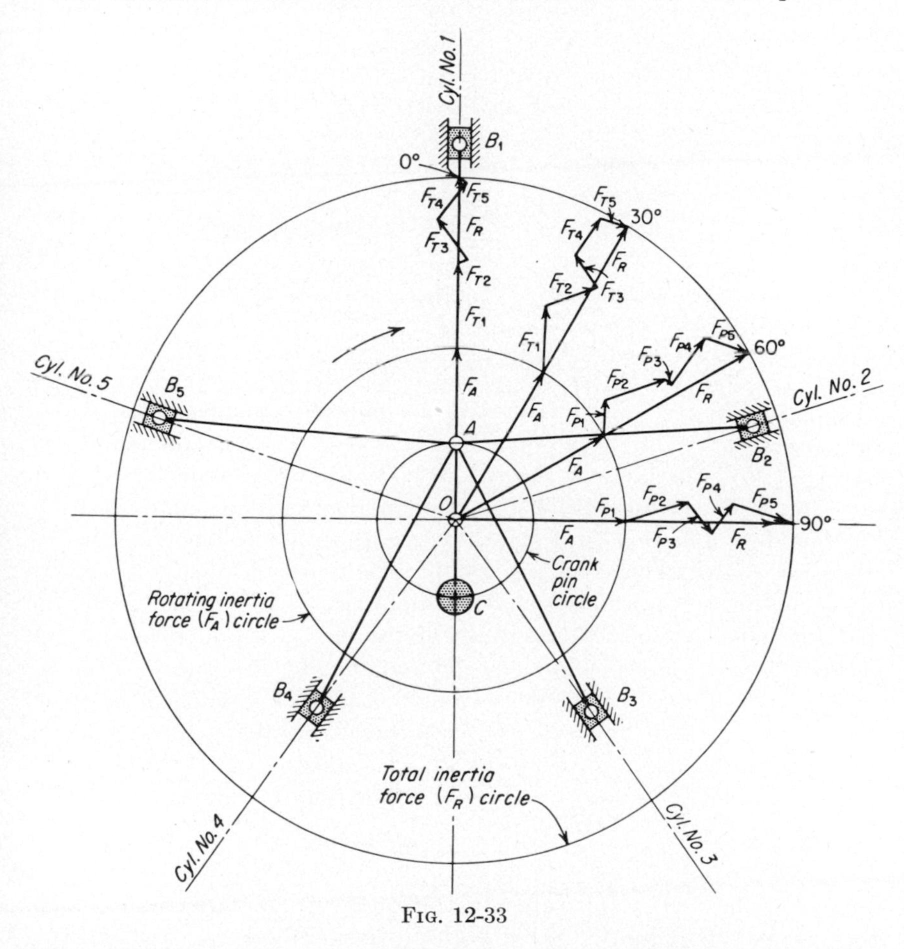

Fig. 12-33

0, 30, 60 deg, etc. Considering first the zero position of the crank, the resultant of the total forces will be obtained by adding to vector F_A (the vector that represents the rotating inertia force) the vectors that represent the five reciprocating inertia forces. These forces are calculated from Eq. (12-34), which for this problem may be reduced to a simplified form as follows:

$$F_T = \frac{4}{32.2} \times \frac{2.6875}{12} \times \left(\frac{2\pi \times 2{,}000}{60}\right)^2 \left(\cos\theta + \frac{2.6875}{10.75}\cos 2\theta\right)$$

which reduces to

$$F_T = 1{,}220(\cos\theta + \tfrac{1}{4}\cos 2\theta)$$

The calculations for the reciprocating inertia forces for the 0-deg crank position, using the above equation, are as follows:

Cyl. No. 1:

$$F_{T_1} = 1{,}220(\cos 0° + \tfrac{1}{4}\cos 0°) = 1{,}525 \text{ lb}$$

Cyl. No. 2:

$$F_{T_2} = 1{,}220(\cos 72° + \tfrac{1}{4}\cos 144°) = 130 \text{ lb}$$

Cyl. No. 3:

$$F_{T_3} = 1{,}220(\cos 144° + \tfrac{1}{4}\cos 288°) = -894 \text{ lb}$$

Cyl. No. 4:

$$F_{T_4} = 1{,}220(\cos 144° + \tfrac{1}{4}\cos 288°) = -894 \text{ lb}$$

Cyl. No. 5:

$$F_{T_5} = 1{,}220(\cos 72° + \tfrac{1}{4}\cos 144°) = 130 \text{ lb}$$

The vectors representing these forces are laid off from the end of vector F_A, giving the resultant shown in Fig. 12-33. In a similar manner the forces F_{T_1}, F_{T_2}, etc., are calculated and laid off for the 30-deg crank position, the resultant force being the same as for the 0-deg crank position. Further determinations will show this resultant to be the same for all crank positions. This condition applies also to other radial engines with the exception of the three-cylinder radial engine—hence the conclusion that for radial engines the resultant of the inertia forces is constant in magnitude (for a given engine speed) and that it always acts along the crank-arm center line. Thus the engine can be completely balanced by a rotating weight opposite the crank arm. If the counterbalancing weight is placed opposite the crank at a distance R, equal to the length of the crank arm, its magnitude is obtained from the equation

$$F_C = m_C R\omega^2 \tag{12-37}$$

where F_C, the inertia force due to the rotating counterbalancing mass m_C, is equal and opposite to the resultant inertia force F_R. Substituting in this equation

$$6{,}100 = \frac{W}{32.2} \times \frac{2.6875}{12} \times \left(\frac{2\pi \times 2{,}000}{60}\right)^2$$

from which $W = 20$ lb.

It can be shown (except in the case of the three-cylinder radial engine) that for radial engines the secondary forces as obtained from Eq. (12-36) are in balance among themselves. It is evident, therefore, that in Fig. 12-33, the same resultant would have been obtained by using the primary-force vectors obtained from Eq. (12-35). This has been done in the case of the 60- and 90-deg crank positions in Fig. 12-33, where the primary-force vectors F_{p_1}, F_{p_2}, etc., are shown to give the same resultant as the total force vectors at the 0- and 30-deg crank positions. For fixed radial engines of K cylinders with all the connecting rods assumed to be of equal length and pivoted on one crankpin axis as in Fig. 12-33, the inertia forces have a constant resultant of $(K/2)mR\omega^2$, approximately, along the crank. This resultant can be balanced completely by a counterbalancing mass of $(K/2)m$ at a radius R opposite the crankpin: If there are two banks of cylinders, each with K cylinders and with cranks at 180 deg, there will be an unbalanced primary couple of $(K/2)mR\omega^2L$, where L is the distance between the rows of cylinders. However, this couple will not exist if each row is counterbalanced in the manner indicated above.

It may be observed that the results obtained above by graphical methods could also be obtained by analytical methods by applying Eq. (12-15) of Art. 12-12 to each of the five slider-crank units and solving for the horizontal and vertical components of the primary and secondary forces with reference to the axis of the crankshaft.

12-24. Other Types of Engines and Reciprocating Machines. Space limitations preclude the discussion of all the possible arrangements of cylinders and schemes for balancing which have been used and proposed for engines and other reciprocating machines such as compressors. For example, there are the V-6 (and other numbers of cylinders in V engines), X-type, a great variety in number of cylinders used in line, opposed-piston types with two pistons in each cylinder, engines with cylinder center lines on circular arcs, and even a radial engine in which the crank was stationary and the cylinders revolved around the crankshaft. Combinations of engines and compressors are used in V and W arrangements, and often in these machines the masses and sizes of pistons, rods, etc., are not all identical. The basic approaches employed in this and previous chapters should be sufficient to allow an approach to the problems of balancing these various types.

An example of an engine using a modification of the conventional crank and connecting-rod arrangement to transform the reciprocating motion of the piston to the rotary motion of the output shaft (or main journal) is shown in Figs. 12-34 and 12-35. This engine has a bore of 14 in., a stroke of 16 in., operates on the two-stroke-cycle principle, and is a single-row radial stationary engine with cylinders in a horizontal plane.

FIG. 12-34. (*Courtesy of the Nordberg Manufacturing Co.*)

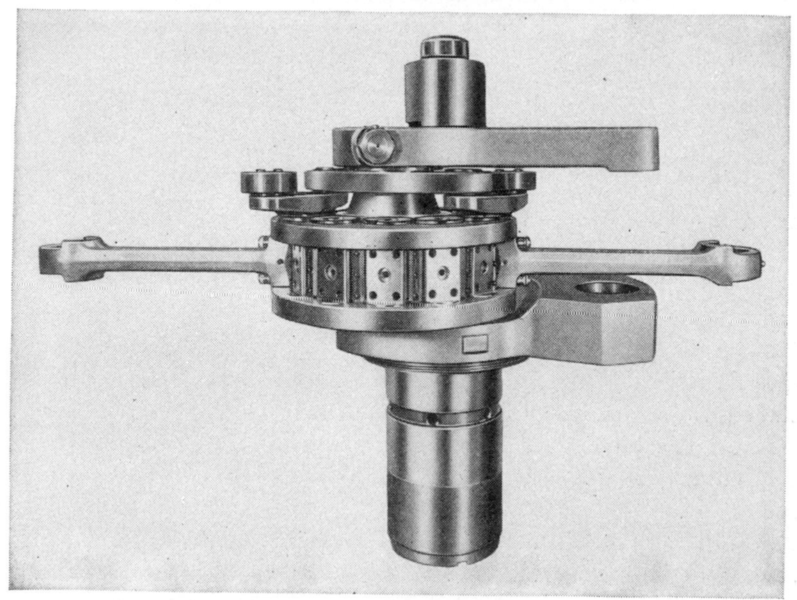

FIG. 12-35. (*Courtesy of the Nordberg Manufaciuring Co.*)

As the pictures show, all connecting rods are identical and connect to the master bearing in the same manner; the master connecting rod of the conventional radial engine is no longer present. Instead, the restraining link and two restraining cranks are used, resulting in identical circles of motion for the knuckle pins. The master bearing *gyrates* but does not rotate completely, while the main journal and crank *do* rotate as usual. In this design, all pistons execute identical motions, which is not the case in the conventional radial engine with master rod, and which is an obvious advantage when considering such things as balance and engine timing. The counterbalancing weights opposite the crank throw are evident in Fig. 12-35.

12-25. Balancing the Crankshaft. The crankshaft is in general an unbalanced system of rotating masses in which the planes of rotation and the crank angles are fixed by the reciprocating system that it operates. The crankshaft may be balanced in either of two ways:

1. By the addition of two balancing masses in two separate planes of rotation, in accordance with the method outlined in this chapter.

2. By the extension of the arms of each crank to form balancing masses for the rotating masses of that crank. In this case there are as many balancing masses as there are crank arms. This method, however, has the advantage that the intermediate parts of the crankshaft do not have to transmit force from one crank to another, since each crank is balanced in its own plane of rotation. The crankshaft is thus relieved of any bending moment due to rotating masses.

In addition to the lateral bending to which the crankshaft is subjected because of unbalance or applied periodic forces, it is also susceptible of vibrating about its axis. This action is known as torsional vibration. A periodic twisting couple agreeing with the natural period of the shaft about its axis is capable of exciting large vibrations of that period, even though the magnitude of the couple producing it is extremely small. Automobile and aircraft engines are now equipped with devices for eliminating the torsional vibration of the crankshaft.

12-26. Balancing the Connecting Rod. The connecting rod has two distinct effects:

1. It does not give harmonic motion to the reciprocating masses, thus causing secondary piston inertia forces. This effect on the balance of the engine has been dealt with in Art. 12-15.

2. Its motion is one of periodic acceleration, and hence inertia forces are set up that have periodic variations. These forces ultimately appear as reactions on the frame or foundation and because of their periodic nature tend to cause vibration. In balancing an engine a commonly used approximate method of taking care of the inertia effect of the connecting rod is to assume the mass of the rod to be distributed between the

crankpin and wrist pin inversely as the mass center divides the rod, as discussed in Art. 12-13. Balancing masses are then added in accordance with this assumed distribution.

12-27. Vibrations Due to Impulse and to Fluctuating Pressure of the Working Fluid. When the forces and couples due to the motion of the parts of an engine have been balanced, the engine is not necessarily free from vibration, because balance of mass does not take into account the fluctuating pressures in the cylinders, with the accompanying effort of each succeeding impulse to rotate the engine frame backward around the crankshaft. Hence, after the inertia forces are balanced, there still remains a couple acting on the frame equal and opposite to the turning couple on the crankshaft, and this couple may have a periodic variation sufficiently great to cause vibration. In order to have an engine entirely free from vibration, therefore, it must be arranged not only for balance among the moving parts, but so that the turning effort on the crank is uniform.

12-28. Vibrations of Foundations and Supports. In the preceding articles, the links of all the mechanisms considered have been regarded as rigid. Although the forces acting on the frame of a machine due to a set of moving masses in balance may form a system of forces in equilibrium, the individual forces of the system cause elastic deformation of the frame at the places where they act. In general, machine frames are stiff enough to limit the elastic deformations to negligibly small amplitude, but in some cases, notably in revolving shafts, there may be appreciable deflections, and these deflections, being of course periodic, may become of sufficient amplitude at certain *critical speeds* to cause vibrations of a serious nature.

The foundation or support of an engine or machine is usually an elastic system susceptible of vibrating in a variety of ways. If the engine or machine is unbalanced, it applies a periodic force or couple to its supports, resulting in forced vibrations that may, in general, be of little consequence, even though a large force or couple may be acting. If, however, the support happens to have among its natural modes of vibration one whose natural period is equal or approximately equal to the time of revolution or other motion cycle of the machine, then the amplitude of the forced vibration may be large, even though the magnitude of the force producing it is extremely small. This phenomenon may be explained by the fact that the work done in displacing the system from its position of rest appears as the energy of the vibration. If the succeeding applications of the force are so timed as to begin to act just at the instant that the vibrations are about to repeat themselves, other small amounts of energy are communicated to the system without interfering with each other. Energy is thus gradually accumulated in the system to such an extent that the amplitude of the forced vibration is all out of proportion to the force producing it.

It is in this way that the engines of a power plant or the machines in a factory, though firmly secured to massive concrete foundations, may set up vibrations in the earth in which the foundation is embedded, which may be transmitted all around the neighborhood. Within the building itself, irrespective of neighboring properties, the vibrating action of machinery is often felt and has been known to cause a great deal of annoyance.

In addition to engines, the types of machinery usually likely to cause annoyance from vibration are printing presses, pumps, ice machines, air compressors, and various types of purely rotative machines such as blowers, centrifugal pumps, etc.

12-29. Critical Speeds. In some unbalanced rotating members, notably in revolving shafts, the deflections and vibrations at certain *critical speeds* become extremely severe. The peculiar behavior of an unbalanced rotating shaft has been a matter of common knowledge for many years, and has received increasing attention as reciprocating machinery has been replaced by high-speed rotative machinery. The most striking phenomena are as follows:

1. Increasing deflection of the shaft as the speed of rotation increases from zero to the critical speed.
2. Dangerous deflection at the critical speed unless some damping influence, usually in the form of friction, comes into play.
3. Gradual return to a condition of stability as the speed increases above the critical value.

The vibration phenomena of a shaft while passing through its critical speed are very interesting from a theoretical point of view, and, at the same time, they are of great practical importance in the design of steam turbines and other high-speed machines. Although the subject merits extended discussion, only the brief treatment given in Chap. 14 is possible in this book.

12-30. Balancing Machines. It is not in general possible to secure the satisfactory balance of a machine member merely by giving it the proper proportions when it is being laid out on the drawing board; and this is particularly true of members intended to rotate at high speeds. This is due to the ordinarily permissible variations in the dimensions of the member that arise from imperfections in the manufacturing process, and also to the fact that the material of which the member is made will usually vary slightly in density from point to point. There is, therefore, a real need for a means of adjusting the balance of a rotating member after it has been made, and this need is met by the various types of balancing machines.

The function of any dynamic balancing machine is to determine where metal should be added or taken away from a rotating member in order to secure its dynamic balance. This is usually accomplished by mounting the member in two bearings on an elastically supported frame. The member is then rotated, and any lack of balance is at once evidenced by the resulting vibration of the frame. The magnitude and angular position of this unbalance are determined by methods peculiar to the type of balancing machine being used. The body is then dynamically balanced by the addition or removal of weight in two arbitrarily selected planes of correction perpendicular to the axis of rotation.

Static balancing machines are used for the measurement and correction of unbalance in narrow-faced parts in which the possibility of dynamic unbalance is practically nonexistent. Such parts are flywheels, clutches, road wheels, narrow fans, pulleys, airplane propellers, etc. These parts were formerly checked for static unbalance on parallel bars or knife-edges. They are now balanced to a high degree of accuracy on balancing machines of the newer types.

A number of different types of balancing machines have been developed for the purpose of measuring static and dynamic balance or combined static and dynamic balance. The earlier machines required that by a trial-and-error process the operator add or remove unbalance from rather poorly indicated positions, the results depending upon the patience and skill of the operator. With successive developments and improvements, balancing machines are now made so that there is a minimum of operator skill required. These newer machines also interpret reading of unbalance into actual practical correction values. For example, the operator reads exactly the depth to drill a given size of hole at a given point or points in the work to produce a condition of balance, or the operator may read directly the number of sixty-fourths of an inch of length of strip steel to be added to a part at a definitely indicated position in order to produce balance in the part.

The latest developments in balancing machines provide an electrical indicating means for obtaining the measurements of the effects of unbalance. The basic principle of operation of one of these machines, typical of the group, may be explained as follows:

Indications on the machine are dependent upon vibrations caused by unbalance in the rotor which is to be balanced when it is rotated at a constant speed. These vibrations actuate magnetic pickups, which produce an alternating current proportional to and in phase with the vibrations. Means for rectifying the alternating-current output of the pickups and obtaining its phase relation to the rotation of the part are provided. Thus the angular location of unbalance is determined by the phase relation between pickup output and machine rotation, and amount of unbalance in specified correction planes is indicated by a meter which registers the rectified current.

Another development is embodied in *portable* dynamic balancing equipment used to measure vibration caused by unbalance in rotating masses. This equipment provides a means of balancing rotors in the field, which may be too large to be handled in a commercial balancing machine and which for this or other reasons must be balanced while they are running in their own or in substitute bearings. Such equipment is particularly applicable where balancing may be needed during the initial manufacture or installation, following subsequent servicing, or as a result of unbalance caused by such action as pitting or corrosion. Typical of such portable dynamic equipment is one type in which the apparatus consists essentially of two magnetic pickups or generators, a contactor or mechanical rectifier, a microammeter, and the necessary mounting fixtures, adapters, and connecting cables. In addition, an electronic amplifying and filtering unit is provided to improve the sensitivity and operation.

12-31. Description of a Balancing Machine. Typical of the balancing machines of the type mentioned in Art. 12-30 is that illustrated in schematic outline in Fig. 12-36. This diagram and the information that follows were supplied by the Gisholt Machine Company, Madison, Wis., manufacturers of the machine. The operation of this machine is based upon a fundamental principle of mechanics that involves a very interesting application of the center of percussion.

As stated in Art. 12-30 and earlier, a body may be dynamically balanced by the addition or removal of weight in two arbitrarily selected planes of correction perpendicular to the axis of rotation. In the schematic drawing, W_1 and W_2 represent the unbalances in the arbitrarily selected correction planes 1 and 2 that will give the same vibration effect at bearings A and B as is produced by the unbalances existing in the rotor. If it is desired that bearing vibration at A and B be reduced to zero, a means must be found for determining the magnitudes of the unbalances W_1 and W_2.

The method whereby this is accomplished may be understood by reference to the schematic diagram (Fig. 12-36). The rotor is carried on two independent, flexibly supported lightweight structures which permit of vibration only in a horizontal plane. The rotor is supported at the bearings A and B where it is desired to reduce vibration to zero. It is driven by the motor through a light, flexible belt.

Before proceeding further with a discussion of the elements of the balancing machine, consider the effect of a horizontal periodic force P applied to the axis LL of the rotor in correction plane 2. This force P

will cause a periodic oscillation of the axis of the rotor in a horizontal plane (the axis of the rotor will move between the lines L_1L_1 and L_2L_2) so that there will be a point X_2 on the axis of the rotor that will have no motion. A vertical axis through X_2, then, will be the axis of suspension for a center of percussion at the intersection of the line of action of the force P with the axis LL of the rotor. Then any point in a plane perpendicular to LL containing X_2 will have no motion due to the periodic force P.

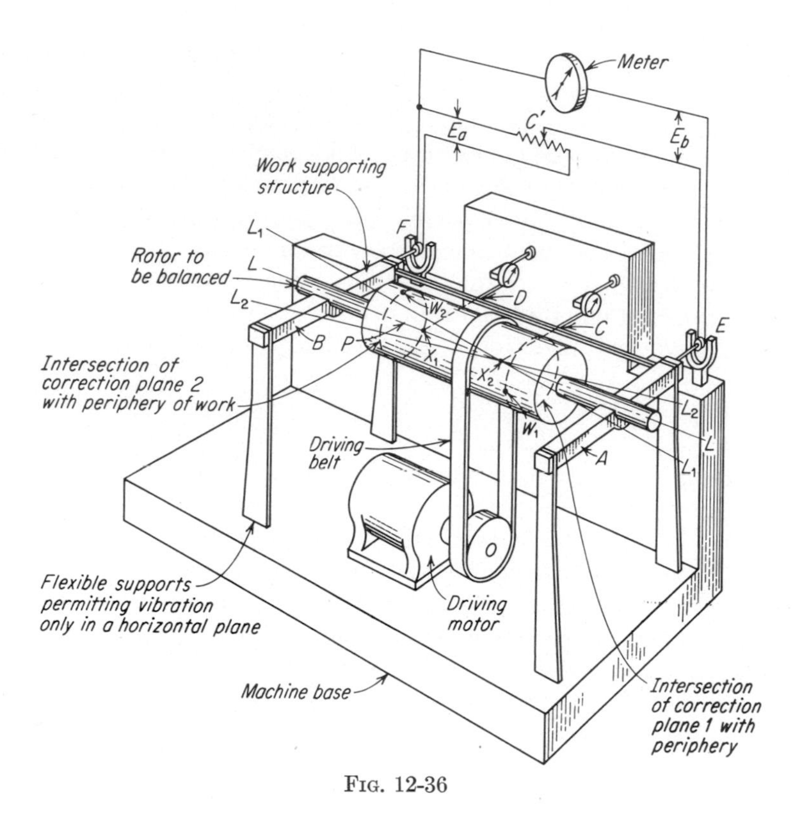

FIG. 12-36

Under rotation, the unbalance W_2 in the rotor will provide the periodic force P because the supporting structure has freedom only in a horizontal plane. Further, the point C, on a bar connecting the two work-supporting structures and in a plane perpendicular to the axis of rotation that contains the point X_2, will have no motion due to the unbalance W_2 in the correction plane 2. However, this point C will have motion due to a force in any other correction plane such as the force produced by unbalance W_1 in correction plane 1. Therefore, an indicator between the machine base and point C will provide a means for measuring the unbalance effect W_1 without any effect from unbalance W_2.

In a similar manner, there can be found a point X_1, which will be on a vertical line that is the axis of suspension for the center of percussion in correction plane 1. Then there will likewise be a point D on the bar connecting the two work-supporting structures, which will have no motion due to unbalance W_1 and will, therefore, serve as a means of measuring the unbalance effect W_2.

In this machine an electrical indicating means is used for obtaining the measurements of the effects of the unbalances. By attaching lightweight generating coils E and F to the work-supporting structures and permitting these coils to vibrate in the fields of powerful permanent magnets, voltages will be generated in the coils that will be proportional to the movements of the bar connecting the work-supporting structure. The motion of point C on this bar is a function of the movements of the ends of the bar, and at a certain position C' on a potentiometer a resultant voltage on the meter will be present which is the same function of the voltages generated at E and F. Effectively, the fraction of the total voltage which is caused by the unbalance in plane 2 is canceled electrically, leaving a meter reading which indicates only unbalance in plane 1. Two such networks must be used, of course—one for each plane. Electrical means of this sort are used because tremendous amplification of the generated voltages is possible and because it is also possible by these electrical means to determine the angular positions of the unbalances W_1 and W_2, for example, by means of a stroboscopic lamp. Such a lamp is caused to flash periodically by the alternating voltages produced by the vibration pickup coils, and the flashes illuminate the position of unbalance on the periphery of the rotating part. Another electrical means for determining angular location of unbalance employs a wattmeter which measures the average value of the product of an instantaneous voltage (from the vibration pickup coils) and an instantaneous current (supplied by some sort of alternating-current generator coupled to the rotating workpiece). When the voltage and current are 90 deg out of phase, the wattmeter reading is zero, and when in phase, the reading is a maximum. The phasing depends, of course, upon the position of the unbalance, which determines the instantaneous value of the voltage output of the pickup coils.

The ordinary dial indicators at C and D, reading in thousandths of an inch, are shown in the illustration merely to indicate how the unbalances W_1 and W_2 might be measured mechanically. A micrometer and spark gap have been used in some machines.

Optical amplification and indication also have been used. A beam of light from a rotating mirror is made to draw a sine wave on a ground-glass screen as the workpiece rotates and vibrates in the cradle of the balancing machine. The height of the wave is a measure of the amount of unbalance, and the position of the crest of the wave along an angularly graduated scale is an indication of where the correction should be applied. This scheme of using a sine-wave display also has been used in electrically indicating balancing machines by employing a cathode-ray oscilloscope.

CHAPTER 13

FORCE ANALYSIS OF A GASOLINE ENGINE

The following analysis provides an excellent means for applying the principles of dynamics, as outlined in the foregoing chapters, in the solution of problems pertaining to high-speed machinery.

This problem may be adapted to classroom work by distributing the routine work among a group of students. In this manner the complete analysis has been made in a period of 6 weeks, working 6 hr per week.

Figures 13-1 to 13-11 and Tables 13-1 to 13-4 give the results of an investigation of the forces acting in a six-cylinder engine of the automotive type.

The following data were taken from information furnished by the manufacturers and from actual measurements of the engine.

Dimensions

Rpm, 2,800.
Number of cylinders, 6.
Diameter of cylinders, 3⅜ in.
Stroke, 5½ in.
Length of connecting rod, center to center, 11 in.
Distance of center of gravity of connecting rod from crankpin end, 2.66 in.
Principal radius of gyration (to be calculated).
Diameter of wrist pin, 1 in.
Diameter of crankpin, 2 in.
Diameter of main bearing, 2 in.
Center of gravity of unbalanced weight of crank from center line of crankshaft, 2⅝ in.

Weights

Connecting rod, complete (including bushings), 2.88 lb.
Piston, complete (including wrist pin, rings, etc.), 1.54 lb.
Estimated unbalanced weight of crank at crank radius, 4.75 lb.

Timing

Inlet valve opens 16 deg after head-end dead center.
Inlet valve closes 44 deg after crank-end dead center.
Exhaust valve opens 46 deg before crank-end dead center.
Exhaust valve closes 10 deg after head-end dead center.

Indicator Diagram

Compression volume, 23.7 per cent of piston displacement.
Maximum explosion pressure, 300 psia.
Pressure rise completed at 2 per cent of stroke.
Development of expansion curve begins at 3.2 per cent of stroke.
Mean suction pressure, 12 psia.
Mean exhaust pressure, 16 psia.
Atmospheric pressure, 14.7 psia.
Pressure at beginning of compression, atmospheric.
Exponent for expansion curve, 1.3.
Exponent for compression curve, 1.28.

13-1. Indicator Diagram. With the given timing data, the crank and piston positions showing the succession of events in the cylinder were laid out as shown in Fig. 13-1. On the stroke of the piston as a base line, the indicator diagram was next drawn. The ordinates used in constructing the expansion and compression curves, corresponding to piston positions laid out for each 15-deg crank position, are given in Table 13-1. In the determination of these ordinates the equation $PV^n = C$ was used. For convenience in making calculations, the equation may be put in the form $\log P = -n \log V + \log C$. Since this is a straight-line equation, the labor in determining the points on the expansion and compression curves is considerably reduced by making use of logarithmic cross-section paper as shown in Fig. 13-2. Points on the expansion curve were obtained by drawing the expansion line in accordance with the form $\log P = -1.3 \log V + \log C$ where pressures are ordinates and percentages of volume are abscissas. The pressure at the beginning of expansion is given as 300 psi, and the expansion curve begins at 3.2 per cent of the stroke, or at $23.7 + 3.2 = 26.9$ per cent of volume. Thus with the coordinates for one point and the slope (n) known, the expansion line is completely determined, and other points on the expansion curve may readily be found. Points for the compression curve are determined in like manner from the compression line in Fig. 13-2.

Since the ordinates in the indicator diagram represent absolute pressures in pounds per square inch, the total effective pressures on the piston (Table 13-1) are obtained by deducting the atmospheric pressure and multiplying by the area of the piston in square inches. In the original diagram the pressures were drawn to a scale of 1 in. = 50 psi.

13-2. Inertia Forces. Having obtained from the indicator diagram the gas pressures acting on the piston throughout the cycle, the next step in the analysis is to determine the inertia forces of the piston, connecting rod, and crank. Although these forces can be determined by analytical methods, the labor is greatly reduced by graphical methods such as have been used in this analysis.

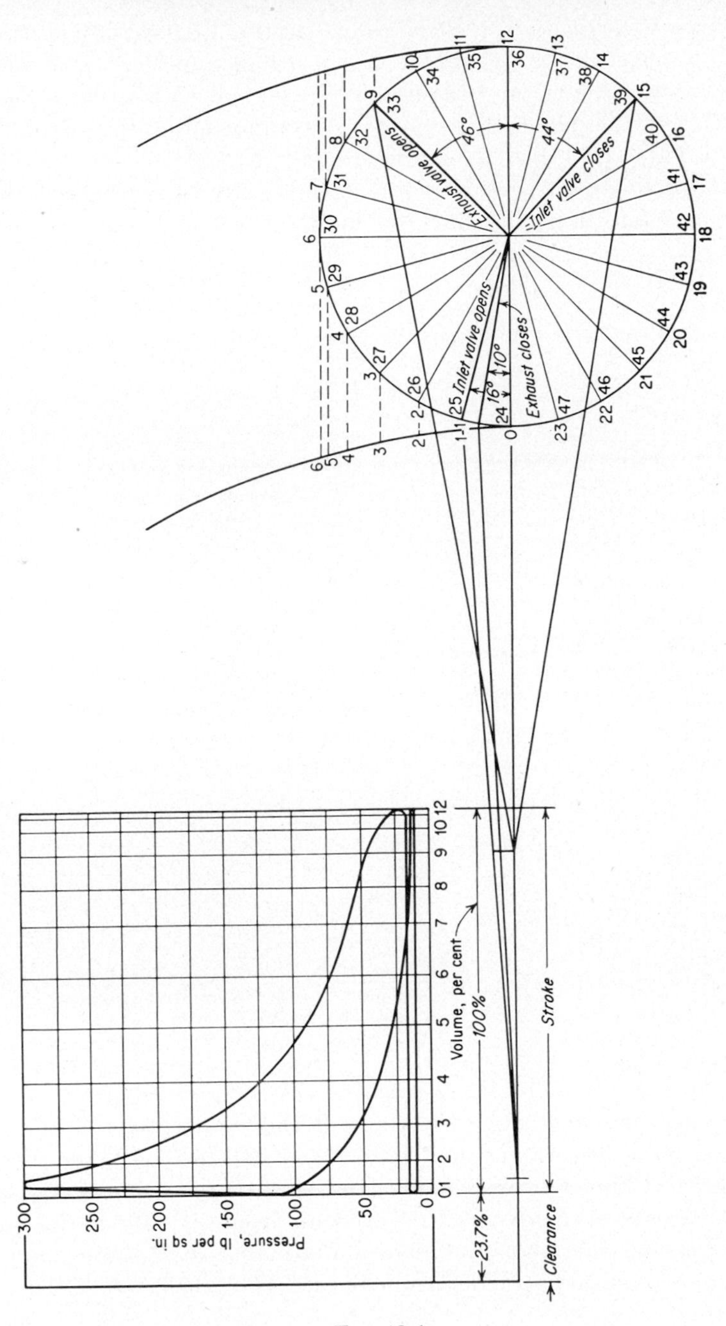

Exhaust valve opens
46°
44°
Inlet valve closes
Inlet valve opens
16°
10°
Exhaust closes
Pressure, lb per sq in.
300
250
200
150
100
50
0
Volume, per cent
0 1 2 3 4 5 6 7 8 9 10 12
100%
23.7%
Stroke
Clearance

FIG. 13-1

1. *Piston Inertia Forces.* By means of the Ritterhaus construction (Chap. 9), the vector *Ob* (Table 13-2), which represents to scale the acceleration of the piston, was obtained for each 15-deg crank position. From this the actual acceleration was determined, and the piston inertia force f_4 (Table 13-3) was then computed.

2. *Connecting-rod Inertia Forces.* A kinetically equivalent system was substituted for the connecting rod as described in Chap. 11. It was

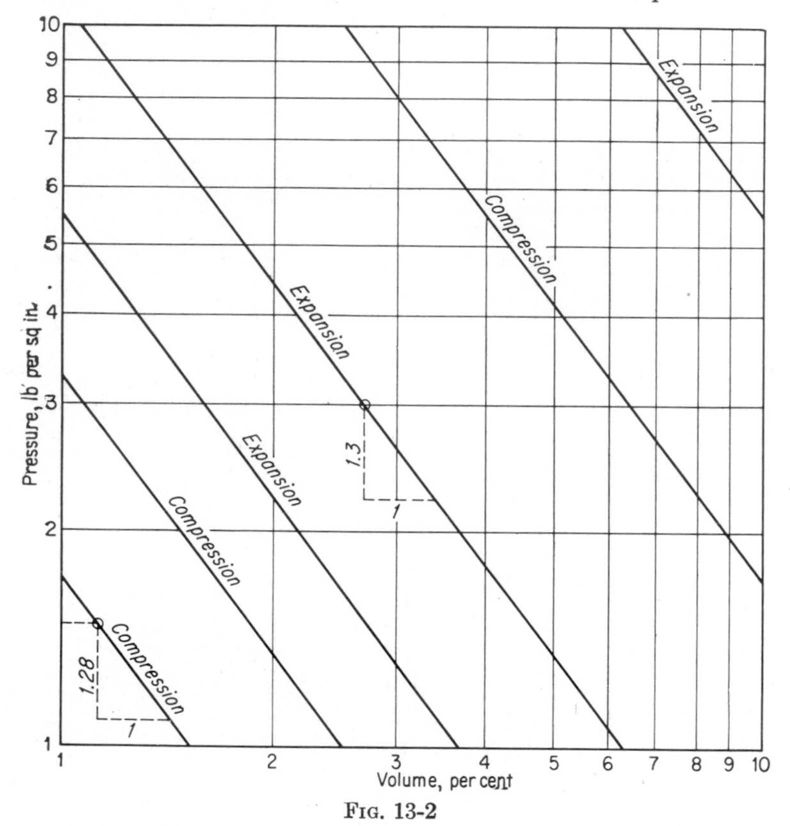

FIG. 13-2

found experimentally that the time for 50 complete swings (over and back constitutes one complete swing) of the connecting rod about the center of the wrist-pin end of the rod was 52.4 sec. By the methods explained in Chap. 11 it was found that with one mass at B (Fig. 13-3) the other mass at E was located 0.25 in. from A. By means of the Ritterhaus construction the acceleration of masses at B and E and of the entire connecting rod were determined, and Table 13-2 completed. The inertia forces f_E and f_B (Table 13-3) were next calculated and the resultant inertia forces of Table 13-3 determined by constructing the inertia-force polygons in accordance with the methods of Art. 11-9.

3. *Inertia Force of Crank.* From the given data the inertia force (centrifugal force) of the crank is

$$mR\omega^2 = 4.75/32.17 \times 2.75/12 \times (2\pi \times 2{,}800)^2/60^2 = 2{,}915 \text{ lb}$$

This force exerts a pressure on the crankshaft bearing but has no effect

TABLE 13-1. INDICATOR-CARD DATA

Expansion curve					Compression curve				
			Gas pressure					Gas pressure	
Piston position	Crank angle, deg	Volume ratio	Absolute pressure, psi, P_a	Total effective pressure $(P_a - 14.7) \times A = P_g \times A$	Piston position	Crank angle	Volume ratio	Absolute pressure, psi, P_a	Total effective pressure, $(P_a - 14.7) \times A = P_g \times A$
0	0	0.237	108	834	36	180	1.237	14.0	−6.3
1	15	0.259	300	2,550	37	195	1.224	14.7	0
e	.....	0.269	300	2,550	38	210	1.185	14.7	0
2	30	0.320	239	2,005	c	...	1.127	14.7	0
3	45	0.415	171	1,370	39	225	1.122	14.8	0.9
4	60	0.535	123	950	40	240	1.034	16.4	15.2
5	75	0.667	92.4	682	41	255	0.926	18.9	37.5
6	90	0.800	72.7	518	42	270	0.800	22.8	72.4
7	105	0.926	60.1	406	43	285	0.667	29.0	127.9
8	120	1.034	52.2	327	44	300	0.535	38.3	211
9	135	1.122	46.9	288	45	315	0.415	53.1	343
10	150	1.185	42.8	251	46	330	0.320	73.5	531
11	165	1.224	35.6	187	47	345	0.258	95.7	742
12	180	1.237	26.0	101	48	360	0.237	108	834
12–24*	180–360	.....	16.0	11.6					
24–36†	0–180	.....	12.0	−24.1					

* Exhaust line.
† Intake line.

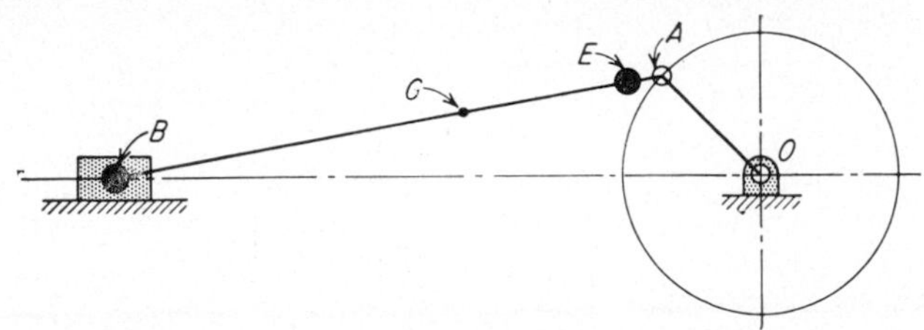

FIG. 13-3

on the piston or the connecting rod. It has therefore been omitted when drawing the inertia-force polygons such as are illustrated in Fig. 11-12 (Art. 11-9). Moreover, if balancing weights are attached to the crankshaft opposite each crank, which is common practice, this force is completely neutralized.

TABLE 13-2. ACCELERATIONS

Crank position				Acceleration of piston $Ob \times k_a = A_B$		Acceleration of center of percussion $Oe \times k_a = A_E$		Acceleration of center of gravity $Og \times k_a = A_G$	
Number		Angle, deg		Vector Ob, in.	Acceleration, ft/sec²	Vector Oe, in.	Acceleration, ft/sec²	Vector Og, in.	Acceleration, ft/sec²
0	24	0	360	6.875	24,630	5.53	19,840	5.83	20,900
1	23	15	345	6.56	23,500	5.53	19,840	5.70	20,430
2	22	30	330	5.46	19,600	5.46	19,600	5.34	19,150
3	21	45	315	3.90	13,980	5.43	19,500	4.88	17,520
4	20	60	300	2.07	7,420	5.36	19,220	4.46	16,000
5	19	75	285	0.21	753	5.33	19,130	4.17	14,950
6	18	90	270	1.44	5,160	5.35	19,190	4.19	15,030
7	17	105	255	2.65	9,500	5.36	19,220	4.40	15,770
8	16	120	240	3.43	12,300	5.40	19,360	4.62	16,560
9	15	135	225	3.86	13,830	5.44	19,500	4.88	17,500
10	14	150	210	4.03	14,460	5.46	19,590	5.04	18,060
11	13	165	195	4.11	14,730	5.47	19,600	5.11	18,310
12	...	180	...	4.125	14,770	5.47	19,600	5.167	18,350

TABLE 13-3. INERTIA FORCES AND THEIR RESULTANTS, IN LB

Crank position				Inertia forces				Resultant inertia forces		
Number		Angle, deg		f_E	$f_E \dfrac{EB}{AB}$	f_B	f_4	Turning force f^t_{32}	Normal force crank f^n_{32}	Side thrust f_{14}
0	24	0	360	1,375	1,343	492	1,178	0	3,000	0
1	23	15	345	1,375	1,343	470	1,124	550	2,880	120
2	22	30	330	1,360	1,399	392	938	860	2,400	190
3	21	45	315	1,353	1,322	280	668	830	1,880	200
4	20	60	300	1,333	1,302	148	355	520	1,460	140
5	19	75	285	1,328	1,298	16	36.0	60	1,300	45
6	18	90	270	1,330	1,300	103	247	350	1,390	55
7	17	105	255	1,333	1,302	190	454	610	1,630	135
8	16	120	240	1,342	1,311	246	588	665	1,900	165
9	15	135	225	1,353	1,322	276	662	570	2,120	160
10	14	150	210	1,358	1,327	289	691	410	2,260	110
11	13	165	195	1,360	1,329	294	704	210	2,340	60
12	...	180	...	1,360	1,329	296	706	0	2,390	0

13-3. Combined or Resultant Forces. In order to obtain the total forces acting on the several links of the mechanism, it is necessary to combine the forces due to gas pressure (static forces) and those due to inertia. The forces due to gas pressure are found by means of force

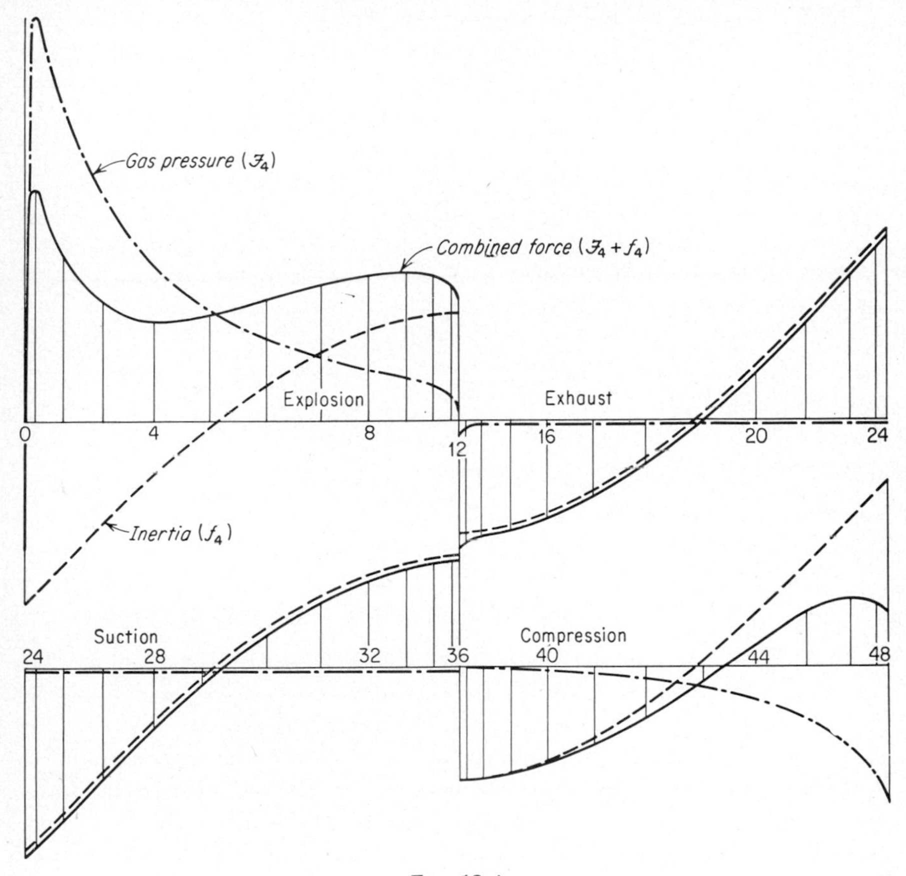

Fig. 13-4

polygons such as described in Art. 11-9, and the results are recorded in Table 13-4 as shown. The table is then completed by combining the gas forces with the inertia forces previously determined.

The investigation up to this point has been very largely concerned with the work of securing and recording all the forces acting on the various members for each 15-deg crank position throughout one complete cycle (two revolutions). The characteristics of these forces and their effects can best be understood from a study of the diagrams (Figs. 13-1 to 13-11) that have been plotted from the data recorded in Tables 13-1 to 13-4.

13-4. Indicator Diagram (Fig. 13-1). This diagram, which has already been discussed, is the starting point in the analysis. It very closely approximates the form of the actual diagram taken from an engine of this design, and the piston pressures obtained therefrom must therefore closely approximate the pressures that actually occur.

13-5. Diagram of Piston Forces (Fig. 13-4). The purpose of this diagram is to show for one cylinder the variation of piston forces that act in the line of stroke, during one cycle. The diagrams have been developed on a base line representing the total motion of the piston (disregarding reversals) during a cycle. The forces due to gas pressure $\mathfrak{F}_4$ (Table 13-1) and the forces due to inertia f_4 (Table 13-3) have been plotted separately and then combined to form the total-pressure diagram $F_4 = \mathfrak{F}_4 + f_4$. Forces aiding the motion of the piston are considered positive and are plotted above the base line, and forces resisting the motion are considered negative and are plotted below the base line.

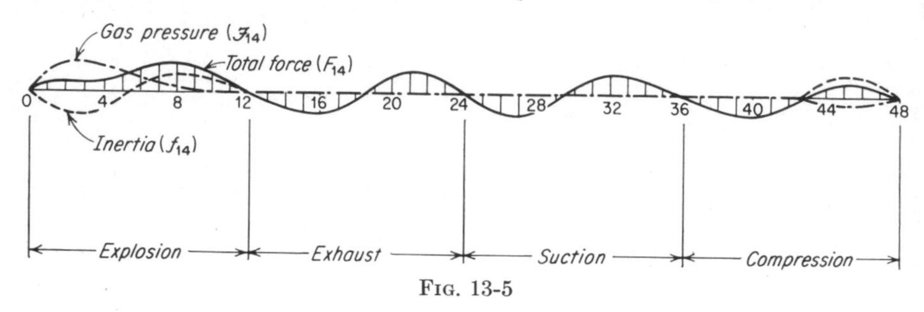

FIG. 13-5

13-6. Side Force of the Piston (Fig. 13-5). This diagram shows the variation of pressure of the piston against the wall of each cylinder during a cycle. From Tables 13-3 and 13-4 the side thrust $\mathfrak{F}_{14}$ due to gas pressure and the side thrust f_{14} due to inertia have been plotted separately and then combined to form the total side-pressure diagram F_{14}. The base line is of arbitrary length, representing the 48 crank positions at 15-deg intervals for one cycle. The scale of the original drawing was 1 in. = 500 lb.

13-7. Turning-effort Diagrams for One Cylinder (Fig. 13-6). The tangential component of the force on the crankpin is the *turning effort*. On an arbitrary length of base line representing the 48 15-deg crank intervals in a cycle, turning effort $\mathfrak{F}^t_{32}$ due to gas pressure, turning effort f^t_{32} due to inertia, and combined turning effort F^t_{32} from Tables 13-3 and 13-4 have been plotted as shown in Fig. 13-6. Turning effort that aids the motion has been plotted above the base line, and that which opposes the motion has been plotted below the base line.

13-8. Combined Turning-effort Diagrams (Fig. 13-7). The total turning effort of the engine is the result of the combined efforts of the six

TABLE 13-4. FORCES DUE TO GAS PRESSURE AND TOTAL FORCES, IN LB

Position		Turning forces due to gas pressure, $\mathfrak{F}^t_{32}$	Total turning forces, F^t_{32}	Gas pressure N, $\mathfrak{F}^n_{32}$	Total along ℄ crank, F^n_{32}	Total force C. rod on crank, F_{32}	Side thrust due to gas pressure, $\mathfrak{F}_{14}$	Total side thrust, F_{14}	Total force C. rod on piston, F_{34}
Number	Angle, deg								
				Explosion Stroke					
0	0	0	0	848	2,155	2,155	0	0	326
1	15	815	290	2,430	460	540	160	50	1,430
2	30	1,240	400	1,600	800	900	260	90	1,100
3	45	1,155	320	810	1,000	1,050	250	70	730
4	60	950	430	300	1,170	1,250	220	80	630
5	75	720	660	15	1,385	1,575	180	120	670
6	90	500	860	130	1,500	1,750	130	180	780
7	105	370	980	210	1,835	2,075	100	240	900
8	120	250	920	235	2,120	2,310	80	250	950
9	135	165	740	240	2,370	2,480	65	225	980
10	150	100	510	230	2,495	2,545	35	150	950
11	165	35	240	182	2,515	2,530	20	80	895
12	180	0	0	110	2,400	2,400	0	0	816
				Exhaust Stroke					
12	180	.....*	0	.....*	2,400	2,400	...*	0	816
13	195	.....	210	.....	2,340	2,350	...	60	705
14	210	.....	410	.....	2,260	2,290	...	110	700
15	225	.....	510	.....	2,120	2,195	...	160	675
16	240	.....	665	.....	1,900	2,015	...	165	610
17	255	.....	610	.....	1,630	1,740	...	135	475
18	270	.....	350	.....	1,390	1,440	...	55	255
19	285	.....	60	.....	1,300	1,300	...	45	55
20	300	.....	520	.....	1,460	1,555	...	140	375
21	315	.....	830	.....	1,880	2,040	...	200	690
22	330	.....	860	.....	2,400	2,545	...	190	945
23	345	.....	550	.....	2,880	2,920	...	120	1,125
24	360	.....	0	.....	3,000	3,000	...	0	1,145
				Suction Stroke					
24	0	.....*	0	.....*	3,000	3,000	...*	0	1,145
25	15	.....	550	.....	2,880	2,900	...	120	1,125
26	30	.....	860	.....	2,400	2,560	...	190	945
27	45	.....	830	.....	1,880	2,050	...	200	690
28	60	.....	520	.....	1,460	1,555	...	140	380
29	75	.....	60	.....	1,300	1,300	...	45	55
30	90	.....	350	.....	1,390	1,440	...	55	255
31	105	.....	610	.....	1,630	1,740	...	135	475
32	120	.....	665	.....	1,900	2,011	...	165	610
33	135	.....	570	.....	2,120	2,195	...	160	675
34	150	.....	410	.....	2,260	2,300	...	116	700
35	165	.....	210	.....	2,340	2,350	...	60	705
36	180	.....	0	.....	2,390	2,391	...	0	707
				Compression Stroke					
36	180	0	0	0	2,390	2,390	0	0	
37	195	.....*	210	.....*	2,340	2,340	...*	60	720
38	210	.....	410	.....	2,260	2,290	...	110	695
39	225	.....	570	.....	2,120	2,190	...	160	680
40	240	.....	665	.....	1,900	2,100	...	165	650
41	255	30	610	16	1,640	1,750	10	140	512
42	270	70	412	19	1,400	1,450	18	65	325
43	285	125	55	0	1,295	1,310	25	20	95
44	300	205	315	70	1,390	1,425	50	85	180
45	315	290	540	195	1,695	1,780	60	130	365
46	330	320	530	410	2,015	2,090	65	130	425
47	345	235	290	680	2,200	2,220	45	60	410
48	360	0	0	848	2,155	2,155	0	0	326

* Negligible.

separate cylinders. In Fig. 12-27 (Art. 12-20) is shown the order in which the events occur in the different cylinders. With the aid of this diagram and data from Tables 13-3 and 13-4, the combined effort for the six cylinders was plotted as shown in Fig. 13-7. Since ordinates represent

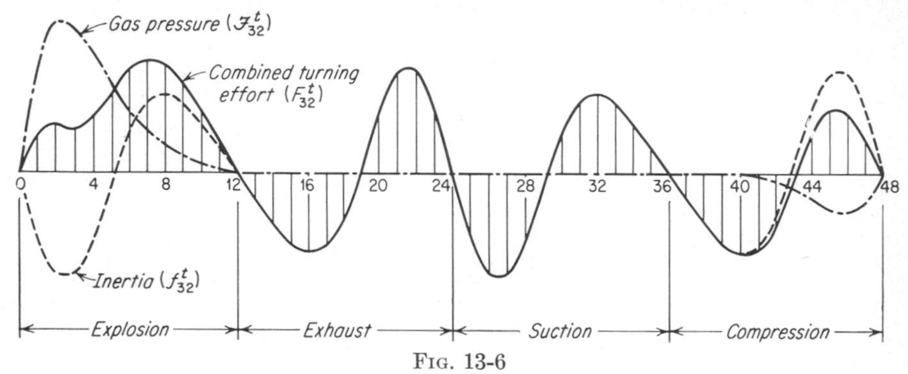

FIG. 13-6

turning effort and abscissas represent movement of the crank, the area under the curve represents the work performed. It is the function of the flywheel to maintain a uniform turning effort of the magnitude represented by the mean ordinate shown in the figure. The areas above the mean

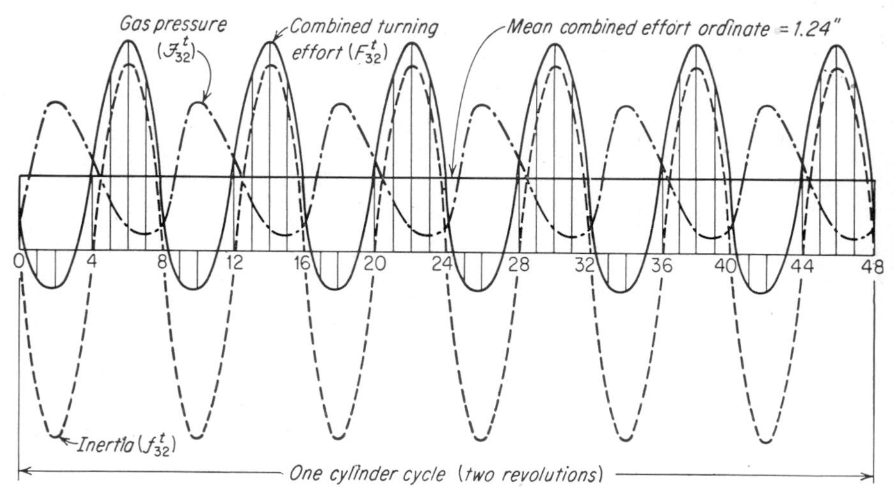

FIG. 13-7

ordinate line therefore represent excess energy that is stored in the flywheel and given up in those parts of the cycle where there is a deficiency of energy, as indicated by the *negative* areas under the mean ordinate line.

Since friction has been neglected in the analysis, the horsepower of the engine when calculated from the turning-effort diagram should be equal to the horsepower calculated from the indicator diagram.

13-9. Pin Forces (Figs. 13-8 to 13-10). In the design of the wrist pin, crankpin, and crankshaft it is necessary to have information regarding the magnitude and distribution of the forces acting on these members and their bearings in order to proportion for strength and wear, and also to determine the best place for the oil to enter the bearing.

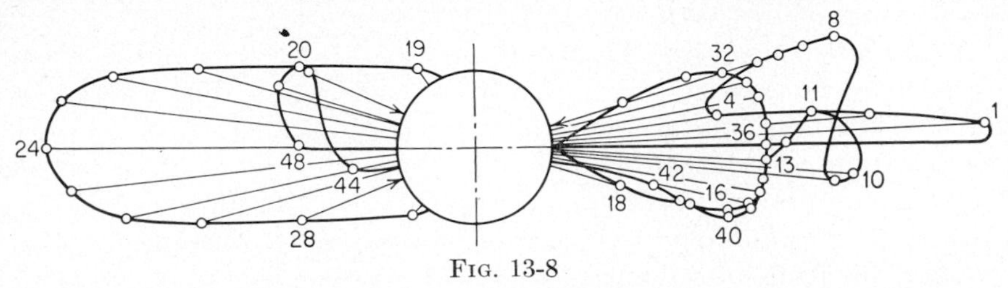

FIG. 13-8

1. *Wrist-pin Forces (Fig.* 13-8). In Fig. 13-8 the wrist pin was drawn full size in the original drawing, and resultant forces F_{34} from Table 13-4 plotted in magnitude and direction as vectors acting against the surface of the pin. The vectors thus represent forces acting on the pin, and the figure shows clearly where wear may be expected to occur on the pin. The pin was assumed to be fixed in the piston, and in the original drawing the vectors were plotted to a scale of 1 in. = 500 lb.

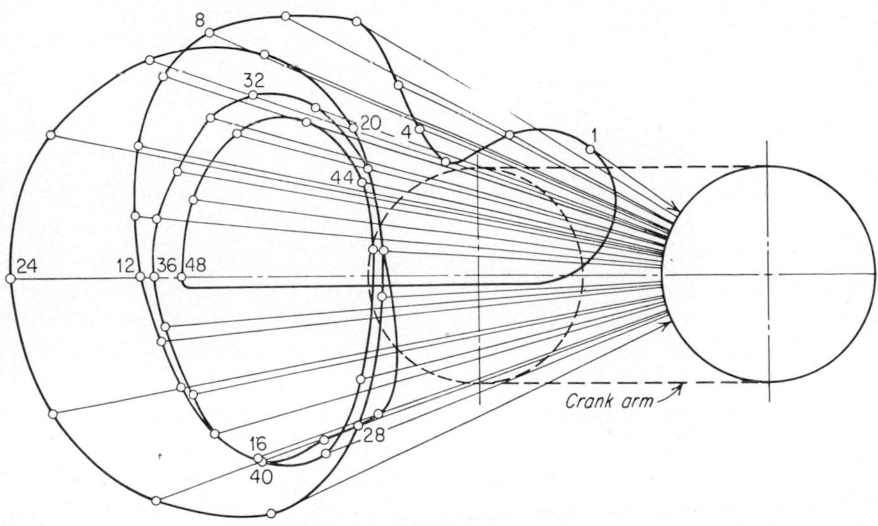

FIG. 13-9

2. *Crankpin Forces (Fig.* 13-9). In the original drawing (Fig. 13-9) the crankpin was drawn full size and the resultant forces F_{32} (Table 13-4) were drawn to a scale of 1 in. = 500 lb as vectors representing forces acting against the surface of the pin. Since the crankpin is rotating, the vectors are drawn, not in their true directions, but in their directions

relative to a line turning with the pin, in this case the center line of the crank. The figure clearly shows that the pressure is always on the same side of the pin, thus demonstrating the cause of uneven wear of the pin. The diagram is also of value in showing where the oil-feed hole in the crankpin should be located, namely, on the side opposite that on which the pressure occurs.

3. *Main-bearing Forces* (*Fig.* 13-10). The force on the crankshaft per cylinder is evidently F_{12} ($= -F_{32}$) from Table 13-4. The reaction or force on the bearing is therefore F_{21} ($= F_{32}$). With this force must be combined the centrifugal force of the crank in order to obtain the resultant force on the bearing if the crank is unbalanced. For a seven-bearing crankshaft the resultant force on each bearing may be assumed as half the forces from each of the pair of adjacent cranks.

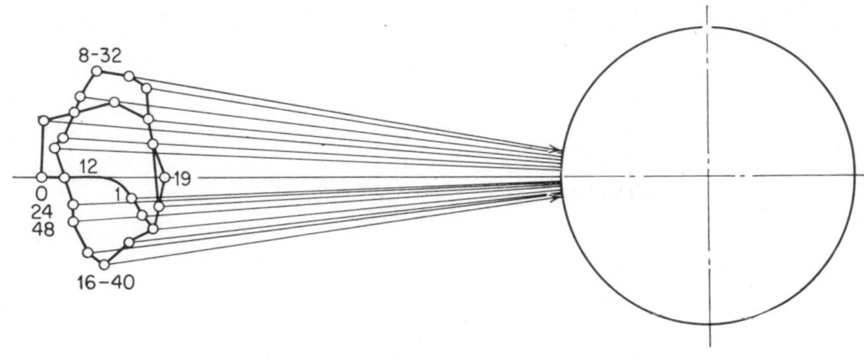

Fig. 13-10

In Fig. 13-10 are shown the magnitude and distribution of forces acting on the crankshaft at the center main bearing. These forces were obtained by combining one-half of F_{12} ($= -F_{32}$) from Table 13-4 for each of the adjacent cranks with one-half the centrifugal force of each of the adjacent cranks. The combined forces are represented by vectors drawn against the surface of the crankshaft. In the original drawing the vectors were drawn to a scale of 1 in. = 1,000 lb.

13-10. Diagram of Shaking Forces (Fig. 13-11). The centrifugal force (inertia force) of the crank arm and pin (when unbalanced), the inertia force of the connecting rod, and the inertia force of the piston all tend to cause vibrations of the engine. The shaking forces for a single cylinder are shown in Fig. 13-11. The vectors in the original diagram were drawn to a scale of 1 in. = 1,500 lb. Force polygons have been drawn for each 15-deg crank position, the end of the resultant vector locating a point on the curve. Referring to crank position 8, from a pole O the vector OA was laid off to represent the centrifugal force of the crank arm and pin. AB represents the inertia force of the connecting rod ($f_E \nrightarrow f_B$) (Table 13-3). BC represents the shaking force due to the mass of the piston. OC represents the resultant shaking force for this crank position, and C is one point on the curve of shaking forces for one revolution of the crank. Now let it be assumed that the crank arm and pin are to be balanced by a rotating weight opposite the crankpin. This introduces a new vector CE ($= OA$) in the force polygon $OABC$, giving a new resultant OE, the point E being a point on the curve of shaking forces with the crank arm and pin balanced. Only the shaking forces of the connecting rod and piston now remain.

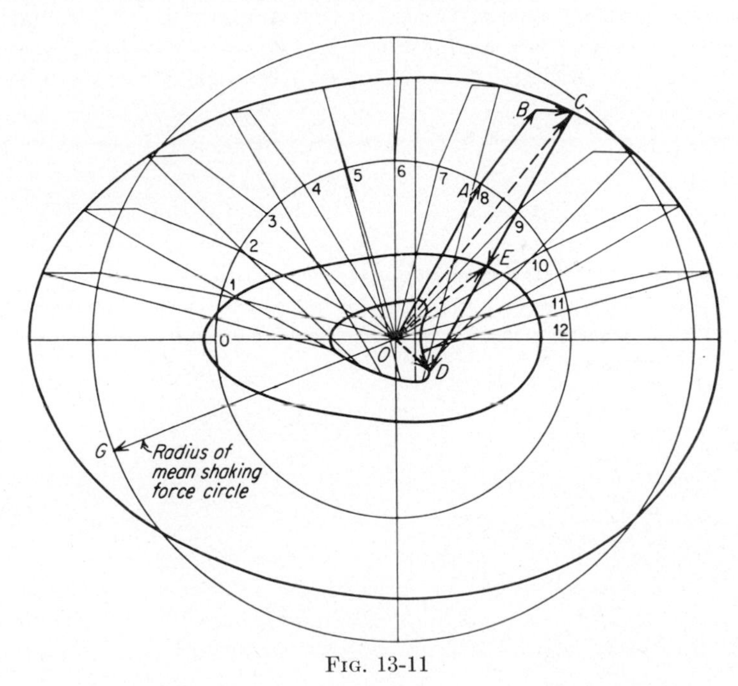

FIG. 13-11

If, instead of balancing the crank arm and pin only, the rotating weight opposite the crankpin is of magnitude such that its centrifugal force is represented by the vector OG, the radius of the mean shaking-force circle, the vector CD ($= OG$), will be brought into the original force polygon $OABC$, giving the resultant OD. D is one point on the curve of remaining shaking forces. This may be considered to represent the best balancing condition, since any further increase in the size of the counterweight, although decreasing the horizontal forces in curve D, would increase the vertical forces.

It should be kept in mind that curves C, E, and D represent the shaking forces for a single cylinder for the conditions stated. This would be the proper procedure in the exact analysis of the balancing conditions for a *single-cylinder* engine. In a six-cylinder engine the shaking forces for the six cylinders combined form a balanced system. In the case of this particular engine, therefore, the diagram (Fig. 13-11) is chiefly of interest in showing what shaking forces are set up in the plane of rotation of each crank.

13-11. Determination of Flywheel Mass. In Art. 13-8 it was shown that the areas above the mean ordinate line (constant load line) in Fig. 13-7 represent the excess of energy that must be stored in the flywheel in order to be available to supply the deficiency of energy represented by the areas below the constant load line. Since in Fig. 13-7 these loops all have the same areas, the area of one of the loops in heavy outline above the constant load line will determine the mass of the flywheel. Therefore, in computing the mass of the flywheel it will be necessary to determine how many foot-pounds of energy this area represents. With the energy that must be stored in the flywheel and the permissible speed fluctuation known, its mass can be computed from the well-known equation for the energy of a rotating mass.

The kinetic energy of a rotating mass m having a moment of inertia I about the center of rotation, a radius of gyration k, and an angular velocity ω is

$$E_k = \tfrac{1}{2}I\omega^2 = \tfrac{1}{2}mk^2\omega^2 \tag{13-1}$$

The change in energy stored in the flywheel during a change in angular velocity from ω_1 to ω_2 is

$$\Delta E_k = \tfrac{1}{2}I(\omega_2^2 - \omega_1^2) = \tfrac{1}{2}m(v_2^2 - v_1^2) \tag{13-2}$$

where ΔE_k = change in rotational kinetic energy, ft-lb

ω_2, ω_1 = angular velocities, radians/sec, at the end and beginning, respectively, of the interval of change: ω_2 and ω_1 represent, in this problem, the extreme values of ω allowable within a complete cycle of engine operation

v_2, v_1 = corresponding magnitudes of velocities, fps, at the radius of gyration at the end and beginning, respectively, of the interval of change

m = mass of flywheel, slugs (= $\text{lb}_m/32.17$)

A coefficient of fluctuation K is defined as the difference between the maximum and minimum angular velocities during an energy cycle divided by the average velocity ω. That is,

$$K = \frac{\omega_2 - \omega_1}{\omega} = \frac{v_2 - v_1}{v} \tag{13-3}$$

Therefore, $\qquad \omega_2 - \omega_1 = K\omega \qquad \text{or} \qquad v_2 - v_1 = Kv \qquad$ (13-4)

If it is assumed that the speed varies by the same amount above and below the average (or nominal) value, then

$$\frac{\omega_2 + \omega_1}{2} = \omega \qquad \text{or} \qquad \frac{v_2 + v_1}{2} = v \tag{13-5}$$

Multiplying (13-4) and (13-5) together,

$$\omega_2^2 - \omega_1^2 = 2K\omega^2 \qquad \text{or} \qquad v_2^2 - v_1^2 = 2Kv^2 \tag{13-6}$$

Substituting these values in Eq. (13-2) gives

$$\Delta E_k = KI\omega^2 \qquad \text{or} \qquad \Delta E_k = Kmv^2 \tag{13-7}$$

Either of these alternative equations can be used.

The kinetic-energy variation ΔE_k is determined as follows. In the original drawing (Fig. 13-7) the length of the diagram was 15 in. and the forces were drawn to a scale of 1 in. = 500 lb. The area of the combined turning-effort diagram is then $1.24 \times 15 = 18.60$ sq in. This area represents the energy or work output per cycle (per two revolutions). The energy per cycle is equal to the mean tangential effort times the distance traveled by the crankpin in one cycle (two revolutions), or $1.24 \times 500 \times \pi \times 5.5/12 \times 2 = 1{,}785$ ft-lb. One square inch of the tangential-effort diagram therefore represents $1{,}785/18.60 = 96$ ft-lb.

The area of one of the loops of the combined turning-effort diagram above the mean ordinate line in Fig. 13-7 is 1.80 sq in. The maximum variation in energy is therefore $1.80 \times 96 = 173$ ft-lb.

It will be assumed that the flywheel has a radius of gyration k of 6.38 in., and that the speed is not to fluctuate more than 5 rpm above or below an average of 2,800 rpm. We have then from Eq. (13-3)

$$K = \frac{\omega_2 - \omega_1}{\omega} = \frac{10}{2{,}800} = 0.0036$$

The average velocity at the radius of gyration is

$$v = r\omega = \frac{6.38}{12} \times \frac{2{,}800}{60} \times 2\pi = 155.8 \text{ fps}$$

Substituting in Eq. (13-7),

$$m = \frac{\Delta E_k}{Kv^2} = \frac{173}{0.0036 \times 155.8^2} = 1.98 \text{ slugs}$$

Expressed in pounds, the mass of the flywheel is 63.6 lb. The flywheel must be proportioned, then, so that it has a mass of 63.6 lb when it has a radius of gyration of 6.38 in. Any other combination of mass and radius of gyration which gives the same moment of inertia I and fits the available space in the engine would be satisfactory, of course.

The foregoing calculation was made for one condition of operation in order to illustrate the procedure. Flywheel dimensions depend not only upon the size and number of cylinders but upon the relative importance attached to such factors as smooth action at low speeds, rapid acceleration, and compactness in design. The coefficients of fluctuation recommended for use in the design of flywheels vary over such a wide range that it is necessary to have in mind all conditions that apply before an intelligent selection can be made.

It should be recognized, too, that the crankshaft and other engine parts have masses which produce a flywheel effect which has not been taken into account. Therefore, the speed variation would probably be less than the selected value of 5 rpm.

As suggested in Art. 13-8, the horsepower of the engine should be the same whether calculated from the indicator diagram or the turning-effort diagram. Obviously, there will be some difference, but they should check closely if the drafting work has been done carefully. The area of the indicator diagram (Fig. 13-1) as found by use of the planimeter is 7.81 sq in. The indicated horsepower is then

$$\text{ihp} = \frac{PLAN}{33{,}000} = \frac{7.81}{5.5} \times 50 \times \frac{5.5}{12} \times \frac{3.375^2\pi}{4} \times 3 \times \frac{2{,}800}{33{,}000} = 74$$

The horsepower from the turning-effort diagram (Fig. 13-7) is

$$\text{ihp} = \frac{(\text{tangential force})(\text{velocity of crankpin, fpm})}{33{,}000} = \frac{1.24 \times 500 \times \pi \times 5.5/12 \times 2{,}800}{33{,}000} = 76$$

It is worth noting that the energy-storage property of a rotating wheel finds application in many machines other than engines. Examples are inertia starters for aircraft engines, punch presses, and even vehicles in which downhill braking occurs by energy storage in a heavy rotating flywheel, with subsequent delivery of the stored energy to help move the vehicle on the level or uphill. The analyses for such applications have much in common with the analysis given for engine flywheels.

CHAPTER 14

VIBRATIONS AND CRITICAL SPEEDS IN SHAFTS

When an external force changes the shape of a body within its elastic limit, internal restoring forces are set up that tend to oppose the external force. If the external force ceases to act, the internal forces will return the body to its original shape, but because of its mass the body will pass through its position of equilibrium (i.e., its original shape) and will be distorted in the opposite direction. Thus there will be brought into being, due to inertia effects alone, restoring forces that in general are directed opposite those set up by the original external force. These pendulumlike movements through the position of equilibrium are called vibrations.

A vibration of this kind, in which, after the initial displacement, no external forces act and the motion is maintained by the internal elastic forces, is termed a *free* or *natural vibration.* In practice, the energy possessed by the system is gradually dissipated in overcoming internal and external resistances to the motion, and the body finally comes to rest. Such a vibration is said to be *damped.* A third type of vibration, which is of great practical importance, is that in which a periodic disturbing force is applied to the body. The vibration then has the same frequency as the applied force and is said to be a *forced vibration.*

When a system is acted upon by an external periodic force having the same frequency as a natural frequency of the system, the amplitude of the system will become very large, and the system is said to be in a state of *resonance.* A *critical speed* exists when the frequency of the disturbing force equals or approaches a natural frequency of the system.

14-1. Characteristics of Shaft Vibration. Shafts are subject to longitudinal, lateral (or transverse), and torsional vibrations. Of these, the last two are the types most commonly encountered. Any rotating shaft carrying heavy masses will be subject to lateral vibrations by reason of the flexure of the shaft. Thus, the chief cause of lateral vibrations is the centrifugal force resulting from unbalance in the rotating masses. Torsional vibrations are caused by periodic forces that tend to twist the shaft about its axis.

When the number of power impulses on an engine shaft or the number of revolutions per minute of a propeller shaft, for example, equals a natural frequency of vibration of the shaft, the system is in a state of resonance, and the amplitude of vibration may become very large. If operation is continued at or near this speed, failure is likely to occur. The critical speed of a shaft may be defined, then, as the speed at which the frequency of the disturbing force equals or approaches a natural frequency of vibration of the shaft.

Many shaft failures due to vibration have occurred in engine crankshafts and propeller shafts, owing to the fact that the matter of vibration was not given sufficient attention in their design. Sometimes, where a machine must operate at variable speeds, it may be difficult or impossible to avoid resonance at all speeds. But if the problem is properly analyzed, objectionable vibrations can generally be eliminated or reduced to the point where they are no longer objectionable. Several methods of eliminating or reducing vibrations are employed. The more important of these methods are the following: balancing to eliminate the disturbing force, designing to avoid resonance, introducing frictional forces that will cause damping, and isolating the vibrations by means of elastic supports.

LATERAL VIBRATIONS IN SHAFTS

14-2. Fundamental Equations for the Lateral Vibration of a Shaft. In Fig. 14-1 is shown a mass m located at some point along an elastic rod (or shaft) which is freely supported at the ends. The rod itself is assumed to be weightless. Suppose that the rod is deflected an amount x from its position at rest and then released. Within the elastic limit the force required to produce this deflection is kx, where k is the spring constant, i.e., the force required to produce a deflection of 1 in.

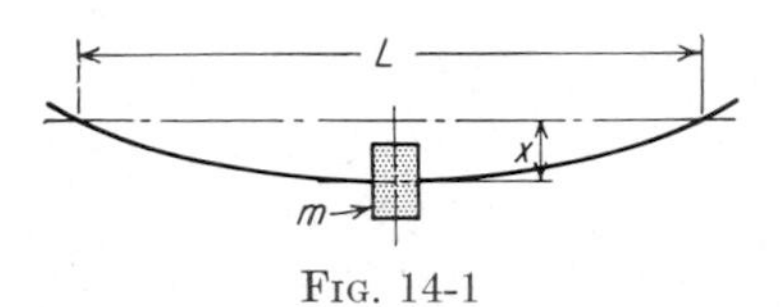

FIG. 14-1

The restoring force exerted on the mass m by the rod, by virtue of its elasticity, is $-kx$. Within the elastic limit, this force is directly proportional to the displacement of the mass from its equilibrium position. Hence, it follows that the acceleration toward the equilibrium position is directly proportional to the displacement from that position, and the vibration is, therefore, simple harmonic. Applying the equation of motion $\Sigma F = ma$, we have

$$m \frac{d^2x}{dt^2} = -kx \tag{14-1}$$

This is the general equation for the free vibration of a body.[1] The negative sign indicates that the acceleration is opposite the displacement.

The reciprocating harmonic motion of a vibrating particle can be represented by the projection of the end of a rotating vector on the diameter of the circle described by the end of the vector. The center of the circle will represent the position of static equilibrium of the particle; the length of the vector will be equal to the amplitude of the vibration; and the projection of the vector on the diameter will be a sine or cosine function of the angle through which the vector has moved. In

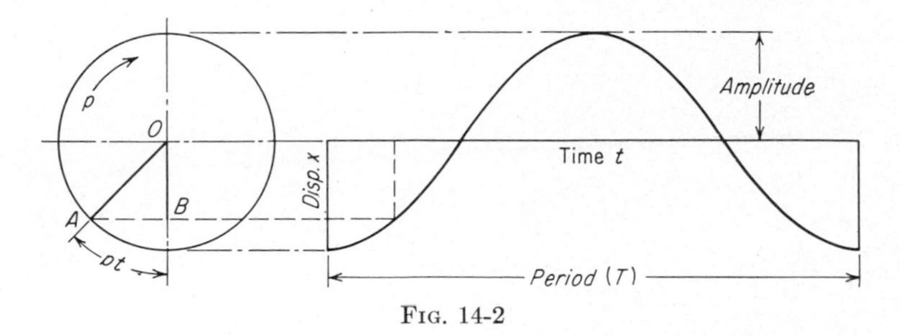

FIG. 14-2

Fig. 14-2 the center O of the circle represents the position of static equilibrium of the vibrating mass m in Fig. 14-1, the vector OA represents the amplitude of the vibration, and the distance OB ($= OA \cos pt$) represents the displacement x. It is evident, therefore, that, as the mass m moves along its vertical path, its position along the path is at the horizontal projection of the end of the vector OA, which rotates at a uniform rate of p radians per second. The angular position of the rotating vector is given by the term pt, where t is time in seconds. If one cycle of the motion (corresponding to one revolution of the vector OA) is completed in time T, it is evident that $pT = 2\pi$ radians. The period of the vibration is then

$$T = \frac{2\pi}{p} \tag{14-2}$$

This is the time in seconds required for a free vibration. The reciprocal of this quantity is the number of free vibrations per second, or the natural

[1] This equation and its solution as indicated in the following article have broader application than to analysis of shafts only. Although space here is not available for detailed treatment, the problems for Chap. 14 in the Appendix include the use of these equations and extensions of them in the solution of simple but practical problems concerned with elastic mountings for machines.

frequency of the vibration. Denoting this frequency by f, we have

$$f = \frac{1}{T} = \frac{p}{2\pi} \tag{14-3}$$

In Fig. 14-2 the displacement-time curve, a cosine wave, was obtained by plotting the harmonic motion against time.

14-3. Solution of the General Equation. Proceed now to the solution of Eq. (14-1) in order to find the value of p in Eqs. (14-2) and (14-3). It will be necessary to find for x a function of time that will satisfy the equation. It will be found that we need for x such a function of time that after twice differentiating it with respect to time we obtain the same function with which we started, multiplied by $-k/m$. It is in this way only that the equation can be satisfied. The functions $x = A \cos pt$ and $x = B \sin pt$ satisfy this requirement. Furthermore, these functions can be added together and still give a function that satisfies this requirement. Thus

$$x = A \cos pt + B \sin pt \tag{14-4}$$

in which A and B are two arbitrary constants. Equation (14-4) gives the general solution of Eq. (14-1), since it contains two constants of integration, as required for a second-order differential equation (14-1). By an appropriate choice of the constants A and B the solution can be adapted to any initial conditions of motion of the mass m.

The general case is that in which the mass is given both an initial displacement and an initial velocity. Assume, for example, that the vibration of the mass is started by displacing it from its equilibrium position an amount x_0 and then giving to it in this displaced position some initial velocity v_0. In this case we have as the initial conditions

$$\text{Displacement } x = x_0 \qquad \text{when } t = 0$$
$$\text{Velocity } v = v_0 = \frac{dx_0}{dt} \qquad \text{when } t = 0$$

Substituting the first condition above into the general solution [Eq. (14-4)], we find $A = x_0$. Now differentiating Eq. (14-4) once with respect to time we have

$$v = \frac{dx}{dt} = -Ap \sin pt + Bp \cos pt \tag{14-5}$$

Substituting the second of the conditions given above into Eq. (14-5), we find that $B = v_0/p$, and the general solution [Eq. (14-4)] becomes

$$x = x_0 \cos pt + \frac{v_0}{p} \sin pt \tag{14-6}$$

in which x_0 is the initial displacement and v_0 the initial velocity of the mass m.

In addition to the general case represented by Eq. (14-6), two particular cases are common in vibration problems: that where the mass m is given an initial displacement x_0 and released without initial velocity, in which case the second term of Eq. (14-6) drops out, and that where the vibration of the mass is started by giving it an initial velocity v_0 at the position of static equilibrium, in which case the first term of Eq. (14-6) drops out.

The displacement-time curve A representing the motion for the particular case of Eq. (14-6), where the initial velocity $v_0 = 0$, is shown in Fig. 14-3. From the displacement-time diagram it is evident that the maximum numerical value of the displacement is x_0 and is the amplitude of the vibration.

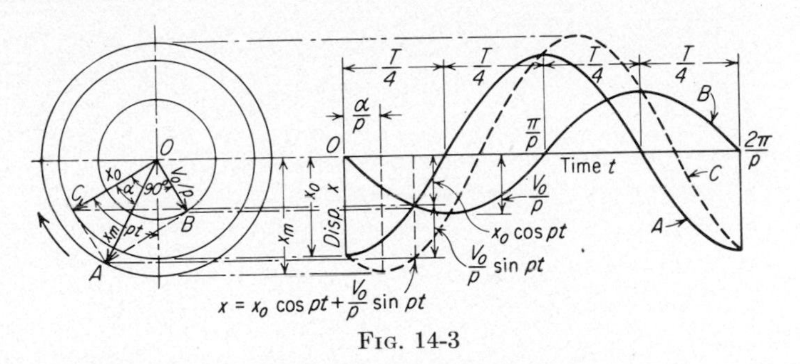

FIG. 14-3

For the particular case of Eq. (14-6) where the initial displacement $x_0 = 0$ the displacement-time curve is shown by curve B in Fig. 14-3. The amplitude of the vibration is in this case v_0/p.

In order to obtain the displacement-time diagram for the general case of Eq. (14-6) it is necessary only to sum up in Fig. 14-3 the ordinates of the displacement-time diagrams A and B for the two preceding particular cases. The result of this summation is again a simple harmonic motion expressed by Eq. (14-6) and represented by curve C.

In proof of the above statement, refer to Fig. 14-3. From point O, which represents the position of equilibrium of the moving mass of Fig. 14-1, draw the vector OC $(= x_0)$ making the angle pt with the positive (downward) direction of the x axis. From the end of OC draw the vector CA $(= OB = v_0/p)$ perpendicular to OC. The projections of these two vectors on the x axis are $x_0 \cos pt$ and $v_0/p \sin pt$, respectively. The projection of their resultant OA, therefore, represents the displacement of the moving mass as given by Eq. (14-6). Hence, it is evident from the figure that, as the mass oscillates along its vertical path, the resultant vector OA is the radius that represents the amplitude of the vibration, and it moves with an angular velocity p. Designating this amplitude by x_m and the angle between vectors OA and OC by α, we have

$$x_m = \sqrt{x_0^2 + \left(\frac{v_0}{p}\right)^2} \tag{14-7}$$

and from Fig. 14-3 it is evident that Eq. (14-6) may be written in the form

$$x = x_m \cos (pt - \alpha) \tag{14-8}$$

Thus it has been shown that the sum of two harmonic motions of a given frequency is again a simple harmonic motion of the same frequency. In any given case the amplitude x_m and the angle α may readily be calculated from the relations shown in Fig. 14-3.

The angle α by which vector OA is behind vector OC in their rotation is called the phase angle. Because of this difference in direction of the vectors their maximum projections on the x axis do not occur at the same time t. Thus, the maximum value of displacement x, which is represented by the maximum ordinate to the curve C in Fig. 14-3, occurs at a time α/p after the ordinate in curve A attains its maximum value. Since the period of the vibration is not affected by the initial conditions of motion, the solution here will be obtained by using the function $x = A \cos pt$ for simplicity.

Let it be assumed that the vibration of the mass m in Fig. 14-1 is started by displacing it an amount x_0 from its equilibrium position and then releasing it without initial velocity. Then we can put $x = x_0 \cos pt$, where x_0 is the amplitude of the vibration in inches, p the angular velocity of the rotating vector x_0 in radians per second, and t the time in seconds. Solving for p in Eq. (14-1), we get from this particular solution of the equation

$$p = \sqrt{\frac{k}{m}} \tag{14-9}$$

Substituting this value of p in Eq. (14-3) we have

$$f = \frac{1}{2\pi}\sqrt{\frac{k}{m}} \tag{14-10}$$

The relation between deflection and spring constant k makes it possible to compute the natural frequency from the static deflection Δ_{st} of the shaft under a gravity load W. When a shaft carrying a load W is placed horizontally and supported at the ends, as in Fig. 14-1, the static deflection $\Delta_{st} = W/k$; and since $m = W/g$, we obtain by substitution in Eq. (14-10)

$$f = \frac{1}{2\pi}\sqrt{\frac{g}{\Delta_{st}}} \tag{14-11}$$

These equations apply whether the shaft is horizontal, vertical, or inclined. It is evident that the period and frequency of a free vibration depend only on the weight carried on the shaft and the stiffness of the shaft, i.e., on mass m and spring constant k, and are not affected by any initial conditions of motion. In any case it is necessary only to compute the static deflection due to the mass m in order to determine the natural lateral frequency of vibration.

14-4. Lateral Vibration Due to a Single Rotating Mass. The development of the equation by means of which the vibrations due to a single rotating mass may be analyzed is quite simple. Consider the shaft (Fig. 14-4) carrying a rotating mass m at some point between the supports. The mass of the shaft is assumed to be negligible compared with the rotating mass m. The center of gravity of the mass may not coincide with the axis of the shaft by a small amount e. In consequence, there will be a centrifugal force which will deflect the shaft a small amount x. The magnitude of this centrifugal force then is $m(x + e)\omega^2$, where ω is the angular velocity of m, in radians per second. The force resisting deflection is kx, where k is the spring constant. Hence, for equilibrium

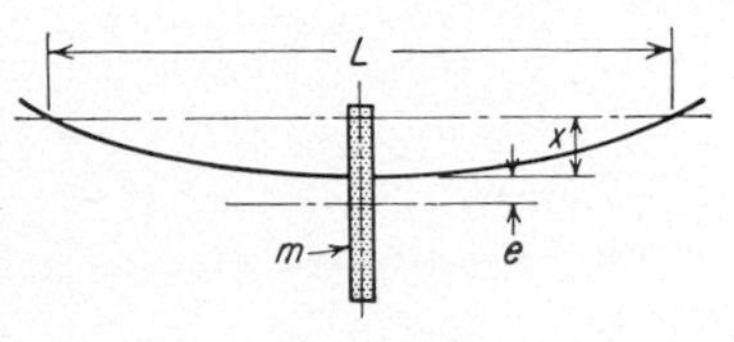

FIG. 14-4

$$m(x + e)\omega^2 = kx \tag{14-12}$$

Solving for x, we have

$$x = \frac{m\omega^2 e}{k - m\omega^2} = \frac{e}{(k/m\omega^2) - 1} \tag{14-13}$$

Unless e is zero, the deflection x becomes infinite when $k = m\omega^2$. The corresponding value of ω is known as the critical speed ω_c. We have then, as the critical speed in radians per second,

$$\omega_c = \sqrt{\frac{k}{m}} \tag{14-14}$$

As in the case of Eq. (14-10), the relation between deflection and spring constant makes it possible to compute the critical speed from the static deflection of the shaft under a weight load equal to that of the rotating mass. If a shaft is deflected an amount Δ_{st} under a static load W, we have $W = k\Delta_{st}$, and from Eq. (14-14) the critical speed in radians per second is

$$\omega_c = \sqrt{\frac{g}{\Delta_{st}}} \tag{14-15}$$

Expressed in revolutions per second, the critical speed is

$$n_c = \frac{1}{2\pi}\sqrt{\frac{g}{\Delta_{st}}} \qquad (14\text{-}16)$$

Note that Eq. (14-16) is identical with Eq. (14-11) which gives the natural frequency of lateral vibration of the same system in cycles per second. Equations (14-15) and (14-16) apply whether the shaft is horizontal, vertical, or inclined. It is necessary only to compute the static deflection under the load equal to the weight of the rotating mass.

Referring again to Eq. (14-13) it is important to note that, if ω exceeds ω_c, the deflection again becomes finite, although *negative*, i.e., in the direction opposite e from the shaft axis. This is known as a phase shift of 180 deg. Equation (14-13) shows that, if ω is very great, k becomes negligible in comparison with $m\omega^2$ and when ω is infinite, $x = -e$. The deflection is then such that the mass m rotates, not around the axis of the shaft, but around its own center of gravity. The shaft is whirling around this center of gravity at the distance e from it, and the operation is stable.

Thus, it is possible to operate smoothly and stably above the critical speed, with the shaft in a flexed condition. Since in the region of the critical speed there is great vibration and danger of breakage, the critical speed of the shaft should be made intentionally low when the running speed is high. This condition is attained by making the stiffness constant k very low; i.e., the shaft should be made very slender. An important early application of this principle was the introduction of flexible shafts in steam turbine design.

Example. The steel shaft, Fig. 14-4, is 1 in. in diameter, 18 in. in length, and carries at its mid-point a rotating disk weighing 50 lb. The bearings are pivoted, and the shaft may be assumed as freely supported at the ends. Determine the critical speed of the shaft.

SOLUTION. The deflection $\Delta_{st} = Wl^3/48EI$. Assume $E = 30{,}000{,}000$ psi, $I = \pi d^4/64 = \pi \times 1^4/64 = 0.0491$ in.^{4} Then $\Delta_{st} = (50 \times 18^3)/(48 \times 30{,}000{,}000 \times 0.0491) = 0.0041$ in. Substituting in Eq. (14-16), the critical speed is

$$n_c = \frac{1}{2\pi}\sqrt{\frac{g}{\Delta_{st}}} = \frac{1}{2\pi}\sqrt{\frac{32.2 \times 12}{0.0041}} = 48.7 \text{ rps}$$

Equation (14-11) would give the natural frequency of lateral vibrations for this same system as 48.7 cps.

While the deflection, theoretically, becomes infinite only at the critical speed, there may be a condition of unsatisfactory operation through a considerable range of speed near it.

14-5. Lateral Vibrations Due to Several Rotating Masses. When a rotating shaft carries *several* masses or has a variable cross section, the exact mathematical solution for the critical speeds (there now may be more than one) becomes quite difficult. Several approximate methods have been developed for the solution of problems of this type. Rayleigh's method which is based on an interchange of energy within the system, gives an approximate solution for the lowest or fundamental frequency. The potential energy at the maximum deflection must equal the kinetic energy as the shaft passes through its equilibrium position. It is assumed that the elastic curve under the static load is the same as that under the dynamic vibration load. It is assumed, also, that the weight of the shaft is neglected. These assumptions are sufficiently accurate for many practical purposes.

Since the potential energy for each mass is $\frac{1}{2}Wx$, that for a number of masses is

$$PE = \tfrac{1}{2}W_1x_1 + \tfrac{1}{2}W_2x_2 + \cdots + \tfrac{1}{2}W_nx_n \qquad (14\text{-}17)$$

The kinetic energy for a single mass is $\frac{1}{2}(W/g)v^2$ and for a number of masses with harmonic motion is

$$KE = \frac{1}{2}\frac{W_1}{g}x_1^2\omega^2 + \frac{1}{2}\frac{W_2}{g}x_2^2\omega^2 + \cdots + \frac{1}{2}\frac{W_n}{g}x_n^2\omega^2 \qquad (14\text{-}18)$$

Equating the potential and kinetic energies and solving first for ω and then for the frequency, which is $f = \omega/2\pi$, we have, in terms of cycles per second,

$$f = \frac{1}{2\pi}\sqrt{\frac{g(W_1x_1 + W_2x_2 + \cdots + W_nx_n)}{W_1x_1^2 + W_2x_2^2 + \cdots + W_nx_n^2}} \qquad (14\text{-}19)$$

where W_1, W_2, etc., are the loads and x_1, x_2, etc., are the total deflections at the loads. Thus, the frequency can be obtained from this equation if the total deflections at each load are known. The equation is sometimes written in the form

$$f = \frac{1}{2\pi}\sqrt{\frac{g\Sigma Wx}{\Sigma Wx^2}} \qquad (14\text{-}20)$$

1. *Solution for a Shaft with Uniform Section.* When the shaft is of uniform diameter, the deflection may be obtained analytically by using the principle of superposition, namely, that the total deflection caused by several weights is equal to the deflection caused by any one weight alone plus the deflections at the same point caused by each of the other weights acting individually. This method is satisfactory for all practical purposes if the deflections are not excessive. The method may be explained by considering the following example:

Example. A shaft of uniform diameter, 48 in. long and freely supported at the ends, carries two pulleys, one weighing 50 lb located 12 in. from the left end and the other weighing 35 lb located 14 in. from the right end. The shaft has a moment of inertia of $I = 0.25$ in.[4] and a modulus of elasticity of $E = 30{,}000{,}000$ psi. It is required to find the lowest natural frequency of this system, neglecting the weight of the shaft.

SOLUTION.[1] The static deflections under each pulley, as obtained from the equations given in books on strength of materials for a deflection due to a concentrated load, are as follows:

At the 50-lb load due to the 50-lb load, $x_a = 0.00830$ in.
At the 50-lb load due to the 35-lb load, $x_b = 0.00533$ in.
At the 35-lb load due to the 35-lb load, $x_c = 0.00733$ in.
At the 35-lb load due to the 50-lb load, $x_d = 0.00762$ in.

The total deflections at the 50- and the 35-lb load are as follows:

$$x_{50} = x_a + x_b = 0.00830 + 0.00533 = 0.01363 \text{ in.}$$
$$x_{35} = x_c + x_d = 0.00733 + 0.00762 = 0.01495 \text{ in.}$$

The natural frequency from Eq. (14-20) is, then,

$$\begin{aligned} f &= \frac{1}{2\pi}\sqrt{\frac{g(W_{50}x_{50} + W_{35}x_{35})}{W_{50}x_{50}^2 + W_{35}x_{35}^2}} \\ &= \frac{1}{2\pi}\sqrt{\frac{g(50 \times 0.01363 + 35 \times 0.01495)}{50(0.01363)^2 + 35(0.01495)^2}} \\ &= 26.3 \text{ cps, or } 1{,}578 \text{ cpm} \end{aligned}$$

Since the critical speed for a rotor is the same as the natural frequency of lateral vibration of the shaft, the above result indicates that a pulley speed at or near 1,578 rpm would not be advisable.

2. *Solution for a Shaft with Nonuniform Section.* When the shaft is not of uniform section, the determination of an equation for the deflection becomes impractical because the value of the sectional moment of inertia, as well as that of the bending moment, varies with the distance along the shaft. The procedure is to obtain first the deflection curve by means of one of the several well-known graphical methods given in books on strength of materials. After the deflections are known, Eq. (14-20) may be used as in case 1 above to determine the lowest natural frequency.

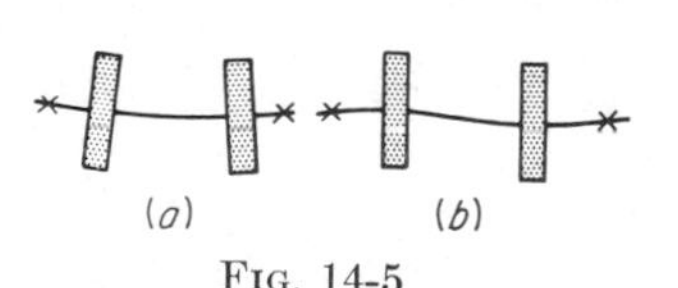

FIG. 14-5

3. *Solution for Shaft with Distributed Load. Higher Critical Speeds.* A shaft may have as many critical speeds as the number of loads that it carries. The first, or lowest, critical speed occurs when the shaft is bent into a single sweep, as in Fig. 14-4. A weightless shaft with only one concentrated mass can deflect in only one way, as in Fig. 14-4. With

[1] For the detailed solution of this problem and for a variety of other problems and solutions pertaining to shaft vibration, the student is referred to the book "Elements of Mechanical Vibration" by Freeberg and Kemler.

two masses there are two possible ways, as in Fig. 14-5. With three masses there are three ways, as in Fig. 14-6, and so on. It is evident that the stiffness is greater for bending into curves with counterflexures than for bending into curves with a single sweep. Hence, for several concentrated masses there will be a first, a second, a third, and perhaps other critical speeds higher than the lowest. A shaft carrying a distributed load, e.g., its own weight, can bend theoretically, at least, in an infinite number of ways and therefore may have an infinite number of critical speeds. Usually it is the first or lowest critical speed that is of most importance.

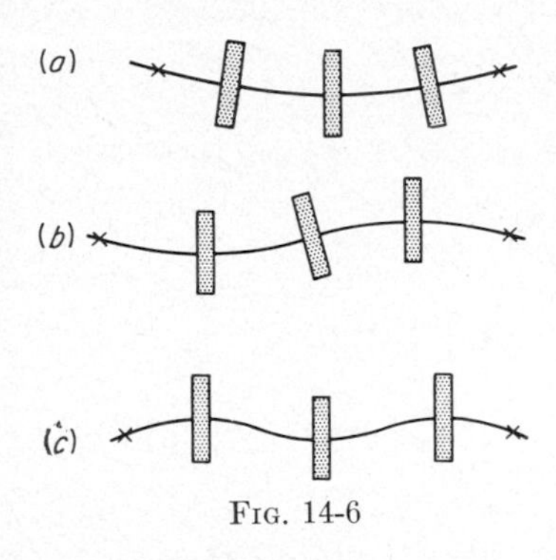

FIG. 14-6

TORSIONAL VIBRATIONS IN SHAFTS

The equations obtained above for lateral vibrations may be used for torsional vibrations if force is replaced by torque, mass by moment of inertia of mass, and linear displacements, velocities, and accelerations by angular displacements, velocities, and accelerations, respectively. In this case, the equation of motion $\Sigma T = I\alpha$ would replace the corresponding equation $\Sigma F = ma$.

14-6. Single Disk on the End of a Shaft. As an example, let it be required to find the frequency of vibration (or oscillation) of the torsional pendulum in Fig. 14-7, which consists of a disk rigidly attached to the slender cylindrical rod, or shaft, of length L. If the disk is given angular displacement θ from its equilibrium position and then released, the disk will vibrate (oscillate) under the influence of the torque exerted by the rod. Within the elastic limit the torque is proportional to the angular displacement and is opposite in sense to the angle θ. Applying the equation of motion $\Sigma T = I\alpha$, we have

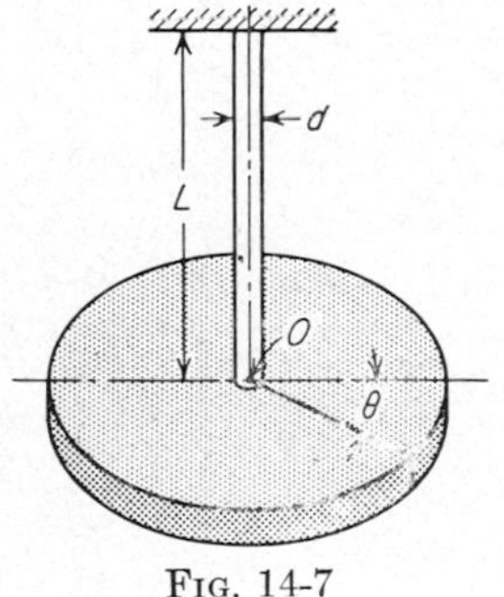

FIG. 14-7

$$I \frac{d^2\theta}{dt^2} = -k\theta \tag{14-21}$$

where k is the torsional spring constant, i.e., the torque required to produce an angle of twist of 1 radian in the rod to which the disk is attached, and I is the mass moment of inertia of the cylindrical disk. This equation indicates that the motion is simple harmonic; it is of the same form as Eq. (14-1); and hence its solution is of the same form. We can therefore proceed with the solution as in the case of Eq. (14-1). In the particular case where the disk is given an initial angular displacement θ_0 from its equilibrium position and then released without initial angular velocity we can put $\theta = \theta_0 \cos pt$, where θ_0 is the amplitude of the vibration in radians, p the angular velocity of the rotating vector θ_0 in radians per second, and t is the time in seconds. Solving for p in Eq. (14-21) we get from this particular solution of the equation,

$$p = \sqrt{\frac{k}{I}} \tag{14-22}$$

The period of oscillation in seconds is, therefore,

$$T = \frac{2\pi}{p} = 2\pi \sqrt{\frac{I}{k}} \tag{14-23}$$

The torsional spring constant for a cylindrical rod, as given in books on strength of materials, is

$$k = \frac{T}{\theta} = \frac{\pi d^4 G}{32L} \tag{14-24}$$

where d is the diameter of the rod, G is the shearing modulus of elasticity of the material in the rod, L is the length of the rod in inches, and θ is the angle of twist in radians. The frequency of the torsional vibration is, therefore,

$$f = \frac{p}{2\pi} = \frac{1}{2\pi}\sqrt{\frac{k}{I}} = \frac{1}{2\pi}\sqrt{\frac{\pi d^4 G}{32IL}} \tag{14-25}$$

Example. The propeller drive on a boat can be represented by the shaft and disk combination shown in Fig. 14-7 because of a large flywheel on the engine drive shaft. The propeller shaft is 5 in. in diameter and 40 in. long. The moment of inertia of the propeller is 1,000 lb-in.-sec². Neglecting the weight of the shaft, what is the natural frequency of this simplified system? Assume the torsional modulus of elasticity of the steel shaft to be 12,000,000 psi.

SOLUTION. From Eq. (14-25)

$$f = \frac{1}{2\pi}\sqrt{\frac{\pi d^4 G}{32IL}} = \frac{1}{2\pi}\sqrt{\frac{\pi \times 5^4 \times 12 \times 10^6}{32 \times 1{,}000 \times 40}} = 21.6 \text{ cps or } 1{,}296 \text{ cpm}$$

Such an approximation as this gives a natural frequency that indicates the lowest speed which would be critical.

14-7. Two Rotating Disks Connected by a Shaft. The above analysis is based on the assumption that one end of the shaft is fixed. It may be applied to a case such as that in the example just given, where the end that is fixed in Fig. 14-7 is connected to a rotating mass such as a very heavy flywheel. In such a case the smaller mass rotates with a velocity that swings periodically above and below the uniform speed of the larger mass, and thus, in effect, the larger mass is fixed with respect to the smaller mass.

The more general case, however, is that in which neither of the two masses carried by a rotating shaft is so large as to be unaffected by the vibration of the other. Consider, for example, the case of the rotors of a large motor and generator connected by a relatively small shaft of uniform diameter d and length L as shown in Fig. 14-8. Whenever the mass at one end tends to twist the shaft forward, that at the other end tends to twist it backward, and vice versa. When vibration is under way, both ends oscillate relative to the mean rotating position, and they oscillate relative to each other.

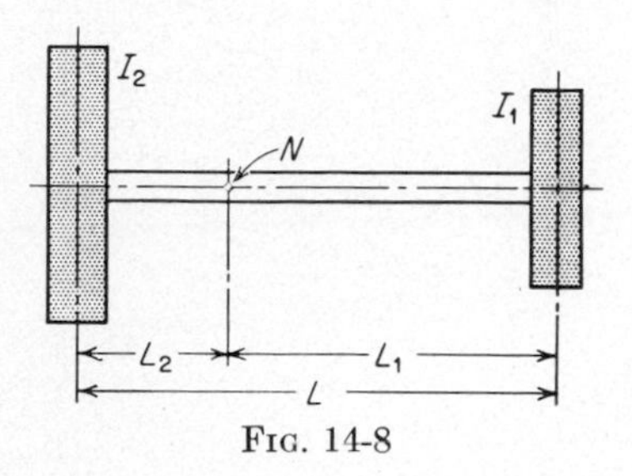

Fig. 14-8

In consequence of this reversal of twist there is one section of the shaft that twists neither forward nor backward during the rotation but rotates with uniform speed. This section can be regarded as the point of anchorage (called the *node* or *nodal point*) for the masses swinging at the ends. Thus, the motion of each disk may be considered as that of a torsional pendulum on a shaft which is fixed at the nodal point N. The position of this point can be determined, since the period of oscillation of both masses is the same. From Eq. (14-25) the frequency is

$$f = \frac{1}{2\pi}\sqrt{\frac{k_1}{I_1}} = \frac{1}{2\pi}\sqrt{\frac{k_2}{I_2}} \tag{14-26}$$

where k_1 and k_2 are the torsional spring constants of the two parts L_1 and L_2 of the shaft, and I_1 and I_2 are the mass moments of inertia of the two disks, respectively. Therefore,

$$\frac{k_2}{k_1} = \frac{I_2}{I_1} \tag{14-27}$$

By substituting the values of k from Eq. (14-24) in Eq. (14-27) we obtain

$$\frac{L_1}{L - L_1} = \frac{I_2}{I_1} \qquad \text{whence } L_1 = \frac{I_2 L}{I_1 + I_2} \tag{14-28}$$

By using the value of L_1 for the length of the shaft in Eq. (14-24) and substituting the resulting value of k in Eq. (14-25), the frequency of free torsional vibration for the system of two masses in Fig. 14-8 is found to be

$$f = \frac{1}{2\pi}\sqrt{\frac{JG(I_1 + I_2)}{I_1 I_2 L}} \tag{14-29}$$

where J is the polar moment of inertia of the area of the cross section of the shaft about the axis of the shaft.

Example. The following data were taken from a preliminary sketch representing the proposed design for a steam turbine driving an electric generator at a speed of 1,800 rpm. The steam turbine rotor and the generator rotor are joined by a 4-in. steel shaft 30 in. long. The steam turbine rotor has an estimated weight of 500 lb and a radius of gyration of 18 in. The generator rotor has an estimated weight of 1,000 lb and a radius of gyration of 16 in. Assume that $G = 11{,}500{,}000$ psi. Is the proposed design satisfactory from the standpoint of vibration? Why?

SOLUTION. The arrangement of the rotors may be represented by Fig. 14-8. Neglecting the weight of the shaft, we have

$$I_1 = m_1k_1^2 = \frac{500 \times 18^2}{32.2 \times 12} = 420 \text{ lb-in.-sec}^2$$

and

$$I_2 = m_2k_2^2 = \frac{1{,}000 \times 16^2}{32.2 \times 12} = 665 \text{ lb-in.-sec}^2$$

Hence, from Eq. (14-29)

$$f = \frac{1}{2\pi}\sqrt{\frac{JG(I_1 + I_2)}{I_1I_2L}} = \frac{1}{2\pi}\sqrt{\frac{\pi \times 4^4 \times 11{,}500{,}000 \times (420 + 665)}{32 \times 420 \times 665 \times 30}}$$
$$= 30.7 \text{ vibrations per second} = 1{,}842 \text{ vibrations per minute}$$

The proposed design would not be satisfactory because the operating speed would be near the condition of resonance.

14-8. Torsional Vibration with Multiple Masses. It is evident from the foregoing analysis that a shaft with two oscillating masses has only one torsional vibration frequency. With three masses there are two, and with four masses, three, and so on. The determination of critical speeds for shafts with several rotating masses is of great importance in the design of machinery with multithrow crankshafts, such as internal-combustion engines, or multicylinder pumps or with machinery having a multiplicity of rotating disks, such as steam and gas turbines, centrifugal pumps, and compressors.

Consider, for example, the case of an engine with its reciprocating masses. The torque transmitted to the crankshaft is not uniform, and the variable part of it may have frequencies that correspond to some of the natural frequencies of torsional vibration of the system represented by the flywheel and reciprocating masses attached to the crankshaft. The speed at which this condition of resonance prevails is a critical speed of the engine, and heavy forced torsional vibrations may result. When a large stationary engine operates in the range of a major critical speed, the noise resulting from the vibration of the whole structure is very annoying. Such engines can be run quickly through the critical ranges to the normal speed. Marine engines, in general, have a long shaft extension between the engine and the propeller shaft which is particularly sensitive to torsional vibration. Many cases of failure of such shafts can be attributed to forced vibrations.

The trouble connected with resonance may be eliminated either by changing the speed of the engine or by changing the dimensions of the shaft in such a way as to make the natural frequency of vibration of the system different from the running speed. If the engine must run in the range of the critical speed, the amplitude of the vibration can be reduced by introducing appropriate damping. In design it is a good rule to avoid entirely a shaft frequency that coincides with the number of power impulses, also one and one-half and twice the number of impulses. Above three times the number of impulses there usually is no likelihood of trouble. In order to avoid critical speeds in a machine, in general, the natural frequency of every part should be considerably higher than the number of impulses that the part receives during the operation of the machine. These conditions can be revealed only by analysis.

14-9. Reduction to an Equivalent System. In many of the cases mentioned in the preceding article, the shafting with all its cranks, pistons, flywheel, and driven machinery is too complicated a structure to attempt an exact determination of its torsional natural frequency of vibration. It is necessary first to simplify or "idealize" the machine to some extent by replacing the pistons, etc., by equivalent disks of the same moment of inertia and by replacing the crank throws by equivalent pieces of straight shaft of the same torsional flexibility. In other words, the machine has to be reduced to an equivalent system consisting of a straight shaft of uniform cross section carrying a number of rotating masses, as illustrated in Fig. 14-9.

In the reduction to an equivalent system a shaft of several cross sections can always be replaced by an equivalent shaft of constant cross section, noting only that a portion of a shaft of length L and diameter d can be replaced, without changing the angle of twist of the shaft, by a portion of length L_0, provided that [note Eq. (14-24)]

$$L_0 = \frac{Ld_0^4}{d^4}$$

An irregular part of a shaft, such as a crank, for example, can be replaced by an equivalent length of straight shaft of constant cross section such that the equivalent straight portion has the same torsional rigidity as the combined torsional rigidity of the journal, the flexural rigidity of the web, and the torsional rigidity of the crankpin. A number of convenient formulas have been developed for this purpose which take into account the clearance in the bearings, etc.

Even after the reduction to the simplified system the solution may become quite involved. Approximate numerical and graphical methods are usually applied in calculating the natural frequencies of vibration. Often, good results in approximating the lowest natural frequency can be obtained by grouping the several rotating masses so as to form a two-mass system. For example, in finding the lowest natural frequency of vibration in an eight-cylinder diesel engine driving the propeller of a ship, the engine and its flywheel will act practically as a solid body and with the propeller may be assumed to form a two-mass system, thus simplifying the solution. There is an extensive literature on the subject of torsional vibrations in engine shafts in which detailed discussions of these problems can be found.

Example. A six-cylinder, four-cycle, diesel engine with a flywheel is directly coupled to a ship propeller through a long propeller shaft as shown diagrammatically in Fig. 14-9. The rpm of the engine is 100. The propeller shaft is 12 in. in diameter

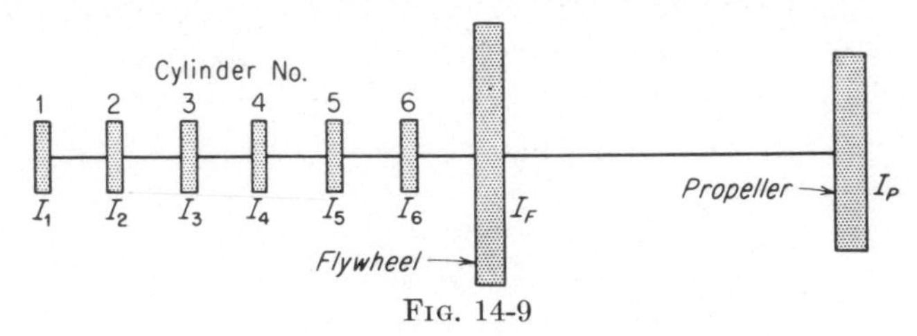

FIG. 14-9

and 150 ft long. The combined moment of inertia of the engine and flywheel mass is 90,000 lb-in.-sec² and the moment of inertia of the propeller is 24,000 lb-in.-sec². It is required to find the lowest natural frequency of the system.

SOLUTION. Let it be assumed first that the engine cranks, pistons, the flywheel, and the propeller have been reduced by the methods suggested above to the equivalent system of rotating masses shown in Fig. 14-9 and that the mass moments of inertia of the disks have been calculated. The lowest natural frequency will be the first mode of vibration with one node. In order to find this frequency the system in Fig. 14-9 can be reduced to a two-disk rotating system such as that shown in Fig. 14-8. In the first mode of vibration with one node the engine with its flywheel will act practically as a solid body (disk 2) and with the propeller (disk 1) will form a two-mass system on the ends of a shaft as in Fig. 14-8. This grouping together of the several rotating masses representing the engine and its flywheel into a single mass for this purpose is an approximation, but it gives a result fairly close to that obtained by an exact analysis.

From Eq. (14-29) the lowest natural frequency is, then,

$$f = \frac{1}{2\pi}\sqrt{\frac{JG(I_1 + I_2)}{I_1 I_2 L}} = \frac{1}{2\pi}\sqrt{\frac{2{,}036 \times (24{,}000 + 90{,}000) \times 12 \times 10^6}{24{,}000 \times 90{,}000 \times 150 \times 12}}$$

$$= 4.25 \text{ cps or } 255 \text{ cpm}$$

where $J = \frac{\pi d^4}{32} = \frac{\pi \times 12^4}{32} = 2{,}036 \text{ in.}^4$ and $G = 12{,}000{,}000$ psi

CHAPTER 15

GYROSCOPIC FORCES

A gyroscope may be defined as a body that is rotating with high speed about an axis, called the spin axis, and is partially free to move in other directions. The spinning top probably is the most familiar example. The wheels of a locomotive when rounding a curve, the engine flywheel in a motor car, the armature of the motor in an electric car, the propeller of an airplane when making a turn are all examples in which gyroscopic action occurs. In many cases the forces that are developed as a result of gyroscopic action may be undesirable. On the other hand, the gyroscope may be used to introduce desirable forces, as, for example, in the stabilizer in a monorail car, ship, or airplane. Another useful application is that in instruments for maintaining direction, of which the gyrocompass is perhaps the most notable example. This same characteristic of a spinning body, namely, that of retaining the direction of the axis of rotation in space, is utilized when a high angular velocity of spin about the longitudinal axis is given to a projectile by rifling the barrel of a gun.

15-1. Vectorial Representation of Angular Motion. In discussing the characteristics of the gyroscope it is advantageous to make use of the idea of vectorial representation of angular motion. Although this principle may be found in any textbook on mechanics, it will be reviewed briefly in this article.

Let the line OA (Fig. 15-1) rotate counterclockwise in the plane of the paper about the center O, and at a given instant let its inclination to the fixed line OX be θ deg. If at the end of a short interval of time the line has moved to the position OB, the angle $\Delta\theta$ is the angular displacement of the line. Angular displacement is a vector quantity, since it has both magnitude and direction. In order to specify completely an angular displacement by a vector, the vector must fix (1) the direction of the axis of rotation in space, (2) the magnitude of the angular displacement, and (3) the sense of the angular displacement, i.e., whether clockwise or counterclockwise. To fix the direction and magnitude, the vector may be drawn at right angles to the plane in which the angular displacement takes place, say along the axis of rotation, and its length may be made to represent the magnitude of the angular displacement to some convenient scale. The conventional way of representing the sense is to use the right-hand screw rule; the arrowhead points along the vector in the same direction as a right-hand screw would move, relative to a fixed nut, if given an angular displacement of the same sense. According to the above convention, the angular displacement $\Delta\theta$ (Fig. 15-1) would be represented by a vector perpendicular to the plane of the paper. The length of the vector would represent the magnitude of $\Delta\theta$ to some convenient scale and the arrowhead would point upward from the paper, since the sense of the displacement is clockwise, viewed from below. This representation can be seen in Fig. 15-2 where the lines in Fig. 15-1 are shown in isometric projection with the vector added.

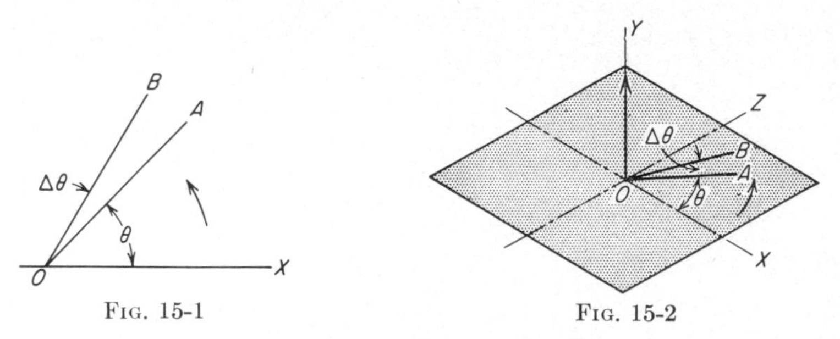

Fig. 15-1 Fig. 15-2

Angular velocity is angular displacement per unit of time. It also has both direction and magnitude, and it may be represented by a vector if the same convention is followed as that just described for angular displacement. Similarly, angular acceleration and angular momentum have both direction and magnitude and are thus vector quantities. The gyroscopic action of a rotating body whose plane of motion is changing direction (as, for example, in the case of the propeller of an airplane when making a turn) may be explained by use of either the principle of angular acceleration or the principle of angular momentum. The method employing the principle of angular momentum is generally considered the more useful, and it is this concept which will be used in this chapter.

The momentum of a particle is the product of the mass of the particle and its velocity at any instant; i.e., momentum $= mv$. Having both magnitude and direction, momentum is a vector quantity. The momentum of a particle is called *linear* momentum in contrast to the moment of momentum of a particle, which is called *angular* momentum. A *body* of mass m, when rotating, behaves as if all its mass were concentrated in a ring at a distance k (the radius of gyration) from the axis of rotation. Therefore, the moment of momentum, or angular momentum, of a rotating body about its axis of rotation is

$$mv \cdot k = mk\omega \cdot k = mk^2\omega = I\omega$$

where I is the mass moment of inertia of the body about its axis of rotation and ω is its angular velocity. If the body is in the form of a disk rotating clockwise about axis OX viewed from the right (Fig. 15-3), the angular momentum $I\omega$ is represented by vector OA.

15-2. Gyroscopic Couple. In Fig. 15-4 is shown a disk that is rotating at high speed about axis OX (called the spin axis) and is free to move in any direction. The disk will offer the same resistance to any translatory force or to any torque about the spin axis that it would if not rotating. If, however, a torque, or couple, is applied to rotate it about any other axis, as the vertical axis OY, for example, it will not rotate about the vertical axis but will rotate about a horizontal axis OZ perpendicular

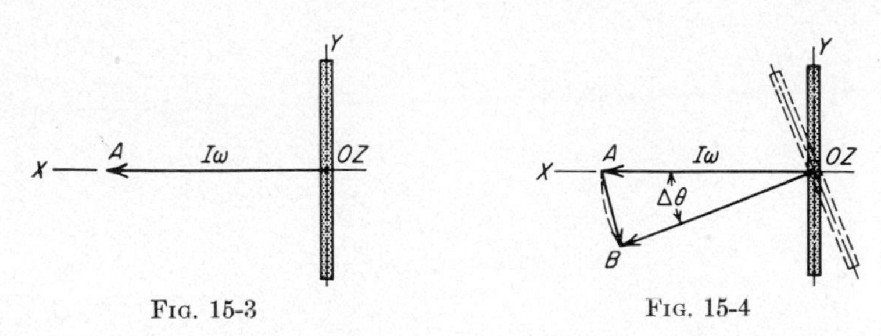

FIG. 15-3 FIG. 15-4

to the other two axes. Conversely, if the disk is rotated about the OZ axis a torque will be set up around the OY axis, perpendicular to the other two axes. This seemingly contradictory behavior of the rotating disk, although it is strictly in accordance with the fundamental principles of mechanics, will now be explained.

Let it be assumed that at a given instant this disk (Fig. 15-4) whose plane is at right angles to the plane of the paper is spinning at high speed with constant angular velocity ω, clockwise about axis OX (in the plane of the paper) when viewed from the right. Assume that at the same time the axis of spin OX is slowly rotating counterclockwise with a constant angular velocity ω_p about axis OZ perpendicular to the plane of the paper. Applying the right-hand screw rule, the angular momentum $I\omega$ of the disk, when in the position shown by full lines, may be represented by the vector OA and, when in the position shown by dotted lines, by the vector OB. The change of angular momentum in time Δt during which the axis of spin moves through the angle $\Delta\theta$ is, therefore, represented by the vector AB. In the limit, when $\Delta\theta$ becomes very small (equal to $d\theta$) and the small interval of time required for the axis of spin to rotate through this angle is dt, the change in angular momentum is

$$AB = OA\, d\theta = I\omega\, d\theta$$

The rate of change of angular momentum, i.e., change of angular momentum in interval of time dt, is

$$\frac{I\omega\, d\theta}{dt} = I\omega\omega_p$$

In the limit the direction of the vector AB, which may represent (to different scales) both the change and rate of change in angular momentum, is perpendicular to OA. This is more clearly shown in Fig. 15-5, where both vectors $I\omega$ and $I\omega\, d\theta$ and their resultant OC are drawn from the same point O. But a torque, or couple, is always required to produce a change in the angular momentum of a body. The plane in which it acts is perpendicular to the vector that represents the change in angular momentum, and the sense of rotation of the couple is such that it would cause a right-hand screw to advance, in the direction of the arrow, along the vector that represents the change in angular momentum. Therefore, a couple must act *on the disk* (Fig. 15-5) in a plane perpendicular to the vector OB (plane of OX and OZ), with a clockwise sense of rotation about the OY axis as viewed from above. Note that, if the thumb of the right hand is pointed in the direction of the change of angular momentum (along OB), the fingers will naturally wrap the proper axis in the proper direction and sense to indicate the direction and sense of the torque or couple which must be applied *to the disk* if the given change of momentum is to take place. The magnitude of the couple is equal to the time rate of change of angular momentum.

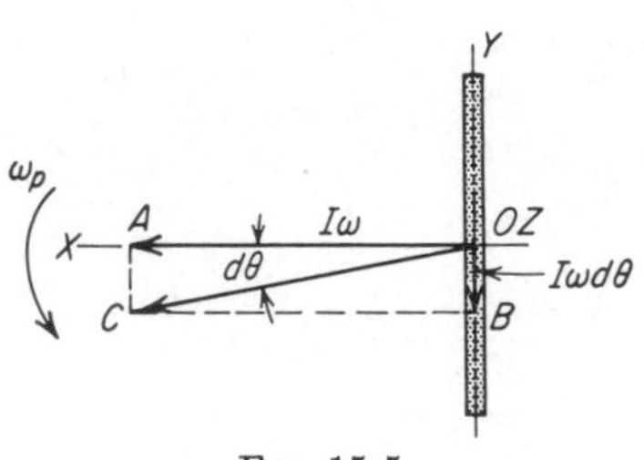

Fig. 15-5

$$T = I\omega\omega_p$$

This couple is called the *gyroscopic couple.* The change in direction of the spin axis of the disk (OX axis) is called *precessional motion.* The angular velocity ω_p which is maintained by the couple is called the velocity of precession, and the corresponding axis (OZ axis) is called the *precession axis.* The axis about which the applied couple acts (OY axis) is called the *torque axis.* Hence, the disk when spinning about the OX axis with angular velocity ω is said to precess about the OZ axis when acted upon by a couple having a moment $I\omega\omega_p$ about the OY axis. The $I\omega$ vector is called the spin vector, and the $I\omega\omega_p$ vector the torque vector. As an aid to visualization the axes and vectors in Fig. 15-5 are represented in Fig. 15-6 in isometric projection.

Referring now to Fig. 15-6, it will be seen that the sense of rotation about the axis of precession is in accordance with the following rule: *The sense of precession is such as to turn the spin vector toward the torque vector by the shortest possible route; i.e., the spin axis tends to become coincident with the torque axis.*

An important characteristic of the gyroscope should be noted at this point. Owing usually to its high rate of spin, the gyroscope possesses a certain "rigidity." That is, it will not precess quickly if the torque tending to turn the spin axis is small. Also, it should be kept in mind that the vector ($I\omega\, d\theta$) representing the change in angular momentum is exaggerated in length in the illustrations shown but actually is very small compared with the angular momentum vector ($I\omega$).

The preceding discussion will be sufficient to explain the main property of the gyroscope and make it possible to apply its principle of operation in solving problems involving gyroscopic action. For an analysis of the forces in the gyroscope the student is referred to textbooks on analytical mechanics, such as "Analytical Mechanics for Engineers" by Seely and Ensign, or to books written specifically about gyroscopes.

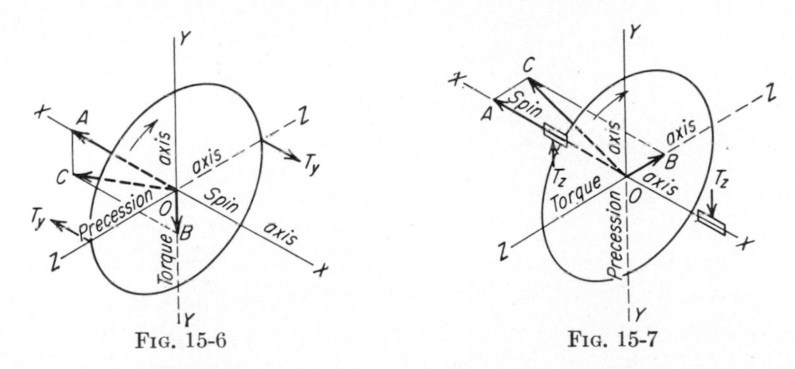

Fig. 15-6 Fig. 15-7

15-3. Some Typical Cases Illustrating Gyroscopic Action. It has now been established that, whenever the axis of spin of a rotating body changes direction, a gyroscopic couple exists. The couple is usually applied through the bearings that support the shaft on which the body rotates. The reaction of the shaft on each bearing is, of course, equal and opposite to the action of the bearing on the shaft. Hence, the precession of the axis of rotation (spin axis) causes a gyroscopic reaction couple (opposite the applied gyroscopic couple) to act on the frame to which the bearings are fixed. The examples that follow will make this clear.

1. A simple experiment for demonstrating the existence of a gyroscopic couple and its behavior in accordance with the above rules may be made by holding a bicycle wheel, Fig. 15-7 (dismounted from its frame), with one hand on each end of the projecting (horizontal) axle OX. If the wheel is spinning in the vertical plane about the axle OX which is held in the hands, any attempt to turn the axle (and hands) in the vertical plane about horizontal axis OZ will cause the wheel (and hands) to turn

(precess) about a vertical axis OY perpendicular to the axis of the wheel. The sense of the precession is determined as follows:

Referring again to Fig. 15-7 and applying the right-hand screw rule, it can be seen that since the rotation about axis OX is clockwise, viewed from the right, the angular momentum vector OA must be directed along the axis OX in the sense indicated. The couple exerted by the hands is the applied couple, or gyroscopic couple, T_z, and for the direction T_z shown, the torque vector OB will point away from the observer

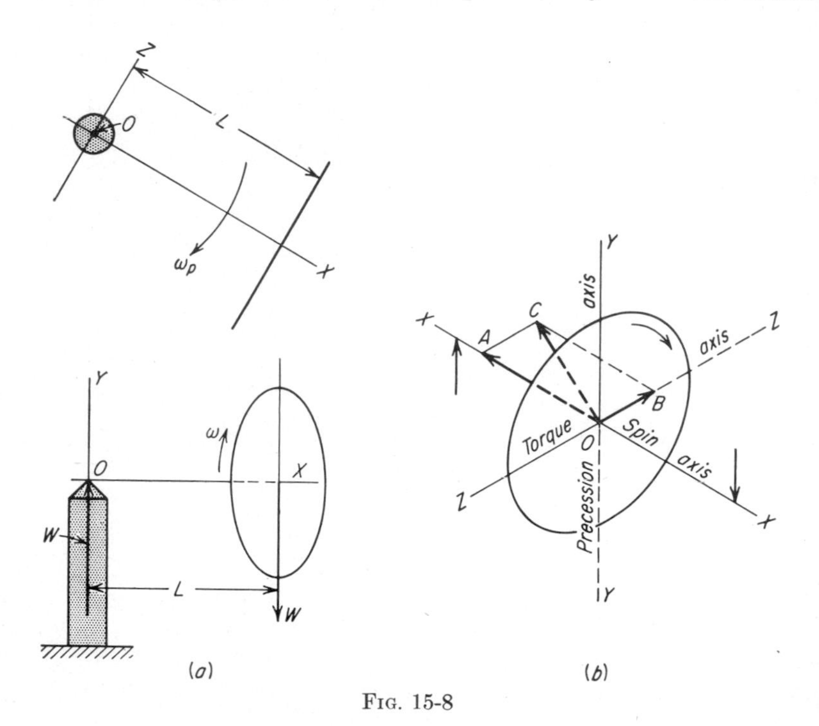

FIG. 15-8

along the axis OZ as indicated. The position of the resultant angular momentum vector OC (after a small change) shows that the precession must be clockwise about the axis OY, viewed from above.

2. Another case of interest is that illustrated at *a* in Fig. 15-8. If a disk of weight W is given an angular velocity ω about the OX axis and then one end of the axis is placed on the vertical post at O, the couple, having a moment WL, will cause the disk (and OX axis) to rotate (precess) with angular velocity ω_p about the axis of the post (OY axis). If the sense of ω is as represented in Fig. 15-8, the sense of precession about the OY axis will be clockwise as viewed from above. It is necessary that the disk shall precess if it is to develop a resistance to the couple WL and hence prevent the disk from falling, as actually happens. Since there

are no bodies to develop or supply a resisting couple, the disk turns (precesses) in the horizontal plane, and thus the couple necessary to keep the OX axis horizontal is developed from the inertia of the disk. If the precession is not allowed, the disk drops immediately. The procedure in determining the sense of precession (sense of ω_p), by applying the rule that has been established, is illustrated at *b* in Fig. 15-8.

3. Consider next the case in Fig. 15-9, where the wheel, shown with its bearings, rotates with a constant angular velocity ω about its axis (axle) OX as it rolls around the curved track (a forced precession about axis OY) with a constant angular velocity ω_p. As shown at *b* in the

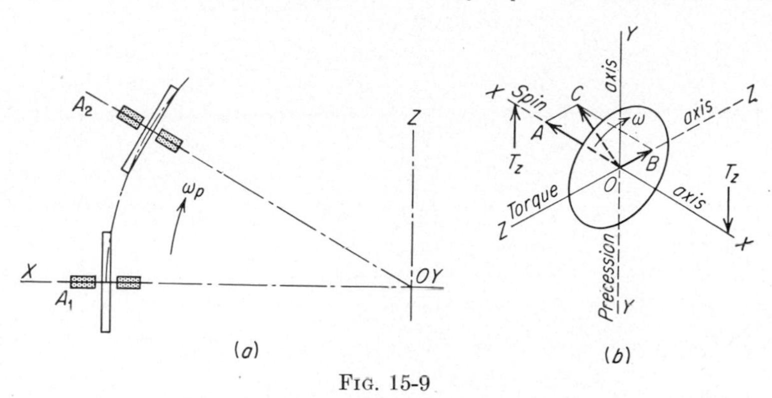

FIG. 15-9

figure the angular momentum vector OA $(= I\omega)$ is forced to rotate in the horizontal plane with spin axis OX toward axis OZ. The vector OB $(= I\omega\, d\theta)$ representing the change of angular momentum will then have to be directed along axis OZ as shown. The torque, or couple, T_z $(= I\omega\omega_p)$ required to produce this change in angular momentum must then act in a plane at right angles to the torque vector OB and, according to the right-hand screw rule, must have a clockwise sense around the torque axis as viewed from behind the wheel as it rolls forward. This is the gyroscopic couple and is the couple exerted *by* the bearings on the axle of the wheel. It is evident that the axle exerts a counterclockwise couple *on* the bearings and that the wheel would turn over counterclockwise (outward) unless the clockwise couple exerted by the bearings acted to prevent it.

4. A case with some interesting practical aspects is that relating to the gyroscopic effect of the propeller of an airplane. Consider the case where the rotation of the propeller is counterclockwise viewed from the pilot's seat and the airplane is making a right-hand turn. In Fig. 15-10 OX is the spin axis of the propeller, and in accordance with the right-hand screw rule OA $(= I\omega)$ is the angular momentum vector. Since the airplane is turning to the right, there is a forced precession of the

propeller clockwise about the OY axis viewed from above. The angular-momentum vector will therefore change direction in a short interval of time from OA to OC, so that the change in angular momentum ($I\omega\, d\theta$), and therefore the applied couple, is represented by the vector OB. The plane of the applied couple is at right angles to OB and therefore vertical, and its sense is clockwise viewed from the right-hand side of the airplane. This is the couple *applied* to the "axle" of the propeller *by* the frame bearings. The reaction couple of the rotating "axle" *on* the frame bearings is opposite to the *applied* couple and therefore counterclockwise, as viewed from the right-hand side of the airplane, and thus tends to raise the nose and depress the tail of the machine.

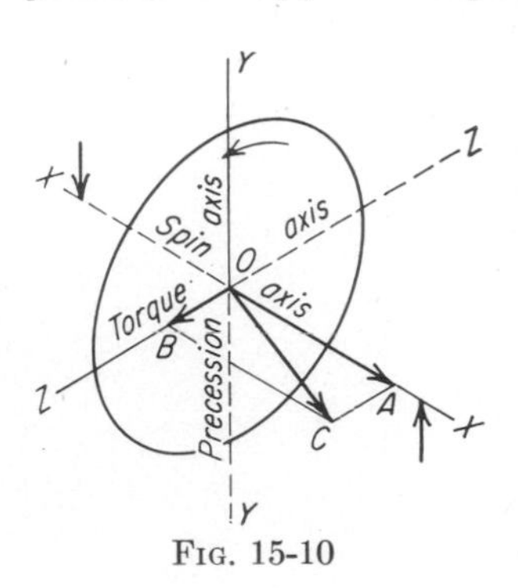

FIG. 15-10

5. The action of one type of gyroscopic stabilizer used in reducing ship roll is illustrated in Fig. 15-11. The device consists of a heavy rotor 1, rotating at high speed about a vertical axis OY. The rotor bearings AA are mounted in a frame 2, which is suspended in two bearings BB so that the frame is capable of rotating about an axis OX across the ship. Let the sense of spin of the rotor be clockwise when viewed from below, so that the angular momentum vector OA ($= I\omega$) points upward as

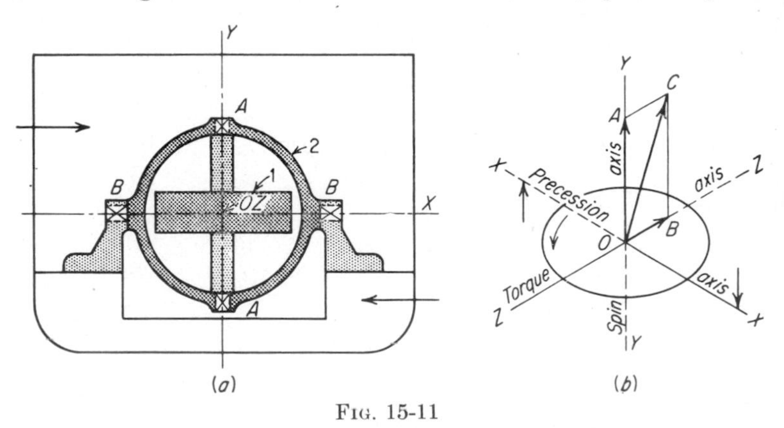

FIG. 15-11

shown at b in the figure. If a wave strikes the ship, starting a clockwise rolling action, viewed from behind, a precession motor starts the top of the frame carrying the rotor moving toward the front of the ship; i.e., the rotor is given a hurried precession in a clockwise sense viewed from the right. The angular-momentum vector OA ($= I\omega$) is therefore tipped toward the front, and the vector OB ($= I\omega\, d\theta$) representing the change in angular momentum must be directed toward the front along axis OZ. This means a clockwise torque exerted by the bearings BB on the rotor frame viewed from the rear along the OZ axis. The rotor frame thus exerts a counterclockwise torque on the bearings BB and thus on the ship. It is thus that the hurried precession of the rotor of the gyroscope creates a torque on the ship that is in opposition to the torque produced by the wave and in that manner counteracts the roll.

6. The gyrocompass was designed to overcome the defects inherent in the magnetic compass. The chief working parts of this instrument of interest here are shown in schematic outline in Fig. 15-12. The rotor 1, spinning on axis OX, may be driven electrically or by an air jet. The ends of its shafts are in bearings BB in a ring 2 which is itself pivoted

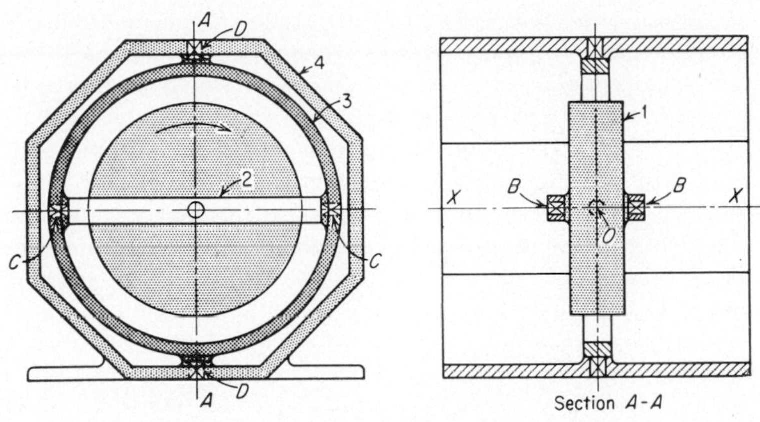

FIG. 15-12

on bearings CC in a vertical ring 3 which, in turn, is pivoted on bearings DD in the outer frame or casing 4. With this sort of mounting the axis OX of the rotor is free to point in any direction, horizontal or oblique, within reasonable limits of obliquity. Or, conversely, since it is usually horizontal, it can stay so and hold a given direction in space, while, if mounted in an airplane, for example, the airplane banks or turns. It actually does stay this way because the parts are delicately balanced and the rings turn easily in their bearings and apply only small forces that tend to turn the axis. As already noted, this is the condition for gyroscopic "rigidity." That is, the instrument works on the principle that a gyroscope will not precess quickly if the forces acting on it are small. It is not *exactly* "rigid," but practically so.

There are auxiliary attachments not shown in the figure to provide for adjustments to compensate for the rotation of the earth, for example, and for other settings and adjustments that may be required.

Although this instrument is used like a compass, it does not get any directive force from the earth's magnetic field. So before it will work as a compass, it must be set to agree with the magnetic compass, its great advantage being that it is not subject to the fluctuation inherent in the magnetic compass. It is, however, affected by the earth's rotation. Suppose, for example, it were at the North Pole. Then it would not require any precession to keep the axis OX horizontal, and the axis would hold a fixed direction in space while the earth rotated under it. To the observer it would appear as though the axis OX were turning at the rate of 15 deg/hr. At the equator there would be no such effect. If the axis pointed along a meridian (north and south), it would never try to point anywhere else. So at the equator the instrument does not need any resetting at all. In this country, about halfway between the equator and the pole, the rate at which the axis OX seems to change its direction varies with the latitude, from about 7 to 11 deg/hr. Therefore, if the gyrocompass is reset every 15 min it will be within 2 or 3 deg of correct; and unlike the magnetic compass, it will be steady.

7. *The Earth as a Gyroscope.* We are accustomed to thinking of the gyroscope as a rotating body having a high rate of spin. Since the gyroscopic effect is proportional to the product $I\omega$, this effect may be produced by a large value of ω and a small value of I or a small value of ω and a large value of I. At one extreme is the spinning top, which has a small mass and a high rate of spin. At the other extreme is the earth, with a large value of I but a small value of ω. Although the linear velocity of a point on the earth's surface at the equator is about 1,000 mph, the angular velocity of the earth is relatively small, one rotation on its axis requiring about 24 hr. However, its mass and radius of gyration are so very large that it has a large angular momentum $I\omega$. It becomes evident, therefore, that the spinning body on which we live must behave in the same manner as the spinning top and will have similar gyroscopic behavior.

It has been said that the fact of the earth's being a spinning body is one of the greatest and most persistent causes of many of the phenomena that occur around us. The equator of the earth makes an angle of 23½ deg with the ecliptic, which is the plane of the earth's orbit around the sun. Or the axis of spin of the earth is always at an angle of 23½ deg with a perpendicular to the ecliptic. This axis of spin points very nearly to the pole star, almost infinitely far away. But because the earth is not a perfect sphere, being somewhat flat at the poles, the attraction of the sun and moon causes a slow conical motion, or precession, of the axis, the complete cycle requiring about 26,000 years. This phenomenon is known in astronomy as the *precession of the equinoxes*, so called because the sun is observed to alter its position relative to the stars at the equinoxes and solstices. The ancient astronomers and philosophers, as well as those of the Middle Ages, noticed this phenomenon. Thus they were very near to the actual knowledge of the earth's rotation and precession.

EXAMPLES

Example 1. The radius of the disk (Fig. 15-8) is 6 in., and its weight is 10 lb. The distance L is 2 ft. If the disk rotates about the horizontal axis OX with a speed of 300 rpm, with what speed will it precess about the vertical axis OY?

SOLUTION. The angular velocity of spin is

$$\omega = 2\pi n = 2\pi \times 300/60 = 31.416 \text{ radians/sec}$$

The gyroscopic couple is

$$T = I\omega\omega_p = 10 \times 2 = 20 \text{ lb-ft}$$

The moment of inertia of the disk is

$$I = \frac{1}{2}mr^2 = \frac{1}{2} \times \frac{10}{32.2} \times \left(\frac{6}{12}\right)^2 = 0.0388 \text{ slug-ft}^2$$

The angular velocity of precession is therefore

$$\omega_p = \frac{T}{I\omega} = \frac{20}{0.0388 \times 31.416} = 16.4 \text{ radians/sec}$$

or

$$n = \frac{\omega_p}{2\pi} = \frac{16.4}{2\pi} \times 60 = 157 \text{ rpm}$$

Example 2. The disk in Fig. 15-9 is 20 in. in diameter and weighs 300 lb. The axle of the disk is supported by bearings 12 in. apart and is constrained to roll around a track of 10 ft radius. The velocity of the center of the disk is 20 fps. What is the force on each bearing due to gyroscopic action alone and in what direction is the force?

SOLUTION. The moment of inertia of the disk about the axis of spin OX is

$$I = \frac{1}{2}mr^2 = \frac{1}{2} \times \frac{300}{32.2} \times \left(\frac{10}{12}\right)^2 = 3.24 \text{ slug-ft}^2$$

The angular velocity of the disk about the axis of spin is

$$\omega = \frac{v}{r} = \frac{20}{10/12} = 24 \text{ radians/sec}$$

The angular velocity of the axis OX of the disk about O is

$$\omega_p = \frac{v}{R} = \frac{20}{10} = 2 \text{ radians/sec}$$

Hence, the gyroscopic couple is

$$T = I\omega\omega_p = 3.24 \times 24 \times 2 = 155.75 \text{ lb-ft}$$

The sense of this couple has been determined as indicated in Fig. 15-9*b* in accordance with the rules of Art. 15-2. The forces constituting the gyroscopic couple are the forces of the bearings on the axle, and hence the force of the axle on the bearing at the outer bearing is downward and that at the inner bearing is upward. Since the distance between the centers of the bearings is 12 in., the magnitude of each of the forces is 155.75/1 = 155.75 lb.

Example 3. A gas-turbine unit in a jet plane revolves at 8,000 rpm counterclockwise when viewed from behind the airplane. The plane executes a climbing arc at a rate of 45 deg in 5 sec. The moment of inertia of the rotors is 6.2 slug-ft². Find the gyroscopic couple and the effect on the airplane.

SOLUTION.

$$I\omega = 6.2 \times \frac{8,000}{60} \times 2\pi$$

$$= 5,180 \text{ slug-ft}^2/\text{sec}$$

$$\omega_p = \frac{\pi}{4} \times \frac{1}{5} = 0.157 \text{ radian/sec}$$

The gyroscopic couple is

$$T = 5,180 \times 0.157 = 814 \text{ lb-ft}$$

The bearings and engine mounts must be capable of furnishing this torque to the rotors. The torque tends to cause the nose of the plane to move to the pilot's left and the tail to his right.

APPENDIX

APPROXIMATE METHOD FOR CONSTRUCTING INVOLUTE GEAR-TOOTH PROFILES

Grant's Odontograph. It will be found that the exact construction of tooth profiles is somewhat tedious and laborious. In making cutters for cut gears the exact outline is, of course, necessary; but in making patterns for cast teeth and for representing teeth on a drawing, an approximation is usually entirely satisfactory. Tooth curves are so short in any case that the ideal form may be very closely approximated by circular arcs, and most of the approximate constructions which have been devised are circular-arc methods. One of the best known of these methods is *Grant's involute*

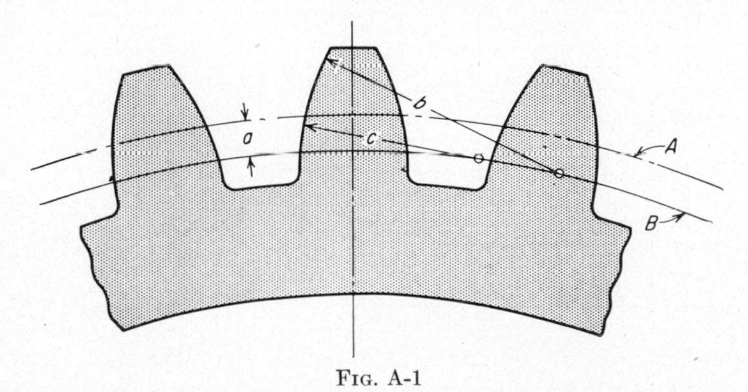

Fig. A-1

odontograph, embodied in Fig. A-1 and Table A-1. The table was devised for a set of interchangeable involute gears of 15-deg pressure angle, but it may be used in drawing the teeth of the 14½-deg standard involute system, the proportions of which are given in Chap. 4.

The procedure in laying out involute teeth by this method is as follows: Draw the pitch, addendum, and dedendum circles in the usual way, as shown in Fig. A-1, and lay off the circular pitch of the teeth on the pitch circle, dividing the latter properly for tooth thickness and tooth space. Next, draw the base circle B at a radial distance one-sixtieth of the pitch diameter inside the pitch circle A. This distance is indicated by a in the figure, and it should be noted that the base circle is the locus of centers for the radii b and c. With face radius b (Table A-1), draw that part of the tooth profile lying between the pitch circle and addendum circle, and with flank radius c draw that part of the profile lying between the pitch circle and base circle. The remainder of the profile, between the base circle and the fillet at the bottom of the spaces, is drawn in as a radial line. Table A-1 gives values of b and c in terms of both *one diametral pitch* and *one circular pitch*, merely as a matter of convenience. The distance a, given as one-sixtieth of the pitch diameter, is simply a convenient means of drawing a base circle that will give a pressure angle of approximately 15 deg.

The profiles of the involute rack teeth are straight lines inclined at an angle of 15 deg with the vertical. In order to avoid interference, the outer half of the addendum is drawn by means of a circular arc having its center on the pitch line of the rack and having a radius of 2.10 in. divided by the diametral pitch, or 0.67 times the circular pitch.

TABLE A-1. GRANT'S INVOLUTE ODONTOGRAPH

Teeth	Divide by the diametral pitch		Multiply by the circular pitch		Teeth	Divide by the diametral pitch		Multiply by the circular pitch	
	Face radius b	Flank radius c	Face radius b	Flank radius c		Face radius b	Flank radius c	Face radius b	Flank radius c
10	2.28	0.69	0.73	0.22	28	3.92	2.59	1.25	0.82
11	2.40	0.83	0.76	0.27	29	3.99	2.67	1.27	0.85
12	2.51	0.96	0.80	0.31	30	4.06	2.76	1.29	0.88
13	2.62	1.09	0.83	0.34	31	4.13	2.85	1.31	0.91
14	2.72	1.22	0.87	0.39	32	4.20	2.93	1.34	0.93
15	2.82	1.34	0.90	0.43	33	4.27	3.01	1.36	0.96
16	2.92	1.46	0.93	0.47	34	4.33	3.09	1.38	0.99
17	3.02	1.58	0.96	0.50	35	4.39	3.16	1.39	1.01
18	3.12	1.69	0.99	0.54	36	4.45	3.23	1.41	1.03
19	3.22	1.79	1.03	0.57	37– 40	4.20		1.34	
20	3.32	1.89	1.06	0.60	41– 45	4.63		1.48	
21	3.41	1.98	1.09	0.63	46– 51	5.06		1.61	
22	3.49	2.06	1.11	0.66	52– 60	5.74		1.83	
23	3.57	2.15	1.13	0.69	61– 70	6.52		2.07	
24	3.64	2.24	1.16	0.71	71– 90	7.72		2.46	
25	3.71	2.33	1.18	0.74	91–120	9.78		3.11	
26	3.78	2.42	1.20	0.77	121–180	13.38		4.26	
27	3.85	2.50	1.23	0.80	181–360	21.62		6.88	

The student should draw an exact profile to a large scale and superimpose an approximate profile on it for comparison. It should be noted that, although the pressure angle in Grant's involute odontograph is 15 deg, the difference is so small that this approximate method may be used for representing the teeth of the 14½-deg standard involute system on drawings, and is commonly used for that purpose.

PROBLEMS

With a few exceptions the problems in the following list have been arranged for solution in pencil on 8½- by 11-in. sheets. A heavy grade of white paper is recommended. The general layout of the sheet is given in Fig. P1-1, where the solution of Prob. 1-1 is shown as a typical example. In the statement of each problem the necessary scales are given. Positions of the mechanisms, diagrams, etc., on the sheet are given by X and Y coordinates measured from the lower left corner of the sheet as the origin. For example, the coordinates designated as $O_2(6, 8)$, Prob. 1-1, are X and Y coordinates that locate the point O_2 in Fig. P1-1, 6 in. from the left-hand edge of the sheet and 8 in. from the lower edge of the sheet. Similarly, the coordinates 0(1¼, 1) locate the point numbered 0 on the displacement diagram, 1¼ in. from the left-hand edge of the sheet and 1 in. from the lower edge of the sheet. All such locating dimensions are in inches.

If larger drawing sheets are preferred, 12 by 18 or 18 by 24 in., for example, these sheets may be blocked out for two or four problems, respectively, or the scale of the drawings, as specified, may be changed as desired.

Problems in greater variety can be obtained by changing the position of the driving or driven link in linkages and by changing the specifications as to data given or required.

Answers to many of the problems are given on pages 497–500.

Chapter 1. Introductory Considerations

1-1. Draw the skeleton outline of the Atkinson gas engine (Figs. 1-15 and P1-1), determine paths of points B and C, and construct the displacement diagram for the piston (link 6), following the general scheme shown in the figure. For scale and dimensions refer to Prob. 8-9. For the mechanism, $O_2(6, 8)$; for the displacement diagram, $o(1¼, 1)$. Length of diagram, 6 in. Suggested proportions: ⅛ in. diameter for the circles that represent the turning joints, and ¼ by 3/16 in. for rectangle that represents the sliding member.

1-2. Referring to Fig. 1-4, draw the skeleton outline of the shaper mechanism to a scale of 2 in. = 1 ft with the crank (link 2) making an angle of 30 deg with the horizontal. Starting with the oscillating arm (link 4) in its extreme right-hand position, divide the crank circle in 12 equal parts and on a base line 6 in. long draw the displacement diagram of the ram (link 6), showing the variation of its motion through the complete cycle. The requirements of the completed drawing as to numbering, lettering, etc., are indicated in Fig. 2-26. For dimensions of the mechanism refer to Prob. 8-20; for the mechanism, $O_4(4½, 4½)$; for the displacement diagram, $o(1⅛, ½)$.

1-3. In Fig. 2-8 the drag-link quick-return mechanism of a slotting machine is shown in skeleton outline. This mechanism is designed to give a slow motion to the cutting tool on the down stroke, or working stroke, and a quick return of the tool on the up stroke, or idle stroke. Draw the mechanism to a scale of 3 in. = 1 ft with the crank (link 2) in the position shown in Fig. 2-8 (30-deg angle with the vertical

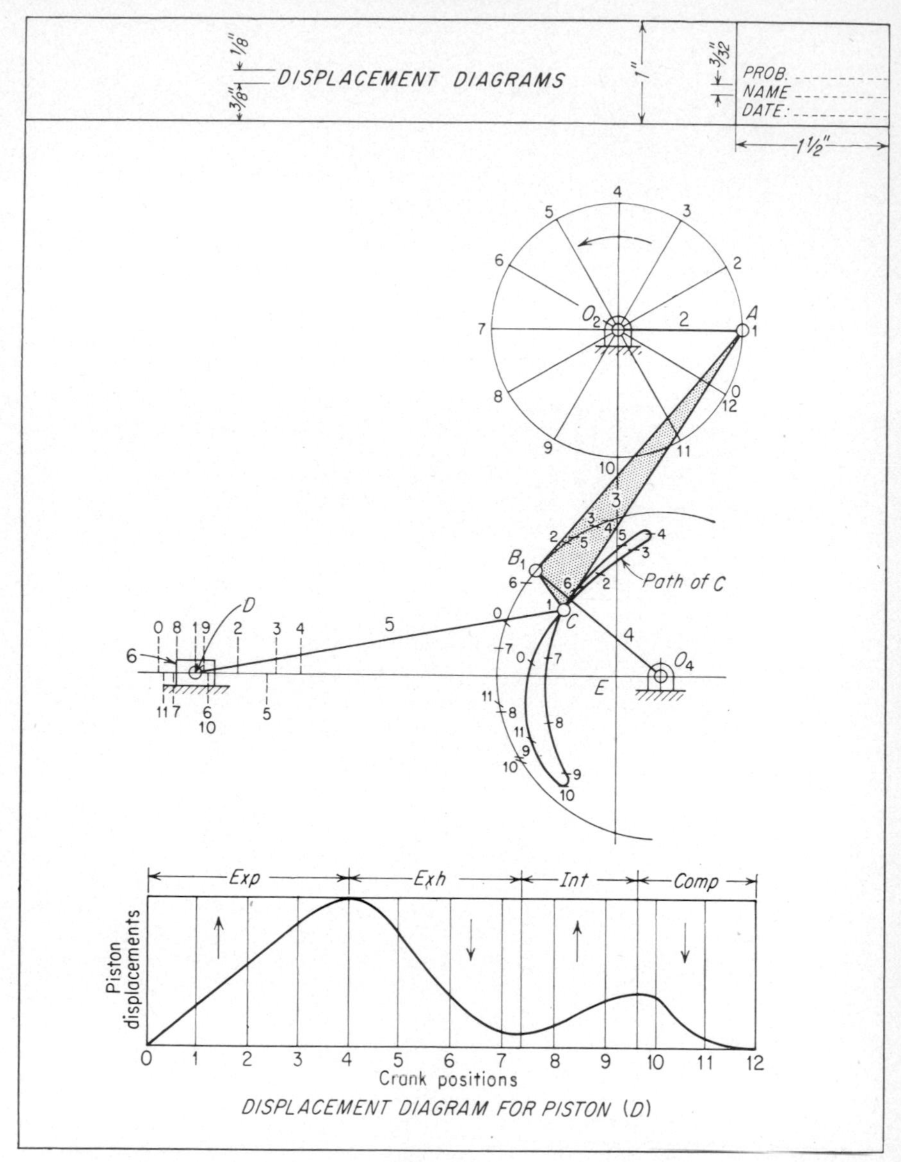

FIG. P1-1

center line). Data: $O_2O_4 = 2\frac{5}{8}$ in.; $O_2A = 5\frac{5}{8}$ in.; $AB = 4\frac{7}{8}$ in.; $O_4B = 6\frac{3}{4}$ in.; $O_4C = 6\frac{3}{4}$ in.; $BC = 9\frac{3}{4}$ in.; $CD = 16\frac{7}{8}$ in.

Starting with the ram (link 6) in its upper extreme position and using 12 equally spaced divisions on the crank circle (since the crank has uniform motion), draw the displacement diagram on a base line $4\frac{1}{2}$ in. long. For the mechanism, $O_2(6\frac{1}{2}, 2\frac{1}{2})$; for the displacement diagram, $o(\frac{3}{4}, —)$.

Refer to Fig. 2-8 for the requirements of the completed drawing as to numbering, lettering, etc.

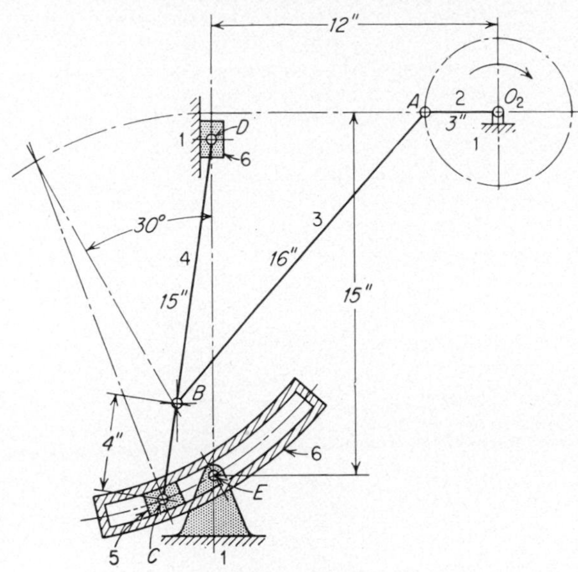

Fig. P1-4

1-4. The mechanism of a variable-stroke boiler feed pump is shown in Fig. P1-4. The drive is from the crank O_2A to the pin B on the connecting rod CBD. The end C of the connecting rod carries a pivoted block which moves along the curved slotted link 6. The radius of curvature of the slot is equal to the length of the connecting rod CD, and the stroke of the pump may be varied by rotating the slotted link 6 about the fixed fulcrum E. The stroke can be varied from 0 to a maximum of 4 in., the position of the slot for maximum stroke being that shown in the figure. Draw the mechanism to a scale of 3 in. = 1 ft. $O_2(6\frac{1}{2}, 7\frac{1}{2})$. It is required to draw a displacement diagram for the piston (point D). Use 12 equal divisions on the crank circle for point A, and start the diagram for the position of the piston in its lowest position. $0(1\frac{1}{8}, 1)$. Length of displacement diagram 6 in.

1-5. In Fig. P1-5 is shown one form of a transport mechanism used for moving materials. The motion gives an intermittent advancement of the material being

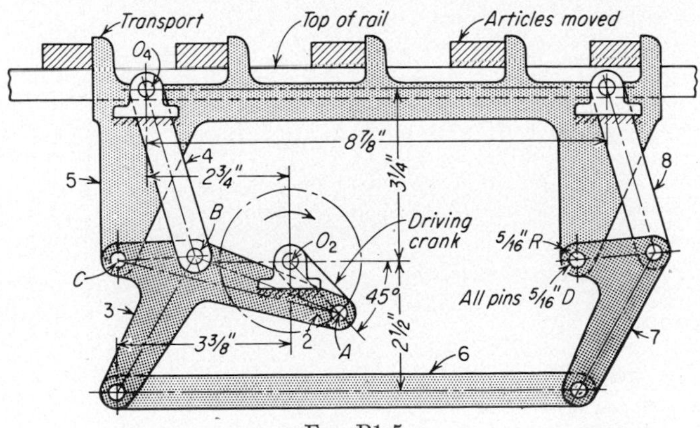

Fig. P1-5

conveyed. The essential characteristic of such a motion is that all points in the moving transport member, link 5, follow similar and equal paths.

a. Draw the skeleton outline of the mechanism, representing only those pairs which are lettered on the figure. $O_4(4, 8)$. Scale full size for the dimensions given (which are proportional to the actual dimensions). $O_2A = 1\frac{3}{8}$ in.; $AB = 3$ in.; $O_4B = 3\frac{5}{16}$ in.; $BC = 1\frac{7}{16}$ in.; $AC = 4\frac{3}{8}$ in.

b. Plot the path of one point C on the transport member. Start with link 2 in the position shown as the zero position, and find the positions of point C corresponding to 24 equally spaced positions of point A on the crank circle.

1-6. Figure P1-6 illustrates diagrammatically the mechanism for adjusting the pusher plate on the front of a bulldozer. Draw the skeleton outline of the mechanism

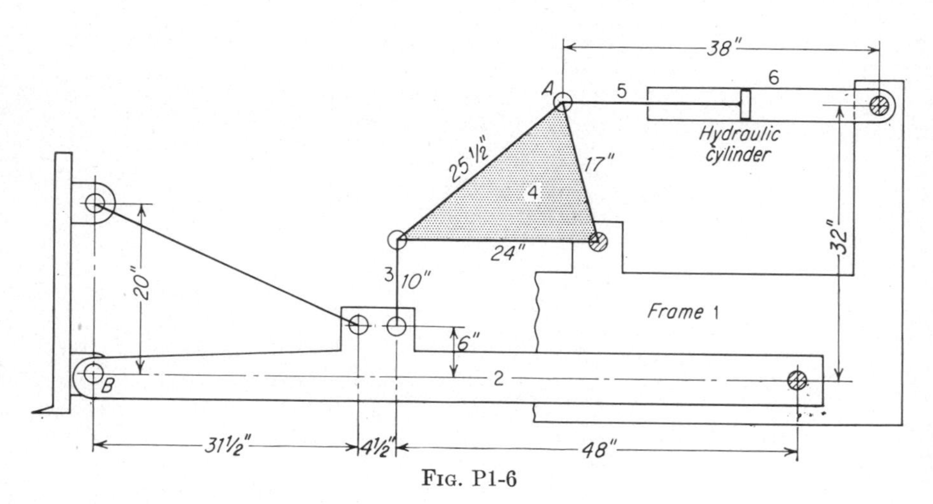

FIG. P1-6

to a scale of 1 in. = 10 in., and locate point B at (1, 3) with the long edge of the sheet horizontal. Determine the position of point B when the piston has moved into the hydraulic cylinder (*a*) 4 in. and (*b*) 8 in.

1-7. A small gas turbine has a 6-in.-diameter rotor which rotates at 30,000 rpm. Find the velocity (in feet per second and miles per hour) and the normal acceleration (in feet per second per second) at a point on the outer periphery.

1-8. A jet airplane executes a turn along a circular arc at 3 deg/sec, thereby requiring a full minute to reverse direction. How many g are applied to the pilot and airplane at (*a*) 500 mph, (*b*) 1,000 mph, (*c*) 1,500 mph? The number of g is the ratio of the actual acceleration to the acceleration of gravity. Also find (*d*) the radii of the arcs along which the turns are made.

1-9. The *stroke* of an automotive engine (see Fig. 1-14) is 4 in. Find the velocity (in feet per second and miles per hour) and the acceleration (in feet per second per second) of the crankpin at a constant engine speed of 3,000 rpm.

1-10. A cam follower moves in a certain interval along a straight line according to the equation

$$s = Bt^2 + Ct^3$$

The follower is to move 1 in. in the first 0.1 sec and a further amount of 5 in. (total of 6 in.) by the end of 0.3 sec. Find the distance moved, the velocity, and the acceleration of the follower at the end of 0.2 sec.

Chapter 2. Linkages and Flexible Connectors

2-1. Figure P2-1 shows the skeleton outline of an air-pump mechanism designed to secure a piston stroke equal to four times the length of crank instead of twice the length, as would be the case when using the ordinary slider-crank mechanism. Plot the paths of points E and C, and draw a displacement diagram $4\frac{1}{2}$ in. long for the piston (link 5) for one complete rotation of the crank (link 2).

In the figure the relative positions of the links are shown when the piston is in its upper extreme position. Draw this configuration of the mechanism to the dimensions given and in the solution of the problem use 12 equal divisions on the crank

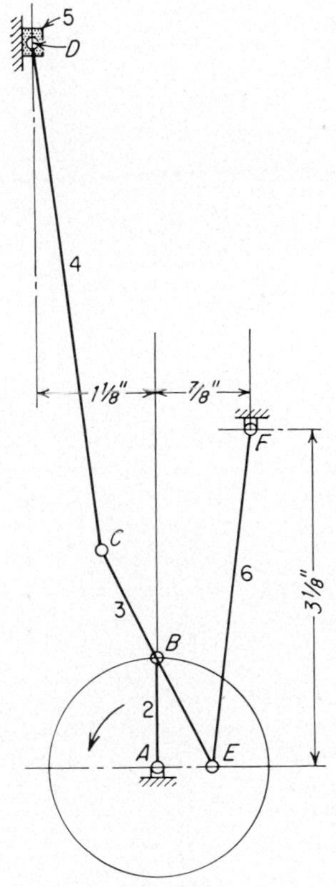

FIG. P2-1

circle. Data: $AB = 1$ in.; $BC = BE = 1\frac{1}{8}$ in.; $FE = 3\frac{1}{8}$ in.; $DC = 4\frac{3}{4}$ in.; scale, full size. For the mechanism, $A(3\frac{3}{8}, 2\frac{1}{2})$; for the displacement diagram, $o(3\frac{5}{8}, —)$.

2-2. In Fig. P2-2 is shown the mechanism for operating the steam valve of a steam engine equipped with the Corliss nonreleasing type of valve gear. It is that part of the valve gear shown at C in Fig. 2-45.

The angular motion imparted to link 6 by the reach rod 7 during one complete rotation of the eccentric is indicated in the figure, the unequal angular divisions shown corresponding to equal angular divisions on the eccentric circle. The steam valve rotates through the same angle as link 3 to which it is rigidly connected.

a. Determine the angular positions of link 3 corresponding to the given angular positions of link 6.

b. Draw a diagram of displacements of B (valve displacements) in Fig. P2-2, corresponding to equal angular positions of the eccentric Fig. 2-45. (Corresponding angular positions of O_6E are shown in Fig. P2-2.) Data: Draw the mechanism to the following dimensions: $O_2O_6 = 6\frac{3}{8}$ in.; $O_2A = 3\frac{1}{4}$ in.; $O_2B = 2\frac{5}{8}$ in.; $BC = 3$ in.; $AC = 1\frac{3}{4}$ in.; $CD = 2\frac{1}{4}$ in.; $O_6D = 4\frac{1}{4}$ in.; $DE = 5\frac{5}{8}$ in.; $O_6E = 6$ in. and is in a vertical position. Note: For an $8\frac{1}{2}$- by 11-in. sheet, draw the mechanism to a scale two-thirds of the dimensions given. Let O_2O_6 run parallel to long edge of sheet. For the mechanism, $O_2(2\frac{1}{4}, 6\frac{3}{4})$; for the displacement diagram, $o(\frac{3}{4}, \frac{3}{8})$; length of diagram = 8 in.

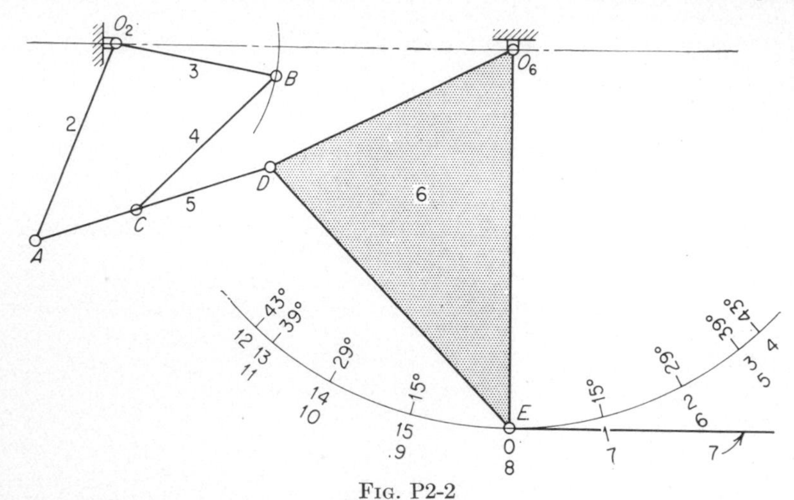

FIG. P2-2

2-3. Figure P2-3 shows the skeleton outline of the Whitworth quick-return mechanism as used in a slotting machine.

a. Draw the skeleton outline to a scale of 3 in. = 1 ft and make a diagram showing ram displacements corresponding to equal angular displacements of the driving link 3. Take ram displacements as ordinates and equal angular displacements of link 3 as abscissas.

b. Determine the time ratio of the mechanism. Data: $O_3O_2 = 3\frac{3}{4}$ in.; $O_3A = 7\frac{1}{2}$ in., and it makes an angle of 60 deg with the horizontal; $O_2B = 4$ in.; $BC = 20$ in. For the mechanism, $O_3(6\frac{1}{8}, 3)$; for the displacement diagram, $o(\frac{5}{8}, —)$. Length of displacement diagram = 6 in.

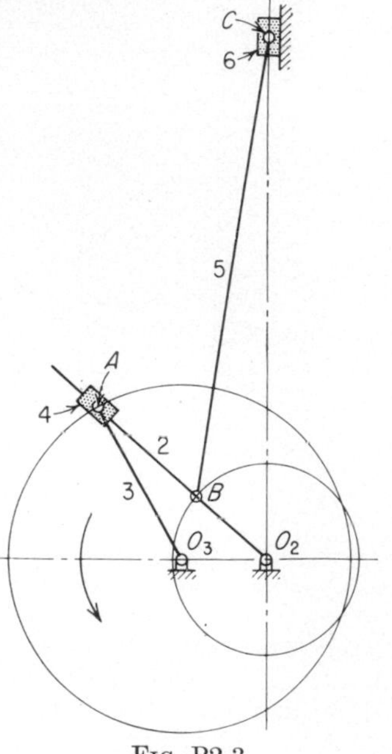

FIG. P2-3

2-4. The Andreau variable-stroke engine is represented in skeleton outline in Fig. P2-4. The circles M and N represent the pitch circles of a pair of Citroën gears (see Art. 5-9) in mesh at P. Links 2 and 5 are rigidly connected to gears M and N, respectively. Diameter of $M = 2\frac{1}{16}$ in.; diameter of $N = 4\frac{1}{8}$ in.

(In the design of a machine or in a study of its operation, particularly where the motions of the members are as complex as in the present case, it is essential to have diagrams showing paths of various points and variations in the piston stroke during a cycle.)

Starting with the member O_5C in the position shown [corresponding to the lowest positions of the piston (link 7)] and placing it successively in 24 equally spaced positions in the cycle (one revolution), plot the path of point B and draw the displacement diagram for the piston (point D). The length of the displacement diagram is 6 in. The first upward stroke of the piston is the exhaust stroke. Indicate along the upper edge of the diagram the distances that represent the exhaust, intake, compression, and power strokes, as in Fig. 1-15. For the mechanism, $O_2(3\frac{7}{8}, 1\frac{1}{2})$; for the displacement diagram, $o(2\frac{1}{4}, —)$. Make the drawing to the dimensions given in the figure.

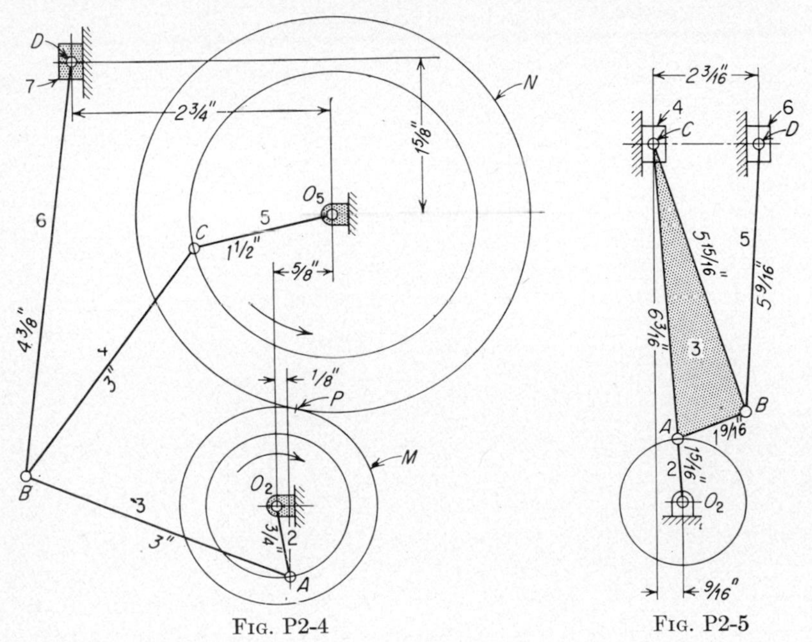

Fig. P2-4

Fig. P2-5

2-5. In Fig. P2-5 is shown a skeleton outline of the Zoller double-piston engine. Draw the mechanism to the dimensions given. Starting with the crank (link 2) in the position indicated in the figure, i.e., with piston 4 in its extreme upper position, and using 12 divisions in the cycle, plot the path of point B. Also, draw the displacement diagrams of the two pistons (points C and D) on the same base line for comparison. For the mechanism, $O_2(5\frac{1}{4}, 1\frac{3}{4})$; for the displacement diagrams, $o(3\frac{1}{4}, 3\frac{1}{4})$. NOTE: Let the base line of the displacement diagrams run vertically on the sheet. Length = 6 in.

2-6. The linkage outlined in Fig. P2-6 is that of the Wellman geared bucket, used in excavating work. The outline of the bucket (one half) in its closed position is shown in dotted lines. Plot the path of the cutting edge of the bucket (point C) between its closed position, where operating link 2 makes an angle of 27 deg with the vertical line through O_2, and its extreme open position, after link 2 has moved through an angle of 105 deg. Draw the mechanism to a scale of $1\frac{1}{2}$ in. = 1 ft: $O_2(3\frac{1}{2}, 6\frac{3}{8})$.

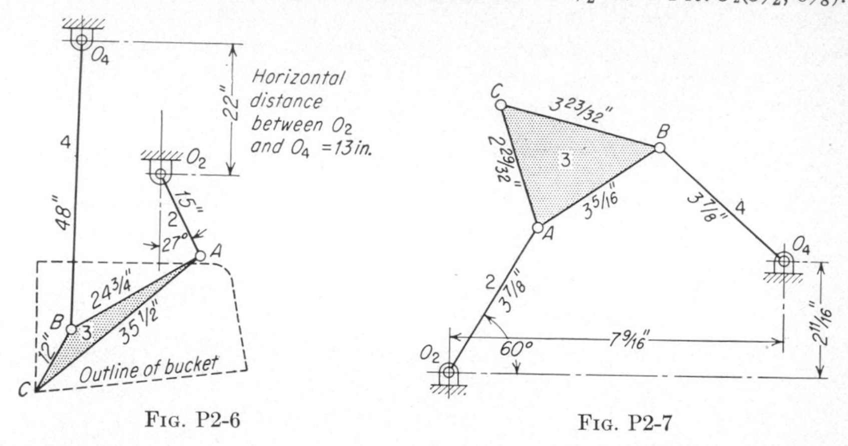

FIG. P2-6 FIG. P2-7

2-7. Figure P2-7 shows in skeleton outline a straight-line mechanism for an oil circuit breaker. The point C moves in a straight line through its operating range.

a. Draw the mechanism to the dimensions given in the figure, $O_2(\frac{1}{2}, 3)$, and plot the paths of points A, B, and C using 1-in. displacements (approximately) of point B, starting with O_2A at an angle of 60 deg with the horizontal. Adopt a numbering scheme to indicate intermediate positions of all points. Determine the complete path of point C, which is a closed curve.

b. Indicate clearly (by showing the mechanism in dotted lines) the dead-point positions when 2 is the driving link and when 4 is the driving link.

2-8. Draw the Crosby indicator mechanism (Fig. P2-8) to a scale of $1\frac{1}{2}$ in. = 1 in. $O_2(1\frac{1}{2}, 4)$; $O_2O_4 = \frac{5}{8}$ in.; O_2A (vertical) $= \frac{15}{16}$ in.; $O_4D = \frac{7}{32}$ in., $BD = \frac{17}{32}$ in.; $BC = 3\frac{1}{8}$ in.; $AB = \frac{21}{32}$ in.

Starting with point C in its lowest position, move it vertically in its straight-line path, and at $\frac{3}{4}$-in. intervals of C plot the path of D. Find the arc of a circle that will pass through the several positions of D (approximately) and thus determine the position of O_5 and the length of link 5.

2-9. The mechanism to be studied here is one that is

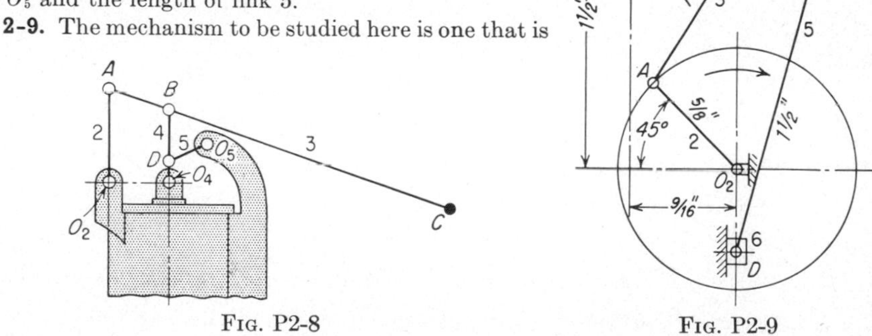

FIG. P2-8 FIG. P2-9

used in the head of a sewing machine to actuate the needle bar. Figure P2-9 shows, in skeleton outline, the relation of the parts and the actual dimensions of the links. Link 2 is the driving crank, which is assumed to rotate clockwise at a uniform rate, and link 6 is the needle bar, which reciprocates vertically. From its lowest position the needle bar moves upward a short distance to create a loop in the thread which is carried by the needle and then moves downward to its lowest position to complete the stitch. After completing the stitch the needle again moves upward and finally returns to its lowest position, thus completing the cycle. It is interesting to note that this complex motion of the needle is accomplished by placing in combination two simple four-link mechanisms.

a. Draw the mechanism to a scale twice full size, in the phase shown in Fig. P2-9. $O_2(2, 4)$, with long edge of sheet horizontal and title space to right.

b. Draw the displacement diagram for point D on the needle bar starting with D in lowest position and using 16 divisions on the crankpin circle. $0(3\frac{3}{4}, —)$. Length of displacement diagram, 6 in.

2-10. Figure P2-10 represents in skeleton outline a part of the mechanism of a wrapping machine. Draw the mechanism in the configuration shown and to the

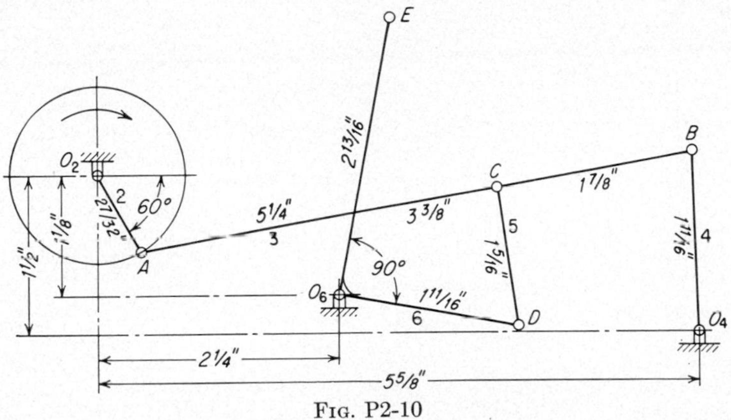

FIG. P2-10

dimensions given on the figure. $O_2(1\frac{3}{4}, 7)$. The point E is in its extreme right-hand position when the crank, link 2, is in the position shown in the figure. Starting with the crank, link 2, in the position shown in the figure as the zero position and using 12 divisions on the crank circle, draw the displacement diagram for point E for one revolution of the crank. Length of displacement diagram 6 in., origin at $o(1\frac{1}{4}, 2)$.

2-11. Figure P8-3 represents in skeleton outline the mechanism of a toggle press. Link 2 is the driving link, and link 8 is the last driven link. Draw the mechanism to a scale of $1\frac{1}{2}$ in. = 1 ft. $O_2(6, 7\frac{1}{2})$. Starting with the ram (link 8) in the lowest extreme position and using 24 equal divisions for A on the crankpin circle, draw the displacement diagram for the ram. Length of displacement diagram 6 in., with origin at $0(1\frac{1}{4}, \frac{3}{4})$.

2-12. A pair of three-step cone pulleys connect a driving and a driven shaft (42 in. apart) by means of a crossed belt. The driving shaft runs at 275 rpm and the smallest step on the driving pulley is 12 in. in diameter. The driven shaft is to run at speeds of 175, 225, and 300 rpm.

a. Determine the size of all steps on both pulleys.

b. Determine the length of belt required.

2-13. Solve the preceding problem if an open belt is used instead of a crossed belt.

2-14. In a belt-driven slotting machine, countershaft A running at 200 rpm is connected with drive shaft B by means of a pair of four-step cone pulleys and a crossed belt. The distance between shafts is 8 ft and the diameter of the largest step on the countershaft pulley is 15 in. The maximum and minimum speeds of the drive shaft are 544 and 68 rpm, respectively, and the intermediate speeds must be in geometric ratio. Calculate the length of the belt required and also the diameters of all the steps of both pulleys.

2-15. Solve the preceding problem if an open belt is used instead of a crossed belt.

2-16. In a block and tackle similar to that shown in Fig. 2-67, the upper block has two sheaves and the lower block has three sheaves. Show the mechanism in diagrammatic form as in Fig. 2-68 and determine the mechanical advantage.

2-17. In a differential chain hoist similar to that shown in Fig. 2-69, the upper block consists of a 22-pocket chain sheave and a 20-pocket chain sheave fastened to the same shaft. The lower, or hook, block is a 21-pocket chain sheave.

a. What is the mechanical advantage of this hoist?

b. If the chain is pulled over the upper sheave at the rate of 50 fpm, at what speed will the load be lifted?

2-18. In Fig. P2-18 is shown the basic mechanism of a portable typewriter. Draw the mechanism full size to the dimensions shown. $O_A(7\frac{3}{4}, 5)$. In order for the type to strike the roller, type arm 6 must swing 90 deg clockwise. Find and label the distance which point B on the type keyboard must be moved.

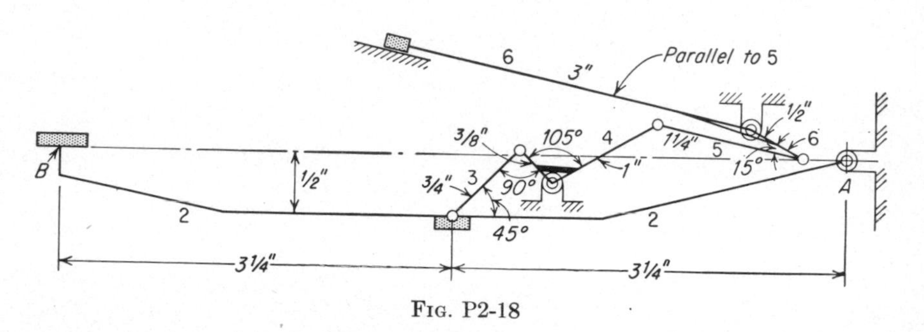

FIG. P2-18

2-19. In Fig. 2-31 of Chap. 2 is shown a pump with thin sliding vanes which move in slots in the rotor. Assume the two circles which represent the rotor and the outer member against which the vanes slide to have diameters of 4 and 5 in., respectively. The centers of the circles are ½ in. apart. Plot a curve of radial displacement (along the slot) of one of the vanes as the rotor revolves. Use at least 12 angular positions of the rotor. Choose the initial position to be that at which the vane is entirely in the slot. Center of rotor at 4¼, 7. Make the drawing full size. Use a 6-in. base line for the curve, and plot ordinates double size.

2-20. Design and draw to scales of your own choosing pantographs which will reduce the size of lettering traced to (*a*) one-fourth of full size and (*b*) one-tenth of full size. Note that adjustable pantographs used on engraving machines allow flexibility in choice of letter sizes without the necessity of having a large number of pattern letter sizes. Sketch a design for such an adjustable pantograph.

2-21. In Fig. 2-49 of Chap. 2 is shown a flyball governor. Suppose such a governor to be attached to and to rotate with a steam turbine shaft which turns at 3,600 rpm. The center of gravity of each flyweight is at a radius of 2 in.

a. Find the velocity in feet per second and acceleration in feet per second per second of the center of gravity of each flyweight, and show the directions on a sketch.

b. Force is proportional to acceleration according to Newton's law $F = (W/g)A =$

$(W/32.2)A$. Find the centrifugal force of a flyweight which weighs 1 lb. Show the direction of the force.

2-22. A centrifugal shaft governor of the type shown in Fig. 2-50 is used on a steam engine which revolves at 350 rpm. At this speed, the center of gravity of the weight W is situated at a radial distance r of $1\frac{1}{2}$ ft. Find the velocity in feet per second and the acceleration in feet per second per second of the center of gravity of the weight and show their proper directions on a sketch. Find also the centrifugal force per pound of weight W. $(F = (W/32.2)A)$.

Chapter 3. Cams

3-1. Construct the outline of a disk cam that will transmit harmonic motion to a roller follower in the following manner: Up 1 in. while the cam turns through 120 deg; rest while the cam turns through 60 deg; down 1 in. while the cam turns through 90 deg; rest while the cam turns through 90 deg. The initial position of the follower relative to the axis of the camshaft and other information regarding the general layout are shown in Fig. P3-1. Coordinates of axis of cam, $(4\frac{1}{4}, 4\frac{3}{4})$.

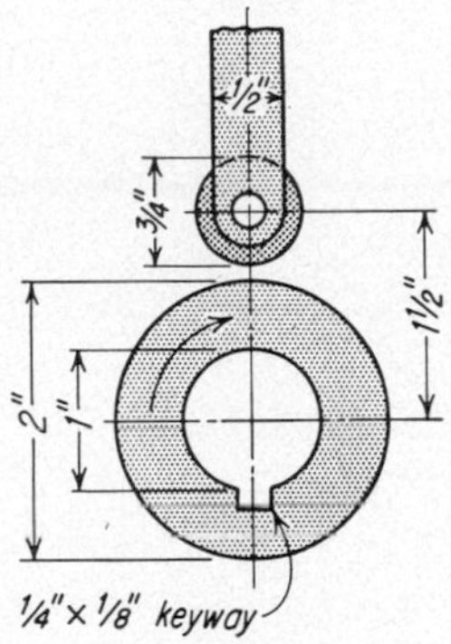

3-2. Same as Prob. 3-1 except that follower has parabolic motion. Coordinates of base curve $o(5, —)$; length $= 3$ in.

3-3. Same as Prob. 3-1 except that follower has modified straight-line motion. Coordinates of base curve, $o(5, —)$; length $= 3$ in.

3-4. Same as Prob. 3-1 except that path of roller follower is in vertical line $\frac{5}{8}$ in. to left of vertical line through axis of cam.

FIG. P3-1

3-5. Same data as in Prob. 3-1, except that: diameter of shaft $= \frac{1}{2}$ in.; diameter of hub $= 1$ in.; keyway $= \frac{1}{8}$ by $\frac{1}{16}$ in.; distance that axis of roller follower is above axis of cam $= 1$ in.; travel of roller follower $= \frac{3}{4}$ in.

a. Construct the outline of the cam and show on the drawing the pitch circle and the maximum pressure angle (refer to Fig. 3-9). Coordinates of axis of cam, $(5\frac{1}{4}, 5)$.

b. Draw a displacement diagram of the follower and show pressure angles (refer to Fig. 3-9). Coordinates of displacement diagram, $o(1\frac{3}{4}, \frac{3}{4})$. (This diagram runs vertically on the sheet.)

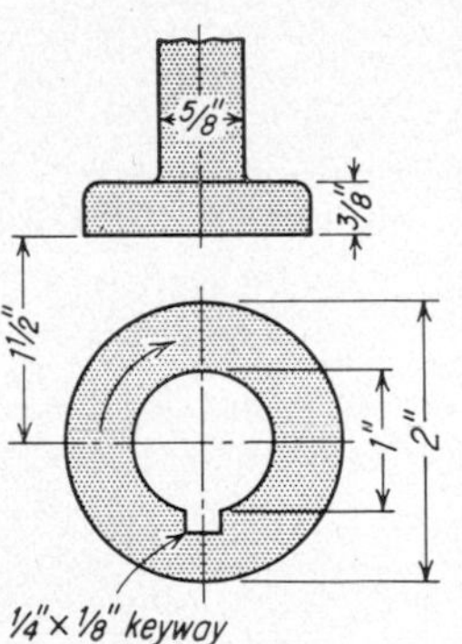

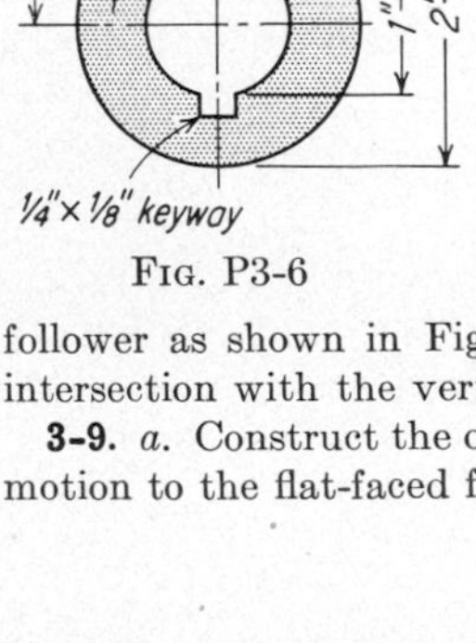

FIG. P3-6

3-6. *a.* Construct the outline of a disk cam that will transmit harmonic motion to a flat-faced follower in the following manner: up 1 in. while the cam turns through 90 deg; rest while the cam turns through 90 deg; down 1 in. while the cam turns through 90 deg; rest while the cam turns through 90 deg.

b. Determine the required length of face of the follower, and then add $\frac{1}{8}$ in. on each end. Information regarding the general layout is given in Fig. P3-6. Coordinates of axis of cam, $(4\frac{1}{4}, 4\frac{3}{4})$.

3-7. Same as Prob. 3-6, except that the motion of the follower is parabolic.

3-8. Same as Prob. 3-6, except that the face of the follower as shown in Fig. P3-6 is to be tipped clockwise 15 deg about its point of intersection with the vertical center line.

3-9. *a.* Construct the outline of a disk cam that will transmit modified straight-line motion to the flat-faced follower in Fig. P3-6 in the following manner: up 1 in. while the cam turns 90 deg, rest while the cam turns through 90 deg, down 1 in. while the cam turns through 90 deg, rest while the cam turns through 90 deg.

b. Determine the required length of face of the follower, and then add ⅛ in. on each end. Coordinates of axis of cam (4¼, 5¾). Length of displacement diagram = 6 in. *o*(1¼, 1½). "Easing off" radius for modified straight-line motion = ½ in.

3-10. A single lobe cam for a gun impulse generator is required to impart a displacement to a roller follower in a radial direction in accordance with the following specifications: The follower is to rise ¼ in. in 90 deg, fall ¼ in. in the next 90 deg, and dwell for 180 deg. The base curve for the follower is to be parabolic for both rise and fall. The maximum pressure angle is to be 20 deg. The cam factor for this pressure angle is 5.5. The diameter of the roller follower is ¾ in.

a. Lay out the parabolic base curve for the rise of the follower to a scale four times full size. *O*(1½, 8½). SUGGESTION: Divide the base line into 10 equal parts.

b. Construct the cam outline, starting with the follower in its lowest position. Clockwise rotation. Locate axis of cam at (4½, 3¾). Scale four times full size.

3-11. The roller follower (Fig. P3-11) is mounted on the end of an arm that is pivoted at *A*. Construct the cam outline that will impart motion to the follower

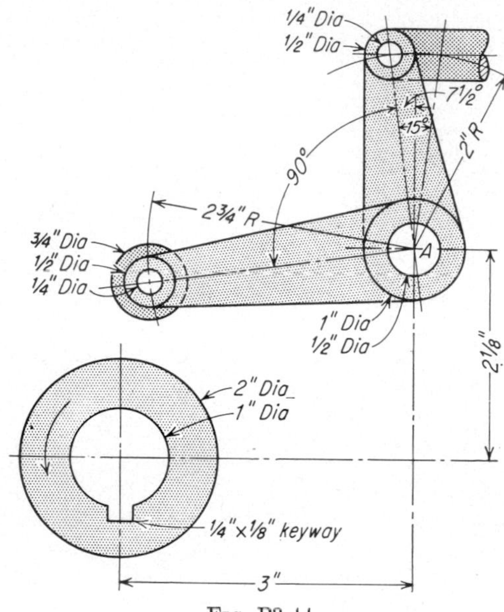

FIG. P3-11

in accordance with the following specifications: Harmonic-motion base curve. Up 15 deg while the cam turns through 120 deg; rest while the cam turns through 150 deg; down 15 deg while the cam turns through 90 deg. The lowest position of the follower and other general information regarding the layout are given in the figure. Coordinates of axis of cam, (4¼, 4½).

3-12. Same as Prob. 3-11 except that up, down, and rest are each 120 deg.

3-13. The flat-faced follower, which is pivoted at *A* in Fig. P3-13, is required to move in accordance with the following specifications: Harmonic-motion base curve. Up 15 deg while the cam turns through 120 deg; rest while the cam turns through 90 deg; down 15 deg while the cam turns through 90 deg; rest while the cam turns through 60 deg.

a. Determine the required cam outline.

b. Determine the limits of contact along the face of the follower. The lowest position of the follower and other general information regarding the layout are given in the figure. Coordinates of axis of cam, (4¼, 5).

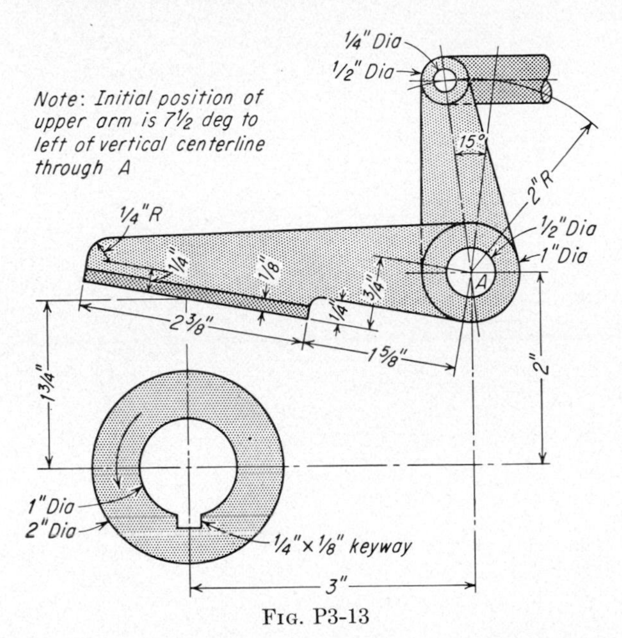

Fig. P3-13

3-14. Same as Prob. 3-13 except that up, down, and rest are each 120 deg.

3-15. Determine the shape of the groove in the cylindrical cam in Fig. P3-15 that will impart motion to the arm that swings about axis *AA* in the following manner: Outward 15 deg with harmonic motion in one-third turn of cam. Rest for one-sixth turn of cam. Return with harmonic motion in one-third turn of cam. The roller is to be ⅝ in. in diameter at large end, ⅜ in. in length, and of such form as to give pure rolling contact. Coordinates for the axis of the cam, (6¼, 3). For the displacement diagram, i.e., the developed cylindrical surface, which will run parallel to left-hand edge of sheet, *o*(2½, ¼).

3-16. In the toe-and-wiper cam mechanism, Fig. 3-25, the face of the follower is shown in its lowest position, which is 1 in. above the axis of the cam. The follower is to have a vertical movement of 1¼ in. while the cam turns counterclockwise through 45 deg at a uniform rate. The positions of the follower corresponding to equal angular positions of the cam are to be ⅛, 7/16, 13/16, and 1¼ in., respectively, above its initial position. Determine the working outline of the cam and draw, in good proportion, the complete outline of both cam and follower. Axis of cam, (2¾, 4¼).

3-17. The dimensions of a stamp-mill cam similar to that shown in Fig. 3-26 are as follows: diameter of camshaft, 5 in.; diameter of hub, 12 in.; center line of stamp shaft, 5 in. to the right of center line of camshaft; the lifting plate is 16 in. above the center line of the camshaft when in its upper extreme position, and from this position it has a fall of 9 in. to its extreme lower position; diameter of lifting plate = 9 in.; height of lifting plate = 10½ in.; diameter of stamp shaft = 3 in. The contact surface of the cam is 2 in. wide, and it passes ¼ in. in front of the stamp shaft. Draw the cam and follower to a scale of 3 in. = 1 ft. Coordinates of axis of camshaft, (4¼, 4⅛). The cam has a counterclockwise rotation of 55 rpm.

a. Assuming 24 time periods per revolution and that the stamp is to rise with uniform motion and rest two-thirds of a time period after it has fallen to its lower extreme position, make a displacement diagram of the follower motion through its complete cycle on a base line 6 in. long. Coordinates, $o(1\frac{1}{4}, \frac{1}{4})$.

b. Determine the required outline of the cam.

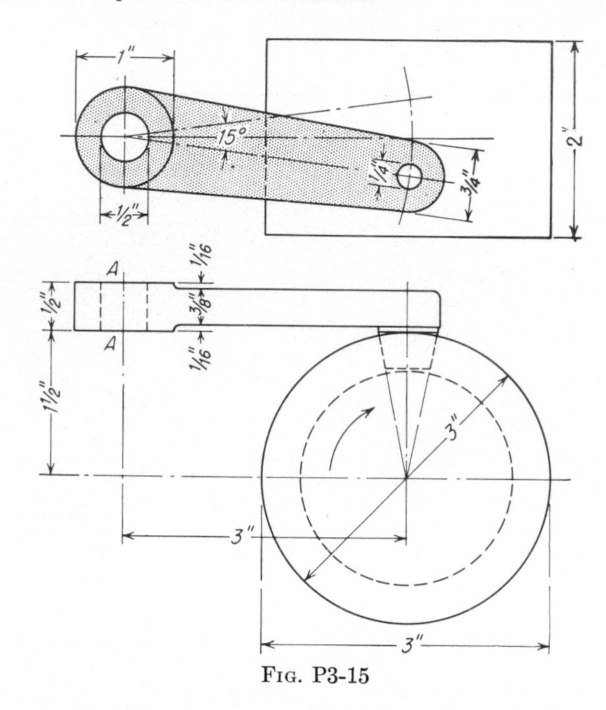

FIG. P3-15

3-18. A cam follower is to be dropped 3 in. from its highest position to its lowest position by simple harmonic motion in 90 deg of cam rotation. If the maximum allowable acceleration of the follower is 59.2 in./sec², what is the maximum allowable cam rpm?

3-19. A cam rotates at 100 rpm. If the follower is to be lifted 2 in. with uniformly accelerated and decelerated motion at 120 in./sec², how many cam degrees will be required?

3-20. A displacement diagram shows a total lift of 1 in. by simple harmonic motion in a total cam angle of 120 deg. The cam is to rotate at 600 rpm. Find the velocity and acceleration of the follower at cam angles of 60 and 80 deg.

3-21. The complete drop of a follower occurs in three stages: constant acceleration for ½ in., constant velocity for 1½ in., and constant deceleration for 1 in. Maximum velocity allowable is 60 ips. Cam speed is 300 rpm.

a. How many cam degrees does each of the three stages require?

b. What are the magnitudes of the acceleration and deceleration?

3-22. Given a follower displacement diagram in which a portion of the motion of the follower conforms to the equation $x = C_0 + C_1t + C_2t^2 + C_3t^3$.

a. Evaluate the "jerk" or rate of change of acceleration.

b. Modify the equation so that the jerk is zero within the interval. Show an accel-time sketch. Is the jerk necessarily zero at the *ends* of the interval?

3-23. In a certain design, shock loading of the follower due to "jerk" (rate of change of acceleration) is to be kept within reasonable limits by choosing the jerk vs. time diagram to be parabolic and symmetrical as shown in Fig. P3-23. The portion of the curve between A and B is given by the equation $d^3x/dt^3 = ct^2$.

a. Determine equations for the acceleration, velocity, and displacement.

b. How can the constants be evaluated so that the displacement diagram can be drawn and the required cam profile constructed?

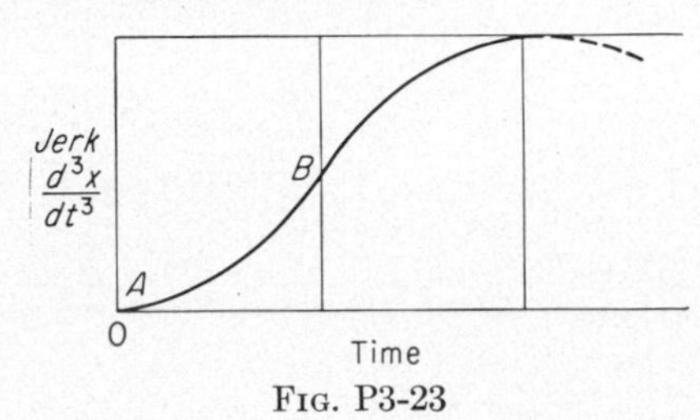

FIG. P3-23

Chapter 4. Toothed Gearing—Spur Gears

4-1. An involute gear with 14½-deg full-depth involute teeth of 1 diametral pitch has a pitch diameter of 12 in. Axis of gear, $o(4\frac{1}{4}, \frac{1}{2})$.

Using the method shown in Fig. 4-13, Art. 4-10, lay out full size one complete tooth. (Refer to Art. 4-26 for tooth proportions.)

4-2. Lay out the teeth of a pair of involute spur gears to conform to the following specifications: 14½-deg full-depth involute teeth. (Refer to Art. 4-26 for tooth proportions.) Diametral pitch = 2½; pitch diameter of pinion = 6 in.; pitch diameter of gear = 12.800 in.; backlash = 0. The pinion is the driver, and its direction of rotation is clockwise. Axis of pinion, $(4\frac{1}{4}, \frac{1}{2})$ (refer to Art. 4-14).

a. Show not more than four or five teeth on each gear, having one pair of teeth in contact at the pitch point.

b. Indicate on the drawing the following: (1) Angles of approach, α_p and α_g; (2) angles of recess, β_p and β_g; (3) line of action; (4) path of involute contact (indicate by heavy line); and (5) pressure angle, θ.

c. Examine the teeth for interference and, if there is any, indicate it in the manner shown in Fig. 4-19.

d. If the interfering tooth tip is to be modified to become cycloidal in form in order to mesh with the radial flank of the meshing tooth, what size describing circle must be used to generate the cycloid?

e. Find the contact ratios for the gears, for involute-to-involute contact.

4-3. Same as Prob. 4-2 except that axis of pinion is located at $(4\frac{1}{4}, 9\frac{3}{4})$.

4-4. Same as Prob. 4-2 except that axis of pinion is located at $(4\frac{1}{4}, 9\frac{3}{4})$ and rotation of pinion is counterclockwise.

4-5. Same as Prob. 4-2 except that rotation of pinion is counterclockwise.

4-6. Statement same as for Prob. 4-2 with the exception that the pitch diameter of the gear is 12 in.

4-7. Same as Prob. 4-6, except that rotation of pinion is counterclockwise.

4-8. A 12-tooth involute pinion with 14½-deg full-depth involute teeth of 1¼ diametral pitch meshes with a rack. Axis of pinion, $O(1, 5\frac{1}{2})$. The pinion is the driver, and rotates clockwise. The rack moves vertically on the sheet. Make drawing full-sized.

a. Show one pair of teeth in contact at the pitch point, and the two adjacent pairs of teeth, one pair on each side of the pitch point. Use the Grant odontograph method given in the Appendix for drawing the profiles.

b. Show the base circle of the pinion and the pitch circles.

c. Show the path of action for involute-to-involute contact by a heavy line with arrows at the ends.

d. Indicate interference by crosshatching the portion of the tooth which would have to be removed to eliminate the interference.

e. Indicate the new minimum pressure angle necessary if interference were to be eliminated by increasing the pressure angle.

f. Indicate the angles of approach and recess for the pinion, for involute contact. What is the angle of action for the rack?

g. Find the contact ratio for the pinion, for involute contact.

4-9. Same as Prob. 4-8, except that the teeth are standard 20-deg involute stub teeth, and axis of pinion is at (8¼, 5½). The rack moves vertically on the left of the sheet. Note that the Grant odontograph method given in the Appendix does not apply to 20-deg teeth. It is suggested that a true cycloidal profile be drawn and a template cut out for use in drawing the profiles required.

4-10. Same as Prob. 4-9, except that the pinion has 16 teeth.

4-11. A drawing and a description of the Root positive blower are given under Art. 2-18. In Fig. P4-11 some general dimensions for such a blower are given. Lay out the cycloidal outlines of the rotors or impellers. Refer to Art. 4-18. The diameter of the generating circle is one-fourth that of the pitch circle. (For a three-lobed wheel it would be one-sixth, etc.) It can be seen that, with the complete curves, the impellers act like a pair of mating teeth on cycloidal gears. There can, however, be real tooth action (in the sense of effective driving) over only certain parts of the revolution. The maintenance of proper rolling is ensured by an external pair of equal gears having the pitch circles indicated in the figure. Axis of lower impeller, (4¼, 3¾). Scale, full size.

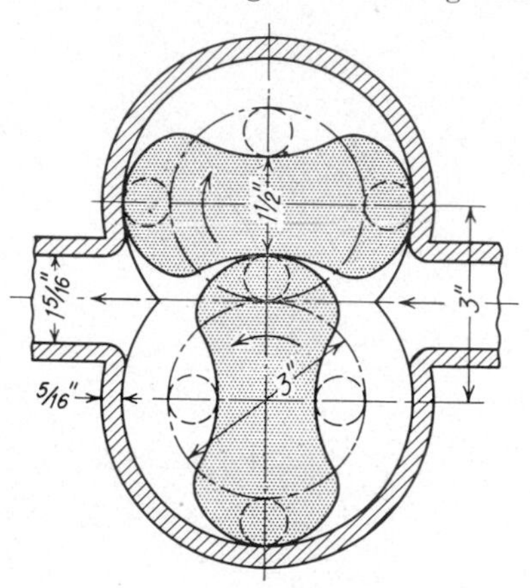

Fig. P4-11

4-12. Statement same as for Prob. 4-11, with the exception that the impellers are to have involute outlines, and the diameter of the pitch circles of the mating gears is 4 in. The pressure angle is 35 deg. Start construction with the long axis of the lower impeller in a 45-deg position in the second quadrant. The end of each impeller is to be rounded off with an arc of a circle having a 19/32 in. radius, the center of this arc to be located on the pitch circle. The same arc, with its center on the pitch circle, will also connect the involute curves of the two lobes inside the base circle. Axis of lower impeller, (4¼, 2⅞). Scale, full size.

Chapter 5. Gear Types and Manufacturing Methods

5-1. Lay out a pair of involute bevel gears to meet the following requirements: 14½-deg composite system (refer to Art. 4-26 for proportions); angle between shafts = 75 deg; velocity ratio = 2:1; diametral pitch = 4; pitch diameter of gear = 8 in.; length of face = 1½ in. Lay out at least two teeth on the developed back cone of each gear, and show sectional view of the gears as in Fig. 5-4. Assume proportions for bores, hubs, webs, etc., using proportions shown in Fig. 5-4 as a guide. Axes of pinion and gear intersect at (4¼, 2, with long edge of sheet horizontal); center line of gear vertical.

5-2. Same as Problem 5-1 except that the teeth are to conform to the Gleason system for bevel gears (Art. 5-15).

5-3 to 5-7. Lay out a pair of straight-tooth bevel gears, in sectional view, to meet the following requirements (select data from accompanying table): Gleason standard teeth; angle between shafts, $A =$; number of teeth in pinion, $B =$; number of teeth in gear, $C =$; diametral pitch, $D =$. The axes of pinion and gear intersect at ($E =$, $F =$) with long edge of sheet horizontal.

Represent the pitch cones and back cones *clearly* by dot and dash lines. Show at least two teeth on the developed back cone of each gear. The involute teeth outlines are to be approximated by using the following radii, with centers on the base circles. For the pinion, $G =$; for the gear, $H =$.

The following proportions are to be used in the layout: Length of face, $I =$; bore of pinion, $J =$; bore of gear, $K =$; length of bore in pinion, $L =$; length of bore in gear, $M =$; the keyway is to fit a square key whose sides are one-fourth the diameter of the bore; the diameter of the hub is twice the diameter of the bore in each case.

Tabulate, on the drawing, the following data for the pinion and gear: Number of teeth, pitch, addendum, dedendum, circular thickness, formative number of teeth.

Prob.	*A*	*B*	*C*	*D*	*E*	*F*	*G*	*H*	*I*	*J*	*K*	*L*	*M*
5-3	90°	16	32	4	4¼	3	¾	1 11/16	1½	1½	1¾	2	2⅛
5-4	90°	14	24	4	4½	3	1 1/16	1 5/16	1¼	1¼	1½	1½	1¾
5-5	90°	14	26	4	4½	3	1 1/16	1 5/16	1¼	1¼	1½	1½	1¾
5-6	90°	14	28	4	4½	3	1 1/16	1 5/16	1¼	1¼	1½	1½	1¾
5-7	90°	14	30	4	4	3	21/32	31/32	1¼	1¼	1½	1½	1¾

All dimensions are in inches.

5-8. Make a layout, similar to that shown in Fig. 5-19, of a worm and worm wheel having a velocity ratio of 18:1. The worm is right-handed and double-threaded and has a linear pitch of ¾ in.

The bore of the worm is 1⅜ in., and that of the worm wheel 1¾ in. Refer to Art. 5-16 for proportions, and use proportions shown in Fig. 5-19 as a guide in drawing hub, web, rim, etc. Coordinates of axis of worm wheel, (4¼, 1½). NOTE: The design illustrated in Fig. 5-19 conforms to the standards of the American Gear Manufacturer's Association (Art. 5-16).

5-9. Same as for Prob. 5-8 except that the worm is single-threaded and the velocity ratio is 36:1.

5-10. The purpose of this problem is to make a comparison of the tooth forms of the 14½-deg composite system, the Fellows stub-tooth system, and the Gleason system for bevel gears (unequal addendum), using the following data and making the drawings to a scale of 2 in. = 1 in.

The 14½-deg composite system: pitch diameter = 4 in.; pitch = 4; tooth outline outside base circle to be approximated by an arc of ¾ in. radius, with center on base circle; radius of fillet = 1/16 in.; axis of gear, (1½, 2) (Arts. 4-23 and 4-26).

The Fellows stub-tooth system: pitch diameter = 4 in.; pitch = 4/5; involute outline of tooth to be approximated by an arc of ¾ in. radius, with center on base circle; radius of fillet = 1/16 in.; axis of gear, (4¼, 2) (Arts. 4-24 and 4-26).

The Gleason system for bevel gears: pitch diameter (pinion) = 4 in.; pitch = 4; addendum = 0.3375 in.; dedendum = 0.2095 in.; tooth thickness = 0.4424 in.; involute tooth outline to be approximated by an arc of ⅞ in. radius with center on base circle; radius of fillet 1/16 in.; axis of gear, (7, 2) (Art. 5-15).

Lay out one tooth of the 14½-deg composite system, one of the Fellows system, and one of the Gleason system. Show on each as dimensions in inches, the addendum, dedendum, and tooth thickness. Also indicate the pressure angle in degrees.

Chapter 6. Intermittent-motion Mechanisms

6-1. Lay out a pair of intermittent gears, similar to the pair shown in Fig. 6-4, in which the driven gear advances one-fourth turn for every turn of the driving gear. Distance between centers = 4 in. Axis of driven gear (4¼, 3⅝). Axis of driving gear on vertical center line through axis of driven gear.

6-2. Lay out a Geneva wheel combination in which the driven wheel makes one-sixth turn for one turn of the driver. Distance between centers = 4 in. Axis of driven wheel (4¼, 3⅝) when directly below driving wheel.

6-3. A circular plate is required to have 17 accurately spaced notches cut in its periphery. Describe fully the method of performing this operation on a milling machine, using the index plate shown in Fig. 6-7.

6-4. Draw to a scale four times that shown in the figure the skeleton outline of the toggle mechanism used to operate the friction clutch (Fig. 6-19) and show the positions of all the links when in their extreme *in* and *out* positions. With the long edge of the paper horizontal, the intersection of the center line of the shaft with the center line of the pulley is (1¼, ¾).

Chapter 7. Trains of Mechanism

7-1. Given the gear train shown in Fig. P7-1 and the following data: gear A, 20 teeth; gear B, 12 teeth, 10 pitch; pinion D 1¾ in. in pitch diameter, 8 pitch; gear E, 30 teeth; worm F triple-threaded, right-hand type; worm wheel G, 45 teeth; gear H, 15 teeth, 6 pitch.

a. How many revolutions of gear A, in the direction indicated by arrow, are necessary to move the rack K a distance of 1 in.?

b. Will the rack move up or down?

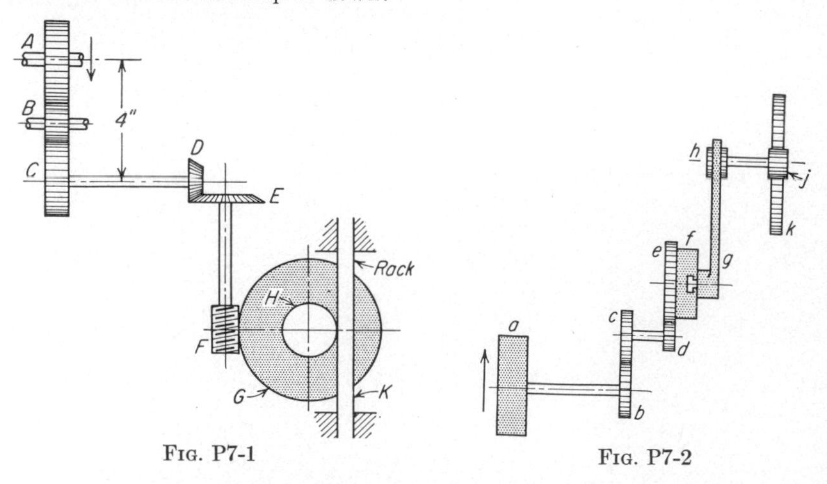

FIG. P7-1

FIG. P7-2

7-2. In Fig. P7-2 is shown a train of mechanism through which power is transmitted to the cutting tools of a bevel-gear-cutting machine. *a* is the driving pulley; *b* and *c* are two of a set of simple change gears by means of which the speed of the tools is controlled; member *f* is a crank plate attached to gear *e*; *g* is a rack in mesh with pinion *h* at one end and pivoted on an adjustable stud in the crank plate on the other end. The adjustable stud provides means for varying the length of tool stroke. The cutting tool is carried on a reciprocating slide to which rack *k* is fastened. Data: drive pulley *a* runs at 500 rpm; the stud in crank plate *f* is set for a throw of $2\frac{1}{2}$ in.; gear *b*, 24 teeth, 8 pitch; gear *c*, 56 teeth, 8 pitch; gear *d*, 15 teeth, 8 pitch; gear *e*, 100 teeth, 8 pitch; gear *h*, 25 teeth, 6 pitch; gear *j*, 25 teeth, 6 pitch.

a. How many strokes per minute does the tool make?

b. What is the average cutting speed in feet per minute?

c. If the tool must run at 50 strokes per minute, figure out the pair of change gears necessary at *b* and *c*.

7-3. A motor running at 1,800 rpm drives a conveyor through the double-stage spur-gear speed reducer shown in Fig. P7-3.

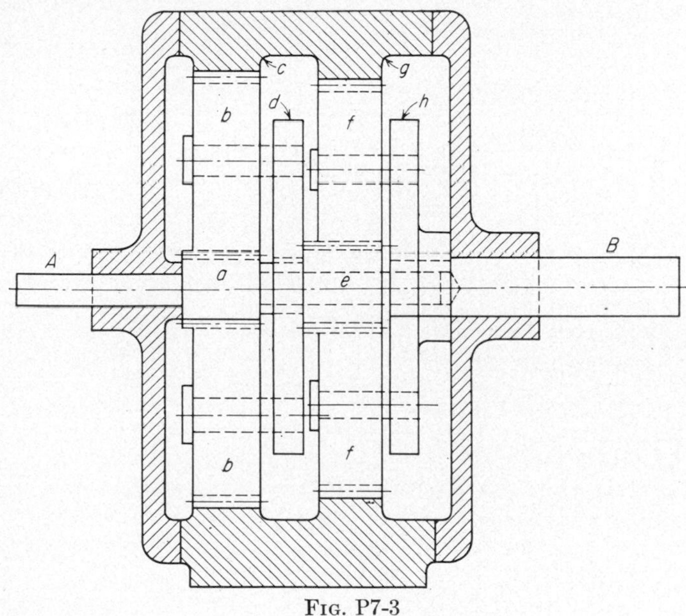

Fig. P7-3

a. Find the total speed reduction from motor to conveyor.

b. Find the rpm of the conveyor shaft.

c. Do the motor and conveyor shafts run in the same direction?

Data: *A* is the shaft to the motor; *B* is the shaft to the conveyor; gear *a*, 12 teeth; gear *b*, 24 teeth; gear *c*, 60 teeth; gear *e*, 14 teeth; gear *f*, 18 teeth; gear *g*, 50 teeth.

7-4. *a.* In the triplex hoist (Fig. 7-14), find the number of turns hand sprocket *a* must make in order to produce one turn of hoisting sprocket *g*.

b. How many feet of hand chain must be moved to raise weight *W* 1 ft?

Data: gear *b*, 13 teeth; gear *c*, 31 teeth; gear *d*, 12 teeth; gear *e*, 49 teeth. Pitch diameter of hand sprocket = $9\frac{3}{4}$ in. Pitch diameter of hoisting sprocket = $3\frac{1}{8}$ in.

7-5. A good example of a spur-gear differential train is shown in Fig. P7-5. The driver (link 2) consists of the shaft and gear *a*, which is keyed to it. The planetary gears b and c, which are keyed together, constitute link 3. Link 4 consists of gear d and the shaft on which it is keyed. The spider that carries the planetary gears b and c is link 5. The motion of the driven shaft (link 4) is the result not only of the continuous motion of the main driving shaft (link 2) but also of the continuous or intermittent motion of link 5, which is driven through the sleeve shown at the left in the figure.

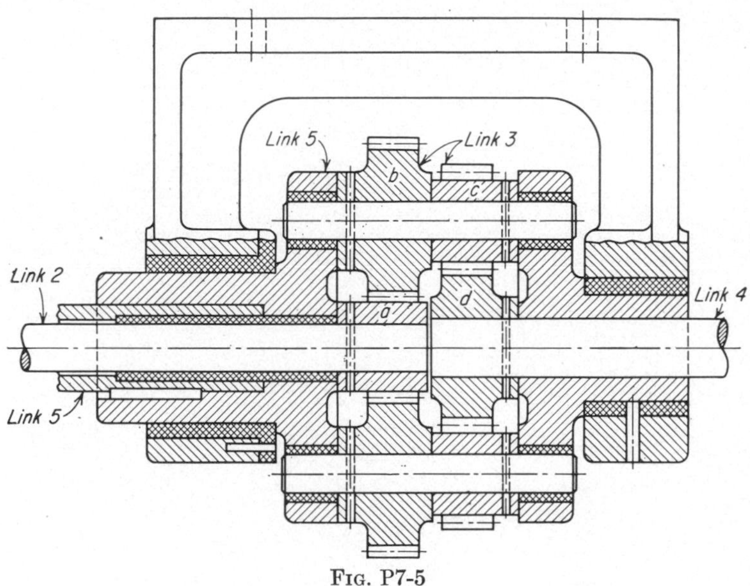

Fig. P7-5

All the gears have the same pitch, and their numbers of teeth are indicated by the letters a, b, c, and d. Determine the angular velocity n_4 of the driven member in terms of the angular velocities n_2 and n_5 of the driving members.

7-6. In Fig. P7-6 is shown a diagram of the arrangement of the gears in the old Ford transmission (model T).

a. For low speed, gear f must be held stationary. In this case, find number of turns of gear d (and hence rear-axle shaft) for one turn of engine drive shaft e.

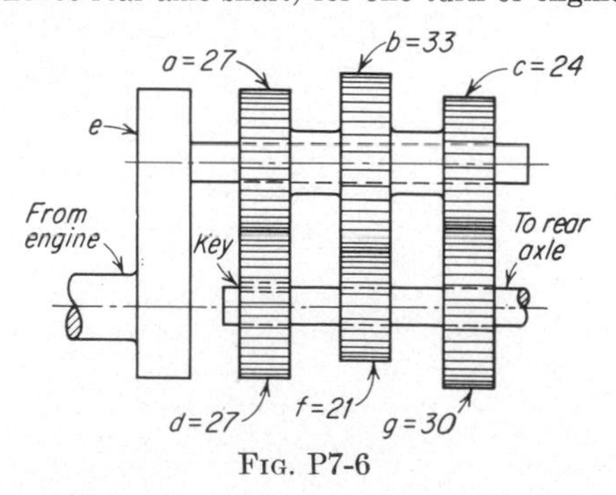

Fig. P7-6

b. For reverse, gear g must be held stationary. In this case, find number of turns of d for one turn of e.

c. For high speed (one turn of d per turn of e), how does the mechanism operate?

7-7. The train of gears shown in Fig. P7-7 includes a planetary-bevel-gear train in the form sometimes used in indexing and differential mechanisms of machine tools. The drive is from *a* and *b* through the planetary train to *d*. Data: gear *a*, 60 teeth, 12 pitch; gear *b*, 60 teeth, 12 pitch; gear *c*, 52 teeth, 8 pitch; gear *d*, 26 teeth, 8 pitch; gear *f*, 24 teeth, 8 pitch; gear *g*, 20 teeth, 8 pitch.

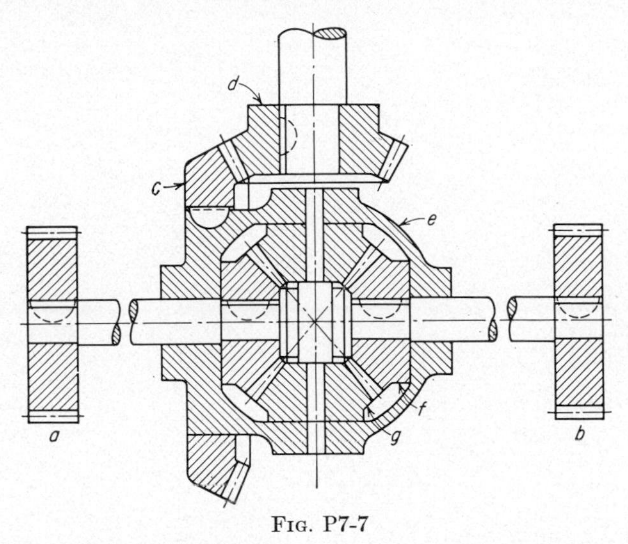

Fig. P7-7

a. If *a* and *b* are running in the same direction at the same speed, how many revolutions does *d* make for one revolution of *a*?

b. If *a* and *b* are running in opposite directions at the same speed, what is the speed of *d*?

c. If *b* is held stationary, how many revolutions does *d* make for one revolution of *a*?

d. If *a* runs at 50 rpm and *b* at 25 rpm in the same direction, what is the rpm of *d*?

7-8. In Fig. P7-8 is shown a planetary boring-bar feed train. Such a train is sometimes used in lathes to cut internal threads. The piece to be bored or threaded is

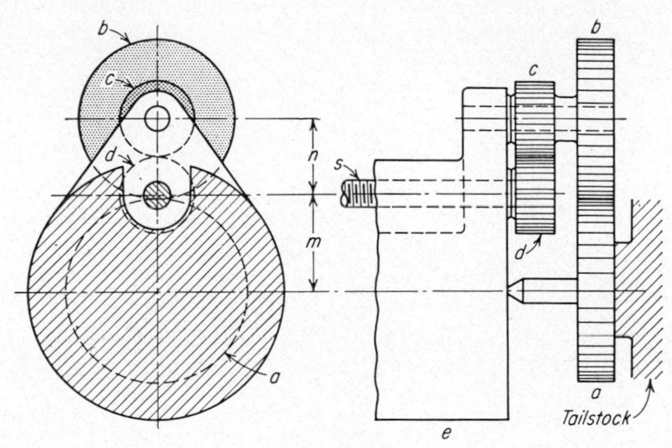

Fig. P7-8

fastened to the bed of the lathe concentric with the boring bar, which is mounted between the lathe centers. The bar rotates with the face plate of the lathe and drives its own feed train as shown in the figure. The boring bar e and the fixed gear a, which is fastened to the tailstock, are concentric with the center line of the lathe. The compound gears b and c rotate on a pin in a lug of the boring bar. Gear d is keyed to the feed screw s, which lies in a horizontal groove in the bar and passes through a threaded lug on the tool head. The direction and magnitude of the feed of the tool head along the bar for each revolution of the bar will depend on the direction and number of revolutions of the feed screw relative to the bar. Assuming that the bar rotates forward and that the lead screw has 8 right-hand threads per inch, determine the number of teeth for gears a and b to cut 6 internal right-hand threads per inch. Gears c and d each have 24 teeth. The ratio of distance n to $m + n$ is 3 to 7.

7-9. Figure P7-9 represents the planetary spur-gear speed reduction from the engine shaft to the propeller shaft in a 1,250-hp, nine-cylinder radial aircraft engine. The engine shaft, which is integral with the internal gear c, drives the pinions b (25 in number), which roll around the fixed gear (sun gear) a. The pinions b are pivoted on a disk d, which is integral with the propeller shaft. The numbers of teeth in the gears are $a = 113$; $b = 15$; $c = 143$. The rpm of the engine is 2,500. It is required to determine the rpm of the propeller shaft. NOTE: The letters that represent the numbers of teeth in the gears, etc., are to be used in the tabular form and in expressing the velocity ratio. The numerical values are to be substituted as a final step in the solution.

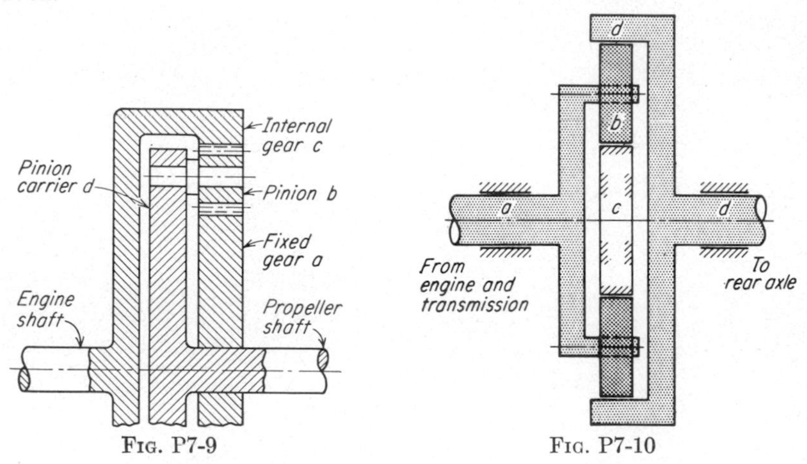

FIG. P7-9 FIG. P7-10

7-10. Figure P7-10 represents the planetary gear set of an automotive overdrive, which is located between the transmission and the rear axle. The shaft from the transmission drives shaft a, which carries the planet pinions b (there are 3). Gear c is stationary, and gear d is fastened to the output shaft d. Numbers of teeth are $b = 12$ and $c = 18$. Find the number of teeth on gear d and the ratio of the speed of the input shaft a to that of the output shaft d.

7-11. Assume that the device of Fig. P7-11 is being studied as a possible part of a computing machine. Numbers of teeth are shown on the drawing. The input rotations to the shafts of B and A are to be multiplied by certain constants, the products added, and the result is to appear as rotations of gear G. Thus, $\theta_G = k_1\theta_B + k_2\theta_A$.

a. If all gears have the same pitch, what must be the number of teeth on gear G?

b. Determine k_1 and k_2.

c. If B rotates at 80 rpm clockwise (plus) when viewed from the left, and A rotates at 100 rpm counterclockwise (minus), what is the rpm (mag. and dir.) of gear G?

d. How could you change k_2 to (-3), retaining the same size of teeth? HINT: Make G an *external* gear.

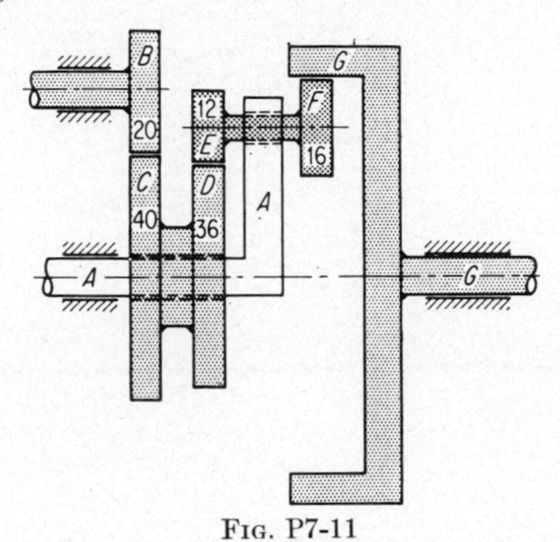

FIG. P7-11

Chapter 8. Velocities in Machines

8-1. Draw the oil-circuit-breaker mechanism (Fig. P2-7) to the dimensions given $O_2(\frac{5}{8}, 2\frac{1}{2})$. Assuming that link 2 has clockwise rotation for the instant and that the velocity of A is represented by a vector $2\frac{1}{2}$ in. long, draw the velocity polygon for the mechanism. $o(3\frac{1}{2}, 3)$.

8-2. In the tool-slide mechanism of a bevel-gear-cutting machine (Fig. P8-2), link 2 is a rotating driving link that imparts a reciprocating motion to a slide attached

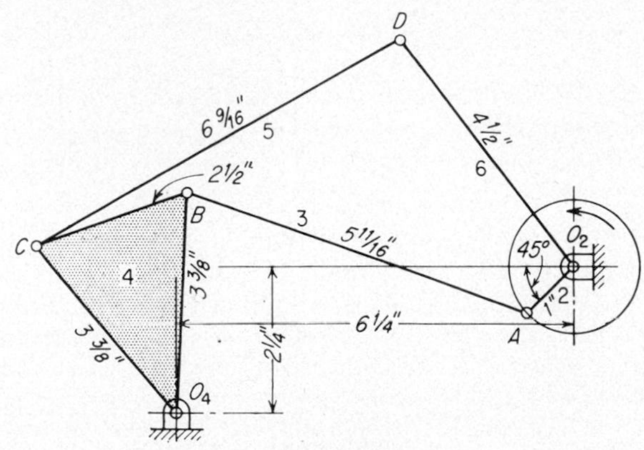

FIG. P8-2

to link 6. The slide carries an adjustable tool head. Draw the mechanism to a scale of 6 in. = 1 ft. $O_4(3, 5\frac{1}{2})$. Assuming counterclockwise rotation of 2 and letting the velocity of point A be represented by a vector $2\frac{1}{2}$ in. long, draw the velocity polygon. $o(3, 3)$.

8-3. Figure P8-3 represents in skeleton outline the mechanism of a toggle press. Link 2 is the driving link, and link 8 is the last driven link. Draw the mechanism to a scale of 1½ in. = 1 ft, with O_2A moved up to 30-deg position with the horizontal. $O_2(6, 7\frac{1}{2})$. If the velocity of A is represented by a vector 1½ in. long, find the length of the vector that represents the velocity of F, by means of the velocity polygon. $O(3, 3)$.

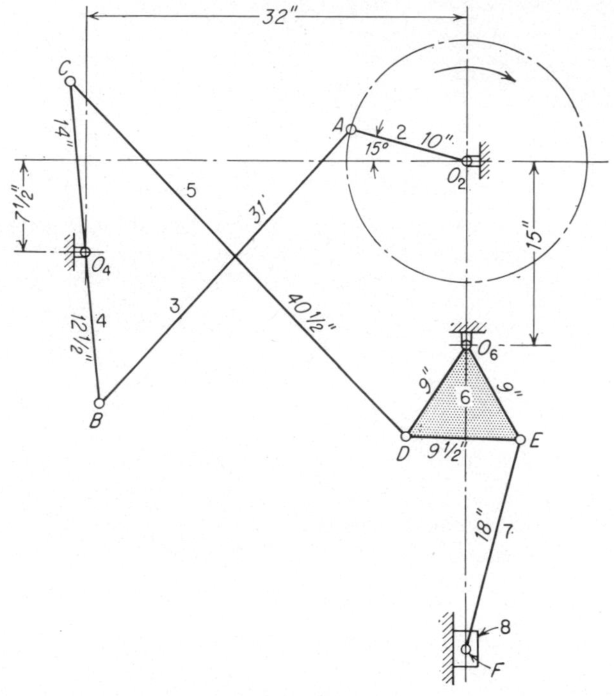

Fig. P8-3

8-4. Draw the drag-link quick-return mechanism (Fig. 2-8) to the dimensions and scale given in Prob. 1-3. $O_2(3\frac{1}{4}, 3\frac{3}{4})$.

The angular velocity of link 2 is 50 rpm counterclockwise. By means of the velocity polygon, find the velocity of the ram (link 6). The velocity scale is 1 in. = 1 fps. $o(5\frac{1}{2}, 7)$.

8-5. Draw the skeleton outline of the Atkinson gas-engine mechanism, given in Fig. 1-15, to a scale of 1½ in. = 1 ft, and by the application of the law of three centers find all the instantaneous centers. Refer to Prob. 8-9 for dimensions. $O_2(6, 7)$.

8-6. Determine all the instantaneous centers of the crank-shaper mechanism referred to in Prob. 8-20. In this case, the crank (link 2) is to make an angle of 15 deg with a horizontal line through O_2 to the right. $O_2(5, 4\frac{1}{2})$.

8-7. Find all the instantaneous centers of the drag-link mechanism shown in Fig. 2-8. Draw the mechanism to the scale and dimensions given in Prob. 1-3. $O_4(4\frac{1}{4}, 4\frac{1}{4})$.

8-8. The Whitworth quick-return mechanism of a slotting machine is shown in Fig. P2-3. Draw this mechanism to the dimensions and scale given in Prob. 2-3 and find all the instantaneous centers. $O_3(6, 3\frac{1}{2})$.

8-9. Given: the skeleton outline of the mechanism of Fig. 1-15 drawn to a scale of $1\frac{1}{2}$ in. = 1 ft. Data: O_2A = 10 in.; O_4B = $12\frac{3}{4}$ in.; AB = $24\frac{3}{4}$ in.; AC = $25\frac{3}{4}$ in.; BC = $3\frac{3}{4}$ in.; CD = $29\frac{1}{2}$ in.; O_2E = $26\frac{3}{4}$ in.; O_4E = $3\frac{1}{2}$ in. $O_2(6, 7\frac{1}{2})$.

a. Find all the instantaneous centers.

b. Draw the velocity polygon. Assume that the velocity of A equals 330 fpm and use a scale of 1 in. = 150 fpm. $o(5\frac{1}{4}, 1\frac{1}{2})$.

c. Check the velocity of D as found in *b* using centers 12, 26, and 16.

d. Same as *c* except that centers 12, 25, and 15 are to be used.

e. Same as *c* except use method of orthogonal components.

8-10. Determine all the instantaneous centers of the shaper mechanism referred to in Prob. 8-20. In this case the crank (link 2) makes an angle of 15 deg with a horizontal line through O_2 in the first quadrant. $O_2(5, 4\frac{1}{2})$. Gear 2 has a uniform velocity of 22 rpm counterclockwise. Draw the skeleton outline to a scale of 2 in. = 1 ft. NOTE: velocity scale, 1 in. = 70 fpm.

a. Find all the instantaneous centers.

b. Find V_C, using centers 12, 25, 15.

c. Find V_C, using centers 12, 26, 16.

d. Find V_C, by means of the velocity polygon. $o(7\frac{1}{4}, 2)$.

e. Find V_C, by use of orthogonal components.

8-11. Find all the instantaneous centers for the shaper mechanism referred to in Prob. 8-20. In this case the crank (link 2) makes an angle of 15 deg with the horizontal in the second quadrant. $O_2(3, 4\frac{1}{2})$. Gear 2 has a uniform velocity of 22 rpm. Draw the skeleton outline to a scale of 2 in. = 1 ft. NOTE: Velocity scale, 1 in. = 70 fpm.

a. Find all the instantaneous centers.

b. Find V_C, using center 25.

c. Find V_C, using center 26.

d. Find V_C, by means of the velocity polygon. $o(6, 2)$.

e. Find V_C, by use of orthogonal components.

8-12. Draw the Corliss valve gear in accordance with the dimensions and scale given in Prob. 2-2. $O_2(2\frac{1}{4}, 8)$.

a. Find all the instantaneous centers.

b. If the velocity of point E on link 6 is represented by a vector 4 in. long, find the vector length that represents the velocity of point B on link 3.

c. Check the velocity found in (*b*) above, by means of the velocity polygon. $o(1, 4)$.

d. Check the velocity in (*b*) using orthogonal components or a combination of this and other methods.

8-13. The drag-link mechanism is to be drawn to the scale and dimensions given in Prob. 1-3. $O_4(4\frac{1}{4}, 4\frac{1}{4})$.

a. Find all the instantaneous centers.

b. If the crank (link 2) rotates at 50 rpm counterclockwise, find the velocity of the ram (link 6), by means of the velocity polygon. Scale, 1 in. = 1 fps. $o(5\frac{1}{2}, 7\frac{1}{2})$.

c. Check the velocity of the ram found in (*b*), using center 26.

d. Check the velocity of the ram using orthogonal components.

8-14. Given: the quick-return mechanism of the slotting machine (Fig. P2-3). Draw the skeleton outline of the mechanism to the dimensions and scale given in Prob. 2-3. $O_3(6, 3\frac{1}{2})$.

a. Find all the instantaneous centers.

b. If link 3, the driving link, rotates at a uniform velocity of 84 rpm, and the velocity scale is 1 in. = 150 fpm, find the velocity of C by means of instantaneous centers, using (1) centers 13, 36, and 16 and (2) centers 13, 37, and 17.

c. Check the above result by means of the velocity polygon. $o(2\frac{1}{2}, 2\frac{1}{2})$.

d. Check the velocity of C using orthogonal components.

8-15. Figure P2-10 represents in skeleton outline the mechanism of a wrapping machine. Draw the mechanism in the configuration shown and to the dimensions given in the figure. $O_2(1\frac{3}{4}, 7\frac{1}{2})$.

a. If the velocity of A is represented by a vector V_A, 2 in. long, determine by means of the velocity polygon the length of vector V_E, which represents the velocity of E. $o(4, 3)$.

b. Find all the instantaneous centers of the mechanism, and check the velocity V_E found in (*a*) above by the method of instantaneous centers.

c. Check the velocity of E using orthogonal components or a combination of this with other methods.

8-16. In Fig. P8-16 is shown the skeleton outline for an oil-circuit-breaker release mechanism. The position of link 2 is 15 deg clockwise from the closed position.

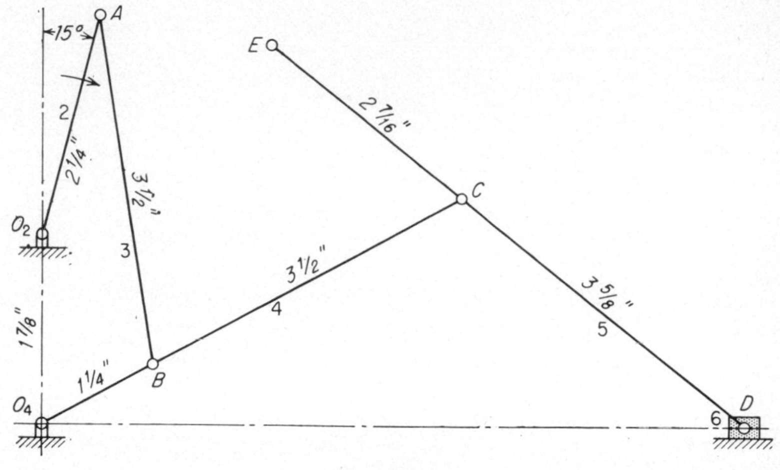

FIG. P8-16

The path of point E on link 5 approximates a vertical straight line through the operating range. Draw the mechanism full size to the dimensions given. $O_4(1, 2)$.

a. Assume that vector V_A is 1 in. in length, and find vector V_E by means of velocity polygon.

b. Locate all the instantaneous centers.

c. Assume that vector V_A is 1 in. in length, and find vector V_E using only the instantaneous-center method.

d. Check V_E using orthogonal components.

8-17. A six-cylinder gasoline engine is operating at 2,500 rpm. Length of crank $O_2A = 2\frac{1}{8}$ in.; length of connecting rod $AB = 8\frac{1}{8}$ in. The engine is to be represented in skeleton outline with center line of cylinder horizontal. Draw the mechanism full size, with the crank in the 45-deg position after head-end dead center, and the piston to the left. Rotation of crank clockwise. $O_2(4\frac{1}{4}, 7\frac{1}{2})$ (see Fig. 9-16, text).

a. Draw the velocity polygon, assuming vector oa equal in length to crank O_2A on the drawing of the mechanism. $o(6\frac{1}{2}, 4\frac{1}{2})$.

b. With pole o at O_2 and oa coinciding with O_2A, superimpose the velocity polygon on the drawing of the mechanism. It will be observed that the intercept formed by the intersection of the connecting rod (extended) with a vertical line through O_2 will represent the velocity of the piston B.

c. Using the above method, determine the vectors that represent the velocity of the piston for one revolution of the crank. Start with the piston at head-end dead center, and make the determination for each 15-deg position of the crank.

d. With the vectors found above, construct a velocity-time diagram for the piston on a base line 6 in. in length. $o(1\frac{1}{2}, 3\frac{1}{8})$.

e. Determine the velocity scale. NOTE: Show calculations for V_A and for velocity scale.

8-18. Given: the quick-return mechanism of a slotting machine drawn to the same dimensions and scale as in Prob. 2-3. Driving link 3 rotates with a uniform velocity of 80 rpm. $O_2(2\frac{3}{4}, 3)$.

a. Construct a velocity-displacement diagram on the path of the ram (link 6) as a base line. NOTE: velocity scale, 1 in. = 150 fpm. Start with the ram in the upper extreme position and make velocity determinations for 12 equally spaced crank positions.

b. Construct a velocity-time diagram on a base line $4\frac{1}{2}$ in. long. $o(6, 4)$. Base line vertical.

8-19. Draw the drag-link mechanism to the dimensions and scale given in Prob. 1-3. $O_4(2, 3)$.

a. Construct a velocity-space diagram for the slide (link 6), using its path as a base line. The crank (link 2) has a uniform velocity of 50 rpm. Velocity scale, 1 in. = 1.6 fps. Start with the slide in its upper extreme position and make velocity determinations for 12 equally spaced crank positions. Determinations may be made for intermediate positions where desirable.

b. Construct a velocity-time diagram on a base line 6 in. long. $o(5\frac{1}{2}, 2)$. Let this diagram run vertically on the sheet.

8-20. In the crank-shaper mechanism (Fig. 1-4), the gear 2 has a uniform velocity of 22 rpm counterclockwise. Draw the skeleton outline of the mechanism to a scale of 2 in. = 1 ft. $O_2A = 5\frac{1}{4}$ in.; $O_4B = 27\frac{7}{8}$ in.; $BC = 6$ in.; $O_2O_4 = 14\frac{3}{4}$ in.; distance from O_2 to path of $C = 12\frac{1}{4}$ in.; $O_2(4\frac{1}{2}, 6\frac{1}{2})$. In the configuration shown, O_2A makes an angle of 30 deg with the horizontal in the second quadrant.

a. Construct a velocity-space diagram on the path of the ram C as a base line. In order to simplify velocity determinations, refer to construction outlined in Art. 8-14. NOTE: Velocity scale, 1 in. = 70 fpm.

Start with the ram in the right-hand extreme position and make velocity determinations for 12 equally spaced crank positions.

b. Construct a velocity-time diagram on a base line 6 in. long. $o(1\frac{1}{4}, 3)$.

8-21. An 18-tooth pinion rotating at 500 rpm in a clockwise direction drives a 32-tooth gear. The teeth are of four pitch and conform to the $14\frac{1}{2}$-deg involute system. By the method of relative velocities, find the rate of sliding between a pair of mating teeth at the last point of contact. Use a graphical method. Velocity scale, 1 in. = 300 fpm. NOTE: It is not necessary to draw in accurate tooth outlines. Axis of pinion, $(4\frac{1}{4}, 2)$.

8-22. Same as Prob. 7-9, except solve by a graphical velocity analysis. As gears all have same pitch (they mesh), gear diameters are proportional to numbers of teeth. Therefore a side view (view from right) can be drawn to scale for use in the solution. Check the answer with that obtained by the tabular method originally used in solving Prob. 7-9.

8-23. Solve Prob. 7-6 by graphical velocity analysis. All teeth are of same pitch; so that gear diameters are proportional to numbers of teeth. Check answers with those obtained by other methods outlined in Chap. 7.

8-24. Solve Prob. 2-17*b* by graphical velocity analysis and check answer with that obtained by other methods outlined in Chap. 2. For making a scale drawing, note that the sheave diameters are proportional to the numbers of chain pockets.

8-25. Draw the device of Fig. 1-11 full size using the following dimensions: AD = 6 in., AB = 1 in., BC = 6 in., CD = 3 in. Link 2 is at 45 deg with the horizontal. If the angular velocity of link 2 is 300 rpm clockwise at the instant shown, find (*a*) the velocity of sliding between 3 and 4 at point B; and (*b*) the angular velocity of link 4. Obtain answers by at least two different methods for a check. (*c*) Can the link AB make full revolutions? (*d*) Show the extremes of position attainable by line BD.

8-26. Refer to Fig. P3-6 and draw the cam device full size. Assume that a cam has been designed, and that it contacts the flat-faced follower ¾ in. to the left of the center line for the position under consideration. The cam is a circular arc of 2 in. radius in the short interval of study, with arc center below the follower. For a cam angular velocity of 180 rpm clockwise, find by at least two methods (*a*) velocity of the follower and (*b*) sliding velocity between the cam and the follower.

8-27. Refer to Fig. P3-1 and draw the cam device full size. Assume that a cam has been designed, and that it contacts the roller follower on a line 30 deg clockwise from a radial line drawn on the roller through the lowest point of the roller. The radius of curvature of the cam curve at the instant is on this 30-deg line, with the center of curvature at a distance of 2 in. (above and to the right) from the roller center. The cam angular velocity is 180 rpm clockwise. Find by at least two methods (*a*) the velocity of the follower rod and (*b*) the angular velocity of the roller.

8-28. In the elevating dump-truck mechanism shown in Fig. P10-9, the body containing the load has angular motion clockwise at a rate of 45 deg in 15 sec for the position given. Draw the mechanism to a scale of 1 in. = 2 ft, $O_A(6, 4)$. Find the velocity of the point of application P of the actuating force F. Suggested velocity scale 1 in. = 0.25 fps.

NOTE: *a*. In drawing a line to or from an instantaneous center not lying on the sheet, the triangle formed by two points of the linkage and the instantaneous center can be drawn to small enough scale to fit the sheet, and *directions* of other lines from the instantaneous center to specific points on the linkage can be taken from images constructed in the small-scale drawing.

b. Do not overlook the possibility that sometimes problems can be worked most easily *in reverse*. That is, *assume* a vector length for the velocity at any convenient point from which it seems easiest to get a solution for velocities elsewhere in the linkage. Then evaluate the velocity *scale* from the given motion and vector length obtained in the solution.

c. Finally, it is emphasized that it is wise to be able to consider the use of *combinations* of the methods of velocity analysis discussed in the text, rather than always attempting to get a solution by one method alone.

8-29. Draw the sewing machine mechanism of Fig. P2-9 twice full size with point O_2 at (4, 4). By two methods, find the velocity of the needle bar (member 6) for the position shown if it completes 480 cycles of motion per minute. Assume the driving crank 2 to rotate at constant speed. Suggested velocity scale 1 in. = 30 in./sec.

8-30. Figure P8-30 depicts a stationary nozzle delivering hot gas to the moving blade of a gas turbine which rotates at 6,000 rpm. The exit gas from the nozzle has a velocity of V_G of 950 fps and the blade moves at 830 fps. In order for the gas to enter the blade smoothly, the velocity of the gas relative to the blade must be tangent to the blade entrance, directed 10 deg counterclockwise from vertical as shown.

a. Find the proper angle θ at which to align the nozzle discharge, and the velocity of the gas relative to the blade.

In reaction blading, the velocity of the gas relative to the blade increases as the gas passes through the blade passages. As shown in the figure, the exit blade angle is 20 deg with the horizontal, and it is desired that the gas leave the blade so that its absolute velocity is directed at 10 deg clockwise from vertical as illustrated.

b. Find the absolute exit velocity and the exit velocity relative to the blade.

c. What is the radius of the blade at the section for which these velocity diagrams have been drawn? Draw the required velocity polygons from poles at (4¼, 6½) and (4¼, 4).

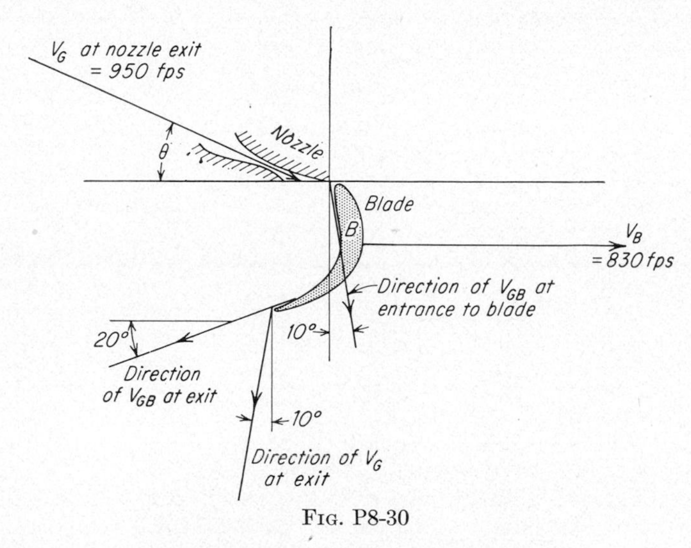

Fig. P8-30

Chapter 9. Accelerations in Machines

9-1. The table of a hydraulically reciprocated grinder is arranged to operate with a ¾-in. stroke and makes 200 complete cycles (over and back) per minute. Except during a very short period at each end of the stroke, the table moves with a uniform velocity of 8 ips. Assuming that during the period of reversal the table is accelerated at a uniform rate, what is the magnitude of the acceleration in feet per second per second?

9-2. At a given instant a 12-in. pulley is turning at a rate of 600 rpm and its speed is increasing at a rate of 1 per cent per revolution.

a. Determine the angular velocity and angular acceleration of the pulley in radians per second and radians per second per second, respectively.

b. Determine the magnitude and direction of the acceleration of a point on the face of the pulley.

c. Determine the magnitude of the acceleration of a point on the straight portion of the belt 8 in. from the point of tangency of the belt and the pulley. Disregard the thickness of the belt.

d. Determine the acceleration of this point relative to the point on the face of the pulley with which it was previously in contact.

9-3. A four-link mechanism similar to that of Fig. 2-1 is to be drawn full size to the following dimensions: link 2 (O_2A) is 2 in. in length and makes an angle of 60 deg with the horizontal. $O_2(2, 6)$. Distance O_2O_4 = 4½ in. Link 4(O_4B), which is in a vertical position, is 2½ in. in length.

a. Draw the velocity polygon. The angular velocity of link 2 is ω_2 = 60 radians/sec (clockwise). Velocity scale, 1 in. = 5 fps. $o(1, 4\frac{1}{2})$.

b. Draw the acceleration polygon. The angular acceleration of link 2 is $\alpha_2 = 900$ radians/sec^2 (clockwise). Acceleration scale, 1 in. = 200 ft/sec^2. $o(7, 5)$.

c. What is the velocity of B in feet per second? What is the acceleration of B in feet per second^2?

9-4. The four-link mechanism of Fig. P9-4 is to be drawn to a scale of 3 in. = 1 ft. Dimensions are shown in the figure. $O_2(2\frac{1}{4}, 6)$. The angular velocity of link 2 is $\omega_2 = 25$ radians/sec (clockwise), and the angular acceleration is $\alpha_2 = 180$ radians/sec^2 (clockwise).

a. Draw the velocity polygon; scale, 1 in. = 5 fps. $o(1, 4\frac{1}{2})$.

b. Draw the acceleration polygon; scale, 1 in. = 100 ft/sec^2. $o(7, 5)$.

c. What is the acceleration of point C in feet per second^2?

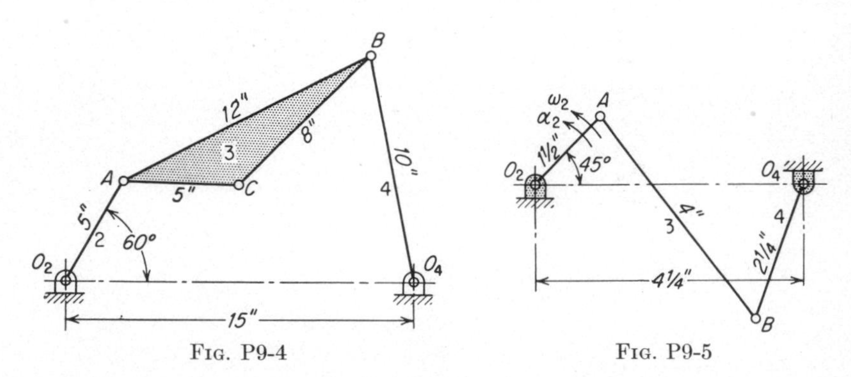

Fig. P9-4

Fig. P9-5

9-5. Lay out the mechanism of Fig. P9-5 to the dimensions given and draw the acceleration polygon, using the *analytical method*, as in Art. 9-5. The angular velocity and angular acceleration of link 2 are $\omega_2 = 40$ radians/sec and $\alpha_2 = 400$ radians/sec^2, respectively, both in the counterclockwise sense.

For the mechanism: scale, full size; $O_2(2\frac{1}{4}, 7)$. For the velocity polygon: scale, 1 in. = $2\frac{1}{2}$ fps; $o(7\frac{3}{4}, 2\frac{1}{2})$. For the acceleration polygon: scale, 1 in. = 100 ft/sec^2; $o(3, 4)$.

9-6. Same as Prob. 9-5 except that the *graphical method* is to be employed, as in Art. 9-8. In this case, let the velocity scale be determined from the conditions of the problem.

9-7. Lay out a four-link mechanism to the following dimensions, using a scale of 3 in. = 1 ft; $O_2(2, 8)$; $O_4(7, 5)$. Link 2 (O_2A) makes an angle of 45 deg with the horizontal in the second quadrant. Link 3 (AB) is 20 in. long, and link 4 (O_4B) is $13\frac{1}{2}$ in. long. The angular velocity and angular acceleration of link 2 are, respectively, $\omega_2 = 60$ radians/sec and $\alpha_2 = 900$ radians/sec^2, both in clockwise sense. Link 2 (O_2A) is 4 in. in length.

a. Draw the velocity polygon; scale, 1 in. = 10 fps. $o(3, 5\frac{1}{4})$.

b. Draw the acceleration polygon; scale, 1 in. = 300 ft/sec.^{2} $o(1, 3\frac{1}{2})$.

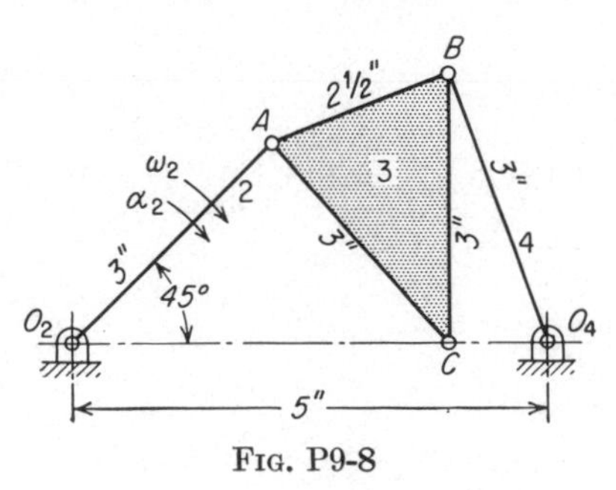

Fig. P9-8

9-8. In the Roberts straight-line mechanism (Fig. P9-8), the path of point C is a straight line. Make a full-size drawing of the mechanism, using the dimensions given on the figure. $O_2(1\frac{1}{2}, 6\frac{1}{2})$. The angular velocity of link 2 is $\omega_2 = 20$ radians/sec, and the angular acceleration is $\alpha_2 = 100$ radians/sec², both clockwise.

a. Draw the velocity polygon. $o(5, 5)$. Scale, 1 in. = 3 fps.

b. Draw the acceleration polygon. $o(4\frac{1}{2}, 4\frac{1}{2})$. Scale, 1 in. = 50 ft/sec².

9-9. Draw the oil-circuit-breaker mechanism to the dimensions given in Fig. P2-7, letting link 2 make an angle of 45 deg with the horizontal. $O_2(\frac{1}{2}, 3\frac{1}{2})$. Scale, full size. The angular velocity and acceleration of link 2 are, respectively, $\omega_2 = 18$ radians/sec and $\alpha_2 = 160$ radians/sec², both in clockwise sense.

a. Draw the velocity polygon. $o(5\frac{1}{2}, 6)$. Scale, 1 in. = 2.5 fps.

b. Draw the acceleration polygon using the analytical method as in Art. 9-5. $o(4\frac{1}{4}, 6)$. Scale, 1 in. = 50 ft/sec².

9-10. Same as Prob. 9-9, except that the graphical method of Art. 9-8 is to be used in the determination of the acceleration polygon. The space scale and the velocity scale are to remain as in Prob. 9-9. Acceleration polygon, $o(4\frac{1}{4}, 5)$.

9-11. Figure P8-3 represents in skeleton outline the mechanism of a toggle press. Link 2 is the driving link, and link 8 is the last driven link. Draw the mechanism to a scale of $1\frac{1}{2}$ in. = 1 ft. $O_2(6, 7\frac{1}{2})$. The crank O_2A is rotating with a uniform velocity, such that $V_A = 1.25$ fps.

a. By means of the velocity polygon, determine the velocity of point F on the ram in feet per second. Scale of velocity polygon, 1 in. = 1 fps. $o(6, 1\frac{1}{4})$.

b. Using the graphical method of Art. 9-8 for the acceleration polygon, determine the acceleration of the point F in feet per second per second. $o(3, 3)$.

9-12. Lay out the control mechanism (Fig. P9-12) to the dimensions given. $O_2(7\frac{1}{4}, 8\frac{1}{4})$. Scale, 3 in. = 1 ft. The angular velocity and angular acceleration of link 2 are, respectively, $\omega_2 = 15$ radians/sec and $\alpha_2 = 100$ radians/sec², both in a counterclockwise sense.

a. Draw the velocity polygon. $o(2\frac{3}{4}, 3\frac{1}{2})$. Scale, 1 in. = 10 fps.

b. Draw the acceleration polygon. $o(1, 6)$. Scale, 1 in. = 100 ft/sec².

c. What is the numerical value of the acceleration of C?

9-13. In the circuit-breaker mechanism represented in Fig. P9-13, the point C moves in a vertical straight line. Draw the mechanism to a scale of 6 in. = 1 ft, using the dimensions given. $O_2(1\frac{1}{2}, 7\frac{1}{2})$. Link 4 is in a vertical position.

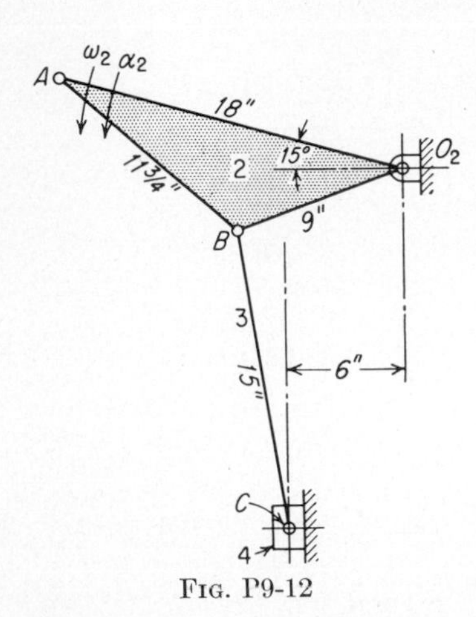

FIG. P9-12

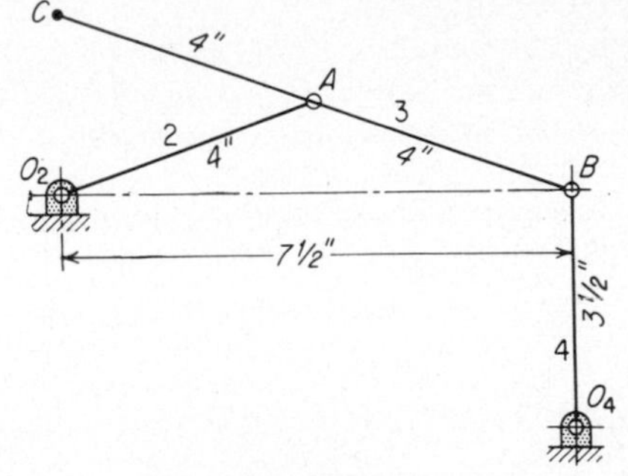

FIG. P9-13

The angular velocity and angular acceleration of link 2 are, respectively, $\omega_2 = 45$ radians/sec and $\alpha_2 = 300$ radians/sec², both in clockwise sense.

a. Draw the velocity polygon. Scale, 1 in. = 10 fps. $o(6\frac{1}{2}, 9)$. Find the velocity of C in feet per second.

b. Draw the acceleration polygon employing the *analytical method*, as in Art. 9-5. Scale, 1 in. = 400 ft/sec². $o(3\frac{7}{8}, 5)$.

c. Draw the acceleration polygon, employing the *graphical method*, as in Art. 9-8. Scale (to be computed). $o(7\frac{1}{2}, 4)$. Find the magnitude of the acceleration of C in feet per second per second. NOTE: Arrange on the sheet, in neat form, all the computations made in connection with the solution of this problem.

9-14. The drag-link mechanism of Fig. 2-8 is to be drawn to the dimensions given in Prob. 1-3, with link 2 making an angle of 30 deg with the vertical center line, in the fourth quadrant. Scale, 3 in. = 1 ft. $O_2(2\frac{3}{4}, 3\frac{7}{8})$. Link 2 has a uniform angular velocity of ω_2 = 50 rpm, counterclockwise.

a. Draw the velocity polygon. Scale, 1 in. = 1.25 fps. $o(5\frac{1}{4}, 8\frac{1}{2})$.

b. Draw the acceleration polygon. Scale, 1 in. = 5 ft/sec². $o(6, 4)$.

c. What is the velocity of the slide (link 6) in feet per second? What is the acceleration in feet per second²?

d. Assuming that the space and velocity scales are the same as in the analytical solution above, draw the acceleration polygon by means of the graphical method, as in Art. 9-8. $o(6, 1)$.

9-15. Using the dimensions given, draw the Watt straight-line mechanism (Fig. P9-15) to a scale of 6 in. = 1 ft. $O_2(1\frac{1}{2}, 7\frac{3}{4})$. The angular velocity and angular acceleration of link 2 are, respectively, ω_2 = 60 radians/sec, and α_2 = 900 radians/sec² both in a clockwise sense.

a. Draw the velocity polygon. Scale, 1 in. = 10 fps. $o(2, 6)$.

b. Draw the acceleration polygon. Scale, 1 in. = 600 ft/sec². $o(6, 3)$.

c. Find acceleration of point C in feet per second per second.

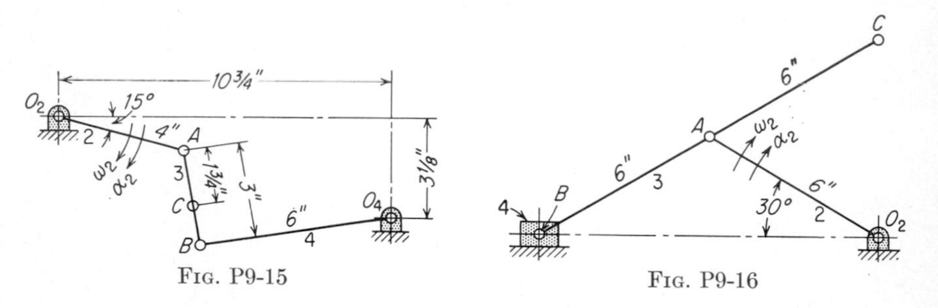

FIG. P9-15

FIG. P9-16

9-16. The Scott-Russell straight-line mechanism is represented in Fig. P9-16. Draw the mechanism to the dimensions given and to a scale of 6 in. = 1 ft. $O_2(7, 6)$. If the angular velocity of link 2 is ω_2 = 40 radians/sec, and the angular acceleration is α_2 = 400 radians/sec², both in clockwise sense, determine the following:

a. The velocity polygon. Scale, 1 in. = 10 fps. $o(1\frac{1}{4}, 1\frac{1}{4})$.

b. The acceleration polygon. Scale, 1 in. = 400 ft/sec². $o(3\frac{3}{4}, 4\frac{3}{4})$.

9-17. Figure P2-6 represents the operating mechanism of the Wellman geared bucket. Link 3 as shown is a skeleton outline representation of the excavating bucket, the point C representing the cutting edge of the bucket. The mechanism is to be drawn to a scale of $1\frac{1}{2}$ in. = 1 ft, with link 2 in a 45-deg position with a vertical line through O_2 in the fourth quadrant. $O_2(2\frac{1}{2}, 6)$. The angular velocity of link 2 is ω_2 = 40 radians/sec, clockwise, and the angular acceleration is α_2 = 400 radians/sec², counterclockwise.

a. Draw the velocity polygon. Scale, 1 in. = 25 fps. $o(7\frac{1}{2}, 3)$.

b. Draw the acceleration polygon. Scale, 1 in. = 500 ft/sec². $o(5\frac{1}{2}, 4\frac{3}{4})$.

9-18. In the tool-slide mechanism of Fig. P8-2, link 2 is the driving member that operates a reciprocating tool block attached to a sliding member that receives its motion through oscillating link 4. Draw the mechanism to the dimensions given with link 2 in the 45-deg position shown. Scale, 6 in. = 1 ft. $O_2(6, 7)$.

The angular velocity of link 2 is $\omega_2 = 96$ radians/sec, and the angular acceleration is $\alpha_2 = 2{,}400$ radians/sec², both counterclockwise.

a. Draw the velocity polygon. Scale, 1 in. = 4 fps. $o(1, 3\frac{1}{2})$.

b. Draw the acceleration polygon. Scale, 1 in. = 200 ft/sec². $o(4\frac{1}{4}, 3\frac{1}{2})$.

9-19. The mechanism shown in Fig. P2-9 is used in the head of a sewing machine to actuate the needle bar. Link 2 is the driving crank, which is assumed to rotate clockwise at a uniform rate of 36 radians/sec. Link 6 represents the needle bar, which reciprocates vertically. The actual dimensions of the links are shown on the figure.

a. Draw the mechanism in the phase shown to a scale twice full size. $O_2(2, 6)$.

b. Draw the velocity polygon. Velocity scale: 1 in. = 1 fps. $o(5\frac{1}{2}, 6)$.

c. Draw the acceleration polygon. Acceleration scale: 1 in. = 40 ft/sec². $o(6, 3\frac{1}{4})$.

9-20. Lay out the mechanism of Fig. P9-5 to a scale of 3 in. = 1 ft, using the following dimensions in place of those given in the figure: $O_2A = 6$ in.; $AB = 24$ in.; $BO_4 = 9$ in.; $O_2O_4 = 23$ in. Extend link 4 to a point C $4\frac{1}{2}$ in. beyond O_4. The centers of gravity G_2, G_3, and G_4 are at the mid-points of links 2, 3, and 4, respectively. O_2A is to be drawn at an angle of 30 deg with the horizontal in the first quadrant. $O_2(1, 8)$.

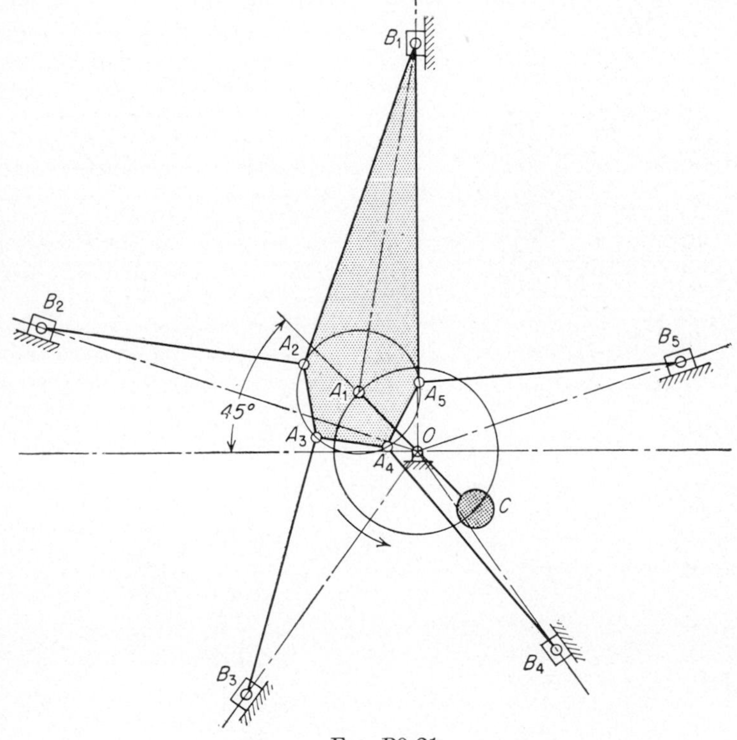

FIG. P9-21

a. Draw the *complete* velocity polygon. $\omega_2 = 20$ radians/sec clockwise. Velocity scale: 1 in. = 5 fps. $o(4, 3\frac{1}{2})$.

b. What is the velocity of point C in feet per second?

c. Draw the *complete* acceleration polygon. $\omega_2 = 20$ radians/sec clockwise; $\alpha_2 =$ 150 radians/sec² clockwise; acceleration scale: 1 in. = 100 ft/sec². $o(3, 4\frac{1}{2})$.

d. What is the acceleration of point C in feet per second per second?

9-21. The following data apply to the five-cylinder radial aircraft engine shown in Fig. P9-21. Rpm = 2,200 (uniform); bore = 4½ in.; stroke = 5½ in.; length of link connecting rods A_2B_2, A_3B_3, etc. = 8⅞ in. Link pins A_2, A_3, etc., are evenly spaced on a circle 4 in. in diameter.

a. Draw the mechanism to a scale of 3 in. = 1 ft in the configuration shown in the figure.

b. Draw the velocity polygon. Scale: 1 in. = 20 fps. $o(3, 3\frac{1}{2})$.

c. Draw the acceleration polygon. Scale: 1 in. = 4,000 ft/sec². $o(4\frac{1}{2}, 4)$. Label polygons, give all scales, and show necessary calculations in available space on sheet.

9-22. In Fig. P9-22 is shown a 12-cylinder V-type aircraft engine. Rpm = 2,000 (uniform). O_2A = 3 in.; AB = 10⅛ in.; AC = 3⅜ in.; BC = 9$\frac{3}{16}$ in.; CD = 7¾ in.

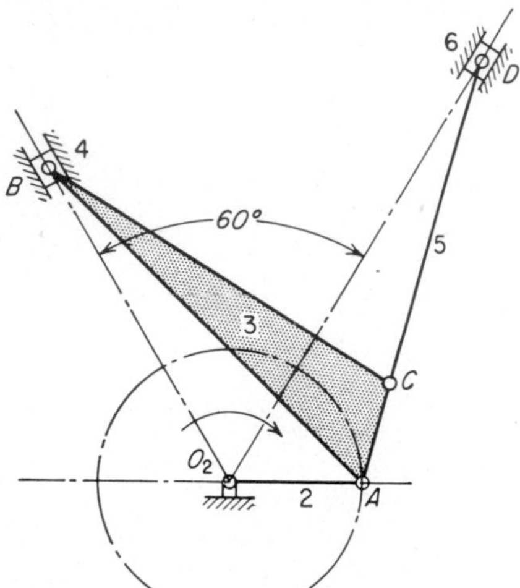

FIG. P9-22

a. Draw the mechanism to a scale of ⅜ in. = 1 in. $O_2(2\frac{1}{4}, 6)$. Crank O_2A in horizontal position, as shown.

b. Draw the velocity polygon. Scale: 1 in. = 20 fps. $o(6\frac{1}{2}, 9)$. Find the velocity of each piston in feet per second.

c. Draw the acceleration polygon. Scale: 1 in. = 2,000 ft/sec². $o(7\frac{1}{2}, 2\frac{3}{4})$. Find the acceleration of each piston in feet per second per second. Find the angular acceleration of link 3 in sense and magnitude.

9-23. A gasoline engine runs at a speed of 2,800 rpm. Length of crank OA = 2¾ in., clockwise rotation. Length of connecting rod AB = 11 in.

a. Draw the mechanism to a scale 6 in. = 1 ft, with the crank in the 30-deg position before crank-end dead center, and construct the velocity and acceleration polygons by means of the Ritterhaus construction. $O_2(6, 7\frac{1}{2})$.

b. Determine the velocity and acceleration scales.

c. Construct the velocity and acceleration polygons, using the scales determined in (*b*) and the graphical method of Art. 9-8. Compare these polygons with those obtained in (*a*). Velocity polygon. $o(2\frac{1}{4}, 2)$. Acceleration polygon. $o(6\frac{1}{2}, 1\frac{1}{2})$. Redraw mechanism for part *c*. $O_2(6, 2\frac{3}{4})$.

9-24. A six-cylinder gasoline engine is operating at 2,500 rpm. Length of crank O_2A = 2⅛ in.; length of connecting rod AB = 8⅛ in. The engine is to be represented in skeleton outline with center line of cylinder horizontal. Draw the mechanism full size, with the crank in the 45-deg position after head-end dead center and the piston to the left. Rotation of crank clockwise. $O_2(4\frac{1}{4}, 7\frac{1}{2})$ (see Fig. 9-16, text).

a. Using the Ritterhaus construction, determine the vectors that represent the acceleration of the piston for one revolution of the crank. Start with the piston at head-end dead center and make the determination for each 15-deg position of the crank. [For crank positions 0 and 180 deg use Eq. (9-33).]

b. With the vectors found above, construct an acceleration-time diagram for the piston on a base line 6 in. in length. $o(1\frac{1}{2}, 3\frac{1}{8})$.

c. Determine the acceleration scale. NOTE: Show calculations for V_A, A_A, velocity scale, acceleration scale; also for A_B at crank positions 0 and 180 deg.

9-25. A six-cylinder gasoline engine is operating at a constant speed of 2,500 rpm. Length of crank $O_2A = 2\frac{1}{8}$ in.; length of connecting rod $AB = 8\frac{1}{8}$ in. The engine is to be represented in skeleton outline with center line of cylinder horizontal. Draw the mechanism full size, with the crank in the 45-deg position after head-end dead center and the piston to the left. Rotation of crank clockwise. $O_2(4\frac{1}{4}, 7\frac{1}{2})$ (see Fig. 9-16).

Displacement diagram: a. Using the above data and the method suggested in Fig. 2-15, plot a displacement-time curve on a base line 6 in. in length representing the time of one revolution of the crank. $o(1\frac{1}{4}, 3\frac{1}{8})$. Start with the piston at head-end dead center and make the determination for each 15-deg position of the crank. Divide the ordinates by 2 when plotting.

Velocity diagram: b. Draw the velocity polygon, assuming vector *oa* equal in length to crank O_2A on the drawing of the mechanism. $o(6\frac{1}{2}, 4\frac{1}{2})$.

c. With pole *o* at O_2 and *oa* coinciding with O_2A, superimpose the velocity polygon on the drawing of the mechanism. It will be observed that the intercept formed by the intersection of the connecting rod (extended) with a vertical line through O_2 will represent the velocity of the piston *B*.

d. Using the above method, determine the vectors that represent the velocity of the piston for one revolution of the crank. Start with the piston at head-end dead center, and make the determination for each 15-deg position of the crank.

e. With the vectors found above, construct a velocity-time diagram for the piston on the same base line as in (*a*).

f. Determine the velocity scale. NOTE: Show calculations for V_A and for velocity scale.

Acceleration diagram: g. Using the Ritterhaus construction, determine the vectors that represent the acceleration of the piston for one revolution of the crank. Start with the piston at head-end dead center and make the determination for each 15-deg position of the crank. [For crank positions 0 and 180 deg use Eq. (9-33).]

h. With the vectors found above, construct an acceleration-time diagram for the piston on the same base line as in (*a*).

i. Determine the acceleration scale. NOTE: Show calculations for V_A, A_A, velocity scale, acceleration scale; also for A_B at crank positions 0 and 180 deg.

9-26. The flywheel of an engine (Fig. P9-26) rotates with an angular velocity $\omega_2 = 30$ radians/sec and an angular acceleration $\alpha_2 = 20$ radians/sec². The governor weight is moving outward relative to the wheel with a uniform velocity of 10 fps. Find the acceleration of *G*, the center of gravity of the governor weight.

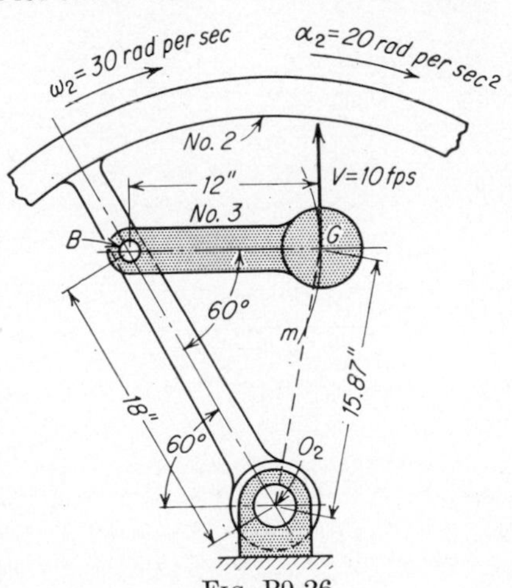

FIG. P9-26

Calculate each component of the acceleration and find the resultant graphically. NOTE: Apply Coriolis' law.

9-27. Wheel A of the Geneva motion shown in Fig. P9-27 rotates at a uniform rate of 100 rpm. Determine the angular acceleration of wheel B when $\theta = 60$ deg. NOTE: Apply Coriolis' law.

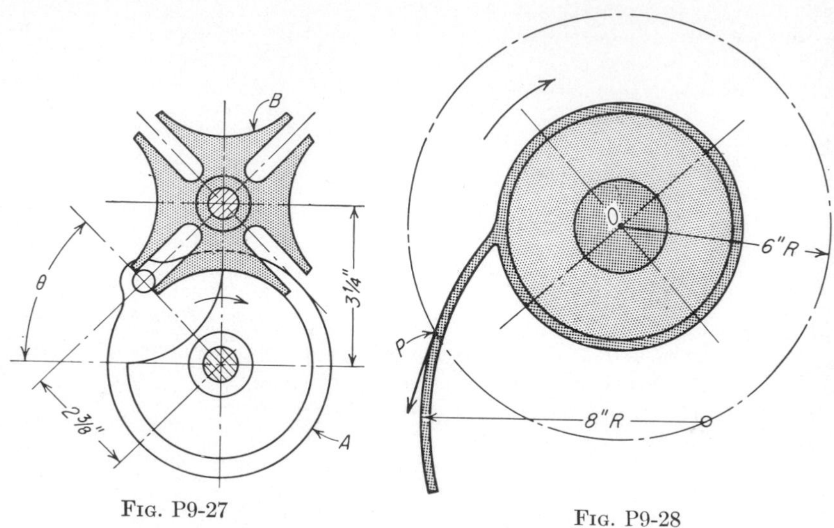

FIG. P9-27

FIG. P9-28

9-28. The vane of a centrifugal pump (Fig. P9-28) rotates with a uniform angular velocity of 30 radians/sec. Using the dimensions given in the figure, find the acceleration of a particle of water P, which moves with a velocity along the vane of 20 fps and an acceleration along the vane of 100 ft/sec². Calculate each component of the acceleration and find the resultant graphically. NOTE: Apply Coriolis' law.

9-29. The driving link 3 of the Whitworth quick-return mechanism (Fig. P2-3) runs at a uniform speed of 80 rpm. Draw the mechanism to a scale of 3 in. = 1 ft, using the dimensions given in Prob. 2-3.

By the application of Coriolis' law, find the acceleration of the piston (link 6).

9-30. The bar, link 2, Fig. P9-30, rotates about O_2 with an angular velocity of 2 radians/sec (clockwise) and an angular acceleration of 5 radians/sec² (clockwise). At the same time the bar, link 3, which is pivoted on link 2, rotates with a constant angular velocity relative to link 2 of 3 radians/sec (clockwise).

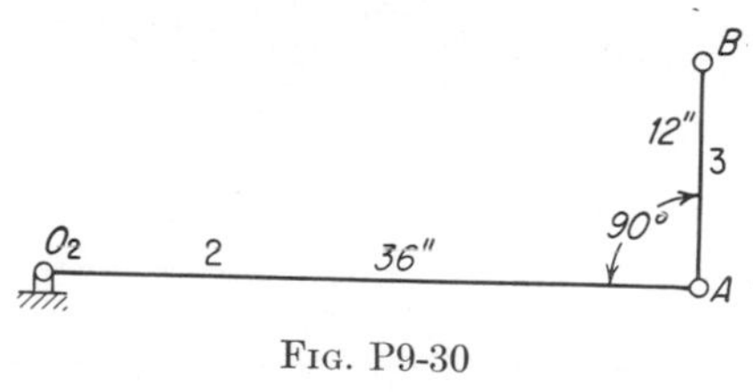

FIG. P9-30

a. By means of the acceleration polygon, determine the acceleration of B when AB is perpendicular to O_2A. Scale of drawing: 1½ in. = 1 ft. O_2(1¼, 7½). Scale of velocity polygon: 1 in. = 3 fps. o(1½, 6½). Scale of acceleration polygon: 1 in. = 8 ft/sec². o(7½, 6½).

b. Check the result obtained in (*a*) above by means of Coriolis' law. Start with the same pole o, and superimpose this solution on the first one, using light lines for contrast.

9-31. Refer to Prob. 7-1. Suppose gear A to be rotating at 1,400 rpm and gaining speed at the rate of 2 rpm per revolution. Find the velocity and acceleration of the rack K.

9-32. Same as Prob. 8-26, except assume cam rpm to be constant, and find the acceleration of the follower as well as the velocity.

9-33. Same as Prob. 8-27, except assume cam rpm to be constant, and find the acceleration as well as the velocity of the follower.

9-34. Same figure as that specified in Prob. 8-25. Angular velocity of crank 2 is constant at 300 rpm clockwise. Find the angular acceleration of link 4 by using Coriolis' law, and check it by an equivalent-link analysis. Also find the sliding acceleration between links 3 and 4 at point B.

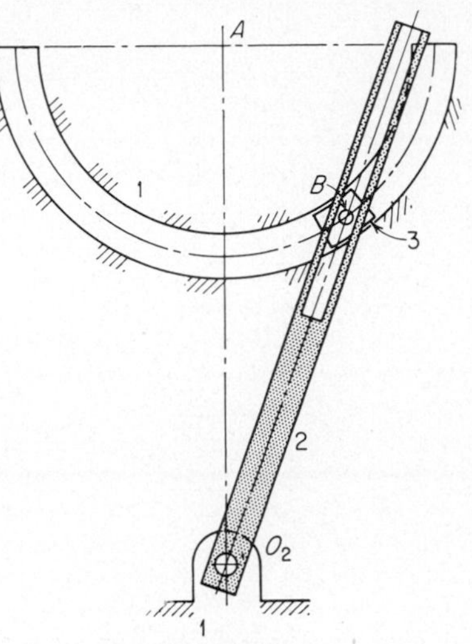

Fig. P9-35

9-35. Draw the skeleton outline of the device of Fig. P9-35 half size. Place O_2 at $(4, 3\frac{1}{2})$. Essential actual dimensions are $O_2A = 8\frac{1}{2}$ in., $O_2B = 6$ in., $AB = 3\frac{1}{2}$ in. At the instant shown, the velocity of point B is along the curved path toward the right and upward, of magnitude 4.5 fps. This magnitude is not changing. Find the angular velocity and angular acceleration of link 2, and also the sliding velocity and sliding acceleration between 2 and 3 at point B. Suggested velocity scale (for graphical solution) 1 in. = 2 fps. For acceleration polygon, $o(6\frac{1}{8}, 1\frac{1}{2})$.

9-36. Draw the skeleton outline of the linkage of Fig. P9-36 to one-sixth full size, placing O_2 at (4, 5). Essential actual dimensions are $AO_2 = 16$ in., $O_2C = 19\frac{1}{2}$ in., $CB = 11\frac{1}{2}$ in. A part of the machine not shown drives link 2 clockwise, imparting a velocity to point A of 5 fps. This velocity is increasing at a rate of 6 ft/sec² at the instant shown. Find the velocity and acceleration of sliding member 3, and also the sliding velocity and sliding acceleration between members 2 and 3 at point B. Suggested velocity scale 1 in. = 2 fps. For acceleration polygon, $o(6, 2\frac{1}{2})$.

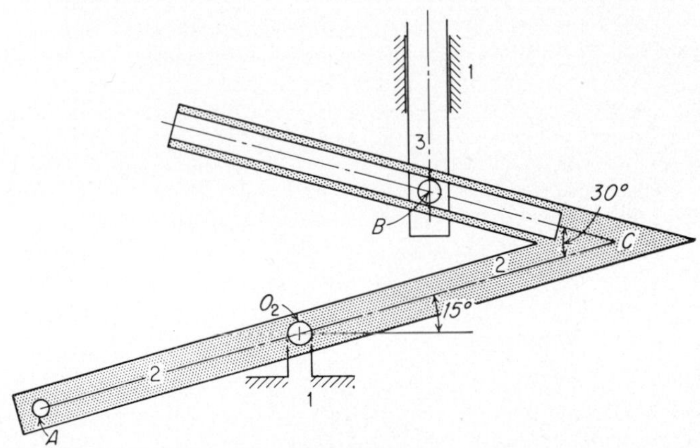

Fig. P9-36

9-37. A research airplane at 30° north latitude flies south at 2,000 mph exactly along a north-south line. Thus there is velocity along a path traced on the surface of the earth, while the earth has angular velocity about its axis. Find the acceleration of the airplane due to the Coriolis effect. What force produces this acceleration?

Chapter 10. Static Forces in Machines

10-1. In the Atkinson gas-engine mechanism, Fig. 1-15 (see Prob. 8-9 for dimensions), the force on the piston (link 6) is represented by a vector 2 in. long directed to the right. By means of the force polygon, determine the vector length that represents the tangential force on the crankpin at A. Scale of drawing, 1½ in. = 1 ft. O_2(6¼, 7¼). Origin of vector P in force polygon, (1½, 6). Friction is to be neglected.

10-2. Draw the drag-link mechanism, Fig. 2-8, to the dimensions given in Prob. 1-3, with the crank (link 2) making an angle of 30 deg with the vertical, in the fourth quadrant, and having counterclockwise rotation. Scale, 3 in. = 1 ft. O_2(3¼, 4). The gear driving O_2A has its axis at O_2 and a diameter of 16½ in. The pinion has a diameter of 5½ in., and its axis lies on a horizontal line to the right of O_2. The pressure angle is 14½ deg.

If the force Q on the ram is represented by a vector 2 in. long, determine the length of vector that represents the normal force on the teeth of the driving pinion. Origin of vector Q in force polygon, (7¼, 7½). Neglect friction. NOTE: Draw the complete force polygon showing all forces properly labeled.

10-3. Draw the oil-circuit-breaker mechanism (Fig. P2-7) to the dimensions given. O_2A makes an angle of 45 deg with the horizontal. O_2(½, 2½). Scale, full size. Assuming that vector Q (= 2 in.) directed upward represents the resistance at C, determine the length of vector P that represents the force applied in a horizontal direction at A. Origin of vector Q in force polygon, (6, 2). Neglect friction.

10-4. Figure P8-3 represents in skeleton outline the mechanism of a toggle press. Link 2 is the driving link, and link 8 is the last driven link. Draw the mechanism to a scale of 1½ in. = 1 ft. O_2(6, 7½). By means of the force polygon, find the vector F_{23}^T (which represents the tangential force acting at A) required to overcome the known resistance Q acting on the ram at F. The force Q is represented by a vector 2 in. long, origin at (3½, 1½).

10-5. The mechanism of a variable-stroke boiler feed pump is shown in Fig. P1-4. Draw the mechanism with the crank O_2A in a position 60 deg clockwise beyond that shown in the figure. O_2(7, 8½). Scale: 3 in. = 1 ft. The downward pressure of the liquid on the piston is 2,250 lb, represented by a vector Q. Determine the tangential force at A required to produce this pressure. Origin of vector Q in the force polygon: (6, 3). Force scale: 1 in. = 1,000 lb. (Neglect friction.)

10-6. For the steam-engine mechanism shown in Fig. 10-2 the following data are given: piston force P = 7,500 lb; length of crank = 6 in.; length of connecting rod = 30 in.; diameter of crankshaft bearing = 5 in.; diameter of crankpin bearing = 3½ in.; diameter of wrist-pin bearing = 2½ in.

Using the methods of Arts. 10-3 and 10-9, determine the following for the 60-deg crank position, shown in the figure (scale, 1 in. = 5 in.):

a. The turning effort T_o, disregarding friction.

b. The turning effort T, when friction is taken into account. Select coefficients of friction from Art. 10-6(4).

c. The efficiency of the mechanism for the position shown.

NOTE: Because of the small size of friction circles, it is desirable to work with a full-size drawing in the solution of part (*b*) of this problem.

10-7. Figure P10-7 represents a rapid-release latch of the dead-center type. The fixed pins are indicated by the crosshatching. The device is shown in latched position under load Q, with the trigger against the stop pin. If there were no friction, the releasing force P would be zero. If the load Q is represented by a vector 3¼ in. in length, origin at (3½, 7¼), determine the vector that represents the magnitude of the releasing force P. The coefficient of sliding friction is to be assumed as 0.25.

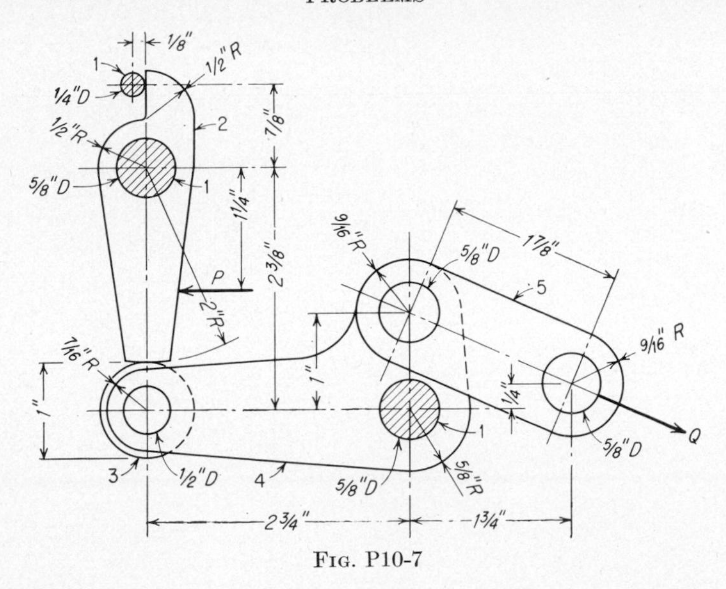

Fig. P10-7

This is a reasonable maximum value obtainable with boundary lubrication. Note that during the releasing operation only the trigger and roller move. The only frictional resistances to be considered are those in the trigger bearing and in the roller bearing.

10-8. Figure P10-8 shows a rapid-release latch of the overcenter toggle type. The device is shown in latched position against the stop pin. The fixed pins are indicated by the crosshatching. If the load Q is represented by a vector $2\frac{1}{2}$ in. long, determine length of vector P, which represents the unlatching force, first without friction, origin of vector $Q(3, 8\frac{1}{2})$; and then with friction, origin of vector $Q(4\frac{1}{2}, 9)$. The coefficient

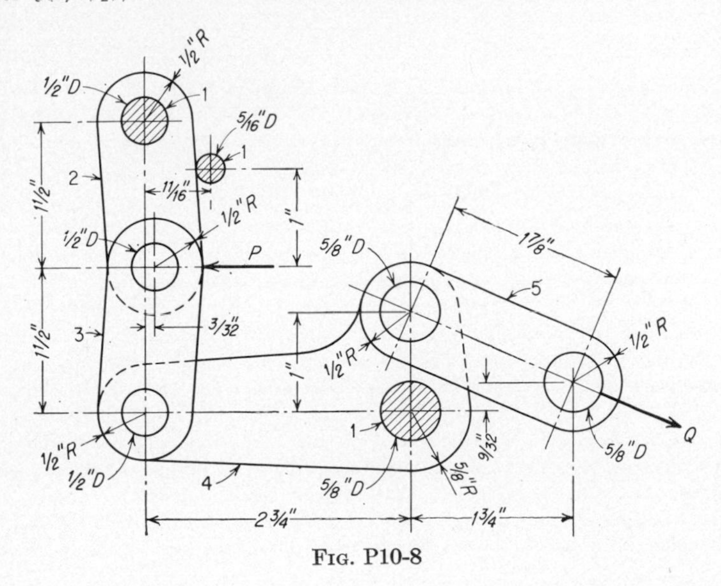

Fig. P10-8

of sliding friction is to be assumed as 0.25. This value is a reasonable maximum obtainable with boundary lubrication.

10-9. One side of an elevating dump-truck mechanism is shown in Fig. P10-9. The full mechanism consists of two such assemblies—one on each side of the body, but only one actuating force F. G indicates the center of gravity of the load and F is the actuating force applied to pin P, which extends also to the mechanism on the other side of the body. Draw the mechanism in the position shown to a scale of 1 in. = 2 ft. $O_A(6, 4)$. For the force analysis, use a scale of 1 in. = 2,500 lb. Find

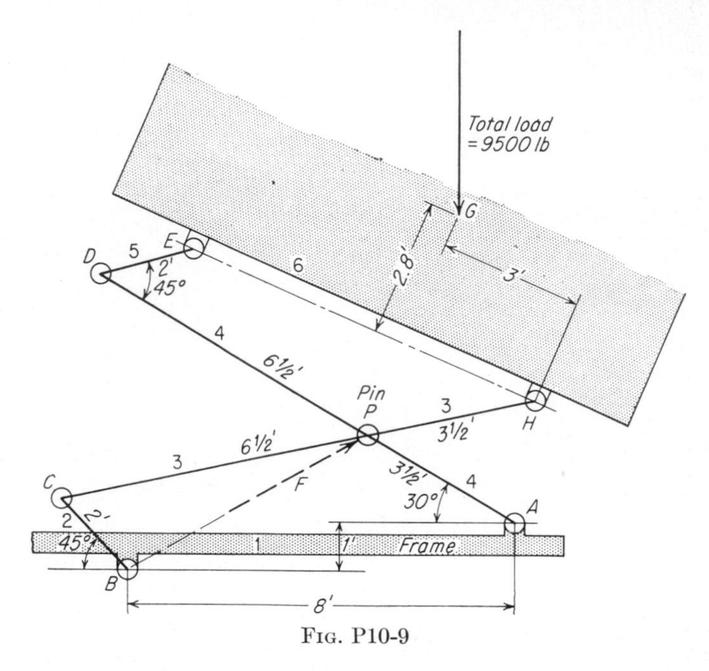

Fig. P10-9

(a) the actuating force F, and (b) the forces on the frame at A and B. (In the analysis, consider the pin P to be a separate member.) Disregard friction and accelerations. Start the force polygon with the tail of the load vector at 3¼, 9¾.

Chapter 11. Inertia Forces

11-1. In this problem an inertia-force analysis of the four-link mechanism of Fig. 11-8 is required. Link 2 makes an angle of 60 deg with the horizontal, ω_2 = 24 radians/sec and α_2 = 240 radians/sec², clockwise. The data are otherwise the same as given in Art. 11-6. Draw the skeleton outline of the mechanism to a scale of 3 in. = 1 ft, using the dimensions given in the figure. $O_2(1\frac{3}{4}, 4)$. In making this analysis, the following procedure is suggested: (a) Draw the velocity polygon (scale, 1 in. = 20 fps). $o(6, 8\frac{1}{2})$. (b) Draw the acceleration polygon (scale, 1 in. = 200 ft/sec²). $o(3, 9\frac{1}{2})$. (c) Determine the inertia forces and draw force polygon with origin of F_4 at $(2\frac{1}{2}, 1)$. Scale, 1 in. = 40 lb. See that all inertia forces and reactions are set up on the figure in magnitude, direction, and line of action, and that necessary computations are indicated in the lower right-hand part of sheet.

11-2. In this problem an inertia-force analysis is required of the four-link mechanism shown in Fig. P11-2. Draw the mechanism in skeleton outline to a scale of

3 in. = 1 ft, using the dimensions given in the figure. $O_2(1\frac{1}{2}, 7)$. $O_4(7\frac{1}{2}, 4)$. In making this analysis the following procedure is suggested:

a. Draw the velocity polygon (scale 1 in. = 10 fps). $o(1\frac{1}{2}, 8\frac{1}{2})$.

b. Draw the acceleration polygon (scale 1 in. = 300 ft/sec²). $o(\frac{3}{4}, 5\frac{3}{4})$.

c. Determine the inertia forces, and draw the inertia-force polygon with origin of vector F_4 at $(7\frac{1}{2}, 3)$. Scale 1 in. = 75 lb. All inertia forces and reactions are to be set up on the figure in magnitude, direction, and line of action, and necessary computations indicated in the space available on the sheet.

NOTE: The driving couple is applied at axis O_2.

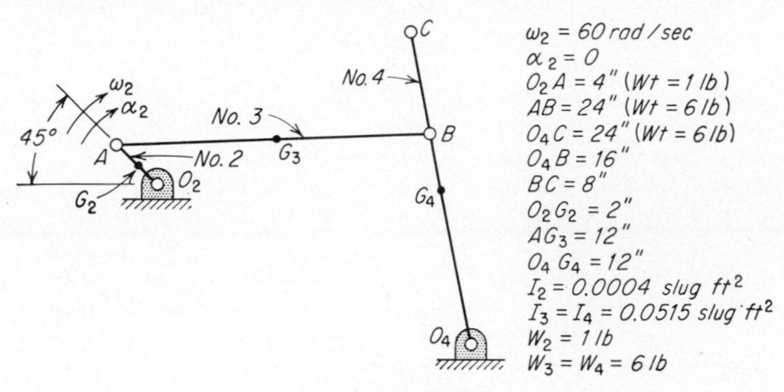

FIG. P11-2

11-3. In this problem an inertia-force analysis is required of the four-link mechanism shown in Fig. P11-3. Draw the mechanism in skeleton outline to a scale of 3 in. = 1 ft, using the dimensions given with the figure. $O_2(2, 8)$. $O_4(7, 5)$. In making this analysis the following procedure is suggested:

a. Draw the velocity polygon (scale 1 in. = 10 fps). $o(1, 3\frac{1}{2})$.

b. Draw the acceleration polygon (scale 1 in. = 300 ft/sec²). $o(1, 6\frac{3}{4})$.

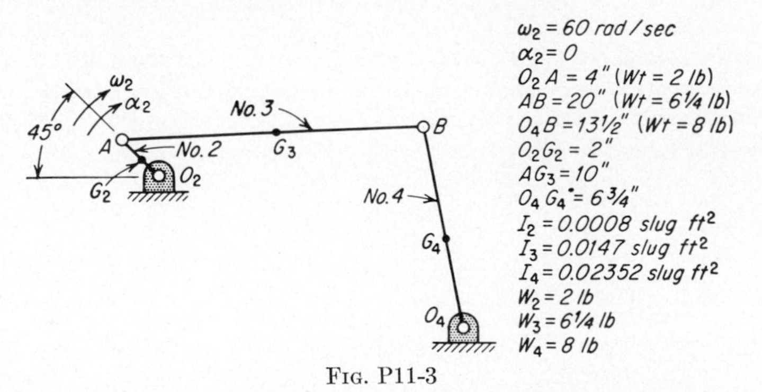

FIG. P11-3

c. Determine the inertia forces, and draw the inertia-force polygon with origin of vector F_4 at $(8\frac{1}{4}, 3\frac{3}{4})$. Scale 1 in. = 75 lb. All inertia forces and reactions are to be set up on the figure in magnitude, direction, and line of action, and necessary computations indicated in the space available on the sheet.

NOTE: The driving couple is applied at axis O_2.

11-4. The following data apply to the five-cylinder radial aircraft engine shown in Fig. P9-21. Rpm = 2,200; bore = $4\frac{1}{2}$ in.; stroke = $5\frac{1}{2}$ in.; length of master connecting rod A_1B_1 = $10\frac{7}{8}$ in.; length of link connecting rods A_2B_2, A_3B_3, etc. = $8\frac{7}{8}$ in. Link pins A_2, A_3, etc., are evenly spaced on a circle of 4 in. diameter. Weight of piston 1 = 3.90 lb. Weight of pistons 2, 3, 4, and 5 = 3.58 lb each. Weight of counterbalance = 9.11 lb. These weights include the connecting-rod weights, which have been distributed at the crankpins and the piston pins in the usual manner (see Art. 12-13).

a. Draw the mechanism to a scale of $\frac{1}{4}$ in. = 1 in. in the configuration shown in the figure (crank O_1A_1 in 45-deg position). $O(4\frac{1}{2}, 6\frac{1}{4})$.

b. Draw the velocity polygon. Scale 1 in. = 20 fps. $o(3, 3\frac{1}{2})$.

c. Draw the acceleration polygon. Scale 1 in. = 4,000 ft/sec². $o(4\frac{1}{2}, 4)$.

d. Tabulate the velocities, the accelerations, and the inertia forces of the five pistons and counterweight.

e. By means of an inertia-force polygon determine the magnitude and direction of the unbalance for this configuration of the mechanism. Origin of inertia-force vector $F_1(7\frac{1}{2}, 3\frac{1}{2})$. Force scale: 1 in. = 1,000 lb. NOTE: Label polygons; give all scales, and show necessary calculations in available space in upper part of sheet.

11-5. A six-cylinder gasoline engine is running at a uniform speed of 2,500 rpm. Length of crank O_2A = $2\frac{1}{8}$ in.; length of connecting rod AB = $8\frac{1}{8}$ in.; distance from A to center of gravity G of connecting rod = $1\frac{13}{16}$ in.; weight of piston = 1.94 lb; weight of connecting rod = 1.75 lb; equivalent unbalanced weight of crank at crankpin radius = 3.25 lb; time for 68 complete swings (over and back constitute one complete swing) of connecting rod about wrist-pin axis is 60 sec.

a. Make an inertia-force analysis of the mechanism when the crank is in the 60-deg position after head-end dead center. Rotation clockwise. All inertia forces and reactions are to be determined, as in Art. 11-6. Procedure: Use two sheets. On sheet 1 make a drawing of the mechanism with center line of engine horizontal and show thereon all dimensions, weights, etc. Scale, $\frac{3}{4}$ in. = 1 in. $O_2(8, 7)$, piston to the left. Place all necessary computations on this sheet. On sheet 2 make another drawing of the mechanism and indicate thereon the inertia forces f_2, f_3, and f_4 in magnitude, direction, and line of action. Scale of drawing, $\frac{3}{4}$ in. = 1 in. $O_2(8, 7)$, piston to left. Locate the several polygons on the sheet as follows: velocity polygon, $o(\frac{1}{2}, 3)$, scale 1 in. = 30 fps; acceleration polygon, $o(\frac{1}{2}, 6)$, 1 in. = 5,700 ft/sec²; inertia-force polygon, origin of vector $f_2(7, 1\frac{1}{2})$, scale, 1 in. = 300 lb.

b. Make a static-force analysis of the mechanism. The gas force on the piston at this position is 750 lb. Origin of vector $\mathfrak{F}_4$, $(\frac{1}{2}, 1\frac{1}{2})$. Scale, 1 in. = 300 lb.

c. By combining the inertia forces and static forces, determine the magnitudes and directions of the following total forces: force on *wrist pin;* force on *crankshaft;* force on *crankpin; turning effort.*

11-6. Same as Prob. 11-5, except that the solution is to be made for some other crank position.

11-7. A six-cylinder gasoline engine is running at a uniform speed of 2,500 rpm. Length of crank O_2A = $2\frac{1}{8}$ in.; length of connecting rod AB = $8\frac{1}{8}$ in.; distance from A to center of gravity of connecting rod = $1\frac{13}{16}$ in.; weight of piston = 1.94 lb; weight of connecting rod = 1.75 lb; equivalent unbalanced weight of crank at crankpin radius = 3.25 lb; time for 68 complete swings (over and back constitute one complete swing) of the connecting rod about the wrist-pin axis is 60 sec. NOTE: The data given and the results required in this problem are the same as in Prob. 11-5. The procedure involves the substitution of a kinetically equivalent system for the connecting rod.

a. Make an inertia-force analysis of the mechanism when the crank is in the 60-deg position after head-end dead center. Rotation clockwise. All inertia forces and reactions are to be determined as in Art. 11-9. Procedure: Use two sheets. On sheet 1 make a drawing of the mechanism with center line of engine horizontal and show thereon all dimensions, weights, etc. Scale, ¾ in. = 1 in., $O_2(8, 7)$, piston to left. Place all necessary computations on this sheet. On sheet 2 make another drawing of the mechanism and indicate thereon the inertia forces f_2, f_e, f_b, and f_4 in magnitude, direction, and line of action. Scale of drawing, ¾ in. = 1 in., $O_2(8, 7)$, piston to the left. Locate the several polygons on the sheet as follows: velocity and acceleration polygons (use Ritterhaus construction); inertia-force polygon, origin of vector $f_2(7, 1\frac{1}{2})$, scale, 1 in. = 300 lb.

b. Make a static-force analysis of the mechanism. The force on the piston due to gas pressure is 750 lb. Origin of vector $\mathfrak{F}_4$, (½, 5). Scale 1 in. = 300 lb.

c. By combining the inertia forces and static forces, determine the magnitudes and directions of the following total forces: force on *wrist pin;* force on *crankshaft;* force on *crankpin; turning effort.*

11-8. Same as Prob. 11-7, except that the solution is to be made for some other crank position.

11-9. A six-cylinder gasoline engine is running at a speed of 2,500 rpm. Length of crank O_2A = 2⅛ in.; length of connecting rod AB = 8⅛ in., distance from A to center of gravity of connecting rod = 1$\frac{13}{16}$ in.; weight of piston = 1.94 lb; weight of connecting rod = 1.75 lb; the crank is balanced; time for 68 complete swings (over and back constitute one complete swing) of the connecting rod is 60 sec. NOTE: The data given in this problem and the results required are the same as in Probs. 11-5 and 11-7. In this case, the crank is balanced. The inertia-force polygon will be superimposed on the Ritterhaus construction on an enlarged drawing of the mechanism.

a. Make an inertia-force analysis of the mechanism when the crank is in the 60-deg position after head-end dead center. Rotation clockwise. This analysis is to be similar to that in Art. 11-10, where the inertia-force polygon is superimposed on the Ritterhaus construction. Procedure: Use two sheets. On sheet 1 make a drawing of the mechanism with the center line of the engine horizontal and show thereon all dimensions, weights, etc. Scale, ¾ in. = 1 in. $O_2(8, 7)$, piston to the left. Place all necessary computations on this sheet. On sheet 2 make another drawing of the mechanism twice full size with the crank in the 60-deg position after head-end dead center. Center line of engine horizontal. Rotation clockwise. $O_2(6\frac{1}{4}, 4\frac{3}{4})$, piston to the left. Determine the inertia forces and reactions by means of the force polygon superimposed on the Ritterhaus construction. Scale of forces, 1 in. = 300 lb.

b. Make a static-force analysis of the mechanism. The force on the piston due to gas pressure is 750 lb. Origin of vector $\mathfrak{F}_4$ is at A on the mechanism. Scale 1 in. = 300 lb.

c. By combining the inertia forces and static forces, determine the magnitudes and directions of the following total forces: force on *wrist pin;* force on *crankshaft;* force on *crankpin; turning effort.*

11-10. Same as Prob. 11-9, except that the solution is to be made for some other crank position.

11-11. *a.* Using the data and methods of Prob. 11-9, plot a diagram of inertia forces of the piston on a base line 6 in. long, located 6⅝ in. from lower edge of sheet. Start with zero position of crank when piston is to the extreme left and make the determinations for each 30-deg crank position during one revolution. For the end positions (0, 6, and 12) use the analytical methods of Art. 11-15. Scale, 1 in. = 300 lb.

b. On a base line 6 in. long, located 2½ in. from lower edge of sheet, plot a diagram showing the variation of the turning effort at the crankpin during one revolution. (NOTE: These diagrams are similar to those shown in Figs. 13-4 and 13-6.) Procedure: Use two sheets. On one sheet make a drawing of the mechanism with piston to the left. $O_2(4\frac{5}{8}, 5\frac{1}{4})$. Scale, 1½ in. = 1 in. Draw in the mechanism for crank positions 1, 3, 5, 8, and 10 and determine the inertia forces and reactions for these positions by means of the force polygon superimposed on the Ritterhaus construction. Scale of forces, 1 in. = 300 lb. (It will be observed that the magnitudes of the inertia forces for crank positions 11, 9, 7, 4, and 2 will be the same as for positions 1, 3, 5, 8, and 10, respectively.)

11-12. In addition to the diagrams plotted in Prob. 11-11, other diagrams such as those described in Chap. 13 may be assigned for problem work as desired.

11-13. The cam shown in Fig. 3-1 is rotating at 180 rpm, and the roller follower receives its maximum displacement of 2 in. while the cam rotates clockwise through the angle of 120 deg. The follower returns to its initial position in the next 120 deg rotation of the cam and remains at rest during the final 120 deg of the cycle. The follower has harmonic motion, and its total weight is 5 lb.

a. On a horizontal base line 6 in. long, 0(1¼, 7½), construct the displacement diagram of the follower for one revolution of the cam. Start with the follower in its lowest position. (Use 8 divisions for each 120 deg.)

b. On a horizontal base line 6 in. long, representing the time for one revolution of the cam, 0(1¼, 5), construct the inertia-force diagram for the follower. Scale 1 in. = 10 lb.

c. On the same base line and to the same scale as in (*b*) construct the diagram representing the weight of the follower.

d. A spring with a scale of 2½ lb/in. set up under an initial load of 0.35 lb will be just sufficient to keep the follower in contact with the cam at all times. On the same base line and to the same scale as in (*b*), construct the diagram representing the force exerted by the spring.

e. Construct the diagram showing inertia force and weight of follower combined.

f. Construct the diagram representing the total force of follower against cam.

g. Mark on the horizontal base line, on the inertia-force diagram, and on the displacement diagram the point at which the follower would leave the cam owing to its acceleration if a spring were not used.

NOTE: In the above force diagrams, plot as ordinates below the base line vertical forces of the follower against the cam and as ordinates above the base line vertical forces of the follower away from the cam. Also, neglect the effect of friction throughout. In Fig. 11-14 of the text is shown another method of plotting these same diagrams.

11-14. A plate cam, similar to that shown in Fig. 3-9, rotates at a speed of 360 rpm and drives its follower in a vertical direction. The follower rises 1½ in. with simple harmonic motion during 180 deg of cam rotation, dwells during 30 deg, and returns 1½ in. to its initial position during the remaining 150 deg of cam rotation. The total weight of the follower is 5 lb.

a. On a horizontal base line 6 in. long, 0(1¼, 8), construct the displacement diagram for the follower for one revolution of the cam. Divide the base line into 24 equal parts, and start with the follower in its lowest position.

b. On a horizontal base line 6 in. long, 0(1¼, 5¼), construct the inertia-force diagram for the follower. Scale: 1 in. = 10 lb.

c. On the same base line and to the same scale as in (*b*), construct the diagram representing the weight of the follower.

d. A spring with a scale of 10 lb/in., set up with zero initial load, is used in order to keep the follower in contact with the cam at all times. On the same base line and to the same scale as in (*b*), construct the diagram representing the force exerted by the spring.

e. Construct the diagram showing inertia force and weight of follower combined.

f. Construct the diagram representing the total force of follower against cam.

g. Indicate on the displacement diagram the position at which the follower would leave the cam because of its acceleration, if a spring were not used.

h. What is the total vertical component of the force of the follower against the cam (1) at the beginning of the upward motion of the follower, (2) at the end of the upward motion, (3) during the dwell period, (4) at the beginning of the downward motion, (5) at the end of the downward motion?

NOTE: In the above force diagrams, plot as ordinates below the base line vertical components of forces of the follower against the cam, and as ordinates above the base line vertical components of forces of the follower away from the cam. Also, neglect the effect of friction throughout.

11-15. The speed of an ammonia compressor is controlled by a gravity-loaded two-ball flyball governor of the type shown in Fig. P11-15. The weight of the central

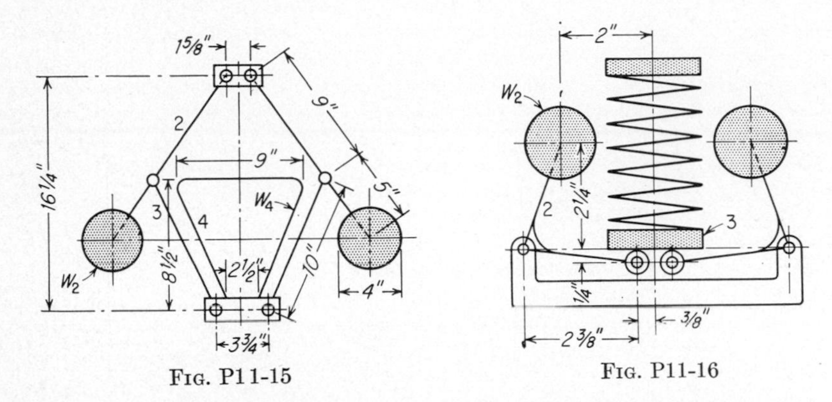

Fig. P11-15

Fig. P11-16

load is W_4 = 74 lb, and the weight of each ball is 9 lb. In the figure the governor is shown in its lowest position (rpm = 0). Lay out the governor to a scale of 3 in. = 1 ft. $O_2(4, 9)$.

a. For this position of the governor (lowest position) determine the speed (rpm) at which it will begin to act.

b. The normal operating speed of the compressor occurs when the central weight is lifted a distance of 2½ in. Determine the speed (rpm) of the governor in this position. NOTE: Use graphical method for the solution, and show free-body diagrams and necessary calculations. Scale of force polygons: 1 in. = 20 lb.

11-16. The spring-controlled governor of a 100-kw steam turbine is shown in diagrammatic form in Fig. P11-16. The governor is mounted on the turbine shaft. Under the action of centrifugal force the two weights W_2 act upon the spring and sleeve as shown in the figure. The weight of each of the two revolving weights is W_2 = 0.15 lb. The normal speed of the turbine is 3,600 rpm, and a total variation of ±2½ per cent from this is permissible. The actual displacement of the sleeve required for this regulation is known to be ⅛ in. (1/16 in. each side of position for normal speed). The position shown in the figure is for the lowest permissible speed. It is required to determine the initial and final load on the spring and its stiffness constant. NOTE: Use analytical method for the solution, and show free-body diagrams and necessary calculations. It will be sufficient to assume for *all* positions that the moment arms of the forces about the pivot point of link 2 are equal to the values given in the figure for the position at lowest speed.

NOTE: In the analysis of governor action, fundamental problems in inertia forces are always involved. The force analysis in the case of the flyball type illustrated in Figs. 2-48 and 2-49 is fairly simple, as illustrated by the solutions of Probs. 11-15 and 11-16. Here it is necessary only to make free-body diagrams showing the forces set up as a result of the centrifugal action of the balls and to solve either analytically or graphically, depending upon which method may be more convenient for the particular case at hand. In the case of the shaft governor (Fig. 2-50) and the inertia governor (Fig. 2-51) the analysis is somewhat more involved. In Fig. 2-50, for example, when ω is constant, we have only the moment of the centrifugal force $F_R = mr\omega^2$ to be balanced by the moment of the spring force S. However, when the wheel receives an angular acceleration α, the weight W develops at G, in addition to the force F_R, a tangential force $F_T = mr\alpha$. At the same time, if the mass of arm 3, which carries the weight, is considered, this entire mass develops a relatively small inertia torque $T = -I\alpha$ with respect to its mass center. However, these secondary effects are relatively small and can be neglected, so that only the effect of the centrifugal force is considered. The force analysis is then reduced to the relatively simple problem of considering the moment about the pivot O_3 of the centrifugal force acting in opposition to the moment of the spring force. Problem 9-26 requires a typical acceleration analysis. A similar situation exists in the case of the inertia governor (Fig. 2-51). Here the inertia torque $T = -I\alpha$ is large. The forces $F_R = mr\omega^2$ and $F_T = mr\alpha$ are relatively small. For a very complete analysis of the forces acting in an inertia type of shaft governor the student is referred to textbooks such as "Analytical Mechanics for Engineers," 3d ed., 1941, by Seely and Ensign.

11-17. A research airplane at 60° north latitude flies north at 2,000 mph along a north-south line. The plane therefore moves along a path which has angular velocity. Determine the magnitude and direction (show on a sketch) of the force which must act on the airplane per thousand pounds of airplane mass because of the Coriolis component of acceleration.

11-18. Assume an airplane to make a turn along a horizontal circular arc such that a full reversal of direction of flight takes 1 min. Evaluate the number of g (actual acceleration/acceleration of gravity) at flight speeds of 500, 1,000, and 2,000 mph. Repeat if the turn takes only ½ min.

11-19. An elevator cage and its load weigh 4,000 lb. The ¾-in. cable is of steel and has a breaking strength of 16.8 tons.

a. For a factor of safety of 4 based on breaking strength, calculate the minimum number of seconds in which it is permissible to stop the elevator from a speed of descent of 500 fpm. Assume constant deceleration.

b. Repeat, using a factor of safety of 8.

c. If deceleration is *not* constant, but the stop is made in the same number of seconds, will the maximum force in the cable be larger or smaller? Justify your answer.

11-20. A drag-line bucket and its load weigh 2,000 lb. The lifting cable is of ½ in. diameter and has a breaking strength of 9.4 tons. What maximum vertical acceleration may be given to the load without breaking the cable? How many g (acceleration/acceleration of gravity) does this represent?

11-21. A machine part is to be made of steel and formed in the shape of a connecting link. A wooden model of the part is made full size and the center of gravity is determined to be at a position 5 in. from one end. An experiment is performed in which the model is pivoted on a knife-edge at a distance of 4 in. from the center of gravity, and allowed to oscillate as a pendulum through a small arc. The frequency of the oscillations is measured as 0.97 per sec. The wooden model weighs 2 lb and is made of wood having a density of 35 lb/cu ft. Density of steel = 489 lb/cu ft. Find the moment of inertia of the proposed *steel* link about its center of gravity. Also indicate how the center of gravity of the model might have been found experimentally.

11-22. A flywheel weighing 500 lb is supported on a knife-edge on the inside of the rim at a distance of $1\frac{3}{4}$ ft from the center and is allowed to oscillate as a pendulum through a small arc. The period of oscillation is measured to be 2.22 sec. Find the moment of inertia of the flywheel about its center.

11-23. A 40-lb package travels by gravity down a chute in a manufacturing plant, emerges at the bottom onto a horizontal section, and strikes a spring-cushioned stop at a velocity of 10 fps. If the package contents will be damaged if the acceleration exceeds $3g$ ($= 96.6$ ft/sec^2), find the proper spring constant for the stop, and the number of inches the spring must compress.

11-24. In a machine, a steel block weighing 1 lb is forced to oscillate with simple harmonic motion along a radial slide in a rotating table. The total motion of the block along the slide is 6 in., with the center point at a radius of 12 in. Each cycle of simple harmonic motion takes place in one-half revolution of the table. The table rotates at a constant speed of 1,200 rpm clockwise with the aid of a very heavy flywheel. Calculate and show on a sketch (*a*) the sideways force on the block (perpendicular to the radius) at the mid-point and at the extremes of travel; and (*b*) the force which connecting linkage must exert on the block along the radial slide direction at the mid-point and at the extremes of travel.

11-25. If the earth's rotation about its axis were to speed up until men and machines at the equator became weightless, how long would a day be? (A trip around the world is about a 25,000-mile journey.)

Chapter 12. The Balancing of Machinery

12-1. Three masses m_1, m_2, and m_3 (Fig. P12-1*a*), which revolve in the transverse planes 1, 2, and 3, are to be balanced by the addition of two revolving masses. Plane 3 is chosen as the plane of one of the masses m_o', and RP_o as the plane of the other mass

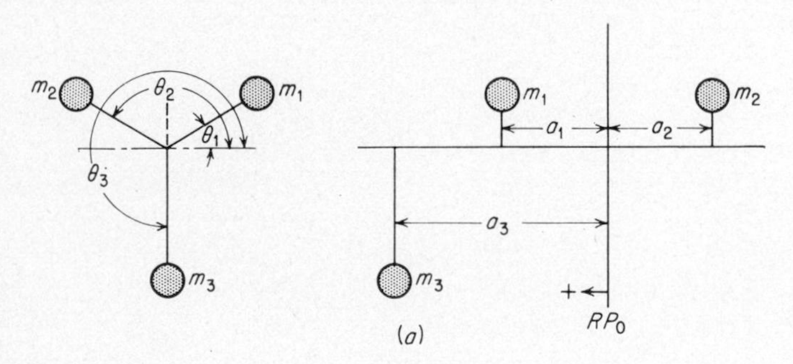

Plane	Mass	Radius	a	θ	mR	mRa
1	15	4	4	30°		
2	30	4	−4	150°		
3	15	5	8	270°		
	Assume		Assume			
0′	20		8			
	Assume					
RP_0	30					

(*b*)

FIG. P12-1

m_o. Determine by graphical method values of m_o and m_o' and the lengths and directions of the corresponding radii for kinetic balance. The accompanying table (Fig. P12-1*b*) gives the values of the masses in slugs, the lengths in feet, the directions of their radii, and their distances in feet from the reference plane *o*. Scale of drawing, ¼ in. = 1 ft. End elevation of axis, *o*(2½, 8). Scale of *mRa* polygon, 1 in. = 300 units. Scale of *mR* polygon, 1 in. = 50 units. NOTE: For actual values of moments and forces for any given speed, the values represented by the vectors in these polygons would have to be multiplied by ω^2.

12-2. Same as Prob. 12-1 except that the solution is to be accomplished by means of two couple polygons.

12-3. Same as Prob. 12-1, except that the solution is to be accomplished by the application of the analytical method.

12-4. Same as Prob. 12-1, except that mass $m_2 = 20$, $R_2 = 5$, and $R_3 = 6$.

12-5. Same as Prob. 12-4, except that the solution is to be accomplished by means of two couple polygons.

12-6. Same as Prob. 12-4 except that the solution is to be accomplished by application of the analytical method.

12-7. The crankshaft of a horizontal stationary gas engine of the center-crank type carries two flywheels *A* and *B*, the planes of revolution of which are 3 ft 8 in. apart. The plane of revolution of the crank is 18 in. from the plane of flywheel *A*. The equivalent weight of the crank arm at a radial distance of 10 in. from the crankshaft and in the plane of revolution of the crank is 110 lb. The crank is to be balanced by weights placed at a radial distance of 24 in. from the crankshaft, one in each flywheel. Find the magnitudes of the weights. Make a diagram of the general setup to a scale of 1½ in. = 1 ft. Start center line of crankshaft at (1½, 6).

12-8. A rotor is unbalanced as specified by the masses shown in Fig. P12-8.

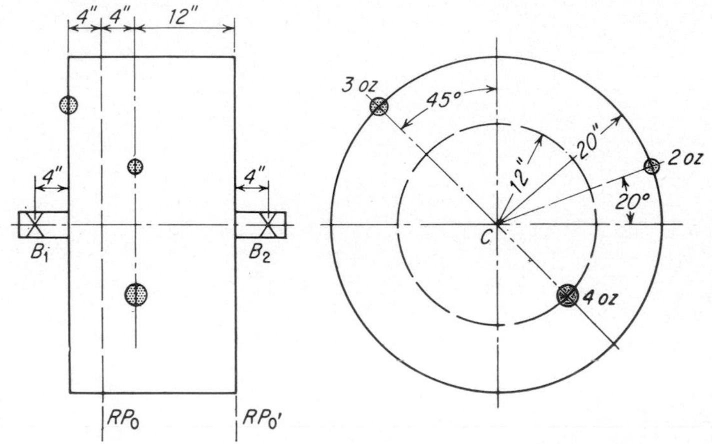

FIG. P12-8

a. Find and show on the figures two masses in the designated reference planes at radii of 20 in. which will completely balance the rotor.

b. Assume that nothing is done to correct the unbalance. What would be the forces on the bearings at B_1 and B_2 at rotor speeds of 3,000 and 6,000 rpm (in addition to forces resulting from the weight of the rotor)? For drawing, place point *C* at (7, 5½), long edge of sheet horizontal; scale, 1 in. = 10 in.

12-9. In a rotor 3 ft long, unbalanced masses are present as follows: (1) in a transverse plane through the left end of the rotor, 4 in.-oz at 30 deg clockwise from vertical (viewed from left end of the rotor); (2) in a plane 1 ft from the left end, 5 in.-oz at 180 deg from vertical; (3) in a plane 2 ft from the left end, 2 in.-oz at 270 deg clockwise from vertical (viewed from left end).

a. Find the two *equivalent* unbalances and their directions in planes at the ends of the rotor, where corrections are to be applied. (These would be the unbalances indicated by a balancing machine.)

b. Find and show on a sketch the amount and location of proper balancing weights, if weights are to be added at radii of 6 in.

12-10. A single-cylinder horizontal gas engine runs at a speed of 250 rpm. Length of crank = 10¼ in. Length of connecting rod = 51¼ in. Distance of center of gravity of connecting rod from crank end = 19.8 in. Equivalent unbalanced weight of crank at 10¼ in. = 159.5 lb. Weight of piston = 234 lb. Weight of connecting rod = 251 lb. Equivalent weight of counterbalance at 10¼ in. radius = 522 lb. Draw the skeleton outline of the mechanism to a scale 1⁄10 size and indicate thereon the dimensions and weights. O_2(1½, 8½). Place all computations on this sheet. Using the methods outlined in Art. 12-14, plot to a scale of 1 in. = 3,000 lb (on another sheet) the following diagrams of inertia forces. o(4¼, 5).

a. Assuming that no counterbalancing weight is used.

b. Assuming that a counterbalancing weight is used that is equal to the sum of all the rotating weight concentrated at the crankpin and all the reciprocating weight concentrated at the piston, the counterbalancing weight to be placed opposite the crank at a distance equal to the crank radius.

c. For the actual conditions existing in this engine, i.e., when the given counterbalancing weight is used.

12-11. A single-cylinder horizontal oil engine runs at a speed of 250 rpm. Length of crank = 9 in.; length of connecting rod = 47 7⁄16 in.; distance of center of gravity of connecting rod from crank end = 16 in.; equivalent unbalanced weight of crank at 9 in. radius = 150 lb; weight of piston = 198.5 lb; weight of connecting rod = 194.6 lb; equivalent weight of counterbalance at 9 in. radius = 410 lb.

Draw the skeleton outline of the mechanism to a scale of 1½ in. = 1 ft and indicate thereon the dimensions and weights. O_2(7¼, 8½). Place all computations on this sheet.

Using the methods outlined in Art. 12-14, plot to a scale of 1 in. = 3,000 lb (on another sheet) the following diagrams of inertia forces:

a. Assuming that no counterbalancing weight is used. o(4¼, 5).

b. Assuming that a counterbalancing weight is used that is equal to the sum of the rotating weight concentrated at the crankpin and the reciprocating weight concentrated at the piston, the counterbalancing weight to be placed opposite the crank at a distance equal to the crank radius.

c. For the actual conditions existing in this engine; i.e., when the given counterbalancing weight is used.

12-12. A two-cylinder 80-hp vertical diesel engine running at 327 rpm has the two cranks in the same axial plane on the same side of the crankshaft. Length of crank = 6⅞ in.; length of connecting rod = 31¾ in., distance of center of gravity of connecting rod from crank end = 11.32 in.; equivalent unbalanced weight of one crank at 6⅞ in. = 133 lb; weight of piston = 194 lb; weight of connecting rod = 146 lb; equivalent weight of counterbalance opposite each crank at 6⅞ in. radius = 215 lb.

Draw the skeleton outline of the engine to a scale 2 in. = 1 ft and indicate thereon the dimensions and weights. O_2(2, 2½) center line of engine vertical. Rotation clockwise. Draw crank in 45-deg position in the first quadrant. Place all computations on this sheet.

Using the methods outlined in Art. 12-14, plot to a scale of 1 in. = 3,000 lb (on another sheet) the following diagrams of inertia forces per cylinder (use 30-deg crank intervals):

a. Assuming that no counterbalancing weight is used. $o(4\frac{1}{4}, 4\frac{3}{4})$.

b. Assuming that a counterbalancing weight is used that is equal to the sum of the rotating weight concentrated at the crankpin and the reciprocating weight concentrated at the piston, the counterbalancing weight to be placed opposite the crank at a distance equal to the crank radius.

c. For the actual conditions existing in this engine; i.e., when the given counterbalancing weight is used.

12-13. A single-cylinder vertical oil engine operates at a speed of 720 rpm. Length of crank = $2\frac{1}{2}$ in.; length of connecting rod = 13 in.; distance of center of gravity of connecting rod from crank end = 4 in.; equivalent unbalanced weight of crank at $2\frac{1}{2}$ in. radius = 9 lb; weight of piston = 13.2 lb; weight of connecting rod = 16.5 lb; equivalent weight of counterbalance at $2\frac{1}{2}$ in. radius = 30 lb.

Draw the skeleton outline of the engine to a scale of 6 in. = 1 ft, and indicate thereon the dimensions and weights $O_2(2\frac{1}{2}, 2)$. Assume clockwise rotation for the crank, and draw it in 45-deg position clockwise from vertical center line of engine. Place all computations on this sheet.

Using the methods outlined in Art. 12-14, plot to a scale of 1 in. = 500 lb (on another sheet) the following diagrams of inertia forces:

a. Assuming that no counterbalancing weight is used. $O(4\frac{1}{4}, 5)$.

b. Assuming that a counterbalancing weight is used which is equal to the sum of the rotating weight concentrated at the crankpin and the reciprocating weight concentrated at the piston, the counterbalancing weight to be placed opposite the crank at a distance equal to the crank radius.

c. For the actual conditions existing in this engine, i.e., when the given counterbalancing weight is used.

12-14. The arrangement of cranks in an eight-cylinder 16 by 20 (bore by stroke) vertical diesel engine is shown diagrammatically in Fig. P12-14, with crank 1 making any angle θ_1 with the vertical center line.

a. Assuming the quantities $mR\omega^2$ and distance between cranks each equal to unity, determine by the methods illustrated in Arts. 12-15 to 12-21 what forces or couples are unbalanced.

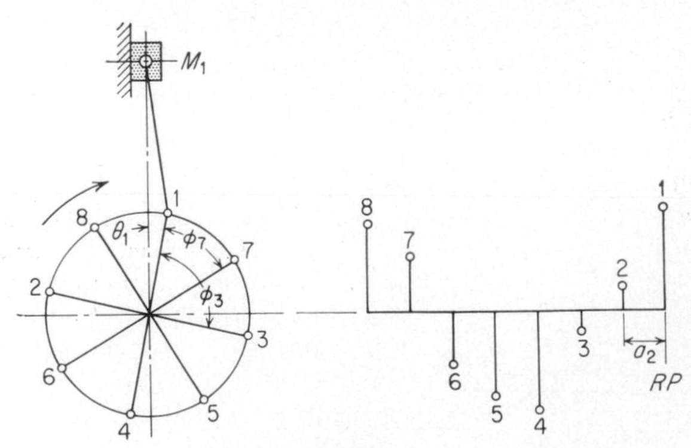

Fig. P12-14

b. The rpm of this engine is 300. Length of connecting rod = 54 in. Total reciprocating weight per cylinder = 984 lb. The distance between cranks is 30 in. Determine the magnitude of the unbalanced couple for positions of $\theta_1 = 0$, 30, 60, etc., to 360 deg and plot a graph showing its variation in magnitude. Scale of ordinates 1 in. = 25,000 lb-ft. (Arrange data on sheet in tabular form, and show sample calculation for $\theta_1 = 60$ deg.)

12-15. The arrangement of cranks in a 90-deg V-8 engine is shown in Fig. 12-31. The rpm of the engine is 2,400. The total reciprocating weight per cylinder is 3 lb. Length of crank = 2½ in. Length of connecting rod = 10 in. The distance between cranks is 5 in. Make a skeleton outline of the engine similar to that shown in Fig. P12-14 with crank 1 making any angle θ_1 (clockwise) with center line of left bank of cylinders, viewed in direction of axis of crankshaft.

a. Assuming that the quantities $mR\omega^2$ and distances between cranks are each equal to unity determine by the methods illustrated in Arts. 12-15 to 12-21 what forces or couples are unbalanced.

b. Determine the magnitude of the unbalanced resultant couple for positions of θ_1 from 0 to 360 deg by 30-deg intervals and from crank center O in the skeleton outline of the engine, plot to a scale of 1 in. = 10,000 lb-in. the polar diagram represented by the ends of the resultant couple vectors.

c. To what extent can this unbalanced couple be eliminated by balancing? Explain. (Arrange data on sheet in tabular form, and show sample calculation for $\theta_1 = 30$ deg.)

12-16. In the 90-deg V-8 engine of Fig. 12-31, consider that the connecting rod has been replaced by an approximate kinetically equivalent system. The total resulting reciprocating mass in each cylinder is m_P, and the total equivalent rotating mass at crank radius is m_C. This includes the equivalent crank mass at crank radius, plus the rotating parts of the masses of the *two* connecting rods attached to the crank. The cylinders and crank throws are spaced uniformly a distance a apart.

a. By a balance check on one bank of cylinders (as in Arts. 12-15 to 12-21), determine what forces or couples are unbalanced.

b. Determine the proper amounts and locations of balancing masses (in terms of m_P and m_C) if each of the rotating masses and primary reciprocating inertia forces is to be counterbalanced in the plane of its crank throw. Show on a sketch.

c. Repeat, but assume that each rotating mass is to be balanced in its own plane and that the reciprocating forces are to be balanced by only two masses—one in each of the end planes (planes of cranks 1 and 4).

d. Repeat, but assume that a total of only two balancing masses will be used—one in each of the end planes.

12-17. Figure 12-32 represents a five-cylinder radial aircraft engine. It is required to make an inertia-force analysis of the engine in order to determine to what extent it is in balance and, if not in balance, to what extent it can be balanced. It can be shown that the accelerations of the articulated-rod pistons do not differ greatly from the accelerations of the master-rod piston. This fact justifies the usual simplifying procedure of assuming the articulated rods to be equal in length to the master rod and that their crank ends are grouped together with the master-rod crank end on the master-rod crankpin axis.

Draw the mechanism to a scale of ¼ in. = 1 in. with piston 1 in extreme head-end position (crank vertical). Location of axis of crankshaft (4¼, 5½). Length of crank = 2¾ in. Length of connecting rods 11 in., with their crank-end axes all coinciding with the crankpin axis of the master rod.

The rpm of the engine is 2,000. The connecting-rod weights are assumed to be distributed at the crankpin and at the piston pins in the usual manner (see Art. 12-13), so that the estimated total rotating weight at the crankpin is 10 lb and total reciprocating weight at each piston pin is 4 lb. Find the resultant of the secondary inertia forces for the 30-deg crank position. Scale 1 in. = 200 lb. $o(3, 1)$. Are the secondary forces in balance for all crank positions? Find the resultant of the rotating and reciprocating primary inertia forces for the 0-, 30-, 60-, and 90-deg positions of the crank. (Calculate the magnitude of each vector; and, starting with the origin of the rotating inertia-force vector at the axis of the crankshaft, find the resultant vector.) Scale of forces 1 in. = 2,000 lb. To what extent is the engine in balance? If not in balance to what extent can it be balanced?

12-18. The data given below apply to a six-cylinder gasoline engine of the automotive type. A complete force analysis of this engine is required, in accordance with the method outlined in Chap. 13.

Dimensions

Rpm at maximum horsepower (80 hp), 3,300.
Number of cylinders, 6.
Diameter of cylinders, $3\frac{5}{16}$ in.
Stroke, 4 in.
Length of connecting rod, center to center, $7\frac{1}{2}$ in.
Distance of center of gravity of connecting rod from crankpin end, $2\frac{3}{8}$ in.
Diameter of wrist pin, 1 in.
Diameter of crankpin, $2\frac{1}{8}$ in.
Diameter of main bearing, $2\frac{3}{16}$ in.
Center of gravity of unbalanced rotating weight of each crank from center line of crankshaft (when cranks are not balanced), 2 in.
Time for 50 complete swings (over and back constitute one complete swing) of connecting rod about center of wrist-pin end is 43 sec.

Weights

Connecting rod, complete (including bushings), 1.99 lb.
Piston, complete (including wrist pin, rings, etc.), 2.36 lb.
Estimated unbalanced weight of each crank (when cranks are not balanced), 2.25 lb.

Timing

Inlet valve opens 4 deg before head-end dead center.
Inlet valve closes 34 deg after crank-end dead center.
Exhaust valve opens 47 deg before crank-end dead center.
Exhaust valve closes 4 deg after head-end dead center.

Indicator diagram

Compression volume, 22.5 per cent of piston displacement.
Maximum combustion pressure, 375 psia.
Pressure rise completed at 2 per cent of stroke.
Development of expansion curve begins at 5 per cent of stroke.
Mean suction pressure, 12 psia.
Mean exhaust pressure, 16 psia.
Pressure at beginning of compression, 12 psia.
Exponent for expansion curve, 1.3.
Exponent for compression curve, 1.3.
Atmospheric pressure, 14.7 psia.

12-19. The data given below apply to an eight-cylinder in-line gasoline engine of the automotive type. A complete force analysis of this engine is required, in accordance with the method outlined in Chap. 13.

Dimensions

Rpm, 3,000.
Number of cylinders, 8.
Diameter of cylinders, $3\frac{3}{16}$ in.
Stroke, $3\frac{1}{2}$ in.
Length of connecting rod, center to center, $7\frac{11}{16}$ in.
Distance of center of gravity of connecting rod from crankpin end, 1.75 in.
Diameter of wrist pin, $\frac{15}{16}$ in.
Diameter of crankpin, 2 in.
Diameter of main bearing, $2\frac{5}{16}$ in.
Center of gravity of unbalanced rotating weight of each crank from center line of crankshaft (when cranks are not balanced), $1\frac{3}{4}$ in.
Time for 50 complete swings (over and back constitute one complete swing) of connecting rod about center of wrist-pin end is 42 sec.

Weights

Connecting rod, complete (including bushings), 2.03 lb.
Piston, complete (including wrist pin, rings, etc.), 2.13 in.
Estimated unbalanced weight of each crank (when cranks are not balanced), 2.00 lb.

Timing

Inlet valve opens 5 deg after head-end dead center.
Inlet valve closes 39 deg after crank-end dead center.
Exhaust valve opens 45 deg before crank-end dead center.
Exhaust valve closes 15 deg after head-end dead center.

Indicator diagram

Compression volume, 21.2 per cent of piston displacement.
Maximum combustion pressure, 375 psia.
Pressure rise completed at 2 per cent of stroke.
Development of expansion curve begins at 5 per cent of stroke.
Mean suction pressure, 12 psia.
Mean exhaust pressure, 16 psia.
Pressure at beginning of compression, 12 psia.
Exponent for expansion curve, 1.3.
Exponent for compression curve, 1.3.
Atmospheric pressure, 14.7 psia.

12-20. The data given below apply to an eight-cylinder 16- by 20-in. two-cycle vertical diesel engine. A complete force analysis of this engine is required, in accordance with the method outlined in Chap. 13.

Dimensions

Rpm, 300.
Number of cylinders, 8.
Diameter of cylinders, 16 in.
Stroke, 20 in.
Length of connecting rod, center to center, 54 in.
Distance of center of gravity of connecting rod from crankpin end, 21.6 in.
Diameter of wrist pin, 7 in.
Diameter of crankpin, 11 in.
Diameter of main bearing, 11 in.
Center of gravity of unbalanced rotating weight of crank from center line of crankshaft, 8.85 in.

Time for 50 complete swings (over and back constitute one complete swing) of connecting rod about center of wrist-pin end is 111.5 sec.

Weights

Connecting rod, complete (including bushings), 590 lb.
Piston, complete (including wrist pin, rings, etc.), 748 lb.
Estimated unbalanced rotating weight of crank, 525 lb.

Timing (two-cycle)

Exhaust port opens 60 deg before crank-end dead center.
Exhaust port closes 60 deg after crank-end dead center.

Indicator diagram

The indicator card, drawn to scale, is given in Fig. P12-20.
Clearance volume is 6.83 per cent of piston displacement.

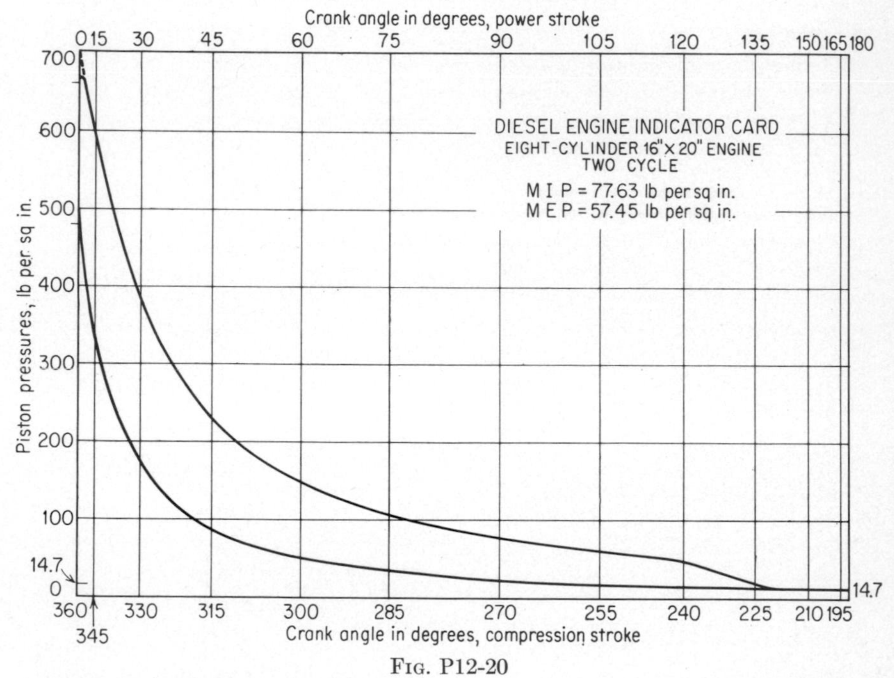

Fig. P12-20

12-21. The purpose of this problem is to determine the flywheel requirements for a single-cylinder horizontal gas engine. This is the same engine as that referred to in Prob. 12-10. It is a four-cycle engine with two flywheels, one on each end of the crankshaft with the crank midway between. It is to be assumed that the engine is to be used for ordinary power purposes and is working against a constant load. The engine specifications are as follows:

Piston diameter, 12 in.
Stroke, 20½ in.
Rpm, 250.
Connecting rod length, 51¼ in.
Compression ratio, 4.57:1.
Flywheel rim speed, 75 fps.
Coefficient of fluctuation, 0.033.

In Fig. P12-21 is shown the general layout that the student would be required to make on an 18- by 24-in. sheet if time permitted. In order to save time the data that would be obtained from this layout are given in Fig. P12-21, and the student will be required to make the computations necessary for obtaining the size and weight of the flywheel.

The layout work may be outlined briefly as follows: First, the crank circle is divided into 24 equal parts, and the engine mechanism OAB is drawn in one of those positions. Then the theoretical indicator card and the inertia-force diagram are plotted. It is usual to construct the tangential effort diagram from a card such as would be obtained at normal or rated engine load and to assume that the engine is working against a constant and uniform torque. The inertia-force diagram includes the full inertia effect of the piston and the reciprocating portion of the connecting rod. One-third of the weight of the connecting rod is considered to be concentrated at the wrist pin. It can be shown that in this problem the inertia effects of the reciprocating masses will not appreciably affect the results, and they will, therefore, be neglected. The inertia-force diagram is given here as a matter of interest in showing the relative magnitudes of the inertia forces and gas forces. All the diagrams are plotted to a scale of 1 in. = 200 psi of piston area. In the indicator diagram the exhaust and intake lines are so near the atmospheric line they are not shown.

The radial tangential effort diagram is plotted next. The effective force GH, for example, at the crosshead pin for any position of the crosshead may be taken off by the dividers and transferred to the position OM along the crank. From point M a line is drawn parallel to the connecting rod, giving vertical line ON, which represents the turning effort for this crank position. Vector ON is next transferred to position Ak, giving k as one point on the tangential effort diagram.

This graphical method of finding the turning effort vector may be explained as follows: In Art. 9-9 it was shown that the triangle OAB in Fig. P12-21 represents the velocity polygon turned through 90 deg from its true position, OA representing the velocity V_A of crankpin A and OB the velocity V_B of the piston B. Since the work done at A is equal to the work done at B, neglecting friction, we have

$$F_A \times V_A = F_B \times V_B$$

where F_A and F_B are the forces acting at A and B, respectively. Therefore

$$\frac{F_B}{F_A} = \frac{V_A}{V_B} = \frac{OM}{ON}$$

The radial tangential effort diagram shows how the tangential force varies during the energy cycle. It should be noted that in this problem an energy cycle is completed in two revolutions of the crank. To determine how the energy varies during the cycle it is necessary to rectify the radial diagram; i.e., the ordinates of the radial diagram are laid off from a straight base line of any convenient length representing the length of the path of the crankpin for the cycle. The area under the rectified diagram will be proportional to the energy delivered to the crankshaft during the energy cycle. The mean ordinate of the rectified diagram will be the distance from the base line to the straight line that represents the load diagram. In getting this mean ordinate Durand's rule may be applied or a planimeter may be used. With the load curve superimposed on the rectified tangential effort diagram it becomes possible to determine when and in what amounts the supply of energy is greater or less than the demand. The crosshatched area above the constant load line represents the excess of energy which must be stored in the flywheel in order to be available to supply the deficiency of energy represented by the area below the constant load line.

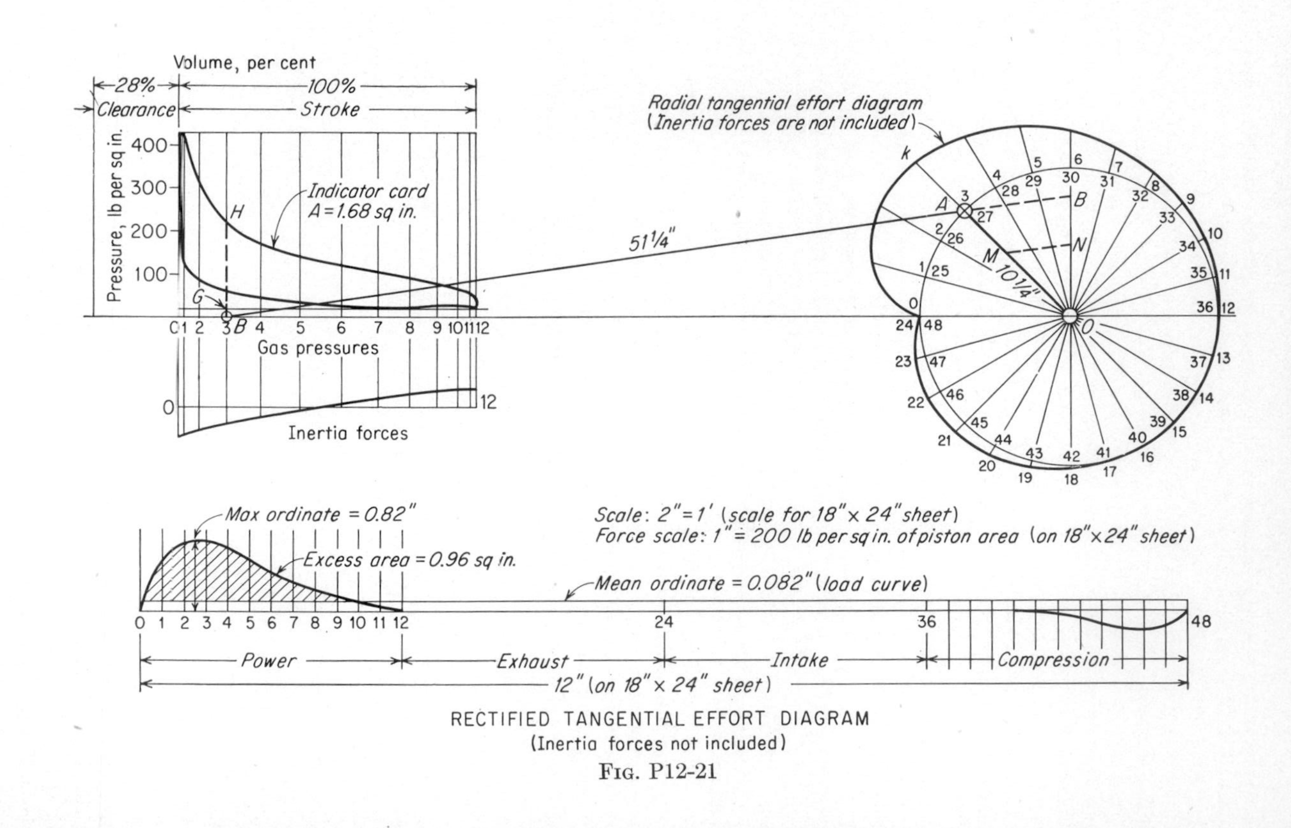

RECTIFIED TANGENTIAL EFFORT DIAGRAM
(Inertia forces not included)

Fig. P12-21

The magnitude in square inches of the excess area, as determined by means of a planimeter, is shown on the drawing. This area will determine the weight of the flywheel. Therefore, in computing the weight of the flywheel it will be necessary to determine how many foot-pounds of energy this area represents. With the energy that must be stored in the flywheel and the permissible speed fluctuation known, its weight can be computed from the well-known equation for the energy of a rotating mass (see Art. 13-11). Tests on the bursting speeds of flywheels have shown that for safety the rim speed of cast-iron flywheels should not exceed about 100 fps. The coefficient of fluctuation for flywheels has been pretty well established from experience gained in the design and operation of various classes of machinery. These coefficients can be found in handbooks and textbooks on machine design. There is a wide variation in values. For gas engines the range is from about 1/50 for constant loads such as in alternating-current generator drives to about 1/30 for ordinary power purposes.

Having available the data from the layout sheet (given in Fig. P12-21), the following items are to constitute the computations for the problem. The computations are to be made in a neat and orderly manner, in the following sequence:

a. I hp from indicator card.

b. I hp from tangential effort diagram.

c. Total energy in foot-pounds represented by 1 sq in. of rectified tangential effort diagram.

d. Total maximum variation of energy per cycle.

e. Maximum torque in inch-pounds on crankshaft from the rectified tangential effort diagram.

f. Weight of the flywheel rim.

g. Mean diameter of flywheel.

h. Dimensions of rim section, assuming for this purpose a square section.

Chapter 14. Vibrations and Critical Speeds in Shafts

NOTE: In the solution of these problems the weight of the shaft is to be neglected.

14-1. A horizontal shaft rotates in bearings at its ends and may be considered freely supported at the ends. A disk weighing 250 lb is keyed to the shaft midway between the bearings, but the center of mass of the disk is located 0.01 in. from the axis of the shaft. In a test to determine the shaft spring constant k, a static force of 1,600 lb applied midway between the bearings deflects the shaft 0.15 in. Calculate the critical speed of rotation.

14-2. Find the critical speed in rpm of the system shown in Fig. 14-4 in which the disk is made of solid steel with a diameter of 6 in. and a thickness of 1 in. The total length of the steel shaft between bearings is 20 in., and its diameter is ½ in. Assume the shaft to be freely supported at the ends.

14-3. Same as Prob. 14-2 except that the disk is located 8 in. from the left-hand bearing.

14-4. A rotating shaft 3 in. in diameter and 48 in. long carries a load of 400 lb 12 in. inside the left-hand bearing and a load of 800 lb 18 in. inside the right-hand bearing. Determine the critical speed in rpm, considering the shaft freely supported at the ends.

14-5. Same as Prob. 14-4 except that the diameter of the shaft between the two loads has been increased to 3½ in.

14-6. A rotating shaft 18 in. long consists of a section 1½ in. in diameter for 8 in. of its length and a section 1 in. in diameter for 10 in. of its length. A load of 300 lb is carried at the point where the section changes. Considering the shaft freely supported at the ends, determine its critical speed in rpm.

14-7. A shaft fixed at one end has on the other end a solid cylindrical disk with a moment of inertia of $I = 8$ lb-in.-sec². It requires 120 in.-lb torque applied at the disk to produce a twist of 1 radian. What is the natural frequency of the system in cycles per minute?

14-8. A steel shaft 2 in. in diameter and 6 ft in length is fixed at one end and carries a steel disk 12 in. in diameter and 2 in. thick at the other end. What is the natural frequency of torsional vibration of the system?

14-9. A diesel engine whose flywheel and other rotating parts (represented by disk I_2, Fig. 14-8) have a combined moment of inertia of 7,200 lb-in.-sec² drives a generator whose rotor (represented by I_1) has a moment of inertia of 1,200 lb-in.-sec². The steel shaft connecting the engine to the generator is 4 in. in diameter and 5 ft long. Assume the shearing modulus of elasticity of the steel to be 12,000,000 psi. Determine the natural frequency of torsional vibration of the system.

14-10. The propeller shaft of a boat (Fig. 14-8) has a length of 12 ft and a diameter of 4 in. On one end of the shaft is a steam turbine rotor of weight $W_1 = 2{,}000$ lb and a radius of gyration of 30 in. On the other end is a propeller of weight $W_2 = 1{,}000$ lb and a radius of gyration of 20 in. The shaft is of steel and has a torsional modulus of elasticity of 12,000,000 psi. Determine the natural frequency of torsional vibration of the system.

14-11. A mass supported on a spring is shown in Fig. P14-11, which represents schematically a machine on an elastic support. A balance of forces in the vertical direction for any instant of time during vertical vibratory motion gives the equation $m(d^2x/dt^2) = -kx$, which is the same as Eq. (14-1). Therefore solutions (14-10) and (14-11) apply. If the mass is 10 lb and is supported on springs having a combined constant of 80 lb/in., (*a*) what is the static deflection Δ and (*b*) what frequency of vertical disturbance applied to the mass will result in resonance with the natural frequency? Assume the motion to take place in a vertical direction only.

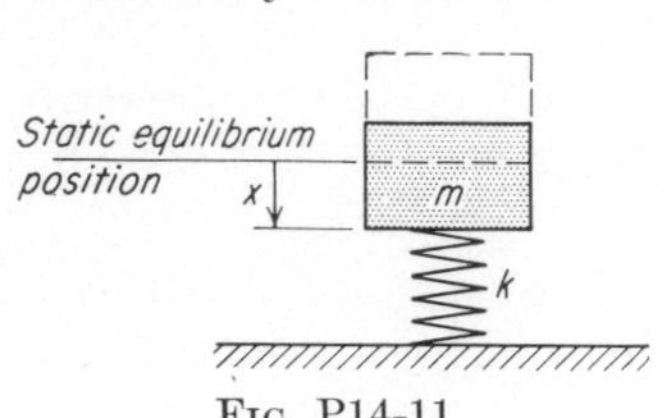

FIG. P14-11

14-12. If the mass of the previous problem is an electric motor which is not perfectly balanced, what static deflections and corresponding spring constants must be avoided for motor speeds of (*a*) 1,200 rpm, (*b*) 1,725 rpm, (*c*) 3,600 rpm?

14-13. Assume a harmonically varying vertical disturbance $F_o \cos \omega t$ to be applied to the mass m of Fig. P14-11. The centrifugal force of a small unbalance of a pump rotor might produce such a force, or the periodic acceleration of a piston of an engine, for example. A balance of forces in the vertical direction gives the equation $m(d^2x/dt^2) = -kx + F_o \cos \omega t$.

a. Verify that a solution of this equation is

$$x = \frac{F_o}{k} \frac{1}{1 - (\omega/\omega_n)^2} \cos \omega t$$

where $\omega_n = (2\pi)$ (natural frequency) $= \sqrt{k/m}$. (Substitute the solution into the differential equation.)

b. In order for the amplitude of the vibration x to be small, what must be true about the ratio ω/ω_n of disturbing frequency to natural frequency?

c. Interpret the meaning of the change of sign of x as ω/ω_n varies from a value less than 1 to a value greater than 1. Also interpret the meaning of F_o/k.

d. Plot a curve of $1/[1 - (\omega/\omega_n)^2]$, usually called the "magnification factor," for ω/ω_n between 0 and 4.

e. From the curve of (*d*), draw conclusions concerning the design of flexible mountings to minimize amplitude of vibration. (Note that, when bringing the machine up to speed, the resonant frequency must be passed through quickly to avoid excessive vibration during this transient period.)

f. What sort of spring constant is indicated for the ratio $\omega/\omega_n = 0$? What does this mean physically?

14-14. An item of importance in considering an elastic mounting for a vibrating machine is the shake felt by the supporting foundation. In Fig. P14-11, this shaking force is the force in the spring, which is kx. The ratio of this shaking force to the disturbing force $F_o \cos \omega t$ at any instant of time gives the fraction of the disturbing force transmitted to the foundation, and is called the "transmissibility."

a. Plot the transmissibility for ω/ω_n between 0 and 4.

b. Draw conclusions concerning the design of flexible mountings intended to minimize shaking force felt by the foundation of a machine.

Chapter 15. Gyroscopic Forces

15-1. A gyroscope arranged as in Fig. 15-8 is composed of a circular rim 24 in. in diameter weighing 10 lb supported on spokes or arms of negligible weight. The distance L is 3 ft. If the wheel is rotating about its own axis with a speed of 600 rpm, clockwise as viewed along its supporting axle from pivot O, in what sense will it precess and with what angular velocity?

15-2. Describe the gyrostatic action of the flywheel of an automobile engine when rounding a curve (*a*) to the right, (*b*) to the left. Rotation is counterclockwise viewed from the rear.

15-3. If an airplane diving at a steep angle changes direction by a sharp curve into the horizontal, what will be the gyroscopic action if the propeller is rotating clockwise when viewed from the rear?

15-4. The flywheel of an engine on a ship weighs 5,000 lb and has a radius of gyration of 3.75 ft. It is mounted on a horizontal axle which is parallel to the longitudinal axis of the ship and has a speed of 400 rpm clockwise when viewed from the rear. Find the gyroscopic couple when the ship is turning to the left with an angular velocity of 0.1 radians/sec. What are the *total* forces on the bearings if the distance between the centers of the bearings is 4 ft?

15-5. The armature of the motor of an electric car weighs 600 lb and rotates in a sense opposite to the rotation of the car wheels. The distance between its bearings is 2 ft, and its radius of gyration is 6 in. The motor makes four revolutions to one revolution of the car wheels which have a diameter of 33 in. If the car is moving forward around a curve of 100 ft radius with a velocity of 20 fps, what are the *total* forces and their directions on the bearings of the motor if the center of the curve is to the right?

15-6. In a stabilizing gyroscope, similar to that shown in Fig. 15-11*a*, a forced precession backward or forward through a given angle about axis BB enables a varying couple to be opposed to the rolling moment due at any instant to a train of waves. The rolling moment at a given instant is 3,000 ft-tons, clockwise when viewed from the rear. The rotor weighs 100 tons and spins at 900 rpm clockwise when viewed from above. The radius of gyration of the wheel about the spin axis is $4\frac{1}{2}$ ft.

a. Calculate the angular velocity of precession required to maintain the ship in an upright position.

b. What is the sense of the precession?

15-7. A locomotive is moving at the rate of 40 mph around a curve of 600 ft radius. The diameter of the drivers is 84 in., and a pair of drivers and axle have a moment of inertia about the axis of rotation of the wheels of 3,000 slug-ft^2. What is the magnitude of the couple introduced by the precessional motion of the pair of wheels, and in what sense does it act? Does it tend to tip the locomotive inward or outward?

15-8. A certain jet-engine rotor turns counterclockwise when viewed from the front of the airplane in which it is mounted. In what direction does the gyroscopic action tend to move the airplane in each of the following maneuvers: (*a*) a turn to the right with the rotor axis always in a horizontal plane, (*b*) a similar turn to the left, (*c*) a pure roll of the airplane about a straight line of flight, (*d*) a climb during which the rotor changes direction from horizontal to vertical, (*e*) a simultaneous climb and turn to the right.

15-9. A certain helicopter rotor turns clockwise in a horizontal plane viewed from above when the helicopter is hovering. The rotor is tilted forward to induce forward flight. Describe the effect of the gyroscopic action on the helicopter.

15-10. A motorcycle rider approaches a banked curve to the right. Describe the gyroscopic action of the front wheel caused by the *bank* (not the *curve*) as the motorcycle is leaned to the right. Is the *steering* affected?

ANSWERS TO PROBLEMS

1-7. 786 fps or 536 mph; 2,470,000 ft/sec^2

1-8. (*a*) 1.19 (*b*) 2.38 (*c*) 3.57 (*d*) 14,010 ft; 28,020 ft; 43,030 ft

1-9. 52.4 fps or 35.7 mph; 16,480 ft/sec^2

1-10. 3.33 in.; 26.67 ips; 33.34 $in./sec^2$

2-12. (*a*) Diameter on driving shaft: 12, 13.9, and 16.1 in.; on driven shaft: 18.9, 17, and 14.75 in. (*b*) $L = 138.18$ in.

2-13. (*a*) Diameter on driving shaft: 12, 14, and 16.2 in.; on driven shaft: 18.9, 17.1, and 14.88 in. (*b*) $L = 132.78$ in.

2-14. (*a*) Diameter on driving shaft: 15, 11.82, 8.3, and 5.22 in.; on driven shaft: 5.52, 8.7, 12.22, and 15.3 in. (*b*) $L = 225.29$ in.

2-15. (*a*) Diameter on driving shaft: 15, 11.86, 8.2, and 5.135 in.; on driven shaft: 5.52, 8.71, 12.07, and 15.12 in. (*b*) $L = 224.25$ in.

2-16. Mechanical advantage = 5

2-17. (*a*) Mechanical advantage = 22 (*b*) 2.272 fpm

2-21. (*a*) 62.8 fps; 23,700 ft/sec^2 (*b*) 736 lb

2-22. 54.9 fps; 2,010 ft/sec^2; 62.4 lb

3-18. 30 rpm

3-19. 154.9 deg

3-20. At 60 deg: 47.1 ips, 0 $in./sec^2$; at 80 deg: 40.8 ips, −2,220 $in./sec^2$

3-21. (*a*) 30; 45; 60 deg (*b*) 3,600 $in./sec^2$; 1,800 $in./sec^2$

3-22. (*a*) $6C_3$ (*b*) Omit term containing t^3

3-23. (*a*) $A = Ct^3/3 + C_1$;
$V = Ct^4/12 + C_1t + C_2$;
$x = Ct^5/60 + C_1t^2/2 + C_2t + C_3$
(*b*) Specify C. Evaluate other constants from initial conditions

4-8. (*e*) 24 deg (*f*) 15, 24, 0 deg (*g*) 1.3

4-9. (*e*) 21 deg (*f*) 21, 18, 0 deg (*g*) 1.3

4-10. No interference (*f*) 18, 14, 0 deg (*g*) 1.42

7-1. (*a*) 7.37 rev (*b*) Down

7-2. (*a*) 32.1 strokes/min (*b*) 13.4 fpm (*c*) $b = 32$ teeth, $c = 48$ teeth

7-3. (*a*) 27.4 to 1 (*b*) 65.7 rpm (*c*) Yes

7-4. (*a*) 10.74 turns (*b*) 33.5 ft

7-5. $n_4 = [1 - (ac/bd)]n_5 + (ac/bd)n_2$

7-6. (*a*) $n_d/n_e = 0.363/1$ (*b*) $n_d/n_e = -0.25/1$ (*c*) All gears locked together

7-7. (*a*) $n_d/n_a = 2/1$ (*b*) Zero (*c*) $n_d/n_a = 1/1$ (*d*) 75 rpm

7-8. $a = 64$ teeth; $b = 48$ teeth

7-9. 1,397 rpm

7-10. 0.70

7-11. (*a*) $N_G = 64$; $k_1 = 0.375$, $k_2 = 1.75$ (*b*) −145 rpm

8-1. $V_C = 1.88$ in.

8-2. $V_D = 2.31$ in.

8-3. $V_F = 0.38$ in.

8-4. $V_D = 1.96$ fps

8-9. $V_D = 225$ fpm

8-10. $V_C = 50.4$ fpm
8-11. $V_C = 54.6$ fpm
8-12. $V_B = 0.28$ in.
8-13. $V_D = 1.93$ fps
8-14. $V_C = 108$ fpm
8-15. $V_E = 0.63$ in.
8-16. $V_E = 2.50$ in.
8-17. (*e*) Velocity scale 1 in. = 21.8 fps
8-21. Rate of sliding = $V_{A_3A_2} = 264$ fpm
8-25. (*a*) 49.5 ips (*b*) 4.5 radians/sec clockwise
8-26. (*a*) 14.1 ips upward (*b*) 28.5 ips
8-27. (*a*) 16.1 ips (*b*) 68 radians/sec counterclockwise
8-28. 0.20 fps
8-29. 53 ips upward
8-30. (*a*) 20 deg; 340 fps (*b*) 340 fps; 950 fps (*c*) 1.32 ft

9-1. 11.9 ft/sec^2
9-2. (*a*) $\omega = 20\pi$ radians/sec; $\alpha = 2\pi$ radians/sec^2 (*b*) $a_P^n = 200\pi^2$ ft/sec^2; $a_P^t = \pi$ ft/sec^2 (*c*) Same as a_P^t
9-3. (*c*) $v_B = 7.5$ fps; $a_B = 440$ ft/sec^2
9-4. (*c*) $a_C = 188$ ft/sec^2
9-5. $a_B = 200$ ft/sec^2
9-6. $a_B = 200$ ft/sec^2
9-7. $a_B = 1{,}050$ ft/sec^2
9-8. (*a*) $v_C = 7.1$ fps (*b*) $a_C = 74$ ft/sec^2
9-9. $a_C = 231$ ft/sec^2
9-10. $a_C = 231$ ft/sec^2
9-11. (*a*) $v_F = 0.7$ fps (*b*) $a_F = 2.63$ ft/sec^2
9-12. (*c*) $a_C = 40$ ft/sec^2
9-13. (*a*) $v_C = 28$ fps (*b*) $a_C = 300$ ft/sec^2 (*c*) $a_C = 300$ ft/sec^2
9-14. (*c*) $v_D = 1.94$ fps; $a_D = 1.15$ ft/sec^2
9-15. (*c*) $a_C = 525$ ft/sec^2
9-16. (*a*) $v_C = 34.5$ fps (*b*) $a_C = 450$ ft/sec^2
9-17. (*a*) $v_C = 76.5$ fps (*b*) $a_C = 1625$ ft/sec^2
9-18. (*a*) $v_D = 7.5$ fps (*b*) $a_D = 580$ ft/sec^2
9-19. (*b*) $v_D = 3$ fps (*c*) $a_D = 42$ ft/sec^2
9-20. (*b*) $v_C = 19.75$ fps
(*d*) $a_C = 235$ ft/sec^2
9-22. (*b*) $v_B = 38$ fps; $v_D = 52$ fps
(*c*) $a_B = 7{,}000$ ft/sec^2; $a_D = 5{,}000$ ft/sec^2; $\alpha_3 = 11{,}380$ radians/sec^2 clockwise
9-23. (*b*) Velocity scale 1 in. = 48.87 fps; acceleration scale 1 in. = 14,325 ft/sec^2
9-24. (*c*) $v_A = 46.6$ fps; $a_A = 12{,}150$ ft/sec^2; velocity scale 1 in. = 21.8 fps; acceleration scale 1 in. = 5,692 ft/sec^2; a_B at 0 deg = 15,300 ft/sec^2; a_B at 180 deg = −8,970 ft/sec^2
9-25. (*f*) Velocity scale 1 in. = 21.8 fps
(*i*) Acceleration scale 1 in. = 5,692 ft/sec^2
9-27. $\alpha_B = 271$ radians/sec^2
9-28. $a_P = 390$ ft/sec^2
9-29. $a_C = 5.25$ ft/sec^2
9-30. $a_B = 40.8$ ft/sec^2
9-31. 190 ipm; 380 in./min^2
9-32. 179 in./sec^2 upward
9-33. 1,490 in./sec^2 upward
9-34. 360 radians/sec^2 counterclockwise; 1,310 in./sec^2
9-35. 5.2 radians/sec clockwise; 190 radians/sec^2 counterclockwise; 3.65 fps; 77 ft/sec^2
9-36. 1.7 fps downward; 6.8 ft/sec^2 upward; 2.7 fps; 1 ft/sec^2
9-37. 0.213 ft/sec^2 eastward

10-1. $F_{32}^t = 1.35$ in.
10-2. $P = 1.10$ in.
10-3. $P = 3.45$ in.
10-4. $F_{23}^t = 1.25$ in.
10-7. $P = 0.25$ in.
10-8. P (without friction) = 0.10 in.; P (with friction) = 0.25 in.
10-9. (*a*) Total F = 13,400 lb
(*b*) 8,750 lb, 2,925 lb

11-1. Resultant shaking force $S = 114$ lb; moment of acceleration couple at $O_2 = 264$ lb-in. counterclockwise
11-2. Resultant shaking force $S = 285$ lb; moment of acceleration couple at $O_2 = 530$ lb-in. counterclockwise

11-3. Resultant shaking force $S = 315$ lb; moment of acceleration couple at $O_2 = 480$ lb-in. counterclockwise

11-5. $F_{34} = 480$ lb; $F_{12} = 1{,}620$ lb; $F_{32} = 520$ lb; $F_{32}^T = 394$ lb

11-7. $F_{34} = 490$ lb; $F_{12} = 1{,}630$ lb; $F_{32} = 535$ lb; $F_{32}^T = 396$ lb

11-9. $F_{34} = 480$ lb; $F_{12} = 525$ lb; $F_{32} = 525$ lb; $F_{32}^T = 400$ lb

11-15. (*a*) 123.6 rpm (*b*) 138.6 rpm

11-16. Initial load = 200 lb; final load = 230 lb; stiffness constant = 240 lb/in.

11-17. 11.5 lb westward

11-18. 1.19; 2.38; 4.76; for 30-sec turn, twice these values

11-19. (*a*) 0.235 sec (*b*) 5.17 sec (*c*) Larger

11-20. 270.5 ft/sec²; 8.4 *g*

11-21. 0.155 slug ft²

11-22. 61.8 slug ft²

11-23. 116 lb/ft; 12.43 in.

11-24. (*a*) At midpoint, 491 lb; at extremes of travel, zero sideways force (*b*) At mid-point, 491 lb toward table center; at position nearest center of table, 123 lb away from the center; at outer extreme position, 1,102 lb toward center

11-25. 1.41 hr/day

12-1. $m_{0'}R_{0'}a_{0'} = 955$ units; $m_0R_0 = 167$ units

12-2. $m_{0'}R_{0'}a_{0'} = 955$ units; $m_0R_0b_0 = 1{,}340$ units

12-3. $m_{0'}R_{0'}a_{0'} = 955$ units; $m_0R_0b_0 = 1{,}340$ units

12-4. $m_{0'}R_{0'}a_{0'} = 975$ units; $m_0R_0 = 137.5$ units

12-5. $m_{0'}R_{0'}a_{0'} = 975$ units; $m_0R_0b_0 = 1{,}100$ units

12-6. $m_{0'}R_{0'}a_{0'} = 975$ units; $m_0R_0b_0 = 1{,}100$ units

12-7. $W_A = 27.1$ lb; $W_B = 18.8$ lb

12-8. (*a*) In RP_0, add 1.905 oz at 269 deg; in $RP_{0'}$ add 1.625 oz at 150 deg (*b*) At 3,000 rpm, on bearing B_1, 399 lb at 78 deg for position shown; on B_2, 393 lb at 354 deg. At 6,000 rpm, four times these force magnitudes

12-9. (*b*) In left end, add 0.223 oz at 264½ deg clockwise from vertical (partial answer only)

12-14. $mR\omega^2 = 25{,}100$ lb; $C_P = 17{,}050$ lb-ft

12-15. (*b*) 19,200 lb-in. (constant) (*c*) Completely, by rotating weights near ends of crankshaft

12-16. (*a*) Primary couples only are unbalanced (*b*) Need $m_C + m_P$ at crank radius opposite each crank throw (*c*) Need m_C at crank radius opposite each crank throw; plus, in plane 1, $\frac{1}{3}m_P$ at crank radius 90 deg counterclockwise from crank 1 and m_P at crank radius opposite crank throw. In plane 4, same as in plane 1, but masses diametrically opposed to those in plane 1. Resultants can be formed, of course (*d*) In plane 1, $(\frac{1}{3}m_P + m_C)$ at 90 deg counterclockwise from crank throw, plus $(m_P + m_C)$ opposite crank (masses at crank radius). In plane 4, same, but all masses diametrically opposed to those in plane 1

12-17. For answers see Art. 12-23

12-21. (*a*) 72.7 (*b*) 75.4 (*c*) 20,300 (*d*) 19,450 (*e*) 190,000 (*f*) 3,000 lb (total for two wheels, or 1,500 lb for each wheel) (*g*) 5.73 ft (*h*) 5.17 by 5.17 in.

14-1. 1,225 rpm

14-2. 1,536 rpm

14-3. 1,585 rpm

14-7. 36.9 cpm

14-8. 816 cpm

14-9. 666 cpm

14-10. 476.4 cpm

14-11. (*a*) 0.125 in. (*b*) 531 cpm

14-12. (*a*) 0.0245 in.; 408 lb/in. (*b*) 0.01186 in.; 844 lb/in. (*c*) 0.00272 in.; 3,672 lb/in.

15-1. Counterclockwise viewed from above. $\omega_P = 1.535$ radians/sec

15-2. Heavier pressure on rear bearing in first case and on front bearing in second case

15-3. Tendency will be to turn the airplane to the right

15-4. On rear bearing, 4,788 lb; on forward bearing, 212 lb

15-5. On left bearing, 273 lb; on right bearing, 327 lb

15-6. (*a*) 0.506 radian/sec (*b*) Clockwise viewed from left side

15-7. 4,927 lb-ft: tends to tip locomotive outward

15-8. (*a*) Tends to dive the plane (*c*) No effect (*d*) Tends to turn plane to pilot's right

15-9. Tends to roll plane to pilot's right

15-10. Wheel tends to turn right

INDEX